Robert Louis Kemp

Super Principia Mathematica

The Rage to Master

Conceptual & Mathematical Physics

The Special Theory

Of

Thermodynamics

Volume 2

Flying Car Publishing Company

Long Beach, California

Flying Car Publishing Company: First edition – June 2010
Flying Car Publishing Company
P.O. Box 91861
Long Beach, CA 90809
310-720-3703

Library of Congress Cataloging-in-Publication Data

Robert Kemp, 2010-
Super Principia Mathematica – The Rage to Master Conceptual & Mathematica Physics - The Special Theory of Thermodynamics / Robert L. Kemp 1st edition
p. cm.
Includes index.
ISBN 978-0-9841518-1-3
Copyright © June 2010

Flying Car Publishing Company: First edition – June 2010
Published by Flying Car Publishing Company
www.Superprincipia.com

Flying Car Publishing Books are available at special discounts for bulk purchases in the United States by corporations, institutions, and other organizations. For more information, contact: Flying Car Publishing Company, P.O. Box 91861, Long Beach, CA 90809
www.Superprincipia.com

Cover Images Courtesy: NASA/JPL-Caltech - http://photojournal.jpl.nasa.gov/gallery

Table of Contents

Acknowledgements

This book is dedicated with sincere gratitude and admonishing to the all wise (Omniscience), Omnipotent, and Omnipresent God the Father of us all; God the Son (Jesus) the Christ, and God the Holy Spirit for providing the wisdom, strength, and insight, and for being the author and finisher of my faith, making this work possible.

Also I want to give my sincere thanks and gratitude to two of the most wonderful women in my life, the Matriarch of our family, my grandmother, Margie Leola Gray; and my mother, Delorise Annese Gray$_{maiden}$.

My Grandmother Margie is a very loving and educated woman that have been such an encouragement to me in life, and growing up have always been a support to me in every way.

My Mother Delorise also a very loving and educated woman is the best mother that any man could have, and was very instrumental in my education as a youth and helping me to become a creative person.

And, last but not least I want to thank my two brothers William T. Kemp, and Thomas E. Kemp for being brothers, Tuskegee University college class mates, and friends.

Robert L. Kemp

Prologue

To the Reader – What is the "Super Principia Mathematica – The Rage to Master Conceptual & Mathematical Physics?"

The work that you are about to engage in, is in general a "Theory of Everything (TOE)," described in several volumes that I developed over a twenty (21) year period starting at the age of 22, in the year 1989; and compiled over a three year period of unrelenting devotion. In this series of works you will find many conceptual and mathematical models which unveil new insights into the physics and mathematics of the universe.

I named the book the "Super Principia Mathematica – The Rage to Master Conceptual & Mathematical Physics", because my goal was to try and present physics in the Super Principia Mathematica similar to the way that Sir Isaac Newton introduced his new concepts of physics in his Principia Mathematica published in the year 1687.

Likewise this work is duly called the "Rage to Master Mathematical & Conceptual Physics" because there was such a rage within, grappling with social life, work life, and physics to bring this work to birth.

My life's work, The Super Principia Mathematica is a body of volumes of work that cultivates mathematics, and conceptual physics into beautiful models of the mechanics of nature.

The main difference between Newton's Principia Mathematica and my Super Principia Mathematica is that Newton conveys physics and mathematics to the reader through the use of laws, axioms, definitions, propositions, and Lemmas.

However, in the Super Principia, I use only laws, definitions and aphorisms to try and convey the same sort of message.

This book was written for anyone that is fascinated by physics, and the incredible way in which nature allows anyone to conceptualize a mechanical process, build a mathematical model, which makes prediction of the process, and finally tests the theory or model with experimentation, to determine the accuracy of the prediction.

Similar to Isaac Newton's "The Principia Mathematica" this book is not designed to be read straight through like a novel, but to be studied and cross referenced with other material, as well as other volumes of this work, based on a specific concept one is investigating.

This book was written to be read by anyone with a minimum high school level physics education, and some basic understanding into algebra, geometry, differential calculus, and integral calculus.

However, this book is not intended to act as a course in basic physics, but to act as a means of giving the reader a deeper understanding of concepts that one is generally familiar; as well as introduces new physics models and new calculations into those generally familiar concepts.

In just about every chapter I introduce new physics models and calculations.

I hope that the reader enjoys this work, I consider it art, as well as science. I earnestly ask that everything be read with an open mind and that the shortcomings in any of the subjects addressed, which are new concepts, may be not so much reprehended as investigated, and kindly supplemented, by new endeavors of my readers.

For me, the mathematics of physics, are the tools that God gave man that he may understand, describe, and predict the great works of God's created universe.

Robert Louis Kemp
June 2010

Epigraph

The following epigraph is taken from the Bible's book of Genesis and modified to fit a Big Bang origin, intelligent design theory.

In the beginning God created the Heavens and the earth; with a very Big Bang, approximately fourteen (14) billion years ago!

And the earth was without form, and void; and there was Aether Dark Matter everywhere. And the Spirit of God moved upon the Aether causing a universe to expand.

And God said; Let there be electromagnetic radiation and heat: and there was light and a Cosmic Force. And God saw the light, that it was good: and God divided the light from the visible matter and the dark matter.

And God called the light Day, and the darkness he called Night. And God caused various places in the universe to rotate and spiral into various orbits, and the evening and the morning were the first day.

And God said; Let there be a firmament in the midst of the expanding universe, and let it divide the Hot Dark Matter from the Cold Dark Matter.

And God made the firmament which is the structure of spacetime, matter, waves, and fields of the universe; and divided the Aether which was internal to the firmament from the Aether which was external to the firmament: and it was so.

And God called the firmament and the contents of the Heavens, Universes. And as the contents of the heavens continued to rotate, evening and the morning were the second day.

And God said; Let the Aether in the various places of the expanding universe be condensed together unto one place, and let the tangible matter appear, and gravitation was created: and it was so.

And God called the condensed matter, electrons, protons, and neutrons, among other self sustaining energy units; and the gathering together of the Aether into elements, He called, atoms and molecules: and God saw that it was good.

And God said; Let the earth bring forth the condensed elements of matter, and the matter combined, and the matter became molecules, after its kind which is evolution, whose self replicating properties is in itself, upon the earth: and it was so.

And the earth brought forth an abundance of elements after its kind which evolved into: hydrogen, helium, carbon, iron, oxygen, and all sorts of solids, liquids, gases, and plasmas; and the atoms and molecules, whose self replication is within itself; and God saw that it was good.

And God said; Let the Hydrogen and Oxygen elements of matter join together, and the waters molecules which formed under the heaven be gathered together unto one place, and let the dry land appear: and it was so.

And God called the dry land Earth; and the gathering together of the waters He called Seas: and God saw that it was good.

And the heavens continued to rotate; evening and the morning were the third day.

And God said; let there be galaxies, stars, and planets, which radiate and consume energy, and be as lights in the firmament of the universe; to divide the day from the night through the rotation of circular and elliptical paths; and let them be for signs, and for seasons, and for days, and years.

And let the stars, quasars, cepheids, and cepheid-like stars be for lights in the firmament of the heaven to give light upon the earth: and it was so.

And God made two great lights to rule the earth; the greater light the sun star to rule the day, and the lesser light the moon which reflects the light and rules the night: He made the stars, the solar system, and the planets also.

And God set the sun, the planets, and the moon into rotational motion in the firmament of the heaven to give light upon the earth, and to rule over the day and over the night, and to divide the light from the darkness: and God saw that it was good.

And the heavens continued to rotate; evening and the morning were the fourth day.

And God said; Let the waters bring forth abundantly the moving creatures that have life, and fowl that may fly above the earth in the open firmament of heaven.

And God created animals that move through the waters and some that fly though the air which evolved after his kind: and God saw that it was good. And God blessed them, saying, be fruitful, and multiply, and fill the waters in the seas evolving, and let fowl multiply evolving in the earth.

And the evening and the morning were the fifth day.

And God said; Let the earth bring forth land animals and bugs of all sorts evolving after his kind: and it was so; and God saw that it was good.

And God said; Let us make man in our image, after our likeness: and let them have dominion over all the earth. And God saw every thing that he had made, and, behold, it was very good. And the evening and the morning were the sixth day.

Thus the heavens and the earth were finished, and all the host of them. And on the seventh day God ended his work which he had made; and he rested on the seventh day from all his work which he had made.

The above intelligent design, origins of the universe theory is lifted from the Bible's origins story (Genesis Chapter 1: verses 1 – 3; and Chapter 2: verses 1 - 2); and modified to fit a Big Bang origin, intelligent design theory posit in this work.

The Bible states that it took God six (6) days to create and finish His entire work of the universe which includes the heavens and the earth. And on the seventh (7) day it states that God rested from all His good work.

However, it can be deduced from the Bible that God completed the physics of the cosmological, atomic, and sub atomic world in four days, and spent the other two days working on earth design details; such as cells, genes, biology, and creatures of all sorts!

If the above is true, this leads one to believe that God the creator of the universe is outside of the universe and therefore is outside of time as we understand it!

This would also mean that God is not subject to the same universal laws of the created physical universe. For example, one day for the creator could be one thousand years for the program.

This would mean that the universe is either a "Big Fish Bowl" filled with water or an aether gas; or the universe is made of pixels in a "Computer Simulation Program" in which the pixels are an aether gas.

And, God is a computer programming genius!

It seems more probable that the universe is a computer simulation program that was created by a Cosmic Computer Genius; God the Creator of the Universe.

This would also give credence for a heaven and a hell as stated in the bible. And, would lead to the conclusion that when we die, we are moved to a new location in the simulation program; a new universe perhaps!

Keep in mind it is all a THEORY!

What I propose is a "***Donut Universe Theory"*** for a model of the universe. This model claims that the universe is an evolving spheroid toroid that expands to a certain size to reach equilibrium. The galaxies that we witness and live in; the milky way are all riding the circulating lines of a vortex toroid which then generate vortex motion on all sub atomic and cosmological scales.

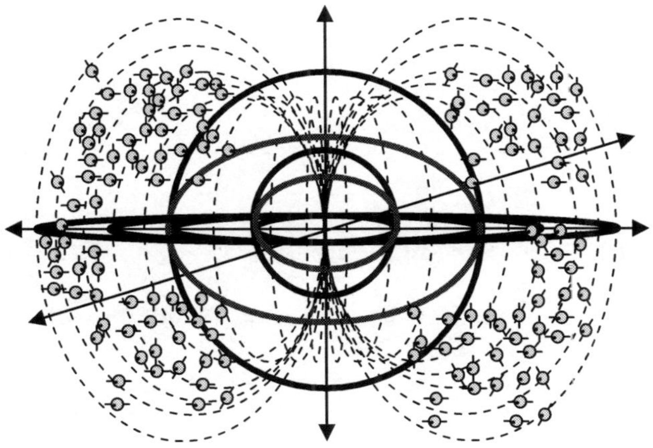

This "Donut Universe" model would explain the expansion of all galaxies which redshift away from each other in all directions. This model explains why all observers in the universe view the universe as being isotropic and on average homogenous. This model also explains why there are fluid dynamic properties to the universe, and why there are cycles in the universe. This model also explains why the universe appears flat and holographic to observers immersed in it. Thus, this model is awkwardly named the ***"Donut Universe Theory"***.

Chapter 1

The Special Theory of Thermodynamics

(The Laws of Thermodynamics)
&
(Adiabatic Specific Heat Capacity Index)

1.1 Introduction

To the reader; this second volume of the, "Super Principia Mathematica – The Rage to Master Conceptual & Mathematical Physics – The Special Theory of Thermodynamics" is dedicated to matter, motion, and thermodynamic energy of systems; independent of rotation.

The "First Law of Motion" describes the energy of motion in general not as being singular but as being dual in nature; the one nature external directional anisotropy motion, and the other nature internal Omni-directional isotropy motion, both working together as one.

The anisotropy is a term to describe directional motion and the isotropy is a term to describe Omni-directional motion for an isolated dynamic conserved system. The two motions are both independent and intertwined; for a body at rest or in uniform motion relative to an inertial frame of reference.

In this second volume of the, "Super Principia Mathematica the subject matter involving thermodynamic system mass bodies in motion will be described in greater detail in the subsequent chapters listed below:

1. The Laws of Thermodynamics & the Adiabatic Specific Heat Capacity Index
2. Thermodynamic Field Irreversible "Free Aether" Heat Energy
3. Hydrodynamic Fluid Resistance & Compressibility Effects
4. Aether Working Fluid Ratio – Locomotion Foreshortening Factor
5. Carnot's Thermal "Anisotropy" Engine Efficiency
6. Carnot's Thermal "Isotropy" Engine Efficiency
7. Thermodynamic Heat Energy Transfer Mechanisms
8. Thermodynamic "Pump" Momentum & Energy
9. Conic Inertial Momentum & the Total Energy Conservation Relation
10. Respiration Spacetime Pump Factor
11. Thermodynamic Spacetime Compressibility

In the study of thermodynamics and the motion of objects, the Super Principia will make use of the concept of an "Isolated Inertial Net Mass System Body" which is a conserved system composed of individual mass units, instead of describing the mythical point particle as a body.

In this work points in space represent locations relative to other locations, not mass bodies.

The isolated system body is considered to be a dense mass object occupying and existing in space, mass, volume, and time in the universe.

To fully describe motion, the concept of the all pervading aether medium must be integrated seamlessly into the study of the motion of objects; this has been attempted since antiquity. And this work is also a model for integrating the aether into the study of matter in motion.

The various concepts of the Aether were described over the centuries by great scientist and physicist such as: Nicolas Copernicus (c. 1500), Johannes Kepler (c. 1600), Rene Descartes (c. 1600), Isaac Newton (c. 1700), Fatio de Duillier (c. 1700), Georges-Louis Le Sage (c. 1700), Michael Faraday (c. 1800), James Clerk Maxwell (c. 1800), Albert Einstein (c. 1900), Hendrik Antoon Lorentz (c. 1900), Alexander Friedmann (c. 1900), and Steven Rado (c. 2000).

In this work the Aether is described as being an all pervading spacetime medium that permeates the entire universe including Baryonic and non Baryonic matter. The constituents of the Aether medium are the Aetherons, which flow and collide with one another; without the need for instantaneous action at a distance forces.

In this work the Aether is revealed and was hidden in the Conservation of Kinetic Energy equation. In this work it is shown that the Aether can be described and measured during any elastic collision.

The Aether is a gas that behaves like an ideal gas that can be described fundamentally with the Kinetic Theory of Gases; and is the fundamental building block of the Kemp Cosmological Principle, which states,

> "**Kemp Cosmological Principle** – In the Universe no single observer at rest or in motion has any special place, and to any universal observer the cosmic force and the speed of light are constants; thus spacetime of the universe behaves like an ideal gas from which matter forms that is homogenous, isotropic, and predictable; whose kinematics and dynamics are governed by physical laws that do not change over time."

The Aether which has a temperature dependence manifest on all orders of magnitude in the universe and on the molecular level is the result of the motion of particles which make up a substance relative to its environment. Many properties of matter are changed with any increases or decreases in temperature.

The Aether is also a form of heat energy where the temperature of that heat is measured on thermodynamic scales, and is a measure of the average kinetic energy of motion of an atmospheric environment and of the particles in translation, rotation, vibration, or the excitation of electron energy levels of individual particles that make up an isolated system Net Inertial Mass body.

The quantity Temperature is a property of a system that determines whether the system will be in thermal equilibrium with other systems. When two systems are in thermal equilibrium, their temperatures are by definition equal.

Temperature is also related to the amount of internal energy of a system. In physics, temperature is a physical property of a system that underlies the common notions of hot and cold.

Any object that is considered hot generally has the higher temperature, and when that same object is considered cold it generally measures a lower temperature.

The Aether as Heat Radiation which is a flow between any high to low temperature difference, is generally thought to travel at the uniform speed of light through the vacuum of spacetime and is slowed down however, in a dense medium such as water, or humid air.

This transfer of aether energy between bodies of different temperatures and kinetic energies happens spontaneously only in the direction of the colder body. The Aether is found in the natural and spontaneous flow from a high temperature reservoir to a low temperature reservoir, in order for the system to establish thermodynamic equilibrium.

The spontaneous aether heat flow from the high temperature to the low temperature is superimposed on the isotropic disordered random Omni-directional energy. Thus aether heat flow is anisotropic ordered directional energy flowing, from hot to cold.

The Aether can also be described as Thermal radiation which is a direct result of the movements of atoms and molecules in a material. Since atoms and molecules are composed of particles (protons, electrons, neutrons), their movements result in the emission of electromagnetic radiation, which carries energy away from a surface body.

Since the amount of emitted radiation increases with increasing temperature, if a surface body is constantly bombarded by thermal radiation from the surroundings, the Aether is found in the transfer of energy to the surface which results in an increase in the average kinetic energy of the body.

The average kinetic energy of an isolated system mass body is comprised of two independent motions that are intertwined in space and time: the external anisotropic translational motion, and the internal isotropic Omni-directional motion; of the individual mass bodies of the system.

Faster molecules which are at a higher temperature striking slower ones at the boundary in elastic collisions will increase the velocity of the slower ones and decrease the velocity of the faster ones, transferring Aether energy from the higher temperature to the lower temperature region.

And, as the Aether heat flows from the isotropic internal environment to the external outside environment, work is done in the form of irreversible heat energy which is radiated away from an internal environment to an external environment, of an isolated system mass body to establish equilibrium with the external environment.

This second volume of Super Principia Mathematica – The Rage to Master Conceptual & Mathematical Physics – The Special Theory of Thermodynamics is art as well measurable science; I hope that my readers enjoy!

1.2 The Zeroth Law of Thermodynamics

The first explicit statement of the first law of thermodynamics was given by Rudolf Clausius (1822 – 1888) a German mathematician and physicist, who in the year 1850 stated,

> "There is a state function E, called 'energy', whose differential equals the work exchanged with the surroundings during an adiabatic process."

The most significant result of this distinction is the fact that one can clearly state the amount of internal energy possessed by a thermodynamic system, but one cannot tell how much energy has flowed into or out of the system as a result of its being heated or cooled, nor as the result of work being performed on or by the system.

Energy transfer by heat can occur between objects either by radiation, conduction and convection. Energy can only be transferred by heat between objects - or areas within an object - with different temperatures (as given by the zeroth law of thermodynamics).

The zeroth law of thermodynamics is a generalization about the thermal equilibrium among bodies, or thermodynamic systems, in contact.

Many systems are said to be in equilibrium if the small exchanges between them do not lead to a net change in the total energy summed over all systems, in accordance with the First Law of Motion.

The "Brownian Motion" gives the description of the motion of bodies in a system where small exchanges between them do not lead to a net change in the total energy of the system.

If a hot high-energy atom comes into contact with a cool low-energy atom, the excited atom will loose some of its energy to the cool atom. The two atoms will settle into an energy level that's creates an equilibrium of temperature between the two. That level is called Thermal Equilibrium.

A system mass body is said to be in thermal equilibrium with its environment when its temperature does not change over time. When Body A, Body B, and Body C are distinct thermodynamic systems or bodies; then, if Body A, and Body C, are in thermal equilibrium with Body B, then Body A is also in thermal equilibrium with Body C.

Law 1.1: **Zeroth Law of Thermodynamics** — a body that is in thermal equilibrium with its environment will remain at that temperature unless another body comes in contact with it, which is of a different temperature than the first body; the result is a temperature difference and heat flow from the hot body to the cold body, and a new thermal equilibrium is established between both bodies and the environment.

$$\Delta T_{Temp} = -\Delta T_{Temp\,\text{Surroundings}} = \left[T'_{Temp\,\text{High}} - T_{Temp\,\text{Low}}\right] \rightarrow K \quad (1.1)$$

The first law of thermodynamics in essence states that a thermodynamic system can store energy and this stored energy, called internal energy, is conserved. The increase in the internal energy of a system is equal to the amount of energy added by heating the system, minus the amount lost as a result of the work done by the system on its surroundings.

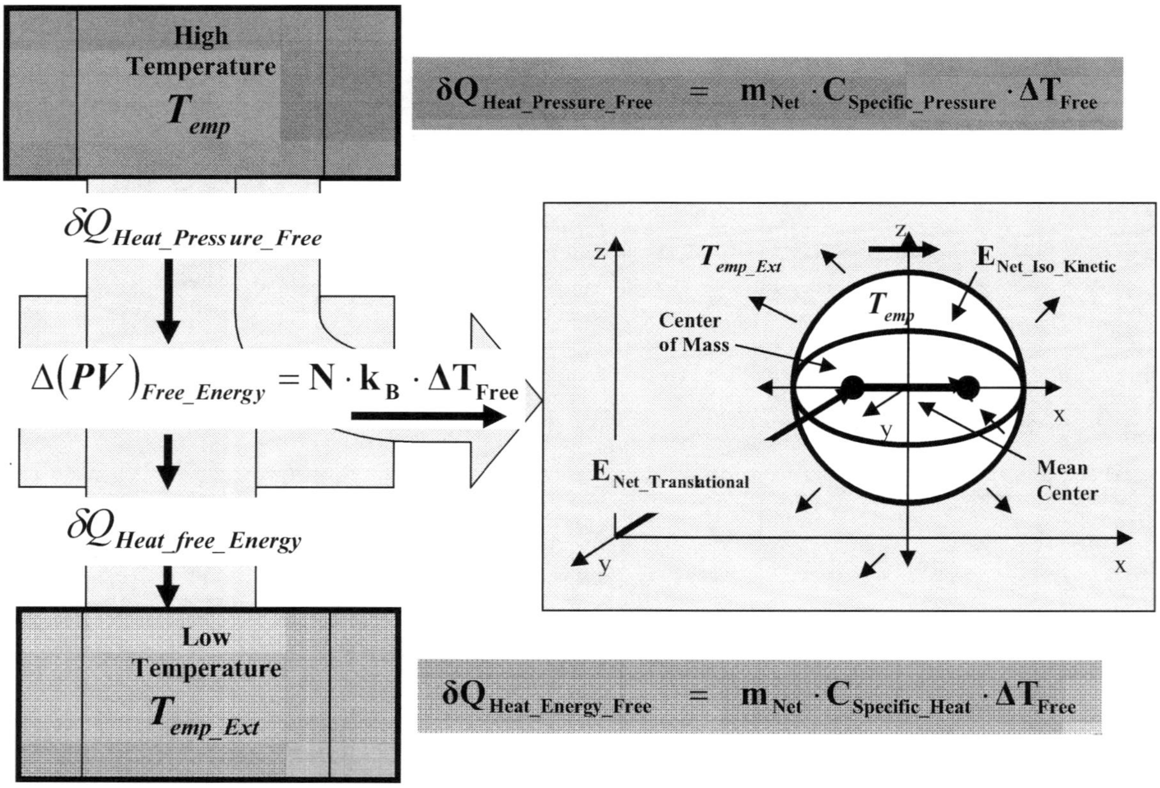

Figure 1.1: Aether Free Work Energy Difference is the measure of the irreversible heat energy that is lost from a system and is gained by the surroundings.

Heat is a process by which energy is either added to a system from a high-temperature source or lost from a system to a low-temperature sink. In addition, energy may be lost by the system when it does work on its surroundings, or conversely, energy may be gained as a result of work done to it by its surroundings.

When an amount of energy is transferred from one body to another solely as a result of the temperature differences between the bodies, we call that amount of energy "heat."

Bodies can transfer energy in this way through three processes: radiation, conduction, or convection. When the hotter body loses thermal energy to a cooler one the process continues until both bodies reach the same "equilibrium" temperature.

In the 18th century scientists thought that heat was something like a fluid which "flowed" from one body to another. Though this model has long been abandoned, much of our terminology still carries remnants of this archaic idea. Some books still say "heat is transferred from one body to another."

This should be understood as a shorthand way of saying "energy is transferred from one body to another through a thermal process." One should *not* say there is "heat in a body." We say instead that the heat given to a body raises that body's *internal energy*.

This usage is similar to the way we use the word *work*. We say that body (A) does work on body (B), but we do *not* say that this work was "*in* body A." Both "work" and "heat" are words which describe and measure the amount of energy transferred from one body to another, but we use other names to represent energy *in* or *possessed by* a body.

For a two body system the interaction between the two bodies produce heat, such that each body has its own absolute temperature; as each mass body approaches equilibrium between the system mass and the surroundings given by the mathematical relationships found later in this section.

The Net Internal Isotropic “Inertial” Omni-directional Kinetic Energy of body (1) and body (2) “Before” the interaction expressed as a function of the initial inertial velocities “Before” the interaction, and the inertial mass of both bodies is given by the following.

Net Internal Isotropic "Inertial" Omni-directional Kinetic Energy

1.1

$$E_{\text{Net\_Iso\_Kinetic}} = \frac{m_{\text{Net}} \cdot \left|v^2\right|_{\text{Iso}}}{2} = \frac{3 \cdot N \cdot k_B \cdot T_{Temp}}{2} \rightarrow kg \cdot m^2 / s^2$$

$$E_{\text{Net\_Iso\_Kinetic}} = E_{\text{Iso\_KE}_1} + E_{\text{Iso\_KE}_2} = \frac{m_1 \cdot \overline{v}_1^2}{2} + \frac{m_2 \cdot \overline{v}_2^2}{2}$$

$$E_{\text{Net\_Iso\_Kinetic}} = \frac{3 \cdot N \cdot k_B \cdot T_{Temp_1}}{2} + \frac{3 \cdot N \cdot k_B \cdot T_{Temp_2}}{2}$$

$$E_{\text{Net\_Iso\_Kinetic}} = E'_{\text{Iso\_KE}_1} + E'_{\text{Iso\_KE}_2} = \frac{m_1 \cdot \overline{v}_1'^2}{2} + \frac{m_2 \cdot \overline{v}_2'^2}{2}$$

$$E_{\text{Net\_Iso\_Kinetic}} = \frac{3 \cdot N \cdot k_B \cdot T'_{Temp_1}}{2} + \frac{3 \cdot N \cdot k_B \cdot T'_{Temp_2}}{2}$$

Net Internal Isotropic "Aether" Omni-directional Kinetic Energy

1.2

$$E_{\text{Net\_Iso\_Kinetic}} = \frac{m_{\text{Net}} \cdot \left|v^2\right|_{\text{Iso}}}{2} = \frac{3 \cdot N \cdot k_B \cdot T_{Temp}}{2} \rightarrow kg \cdot m^2 / s^2$$

$$E_{\text{Net\_Iso\_Kinetic}} = E_{\text{A\_Iso\_KE}_1} + E_{\text{A\_Iso\_KE}_2} = \frac{m_1 \cdot \overline{v}_{A1}^2}{2} + \frac{m_2 \cdot \overline{v}_{A2}^2}{2}$$

$$E_{\text{Net\_Iso\_Kinetic}} = \frac{3 \cdot N \cdot k_B \cdot T_{Aether\_Temp_1}}{2} + \frac{3 \cdot N \cdot k_B \cdot T_{Aether\_Temp_2}}{2}$$

$$E_{\text{Net\_Iso\_Kinetic}} = E'_{\text{A\_Iso\_KE}_1} + E'_{\text{A\_Iso\_KE}_2} = \frac{m_1 \cdot \overline{v}_{A1}'^2}{2} + \frac{m_2 \cdot \overline{v}_{A2}'^2}{2}$$

$$E_{\text{Net\_Iso\_Kinetic}} = \frac{3 \cdot N \cdot k_B \cdot T'_{Aether\_Temp_1}}{2} + \frac{3 \cdot N \cdot k_B \cdot T'_{Aether\_Temp_2}}{2}$$

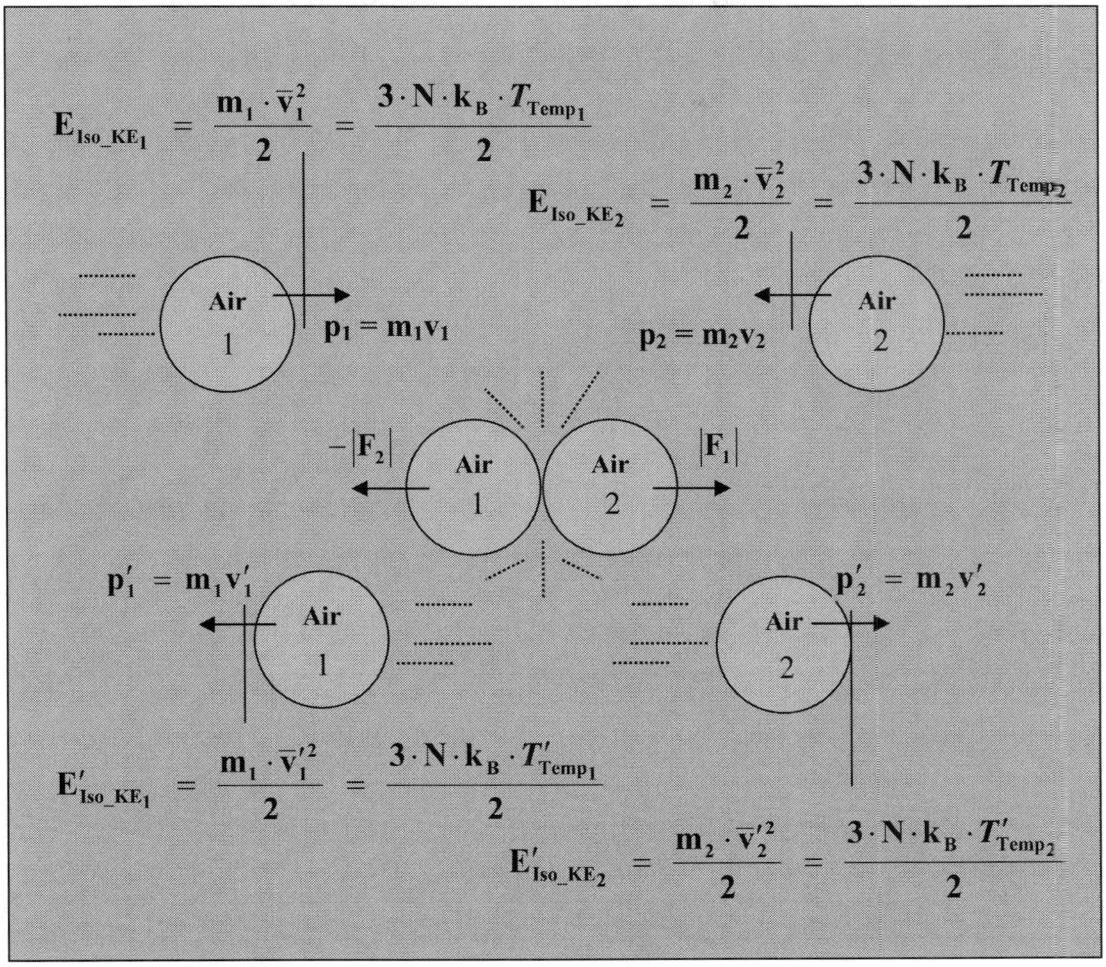

Figure 1.2: According to the Conservation of Energy the mass in motion of an isolated system mass body is constant; and likewise the Internal Absolute Isotropic Temperature is constant.

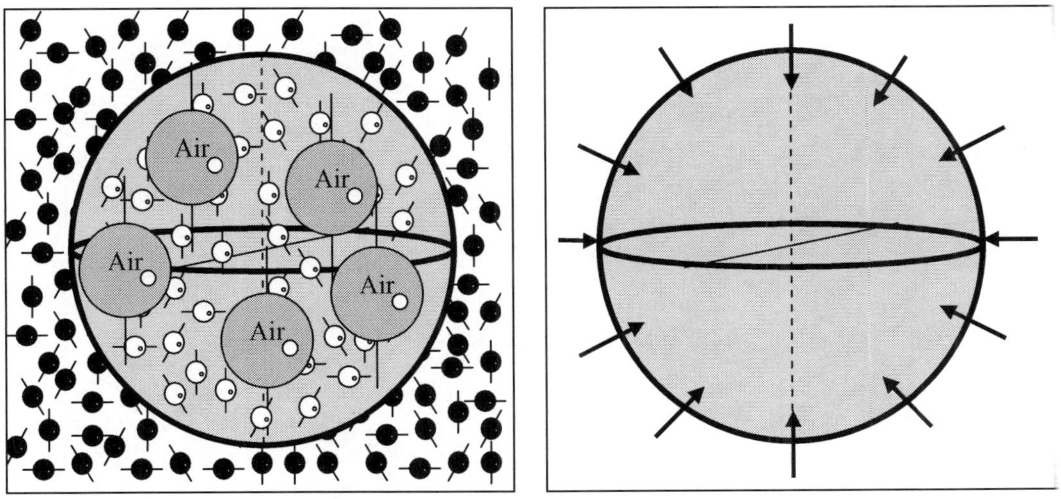

Figure 1.3: The isotropic pressure force can be used to describe Molecular motion as well as Aetheron motion.

Two Body — Net Internal Isotropic "Inertial" Omni-directional Kinetic Energy & Isotropic Absolute Temperature — "Before State" of Body (1)

1.3

$$E_{Iso\_KE_1} = \frac{m_1 \cdot \overline{v}_1^2}{2} = \frac{3 \cdot N \cdot k_B \cdot T_{Temp_1}}{2} \rightarrow kg \cdot m^2/s^2$$

$$T_{Temp_1} = \lfloor T_{Temp} - T_{Temp_2} \rfloor \rightarrow K$$

Two Body — Net Internal Isotropic "Inertial" Omni-directional Kinetic Energy & Isotropic Absolute Temperature — "Before State" of Body (2)

1.4

$$E_{Iso\_KE_2} = \frac{m_2 \cdot \overline{v}_2^2}{2} = \frac{3 \cdot N \cdot k_B \cdot T_{Temp_2}}{2} \rightarrow kg \cdot m^2/s^2$$

$$T_{Temp_2} = \lfloor T_{Temp} - T_{Temp_1} \rfloor \rightarrow K$$

The Net Internal Isotropic "Inertial" Omni-directional Kinetic Energy of body (1) and body (2) "After" the interaction expressed as a function of the initial inertial velocities "Before" the interaction, and the inertial mass of both bodies.

Two Body — Net Internal Isotropic "Inertial" Omni-directional Kinetic Energy & Isotropic Absolute Temperature — "After State" of Body (1)

1.5

$$E'_{\text{Iso\_KE}_1} = \frac{m_1 \cdot \bar{v}_1'^2}{2} = \frac{3 \cdot N \cdot k_B \cdot T'_{Temp_1}}{2} \rightarrow kg \cdot m^2 / s^2$$

$$T'_{Temp_1} = \left[m_1 \cdot \left(\left(\frac{m_1 - m_2}{m_1 + m_2} \right) \cdot \sqrt{\frac{T_{Temp_1}}{m_1}} + 2 \cdot \left(\frac{m_2}{m_1 + m_2} \right) \cdot \sqrt{\frac{T_{Temp_2}}{m_2}} \right)^2 \right]$$

$$T'_{Temp_1} = \left[\begin{aligned} & \left(\frac{m_1 - m_2}{m_1 + m_2} \right)^2 \cdot T_{Temp_1} + 4 \cdot \left(\frac{m_1}{m_1 + m_2} \right)^2 \cdot T_{Temp_2} \\ & + 4 \cdot \left(\frac{m_1 - m_2}{m_1 + m_2} \right) \cdot \left(\frac{\sqrt{m_1 \cdot m_2}}{m_1 + m_2} \right) \cdot \sqrt{T_{Temp_1} \cdot T_{Temp_2}} \end{aligned} \right] \rightarrow K$$

Two Body — Net Internal Isotropic "Inertial" Omni-directional Kinetic Energy & Isotropic Absolute Temperature — "After State" of Body (2)

1.6

$$E'_{\text{Iso\_KE}_2} = \frac{m_2 \cdot \bar{v}_2'^2}{2} = \frac{3 \cdot N \cdot k_B \cdot T'_{Temp_2}}{2} \rightarrow kg \cdot m^2 / s^2$$

$$T'_{Temp_2} = \left[m_2 \cdot \left(2 \cdot \left(\frac{m_1}{m_1 + m_2} \right) \cdot \sqrt{\frac{T_{Temp_1}}{m_1}} - \left(\frac{m_1 - m_2}{m_1 + m_2} \right) \cdot \sqrt{\frac{T_{Temp_2}}{m_2}} \right)^2 \right]$$

$$T'_{Temp_2} = \left[\begin{aligned} & 4 \cdot \left(\frac{m_2}{m_1 + m_2} \right)^2 \cdot T_{Temp_1} + \left(\frac{m_1 - m_2}{m_1 + m_2} \right)^2 \cdot T_{Temp_2} \\ & - 4 \cdot \left(\frac{m_1 - m_2}{m_1 + m_2} \right) \cdot \left(\frac{\sqrt{m_1 \cdot m_2}}{m_1 + m_2} \right) \cdot \sqrt{T_{Temp_1} \cdot T_{Temp_2}} \end{aligned} \right] \rightarrow K$$

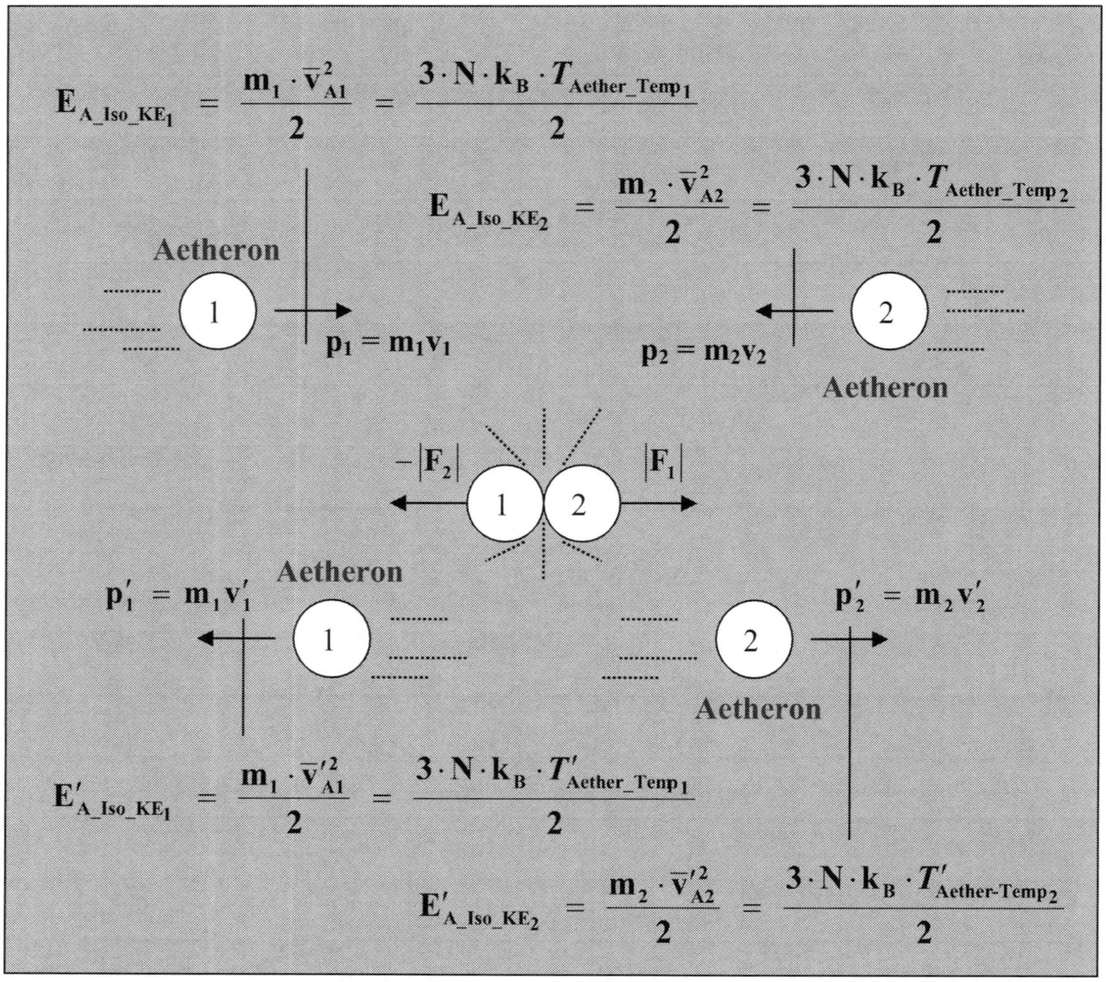

Figure 1.4: According to the Conservation of Energy the Aether mass in motion of an isolated system mass body is constant; and likewise the Internal Absolute Isotropic Temperature is constant.

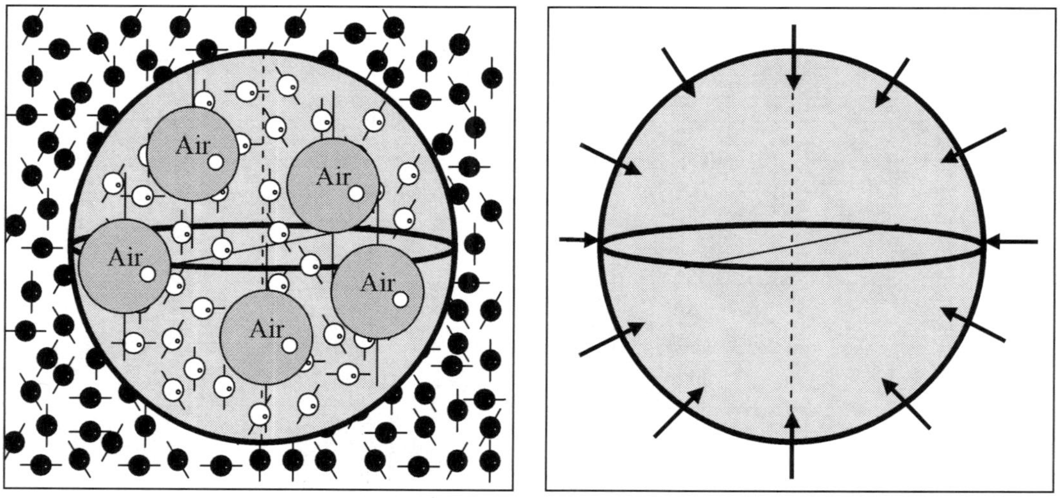

Figure 1.5: The isotropic pressure force can be used to describe Molecular motion as well as Aetheron motion.

The Net Internal Isotropic "Aether" Omni-directional Kinetic Energy of body (1) and body (2) "Before" the interaction expressed as a function of the initial inertial velocities "Before" the interaction, and the inertial mass of both bodies.

Two Body — Net Internal Isotropic "Aether" Omni-directional Kinetic Energy & Isotropic Absolute Temperature — "Before State" of Body (1)

1.7

$$E_{A\_Iso\_KE_1} = \frac{m_1 \cdot \bar{v}_{A1}^2}{2} = \frac{3 \cdot N \cdot k_B \cdot T_{Aether\_Temp_1}}{2} \rightarrow kg \cdot m^2 / s^2$$

$$T_{Aether\_Temp_1} = \left\lfloor T_{Temp} - T_{Aether\_Temp_2} \right\rfloor \rightarrow K$$

$$T_{Aether\_Temp_1} = \left[\left(\frac{m_1 - m_2}{m_1 + m_2}\right) \cdot T_{Temp_1} + 2 \cdot \left(\frac{m_1}{m_1 + m_2}\right) \cdot T_{Temp_2}\right]$$

$$T_{Aether\_Temp_1} = m_1 \cdot \left[\left(\frac{T_{Temp}}{m_{Net}}\right) + \left(\frac{m_2}{m_1 + m_2}\right) \cdot \left(\frac{T_{Temp_2}}{m_2} - \frac{T_{Temp_1}}{m_1}\right)\right]$$

Two Body — Net Internal Isotropic "Aether" Omni-directional Kinetic Energy & Isotropic Absolute Temperature — "Before State" of Body (2)

1.8

$$E_{A\_Iso\_KE_2} = \frac{m_2 \cdot \bar{v}_{A2}^2}{2} = \frac{3 \cdot N \cdot k_B \cdot T_{Aether\_Temp_2}}{2} \rightarrow kg \cdot m^2 / s^2$$

$$T_{Aether\_Temp_2} = \left\lfloor T_{Temp} - T_{Aether\_Temp_1} \right\rfloor \rightarrow K$$

$$T_{Aether\_Temp_2} = \left[2 \cdot \left(\frac{m_2}{m_1 + m_2}\right) \cdot T_{Temp_1} - \left(\frac{m_1 - m_2}{m_1 + m_2}\right) \cdot T_{Temp_2}\right]$$

$$T_{Aether\_Temp_2} = m_2 \cdot \left[\left(\frac{T_{Temp}}{m_{Net}}\right) - \left(\frac{m_1}{m_1 + m_2}\right) \cdot \left(\frac{T_{Temp_2}}{m_2} - \frac{T_{Temp_1}}{m_1}\right)\right]$$

The Net Internal Isotropic "Inertial/Aether" Omni-directional Kinetic Energy of body (1) and body (2) "After" the interaction expressed as a function of the final inertial velocities "After" the interaction, and as a function of the initial inertial velocities "Before" the interaction and the inertial mass of both bodies.

Two Body — Net Internal Isotropic "Aether" Omni-directional Kinetic Energy & Isotropic Absolute Temperature — "After State" of Body (1)

1.9

$$\mathrm{E}'_{\mathrm{A\_Iso\_KE}_1} = \frac{\mathrm{m}_1 \cdot \overline{\mathrm{v}}'^{2}_{\mathrm{A}1}}{2} = \frac{3 \cdot \mathrm{N} \cdot \mathrm{k_B} \cdot T'_{Aether\_Temp_1}}{2} \rightarrow kg \cdot m^2 / s^2$$

$$T'_{Aether\_Temp_1} = \left\lfloor T_{Temp} - T'_{Aether\_Temp_2} \right\rfloor \rightarrow K$$

$$T'_{Aether\_Temp_1} = \left[\left(\frac{\mathrm{m}_1 - \mathrm{m}_2}{\mathrm{m}_1 + \mathrm{m}_2} \right) \cdot T'_{Temp_1} + 2 \cdot \left(\frac{\mathrm{m}_1}{\mathrm{m}_1 + \mathrm{m}_2} \right) \cdot T'_{Temp_2} \right]$$

$$T'_{Aether\_Temp_1} = \mathrm{m}_1 \cdot \left[\begin{array}{c} \left[\dfrac{\mathrm{m}_1^3 - \mathrm{m}_2^3 + \mathrm{m}_1 \cdot \mathrm{m}_2 \cdot (5 \cdot \mathrm{m}_1 + 3 \cdot \mathrm{m}_2)}{(\mathrm{m}_1 + \mathrm{m}_2)^3} \right] \cdot \left(\dfrac{T_{Temp_1}}{\mathrm{m}_1} \right) \\ - 2 \cdot \left[\dfrac{\mathrm{m}_2^3 - \mathrm{m}_1^2 \cdot \mathrm{m}_2}{(\mathrm{m}_1 + \mathrm{m}_2)^3} \right] \cdot \left(\left(\dfrac{T_{Temp_2}}{\mathrm{m}_2} \right) - 2 \cdot \sqrt{\dfrac{T_{Temp_1} \cdot T_{Temp_2}}{\mathrm{m}_1 \cdot \mathrm{m}_2}} \right) \end{array} \right]$$

Two Body — Net Internal Isotropic "Aether" Omni-directional Kinetic Energy & Isotropic Absolute Temperature — "After State" of Body (2)

1.10

$$E'_{A\_Iso\_KE_2} = \frac{m_2 \cdot \bar{v}'^2_{A2}}{2} = \frac{3 \cdot N \cdot k_B \cdot T'_{Aether\_Temp_2}}{2} \rightarrow kg \cdot m^2 / s^2$$

$$T'_{Aether\_Temp_2} = \left\lfloor T_{Temp} - T'_{Aether\_Temp_1} \right\rfloor \rightarrow K$$

$$T'_{Aether\_Temp_2} = \left[2 \cdot \left(\frac{m_2}{m_1 + m_2} \right) \cdot T'_{Temp_1} - \left(\frac{m_1 - m_2}{m_1 + m_2} \right) \cdot T'_{Temp_2} \right]$$

$$T'_{Aether\_Temp_2} = m_2 \cdot \left[\begin{array}{l} \left[\dfrac{m_2^3 - m_1^3 + m_1 \cdot m_2 \cdot (3 \cdot m_1 + 5 \cdot m_2)}{(m_1 + m_2)^3} \right] \cdot \left(\dfrac{T_{Temp_2}}{m_2} \right) \\ - 2 \cdot \left[\dfrac{m_1^3 - m_1 \cdot m_2^2}{(m_1 + m_2)^3} \right] \cdot \left(\left(\dfrac{T_{Temp_1}}{m_1} \right) - 2 \cdot \sqrt{\dfrac{T_{Temp_1} \cdot T_{Temp_2}}{m_1 \cdot m_2}} \right) \end{array} \right]$$

1.3 The First Law of Thermodynamics

The first law of thermodynamics is an expression of the principle of conservation of energy for a thermodynamic system mass body.

For any thermodynamic system the amount of internal energy possessed by a thermodynamic system can easily be determined, however it is not easy to measure how much energy has flowed into or out of the system as a result of its being heated or cooled, nor as the result of work being performed on or by the system. In simple terms, this means that energy cannot be created or destroyed, only converted from one form to another.

Law 1.2: **First Law of Thermodynamics** — the external and internal energy of motion of an isolated system mass body and its surroundings is a total thermodynamic system whose total energy is conserved and can be transformed or changed from one form of energy to another, like in a thermodynamic cycle the net heat supplied to the system equals the net work done by the system; however the system can evolve and devolve into other forms of energy but, the total energy can neither be created nor destroyed.

Internal energy, Enthalpy energy, Available, and Aether Kinematic energy given below are thermodynamic state properties of the system whereas, work done, or heat supplied is added or subtracted from the system such that the system remains conserved.

1.11

$$\mathrm{U}_{\mathrm{Internal\_Energy}} = \left\lfloor H_{\mathrm{Enthalpy\_Energy}} - P_{Iso\_\Pr essure} \cdot \mathrm{V}_{\mathrm{ol}} \right\rfloor \rightarrow kg \cdot m^2 / s^2$$

1.12

$$\mathrm{U}_{\mathrm{Internal\_Energy}} = \left\lfloor H_{\mathrm{Available\_Energy}} + P_{Iso\_\Pr essure} \cdot \mathrm{V}_{\mathrm{ol}} \right\rfloor \rightarrow kg \cdot m^2 / s^2$$

The First Law of Thermodynamics is primarily used to explain how the above thermodynamic state properties are changed by heating, cooling, and work being done on or by the system; and whether the thermodynamic parameter is an exact or inexact differential during a process change.

1.4 The Second Law of Thermodynamics

The second law states that in an isolated mass system body, concentrated energy disperses over time, and consequently less concentrated energy is available to do useful work.

The second law can be stated in various succinct ways, including:

- It is impossible to produce work in the surroundings using a cyclic process connected to a single heat reservoir (Kelvin, 1851).
- It is impossible to carry out a cyclic process using an engine connected to two heat reservoirs that will have as its only effect the transfer of a quantity of heat from the low-temperature reservoir to the high-temperature reservoir (Clausius, 1854).

Law 1.3: **The Second Law of Thermodynamics.** – The entropy of an isolated system which is not in equilibrium will tend to increase over time, approaching a maximum value at equilibrium. The consequence of increasing entropy is that an isolated mass system body, which is concentrated energy, disperses over time, and consequently less concentrated energy is available to do useful work.

Heat cannot transfer from a colder to a hotter body. As a result of this fact of thermodynamics, natural processes that involve energy transfer must have one direction, and all natural processes are irreversible.

This law also predicts that the entropy of an isolated system always increases with time. Entropy is the measure of the disorder or randomness of energy and matter in a system.

Because of the second law of thermodynamics both energy and matter in the Universe are becoming less useful as time goes on. Perfect order in the Universe existed for a short time the instance after the Big Bang when energy and matter and all of the forces of the Universe were unified.

In simple terms, the second law is an expression of the fact that over time, ignoring the effects of self-gravity, differences in temperature, pressure, and density, the energy of the system as a whole tends to even out in a physical system that is isolated from the outside world.

Consider a Heat "Sink" Temperature Reservoir (T_{Death_Temp}) which is cooler than the hot Heat "Source" Temperature Reservoir (T_{emp}). Now consider that there is only one hot Heat "Source" Temperature Reservoir (T_{emp}) if there are no other heat reservoirs around and no other systems at a lower temperature which can be used as heat reservoirs, we cannot use any of the internal energy of the reservoir for the performance of work.

The internal energy of the reservoir is unavailable. To extract some of it and do work requires a lower-temperature reservoir or system to accept the heat rejected. Thus, according to the Second Law of Thermodynamics it is not possible to remove internal energy from a system and convert that energy completely into useful work; because some energy is always lost during this process.

In essence the conversion of internal energy completely into useful work would decrease the entropy of the universe; which is not allowed. Whenever an irreversible process occurs, some energy is made unavailable for doing useful work, which was available before the process took place.

Sadi Carnot (1796-1832) found that the efficiency of an engine operating on the Carnot cycle depends only on the temperatures of the hot and cold reservoirs. Carnot's formula gives the efficiency as $\left[1 - \left(\frac{T_{Death\_Temp}}{T_{emp}}\right)\right]$. Where the cold temperature reservoir is given by (T_{Death_Temp}) and hot temperature reservoir (T_{Temp}) in the Kelvin temperature scale. If (T_{Death_Temp}) is zero, then the efficiency is 1 or 100%.

The second law of thermodynamics states that the efficiency of any process must always be less than 100%. An alternate statement is that no heat engine can have efficiency greater than the efficiency of an engine operating with the Carnot cycle.

Aphorism 1.1: **Syntropy Change "Energy Dispersal" of the Universe** – During any irreversible process in which Syntropy Change "Energy Dispersal" of the Universe occurs, some energy becomes unavailable for the purpose of doing useful work; the amount of energy made unavailable

$$T_{Death\_Temp} \cdot \Delta S_{Syntropy} = \Delta(PV_{ol})_{Total\_Energy} \cdot \left[1 - \left(\frac{T_{Death\_Temp}}{T_{emp}}\right)\right]$$

where (T_{Death_Temp}) is the lowest temperature reservoir available.

If thermodynamic work is to be done at a finite rate, free energy must be expended. Energy dispersal also means that differences in temperature, pressure, and density even out.

If the temperature of the system is not constant, then the relationship becomes a differential equation and the entropy of the system is increased as described by the following

1.13

$$T_{emp} \neq T_{emp\_Ext} \rightarrow K$$

The change in entropy for the surroundings is given by the following

1.14

$$\Delta S_{Entropy\ Free\_Energy} = \left(\frac{\mathrm{V_{ol}}}{T_{emp}}\right) \cdot \left[\int_{P_{Iso\_Pressure}}^{q_{Dynamic\_Pressure}} d\left(P_{Iso\_Pressure}\right) \right]$$

$$\Delta S_{Entropy\ Free\_Energy} = \frac{\Delta(PV)_{\mathrm{Free\_Energy}}}{T_{emp}} \rightarrow kg \cdot m^2 / s^2 \cdot K$$

Then the total change in entropy for the surroundings is given by the following

1.15

$$\Delta S_{Entropy\ Free\_Surroundings} = \Delta(PV)_{\mathrm{Free\_Energy}} \cdot \left[\frac{1}{T_{emp\_Ext}} - \frac{1}{T_{emp}} \right] \rightarrow kg \cdot m^2 / s^2 \cdot K$$

$$\Delta S_{Entropy\ Free\_Surroundings} = \mathrm{N} \cdot \mathrm{k_B} \cdot \left[\frac{\left[T_{emp} - T_{emp\_Ext}\right]}{T_{emp\_Ext}} - \frac{\left[T_{emp} - T_{emp\_Ext}\right]}{T_{emp}} \right]$$

Then the production of entropy is given by the following

1.16

$$\Delta S_{Entropy\ Free\_Production} = \left[\frac{\Delta(PV)_{\mathrm{Free\_Energy}}}{T_{emp\_Ext}} \right] = \left[\begin{array}{l} \Delta S_{Entropy\ Free\_Energy} \\ + \Delta S_{Entropy\ Free\_Surroundings} \end{array} \right] \rightarrow kg \cdot m^2 / s^2 \cdot K$$

1.5 The Third Law of Thermodynamics

The third law of thermodynamics simply states that it is not possible to reach a temperature of absolute zero. It is possible to come arbitrarily close to absolute zero but not possible to reach that temperature by any process or series of processes.

If a temperature of absolute zero is impossible, then it is not possible for a Carnot cycle engine to reach 100% efficiency. The third law of thermodynamics closes the loophole that the Carnot cycle opens in the second law of thermodynamics.

When an object's temperature decreases its internal heat energy causes the random motions of individual atoms and molecules to decrease. Random molecular motions are faster at higher temperatures and slower at lower temperatures.

The lowest possible temperature is when these random atomic and molecular motions are at the minimum possible energy. This temperature is absolute zero, which is the zero point in the Kelvin temperature scale. Absolute zero is -273.15 degrees Celsius or -459.67 degrees Fahrenheit.

Absolute zero, at which all activity would stop if it were possible to happen, is −273.15 °C (degrees Celsius), or −459.67 °F (degrees Fahrenheit) or 0 K (kelvins, formerly sometimes degrees absolute).

As a system asymptotically approaches absolute zero of temperature all processes virtually cease and the entropy of the system asymptotically approaches a minimum value.

The third law of thermodynamics states that if all the thermal motion of molecules (kinetic energy) could be removed, a state called absolute zero or zero point energy would occur.

The Universe will attain absolute zero when all energy and matter is randomly distributed across space. The current temperature of empty space in the Universe is approximately 2.7 Kelvin degrees.

The third law was developed by the chemist Walther Nernst (1864-1941), during the years 1906-1912, and is thus sometimes referred to as Nernst's theorem or Nernst's postulate.

> "*it is impossible by any procedure, no matter how idealized, to reduce any system to the absolute zero of temperature in a finite number of operations*".

Thus the third law of thermodynamics states that the entropy of a system at absolute zero is a well-defined constant. This is because a system at zero temperature exists in its ground state, so that its entropy is determined only by the degeneracy of the ground state.

Law 1.3: **The Third Law of Thermodynamics.** — Every atomic substance has positive entropy and the change in entropy of a homogeneous system undergoing an isothermal reversible process approaches zero as the internal isotropic absolute temperature approaches absolute zero.

The absolute entropy of a substance can be calculated from measured thermodynamic properties by integrating the differential equations of state from absolute zero. For a gas this requires integrating through solid, liquid and gaseous phases.

In simple terms, the Third Law states that the entropy of most pure substances approaches zero as the absolute temperature approaches zero. This law provides an absolute reference point for the determination of entropy. The entropy determined relative to this point is the absolute entropy.

The ground state of a quantum mechanical system is its lowest-energy state; the energy of the ground state is known as the zero-point energy of the system. An excited state is any state with energy greater than the ground state. The ground state of a quantum field theory is usually called the vacuum state or the vacuum.

If more than one ground state exists, they are said to be degenerate. Many systems have degenerate ground states, for example, the hydrogen atom.

According to the third law of thermodynamics, a system at absolute zero temperature exists in its ground state; thus, its entropy is determined by the degeneracy of the ground state.

The Third Law of Thermodynamics can be visualized by thinking about water. Water in gas form has molecules that can move around very freely. Water vapor has very high entropy (randomness).

As the gas cools, it becomes liquid. The liquid water molecules can still move around, but not as freely. They have lost some entropy. When the water cools further, it becomes solid ice.

The solid water molecules can no longer move freely, but can only vibrate within the ice crystals. The entropy is now very low. As the water is cooled more, closer and closer to absolute zero, the vibration of the molecules diminishes.

If the solid water reached absolute zero, all molecular motion would stop completely. At this point, the water would have no entropy (randomness) at all.

Many systems, such as a perfect crystal lattice, have a unique ground state and therefore have zero entropy at absolute zero; because ($l\mathrm{n}(1)=0$).

1.17

$$\left(\int_{T_{emp\,Initial}}^{T'_{emp\,Final}} \frac{dT_{emp}}{T_{emp}}\right) = \left[\left(\int_{\mathrm{V}_{\mathrm{ol}\,Initial}}^{\mathrm{V}'_{\mathrm{ol}\,Final}} \frac{d\mathrm{V}_{\mathrm{ol}}}{\mathrm{V}_{\mathrm{ol}}}\right) + \left(\int_{P_{Iso\_\mathrm{Pr}essure\,Initial}}^{P'_{Iso\_\mathrm{Pr}essure\,Final}} \frac{dP_{Iso\_\mathrm{Pr}essure}}{P_{Iso\_\mathrm{Pr}essure}}\right)\right]$$

$$l\mathrm{n}\left(\frac{T'_{emp\,Final}}{T_{emp\,Initial}}\right) = \left[l\mathrm{n}\left(\frac{\mathrm{V}'_{\mathrm{ol}\,Final}}{\mathrm{V}_{\mathrm{ol}\,Initial}}\right) + l\mathrm{n}\left(\frac{P'_{Iso\_\mathrm{Pr}essure\,Final}}{P_{Iso\_\mathrm{Pr}essure\,Initial}}\right)\right] = \mathrm{Constant}$$

A pure crystal is the substance in which all the molecules are perfectly identical and the alignment of molecules with each other is perfectly uniform throughout the substance.

As per the third law of thermodynamics when such a substance is cooled to zero degree Kelvin, all the movements of all the molecules stop completely and the entropy of the substance becomes zero. This is an ideal condition.

In actuality there is no substance which has all the molecules identical and no movements of the molecules are perfectly uniform, hence in practical case at absolute zero the entropy is not zero, its value is above zero. This also means that the value of entropy can never be negative.

Let us consider another simple example of hot steam. Steam is the gaseous form of water at high temperature. The molecules within it move freely and hence it has high entropy.

If you cool this steam to below 100 degree Celsius it will get converted into water, where the movement of the molecules will be restricted resulting in decrease in entropy of water.

When this liquid is further cooled to below zero degrees Celsius, it gets converted into solid ice, where the movement of molecules is further reduced and the entropy of the substance further reduces.

As the temperature of this ice goes on reducing the movement of the molecules and along with it the entropy of the substance goes on reducing. When this is ice is cooled to absolute zero ideally the entropy should become zero.

But in practical situations it is just not possible to cool any substance to absolute zero temperature, nor does entropy become zero, but it remains always above zero.

The Quantum, Microscopic, and Macroscopic Thermodynamic Constant Entropy

The ground state of a quantum mechanical system is its lowest-energy state; the energy of the ground state is known as the zero-point energy of the system. The ground state is a constant related to the Microscopic Boltzmann Gas Energy-Temperature Constant at Constant Specific Heats as shown below.

1.18

$$\mathrm{N}\cdot\mathrm{k_B} = \mathrm{n_{Moles}}\cdot\mathrm{R} = \mathrm{m_{Net}}\cdot\left(C_{Specific\_Pressure} - C_{Specific\_Heat}\right) \rightarrow {kg\cdot m^2}/{s^2\cdot K}$$

Ideal Gas Energy Equation of State

$$\mathrm{N}\cdot\mathrm{k_B} = \frac{P'_{Iso\_Pressure\,Final}\cdot\left(\mathrm{V}'_{ol\,Final}\right)^{\gamma_{Heat}}}{T'_{emp\,Final}\cdot\left[\mathrm{V}'_{ol\,Final}\right]^{1\cdot\left[\gamma_{Heat}-1\right]}} = \frac{P_{Iso\_Pressure\,Initial}\cdot\left(\mathrm{V}_{ol\,Initial}\right)^{\gamma_{Heat}}}{T_{emp\,Initial}\cdot\left[\mathrm{V}_{ol\,Initial}\right]^{1\cdot\left[\gamma_{Heat}-1\right]}}$$

Entropy Equation of State

$$\mathrm{N}\cdot\mathrm{k_B} = \left[\left[T'_{emp\,Final}\cdot\left[\mathrm{V}'_{ol\,Final}\right]^{1\cdot\left[\gamma_{Heat}-1\right]}\right]\cdot\left[\frac{\left[P'_{Iso\_Pressure\,Final}\right]^{1\cdot\left[\gamma_{Heat}-1\right]}}{\left(T'_{emp\,Final}\right)^{\gamma_{Heat}}}\right]\right]^{\frac{1}{\left[\gamma_{Heat}-1\right]}}$$

$$\mathrm{N}\cdot\mathrm{k_B} = \left[\left[T_{emp\,Initial}\cdot\left[\mathrm{V}_{ol\,Initial}\right]^{1\cdot\left[\gamma_{Heat}-1\right]}\right]\cdot\left[\frac{\left[P_{Iso\_Pressure\,Initial}\right]^{1\cdot\left[\gamma_{Heat}-1\right]}}{\left(T_{emp\,Initial}\right)^{\gamma_{Heat}}}\right]\right]^{\frac{1}{\left[\gamma_{Heat}-1\right]}}$$

The Quantum, Microscopic, and Macroscopic Thermodynamic Constant Entropy

1.19

$$\mathrm{N} \cdot \mathrm{k_B} \;=\; \mathrm{n_{Moles}} \cdot \mathrm{R} \;=\; \mathrm{m_{Net}} \cdot \left(C_{Specific\_Pressure} \;-\; C_{Specific\_Heat} \right) \;\rightarrow\; {kg \cdot m^2}/{s^2 \cdot K}$$

$$\mathrm{N} \cdot \mathrm{k_B} \;=\; \frac{\Delta(PV)_{\mathrm{Field\_Energy}}}{T_{emp\ Initial} \cdot ln\left(\dfrac{T'_{emp\ Final}}{T_{emp\ Initial}} \right)} \;=\; \frac{\Delta(PV)_{\mathrm{Field\_Energy}}}{T_{emp\ Initial} \cdot [1 \;-\; \gamma_{Heat}] \cdot ln\left(\dfrac{\mathrm{V}'_{\mathrm{ol}\,Final}}{\mathrm{V}_{\mathrm{ol}\,Initial}} \right)}$$

$$\mathrm{N} \cdot \mathrm{k_B} \;=\; \frac{\Delta(PV)_{\mathrm{Field\_Energy}}}{T_{emp\ Initial} \cdot \left[1 \;-\; \dfrac{1}{\gamma_{Heat}} \right] \cdot ln\left(\dfrac{P'_{Iso\_Pressure\ Final}}{P_{Iso\_Pressure\ Initial}} \right)} \;\rightarrow\; {kg \cdot m^2}/{s^2 \cdot K}$$

If an object is a black body, the radiation is termed black-body radiation. The emitted wave frequency of the thermal radiation is a probability distribution depending only on temperature and for a genuine black body is given by Planck's law of radiation.

Albert Einstein and Otto Stern carried out an analysis of the specific heat of hydrogen gas at low temperature, and concluded that the data are best represented if the vibrational energy is taken to have the form:

Average - Vibrational Energy of Quantized Linear Harmonic Oscillator

1.20

$$\in_{\mathrm{Thermal\_Energy}} \;=\; \left\lfloor \in_{\mathrm{Zero\_Point\_Energy}} \;+\; \in_{\mathrm{Radiation\_Energy}} \right\rfloor \;\rightarrow\; {kg \cdot m^2}/{s^2}$$

$$\in_{\text{Thermal\_Energy}} = \left[\in_{\text{Zero\_Point\_Energy}} + \in_{\text{Radiation\_Energy}}\right]$$

1.21

$$\in_{\text{Thermal\_Energy}} = \left[\frac{h_{Planck} \cdot f_{Photon}}{\left(e^{\frac{h_{Planck} \cdot \nu}{k_B \cdot T_{Temp}}} - 1\right)} + \frac{h_{Planck} \cdot f_{Photon}}{2}\right] \rightarrow kg \cdot m^2 / s^2$$

$$\in_{\text{Thermal\_Energy}} = \left[\frac{\hbar \cdot \omega_{Photon}}{\left(e^{\frac{\hbar \cdot \omega}{k_B \cdot T_{Temp}}} - 1\right)} + \frac{\hbar \cdot \omega_{Photon}}{2}\right]$$

The second term in the above energy quantization equation is the Thermal Excitation Energy. According to the above expression, an atomic system at absolute zero temperature ($T_{\text{Temp}} = 0$) retains a quantized amount of energy given by the following derivation.

Letting the following ratio be called the Thermal Excitation Energy Ratio and setting the absolute temperature equal to zero ($T_{Temp} = 0$) in the Zero-Point Energy equation yields the following result.

1.22

$$\psi_{\text{Temp\_Quantization}} = \frac{h_{Planck} \cdot f_{Photon}}{k_B \cdot T_{Temp}} = \frac{h_{Planck} \cdot f_{Photon}}{0} = \infty$$

Substituting the above equation into the root of the equation below yields the following result

1.23

$$e^{\frac{h_{Planck} \cdot f_{Photon}}{k_B \cdot T_{Temp}}} = e^{\frac{h_{Planck} \cdot f_{Photon}}{0}} = e^{\infty} = \infty$$

$$\in_{\text{Thermal\_Energy}} = \frac{h_{Planck} \cdot f_{Photon}}{(\infty - 1)} + \frac{h_{Planck} \cdot f_{Photon}}{2}$$

$$\in_{\text{Thermal\_Energy}} = \in_{\text{Radiation\_Energy}} = \frac{h_{Planck} \cdot f_{Photon}}{2}$$

1.6 The Fourth Law of Thermodynamics

The fourth law of thermodynamics simply states that for any isolated system mass body and its atmospheric environment temperature difference causes heat energy to flow in one direction and aether mass density differences causes matter to flow in the opposite direction.

The way in which a system transfers's energy internal and external to the system is via two distinct heat engine mechanisms called ***Inertial & Aether Mass Kinetic Energy Transfer mechanisms.***

The first heat energy transfer mechanism is called an ***Inertial Mass Kinetic Energy Transfer mechanism***, which behaves like a Carnot Thermal Heat Engine does work transferring heat energy between two different temperature reservoirs, with a goal to establish equilibrium between an inertial net mass body and its atmospheric surroundings medium.

The ***Inertial Mass Kinetic Energy – Carnot Heat Engine- Transfer mechanism*** is used to transfer kinetic heat energy, via kinetic pressure, and temperature differences between the Isotropy and the Anisotropy of the isolated system and does work in the process.

The Work done by Inertial Mass Kinetic Energy Transfer mechanism of the system mass body is to transfer kinetic energy in the form of heat energy based on: Temperature difference, Kinetic Energy difference, and Pressure differences; between Internal Isotropy and External Anisotropy reservoirs internal and external to an isolated system mass body.

The second heat energy transfer mechanism inherent to an isolated system mass body is called a **Carnot Thermal Heat Pump** - ***Aether Mass Kinetic Energy Transfer mechanism***, which functions like a Thermal Heat Pump doing work transferring heat energy between two different density reservoirs, with a goal to establish equilibrium between an inertial net mass body and its atmospheric surroundings medium.

The ***Aether Mass Kinetic Energy Transfer*** is also the work of a Rarefaction/Condensing mechanism of the system mass body used to reduce or increase the aether mass density.

The Work done by ***Aether Mass Kinetic Energy –Carnot Heat Pump - Transfer mechanism*** is to transfer kinetic energy in the form of heat energy based on Aether mass-energy difference, Aether density difference , and Aether Pressure differences between Internal Isotropy and External Anisotropy reservoirs internal and external to an isolated system mass body.

Law 1.4: **The Fourth Law of Thermodynamics.** – An isolated system mass body and its environment will establish thermodynamic equilibrium by heating and cooling the system via temperature difference in one direction and condensing and rarefying the system via aether mass density difference in the opposite direction and vice versa depending on the ratio of the anisotropy to isotropy of the system. Mass density difference of a system flows opposite to temperature differences of the system.

Based on the above discussion of work and heat energy transfer, a dynamic system mass body and its atmospheric environment perform work via two distinct heat energy transfer mechanisms.

1. Inertial Mass Kinetic Heat Energy Transfer mechanism
   a. Behaves like a ***Carnot Heat Engine***

2. Aether Mass Density and Kinetic Energy Transfer mechanism
   a. Behaves like a ***Carnot Heat Pump***

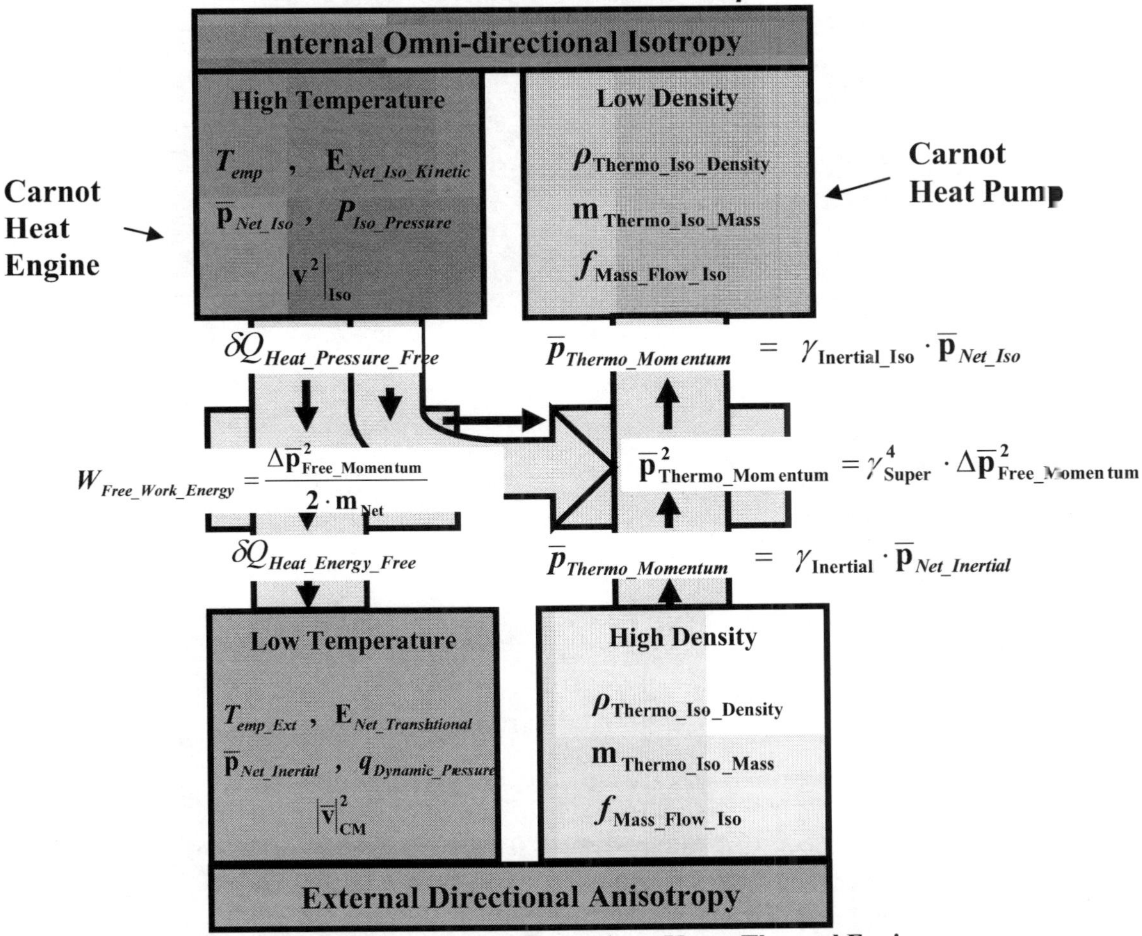

Figure 1.6: Heat Energy Transfer Device – Kemp Thermal Engine.

1.7 Specific Heat Capacity of an Atomic Substance

The Specific Heat Capacity of a system mass body is the capacity of the body to store heat and release heat. The Specific Heat Capacity is the amount of heat required to change the temperature of the Net Inertial Mass or Mole of a substance by one degree.

More heat energy is required to increase the temperature of a substance with high specific heat capacity than one with low specific heat capacity. For example, eight times the heat energy is required to increase the temperature of mass of magnesium than is required for the lead having the same mass. And a tub of water has a greater heat capacity than a cup of water.

The specific heat of virtually any substance can be measured, including chemical elements, compounds, alloys, solutions, and composites.

There are three distinctly different experimental conditions under which specific heat capacity is measured and these are denoted by:

- Specific Heat function of Temperature — $C_{Specific\_Temperature} = \left(\frac{\mathrm{N} \cdot \mathrm{k_B}}{\mathrm{m_{Net}}}\right)$
- Specific Heat function of Pressure — $C_{Specific\_Pressure}$
- Specific Heat function of Volume — $C_{Specific\_Heat}$

The specific heats of substances are typically measured under constant pressure ($C_{Specific\_Pressure}$). However, fluids such as gases and liquids are typically also measured at constant volume ($C_{Specific\_Heat}$). Measurements under constant pressure produce greater values than those at constant volume because work must be performed in energy transfer between Specific Heat at Constant Pressure and the Specific Heat at Constant Volume.

1.24

$$C_{Specific\_Temperature} = \left(\frac{\mathrm{N} \cdot \mathrm{k_B}}{\mathrm{m_{Net}}}\right) = \left(C_{Specific\_Pressure} - C_{Specific\_Heat}\right) = \text{Constant}$$

These differences in specific heats in gases under constant pressure are typically 30% to 66.7% greater than those at constant volume. This means that one needs to know at all times where all parts of the system are, how much mass they have, and how fast they are moving.

This information is used to account for different ways that heat of Adiabatic Isentropic Thermodynamic Fields can be stored as kinetic energy (energy of motion) and potential energy (energy stored in force fields), as an object expands or contracts.

For all real systems, the path through these changes must be explicitly defined, since the value of heat capacity depends on which path from one temperature to another, is chosen. Of particular usefulness in this context are the values of heat capacity for constant volume, ($C_{Specific\_Heat}$), and constant pressure, ($C_{Specific\_Pressure}$). These will be defined below.

Measuring the heat capacity at constant volume can be very difficult for liquids and solids. That is, small temperature changes typically require large pressures to maintain a liquid or solid at constant volume; implying the containing vessel must be nearly rigid or at least very strong.

It is easier to measure the heat capacity at constant pressure (allowing the material to expand or contract as it wishes) and solve for the heat capacity at constant volume using mathematical relationships derived from the basic thermodynamic laws.

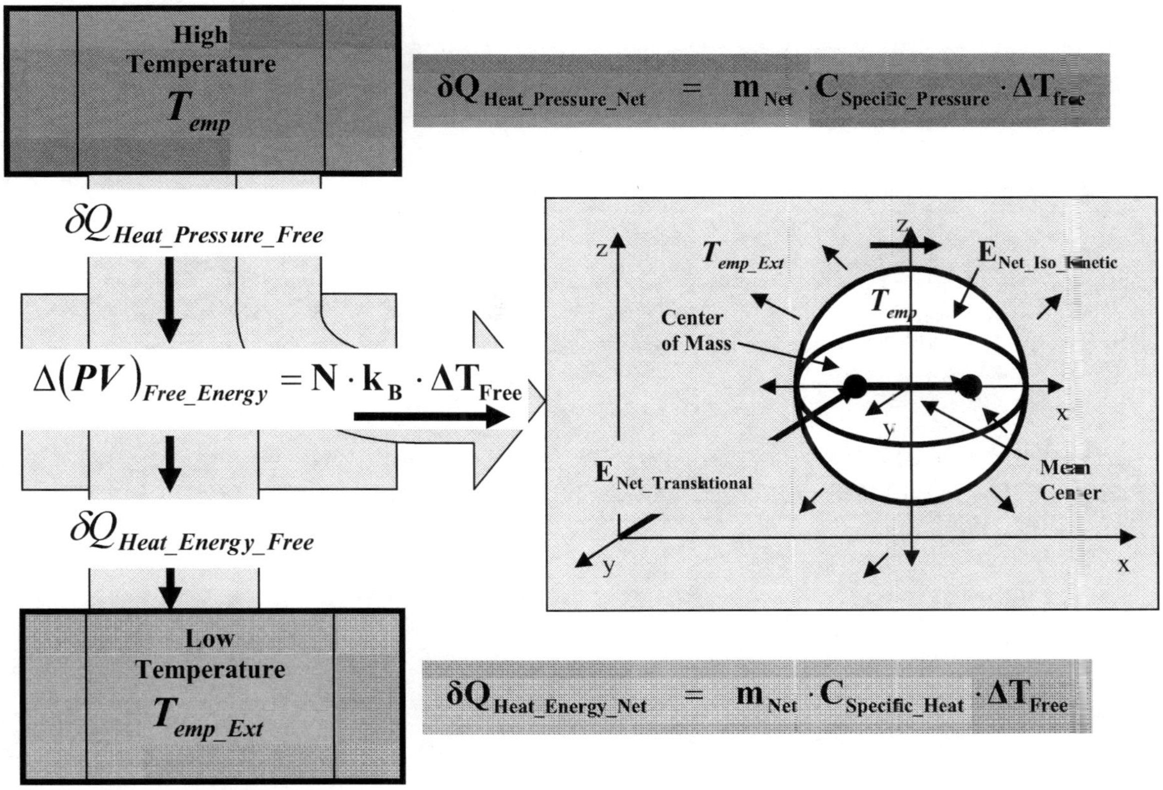

Figure 1.7: The total Aether Kinematic Work Energy transfer is equal to the difference in the high thermodynamic potential subtracted from the low thermodynamic potential.

Note that at constant volume all of the heat absorbed by the gas goes into increasing its internal energy, and, hence, its temperature, whereas at constant pressure some of the absorbed heat is used to do work on the environment as the volume increases.

This means that, in the latter case, less heat is available to increase the temperature of the gas. Thus, we expect the specific heat at constant pressure to exceed that at constant volume, as indicated by the above formula.

The heat capacity ratio or adiabatic index or ratio of specific heats, is the ratio of the heat capacity at constant pressure ($C_{Specific\_Pressure}$) to heat capacity at constant volume ($C_{Specific\_Heat}$). It is sometimes also known as the isentropic expansion factor

As temperature increases, higher energy rotational and vibrational states become accessible to molecular gases, thus increasing the number of degrees of freedom and lowering γ. For a real gas, both ($C_{Specific\_Pressure}$) and ($C_{Specific\_Heat}$) increase with increasing temperature, while continuing to differ from each other by a fixed constant ($C_{Specific\_Temperature} = \left(\frac{N \cdot k_B}{m_{Net}}\right) = \left(C_{Specific\_Pressure} - C_{Specific\_Heat}\right)$) which reflects the relatively constant (P^*V_{ol}) difference in work done during expansion, for constant pressure vs. constant volume conditions.

An example of specific heating for an Adiabatic Thermodynamic Field is to consider the following experiment:

A closed cylinder with a locked piston contains air. The pressure inside is equal to the outside air pressure. This cylinder is heated to a certain target temperature. Since the piston cannot move, the volume is constant, while temperature and pressure rise.

When the target temperature is reached, the heating is stopped. The piston is now freed and moves outwards, expanding without exchange of heat (adiabatic expansion).

Doing this work cools the air inside the cylinder to below the target temperature. To return to the target temperature (still with a free piston), the air must be heated. This extra heat amounts to about 40% of the previous amount added. In this example, the amount of heat added with a locked piston is proportional to ($C_{Specific\_Heat}$), whereas the total amount of heat added is proportional to ($C_{Specific\_Pressure}$). Therefore, the heat capacity ratio in this example is 1.4.

Another way of understanding the difference between ($C_{Specific\_Pressure}$) and ($C_{Specific\_Heat}$) is that ($C_{Specific\_Pressure}$) applies if work is done to the system which causes a change in volume (e.g. by moving a piston so as to compress the contents of a cylinder). Or if work is done by the system which changes its volume (e.g. heating the gas in a cylinder to cause a piston to move).

($C_{Specific\_Heat}$) applies only the work done - is zero. Consider the difference between adding heat to the gas with a locked piston, and adding heat with a piston free to move, so that pressure remains constant. In the second case, the gas will both heat and expand, causing the piston to do mechanical work on the atmosphere.

The heat that is added to the gas goes only partly into heating the gas, while the rest is transformed into the mechanical work performed by the piston. In the first, constant-volume case (locked piston) there is no external motion and thus no mechanical work is done on the atmosphere; ($C_{Specific\_Heat}$) is used.

In the second case, additional work is done as the volume changes, so the amount of heat required to raise the gas temperature (the specific heat capacity) is higher for this constant pressure case Thus, the ratio of the two values, (γ_{Heat}), decreases with increasing temperature.

Specific Heat Capacity Adiabatic Index Ratio

Definition 1.1: The **Specific Heat Capacity Adiabatic Index Ratio** (γ_{Heat}) is a unit-less ratio quantity defined as amount of heat energy of spacetime at constant pressure divided into the amount of heat energy of spacetime at constant volume; and is defined as the ratio of the **Specific Heat Capacity of an Atomic Substance** at constant **Pressure** ($C_{Specific\_Pressure}$) divided by the **Specific Heat Capacity of an Atomic Substance** at constant **Volume** ($C_{Specific\_Heat}$).

Specific Heat Capacity Adiabatic Index Ratio

1.25

$$\gamma_{Heat} = \left(\frac{C_{Specific\_Pressure}}{C_{Specific\_Heat}} \right) = \left(1 + \frac{\mathrm{N} \cdot \mathrm{k_B}}{\mathrm{m_{Net}} \cdot C_{Specific\_Heat}} \right) \rightarrow \text{Unit-less}$$

Specific Heat Capacity Adiabatic Index Ratio

1.26

$$\gamma_{Heat} = \left(\frac{C_{Specific\_Pressure}}{C_{Specific\_Heat}}\right) = \left(1 + \left(\frac{C_{Specific\_Temperature}}{C_{Specific\_Heat}}\right)\right)$$

$$\gamma_{Heat} = \left(\frac{C_{Specific\_Heat} + C_{Specific\_Temperature}}{C_{Specific\_Heat}}\right) \rightarrow \text{Unit-less}$$

The heat capacity ratio (γ_{Heat}) for an ideal gas can be related to the degrees of freedom ($f_{Degrees\_Freedom}$) of a molecule by:

$$\gamma_{Heat} = \left(\frac{f_{Degrees\_Freedom} + 2}{f_{Degrees\_Freedom}}\right)$$

Thus we observe that for a monatomic gas, with three degrees of freedom ($f_{Degrees\_Freedom} = 3$):

$$\gamma_{Heat} = \left(\frac{3 + 2}{3}\right) = \left(\frac{5}{3}\right)$$

The terrestrial air is primarily made up of diatomic gasses (~78% nitrogen (N_2) and ~21% oxygen (O_2)) and, at standard conditions it can be considered to be an ideal gas.

A diatomic molecule has five degrees of freedom (three translational and two rotational degrees of freedom). This result in a value for a diatomic gas, with five degrees of freedom (at room temperature) ($f_{Degrees\_Freedom} = 5$):

$$\gamma_{Heat} = \left(\frac{5 + 2}{5}\right) = \left(\frac{7}{5}\right)$$

Specific Heat Capacity Adiabatic Index Ratio – Table of various gases

Heat Capacity Ratio for various gases

Temp.	Gas	γ_{Heat}	Temp.	Gas	γ_{Heat}	Temp.	Gas	γ_{Heat}
−181°C	H_2	1.597	200°C	Dry Air	1.398	20°C	NO	1.400
−76°C		1.453	400°C		1.393	20°C	N_2O	1.310
20°C		1.410	1000°C		1.365	−181°C	N_2	1.470
100°C		1.404	2000°C		1.088	15°C		1.404
400°C		1.387	0°C	CO_2	1.310	20°C	Cl_2	1.340
1000°C		1.358	20°C		1.300	−115°C	CH_4	1.410
2000°C		1.318	100°C		1.281	−74°C		1.350
20°C	He	1.660	400°C		1.235	20°C		1.320
20°C	H_2O	1.330	1000°C		1.195	15°C	NH_3	1.310
100°C		1.324	20°C	CO	1.400	19°C	Ne	1.640
200°C		1.310	−181°C	O_2	1.450	19°C	Xe	1.660
−180°C	Ar	1.760	−76°C		1.415	19°C	Kr	1.680
20°C		1.670	20°C		1.400	15°C	SO_2	1.290
0°C	Dry Air	1.403	100°C		1.399	360°C	Hg	1.670
20°C		1.400	200°C		1.397	15°C	C_2H_6	1.220
100°C		1.401	400°C		1.394	16°C	C_3H_8	1.130

Figure 1.8: Specific Heat Capacity Adiabatic Index Ratio – Table of various gases.

Heat energy is stored in internal motions of substances, on a per-atom basis; the heat capacity of molecules does not exceed the heat capacity of monatomic gases, unless vibrational modes are brought into play.

In molecules, however, rotational modes may become active due to higher moments of inertia about certain axes. Further internal vibrational degrees of freedom also may become active.

For instance, nitrogen, which is a diatomic molecule, has five active degrees of freedom at room temperature: the three comprising translational motion plus two rotational degrees of freedom internally.

Two separate nitrogen atoms form a molecule that has a total of six degrees of freedom – which comprise the three translational degrees of freedom of each atom. When two or more atoms are bonded together the molecule will still only have three translational degrees of freedom, as the two atoms of the molecule move as one.

However, the molecule cannot be treated as a point object, and the moment of inertia will increased sufficiently about two axes to allow two rotational degrees of freedom to be active at room temperature to give five degrees of freedom. The moment of inertia about the third axis remains small, as this is the axis is passing through the center of the two atoms, and so is similar to that of a monatomic gas.

At higher temperatures, however, nitrogen gas gains two more degrees of internal freedom, as the molecule is excited into higher vibrational modes which store heat energy, and then the heat capacity per volume or mole of molecules approaches seven-thirds that of monatomic gases, or seven-sixths of monatomic, on a mole-of-atoms basis. This is now a higher heat capacity per atom than the monatomic figure, because the vibrational mode enables an extra degree of potential energy freedom per pair of atoms, which monatomic gases cannot possess.

Thus, it is the heat capacity per-mass-of-atoms, not per-mass-of-molecules, which come closest to being a constant for all substances at high temperatures. For this reason, some care should be taken to specify a mass-of-molecules basis vs. a mass-of-atoms basis, when comparing specific heat capacities of molecular solids and gases.

Ideal gases have the same numbers of molecules per volume, so increasing molecular complexity adds heat capacity on a per-volume and per-mass-of-molecules basis, but may lower or raise heat capacity on a per-atom basis, depending on whether the temperature is sufficient to store energy as atomic vibration.

In solids, the limit of heat capacity in general is about $(3 \cdot R_{Gas} = 3 \cdot N_{Avogadro} \cdot k_B)$ per mole of atoms, where (R_{Gas}) is the ideal gas constant. Six degrees of freedom (3 kinetic and 3 potential) are available to each atom. The equation defining the heat capacity at constant temperature is given by the following.

1.27

$$C_{Specific\_Temperature} = \left(\frac{N \cdot k_B}{m_{Net}}\right) = \left(\frac{N \cdot R}{N_{Avogadro} \cdot m_{Net}}\right) = \left(\frac{n_{Moles} \cdot R}{m_{Net}}\right) \to \frac{m^2}{s^2 \cdot K}$$

Each of these 6 contributes $(\frac{1}{2} \cdot R_{Gas} = \frac{1}{2} \cdot N_{Avogadro} \cdot k_B)$ specific heat capacity per mole of atoms. For monatomic gases, the specific heat is only half of this $(\frac{3}{2} \cdot R_{Gas} = \frac{3}{2} \cdot N_{Avogadro} \cdot k_B$ per mole) due to loss of all potential energy degrees of freedom.

For polyatomic gases, the heat capacity will be intermediate between these values on a per-mole-of-atoms basis, and (for heat-stable molecules) would approach the limit of $(3 \cdot R_{Gas} = 3 \cdot N_{Avogadro} \cdot k_B)$ per mole of atoms for gases composed of complex molecules with all vibrational modes excited.

This is because complex gas molecules may be thought of as large blocks of matter which have lost only a small fraction of degrees of freedom, as compared to the fully integrated solid.

Definition 1.2: The amount of **Specific Heat Capacity of an Atomic Substance** at constant **Temperature** **(**$C_{Specific\_Temperature}$**)** is the amount of heat required to change the temperature of the **Net Inertial Mass** or **Mole of a Substance** by one degree, at constant temperature; and is defined as the **Number of particles** multiplied by the **Microscopic Boltzmann Gas Energy-Temperature Constant** divided by the Net Inertial Mass of a substance system mass body; and likewise is equal to the **Specific Heat Capacity of an Atomic Substance** at constant **Pressure** **(**$C_{Specific\_Pressure}$**)** Subtracted from the **Specific Heat Capacity of an Atomic Substance** at constant **Volume** **(**$C_{Specific\_Heat}$**)**.

Specific Heat Capacity of an Atomic Substance — Constant Temperature

1.28

$$C_{Specific\_Temperature} = \left(\frac{\mathrm{N}\cdot \mathrm{k_B}}{\mathrm{m_{Net}}}\right) = \frac{1}{\mathrm{m_{Net}}}\cdot\left(\frac{\Delta(PV)_{\mathrm{Free\_Energy}}}{\Delta T_{Free}}\right) = \text{Constant}$$

$$C_{Specific\_Temperature} = \left(\frac{\mathrm{N}\cdot \mathrm{k_B}}{\mathrm{m_{Net}}}\right) = \left(C_{Specific\_Pressure} - C_{Specific\_Heat}\right) = \text{Constant}$$

$$C_{Specific\_Temperature} = C_{Specific\_Pressure}\cdot\left(1 - \left(\frac{C_{Specific\_Heat}}{C_{Specific\_Pressure}}\right)\right)$$

$$C_{Specific\_Temperature} = \left(1 - \frac{1}{\gamma_{Heat}}\right)\cdot C_{Specific\_Pressure}$$

$$C_{Specific\_Temperature} = \left(\frac{\mathrm{N}\cdot \mathrm{k_B}}{\mathrm{m_{Net}}}\right) = \frac{1}{3}\cdot\frac{\left|\mathrm{v}^2\right|_{\mathrm{Iso}}}{T_{emp}} = \frac{1}{3}\cdot\frac{\left|\bar{\mathrm{v}}\right|^2_{\mathrm{CM}}}{T_{emp\_Ext}}$$

$$C_{Specific\_Temperature} = \left(\frac{\mathrm{N}\cdot \mathrm{k_B}}{\mathrm{m_{Net}}}\right) = \frac{1}{\gamma_{Heat}}\cdot\frac{\mathrm{c^2_{Sound}}}{T_{emp}} \rightarrow \frac{\mathrm{m}^2}{\mathrm{s}^2\cdot K}$$

The **Specific Heat Capacity of an Atomic Substance — Constant Temperature,** in three dimensional Cartesian coordinates x, y, and z, are given by the following,

1.29

$$C_{Specific\_Temperature_x} = \left(\frac{N \cdot k_B}{m_{Net}}\right)_x = \frac{1}{3}\cdot\left(\frac{\left|v^2\right|_{Iso\,x}}{T_{emp}}\right) = \frac{1}{3}\cdot\left(\frac{\left|\overline{v}\right|^2_{CM\,x}}{T_{emp\_Ext}}\right)$$

$$C_{Specific\_Temperature_y} = \left(\frac{N \cdot k_B}{m_{Net}}\right)_y = \frac{1}{3}\cdot\left(\frac{\left|v^2\right|_{Iso\,y}}{T_{emp}}\right) = \frac{1}{3}\cdot\left(\frac{\left|\overline{v}\right|^2_{CM\,y}}{T_{emp\_Ext}}\right)$$

$$C_{Specific\_Temperature_z} = \left(\frac{N \cdot k_B}{m_{Net}}\right)_z = \frac{1}{3}\cdot\left(\frac{\left|v^2\right|_{Iso\,z}}{T_{emp}}\right) = \frac{1}{3}\cdot\left(\frac{\left|\overline{v}\right|^2_{CM\,z}}{T_{emp\_Ext}}\right)$$

1.30

Scalar/Tensor		
Magnitude	Magnitude/Tensor	Units
Specific Heat Capacity of an Atomic Substance — Constant Temperature		
$C_{Specific\_Temperature} = \begin{bmatrix} C_{Specific\_Temperature_x} \\ +C_{Specific\_Temperature_y} \\ +C_{Specific\_Temperature_z} \end{bmatrix}$	$C_{Specific\_Temperature} = \begin{bmatrix} C_{Specific\_Temperature_x} \\ C_{Specific\_Temperature_y} \\ C_{Specific\_Temperature_z} \end{bmatrix}$	$\frac{m^2}{s^2 \cdot K}$

1.8 Carnot's Thermodynamic Engine "Anisotropy" Efficiency Factor

Carnot's theorem, also called Carnot's rule is a principle which sets a limit on the maximum amount of efficiency any possible ***anisotropy heat engine*** can obtain, which thus solely depends on the difference between the hot and cold temperature reservoirs isotropy and anisotropy. Carnot's theorem states:

"No engine operating between two heat reservoirs can be more efficient than a Carnot engine operating between the same reservoirs."

Carnot's ***Anisotropy Heat Engine*** theorem sets essential limitations on the yield of a cyclic heat engine such as steam engines or internal combustion engines, which operate on the Anisotropic Carnot Cycle.

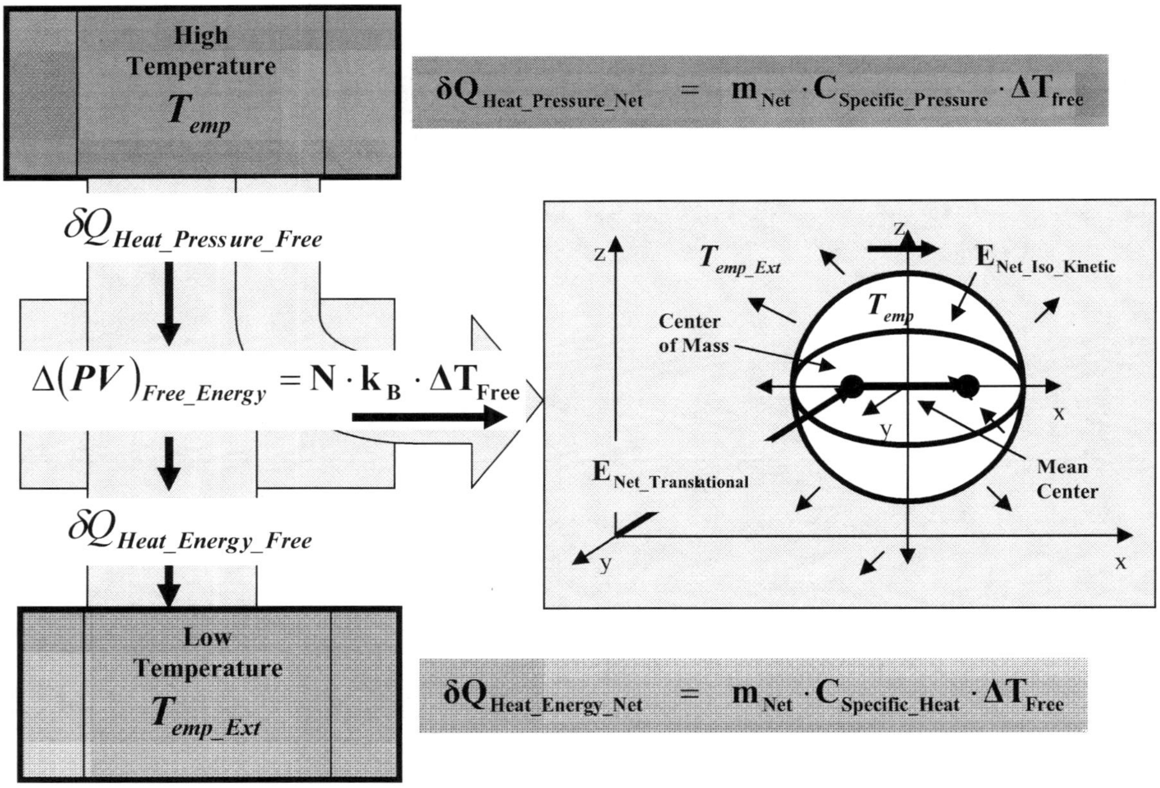

Figure 1.9: The total Aether Kinematic Work Energy transfer is a Heat Engine equal to the difference in the high "isotropic" thermodynamic potential subtracted from the low "anisotropic" thermodynamic potential.

Thus, when transforming thermal energy into mechanical energy, the thermal efficiency of a heat engine is the percentage of energy that is transformed into work *cooling the body and heating the environment*.

Aphorism 1.2: The **Carnot Thermal Engine "Anisotropy" Efficiency Factor** (η_{Carnot}) is the measure of the ***anisotropy heat engine*** efficiency of an a isolated system mass body and its atmospheric environment doing work cooling the mass body as it transfers heat energy from a higher temperature internal isotropic absolute temperature (T_{emp}) reservoir to a lower temperature external anisotropic aerodynamic temperature (T_{emp_Ext}) reservoir; and changes in direct proportion to changes in the **Inertial Motion Foreshortening Factor** ($\phi_{\text{Loco\_Motion}}$).

$$\eta_{\text{Carnot}} \propto \phi_{\text{Loco\_Motion}}$$

Heat Engines that operate on the Anisotropic Carnot cycle can extract only a certain proportion of mechanical energy from the heat of the working fluid, and this maximal amount is realized by the ideal Carnot heat engine.

If we allow the final temperature to be the Aerodynamic Temperature (T_{emp_Ext}) to be lower temperature heat sink of the system then the following condition holds,

1.31

$$\phi_{\text{Loco\_Motion}} \leq 1$$

$$T_{emp\_Ext} \leq T_{emp} \rightarrow K$$

$$\text{E}_{\text{Net\_Translational}} \leq \text{E}_{\text{Net\_Iso\_Kinetic}} \rightarrow kg \cdot m^2 / s^2$$

Starting with the Thermodynamic Entropy equation it can be demonstrated that the maximum efficiency possible by any sort of engine has a limit, defined by the following Carnot "anisotropy" efficiency factor (η_{Carnot}).

Since Heat flows between systems that are not in thermal equilibrium with each other in an attempt to reach equilibrium spontaneously flows from the areas of high temperature to areas of low temperature; in accordance with the Zeroth Law of Thermodynamics which states that a body in thermodynamic equilibrium with its environment will remain at that temperature unless another body comes in contact with it, which is of a different temperature than the first body.

The Entropy associated with the anisotropy of the system is equally essential in predicting the extent of complex chemical reactions, for such applications the total entropy of a system must be incorporated in an expression that include both the system and its surroundings, as described below.

Thermodynamic Free Heat Energy Entropy

1.32

$$\Delta S_{Entropy\ Free\_Surroundings} = \Delta(PV)_{Free\_Energy} \cdot \left[\frac{1}{T_{emp\_Ext}} - \frac{1}{T_{emp}} \right]$$

$$\Delta S_{Entropy\ Free\_Surroundings} = \left(\frac{\Delta(PV_{ol})_{Free\_Energy}}{T_{emp\_Ext}} \right) \cdot \left[1 - \left(\frac{T_{emp\_Ext}}{T_{emp}} \right) \right]$$

$$\Delta S_{Entropy\ Free\_Surroundings} = \left(\frac{\Delta(PV)_{Free\_Energy}}{T_{emp\_Ext}} \right) \cdot \left[\frac{T_{emp} - T_{emp\_Ext}}{T_{emp}} \right]$$

$$\Delta S_{Entropy\ Free\_Surroundings} = \left(\frac{\Delta(PV)_{Free\_Energy}}{T_{emp\_Ext}} \right) \cdot \left[\frac{\Delta T_{Free}}{T_{emp}} \right] \rightarrow kg \cdot m^2 / s^2 \cdot K$$

It can be showed that the maximum efficiency possible by any sort of anisotropic engine has a limit defined by the following efficiency factor:

Definition 1.3: The **Carnot Thermal Engine "Anisotropy" Efficiency Factor (η_{Carnot})** is the measure of the efficiency of a heat engine doing work cooling the mass body and heating the atmosphere environment as it transfers heat energy from a higher temperature isotropic reservoir to a lower temperature anisotropic reservoir of an isolated system mass body; and is equal to one subtracted from the ratio of the Final Low Temperature Reservoir divided by the Initial High Temperature Reservoir ($\left[1 - \left(\frac{T_{emp\_Ext}}{T_{emp}} \right) \right]$).

Carnot Thermal Engine "Anisotropy" Efficiency Factor

1.33

$$\eta_{\text{Carnot}} = \left(\frac{T_{emp\_Ext} \cdot \Delta S_{Syntropy}}{\Delta(PV_{\text{ol}})_{Free\_Energy}} \right) = \frac{\Delta S_{Syntropy}}{\left(\frac{\Delta(PV_{\text{ol}})_{Free\_Energy}}{T_{emp\_Ext}} \right)} \rightarrow \text{Unit-less}$$

$$\eta_{\text{Carnot}} = \left(\frac{\Delta T_{Free}}{T_{emp}} \right) = \left[1 - \phi_{\text{Loco\_Motion}} \right]$$

$$\eta_{\text{Carnot}} = \left(\frac{\Delta T_{Free}}{T_{emp}} \right) = \left[1 - \left(\frac{T_{emp\_Ext}}{T_{emp}} \right) \right]$$

$$\eta_{\text{Carnot}} = \left[1 - \left(\frac{|\overline{\text{v}}|^2_{\text{CM}}}{|\text{v}^2|_{\text{Iso}}} \right) \right] = \left[1 - \frac{\gamma_{Heat}}{3} \cdot \left(\frac{|\overline{\text{v}}|^2_{\text{CM}}}{\text{c}^2_{\text{Sound}}} \right) \right]$$

$$\eta_{\text{Carnot}} = \left[1 - \left(\frac{\text{m}_{\text{Net}}}{\text{m}_{\text{Dark\_Matter}}} \right) \right] = \left[1 - \left(\frac{\overline{\text{p}}^2_{\text{Inertial\_Net}}}{\overline{\text{p}}^2_{\text{Net\_Iso}}} \right) \right]$$

1.34

Scalar	
Magnitude	Units
Carnot Thermal Engine "Anisotropy" Efficiency Factor	
$\eta_{\text{Carnot}} = \left(\frac{\Delta T_{Free}}{T_{emp}} \right) = \left[1 - \left(\frac{T_{emp\_Ext}}{T_{emp}} \right) \right]$	Unit-less

1.9 Carnot's Thermodynamic Engine "Isotropy" Efficiency Factor

Carnot's theorem, also called Carnot's rule is a principle which sets a limit on the maximum amount of efficiency any possible ***isotropy heat engine*** can obtain, which thus solely depends on the difference between the hot and cold temperature isotropy and anisotropy reservoirs. Carnot's theorem states:

"No engine operating between two heat reservoirs can be more efficient than a Carnot engine operating between the same reservoirs."

Carnot's ***Isotropy Heat Engine*** theorem sets essential limitations on the yield of a cyclic heat engine such as steam engines or internal combustion engines, which operate on the Isotropic Carnot Cycle.

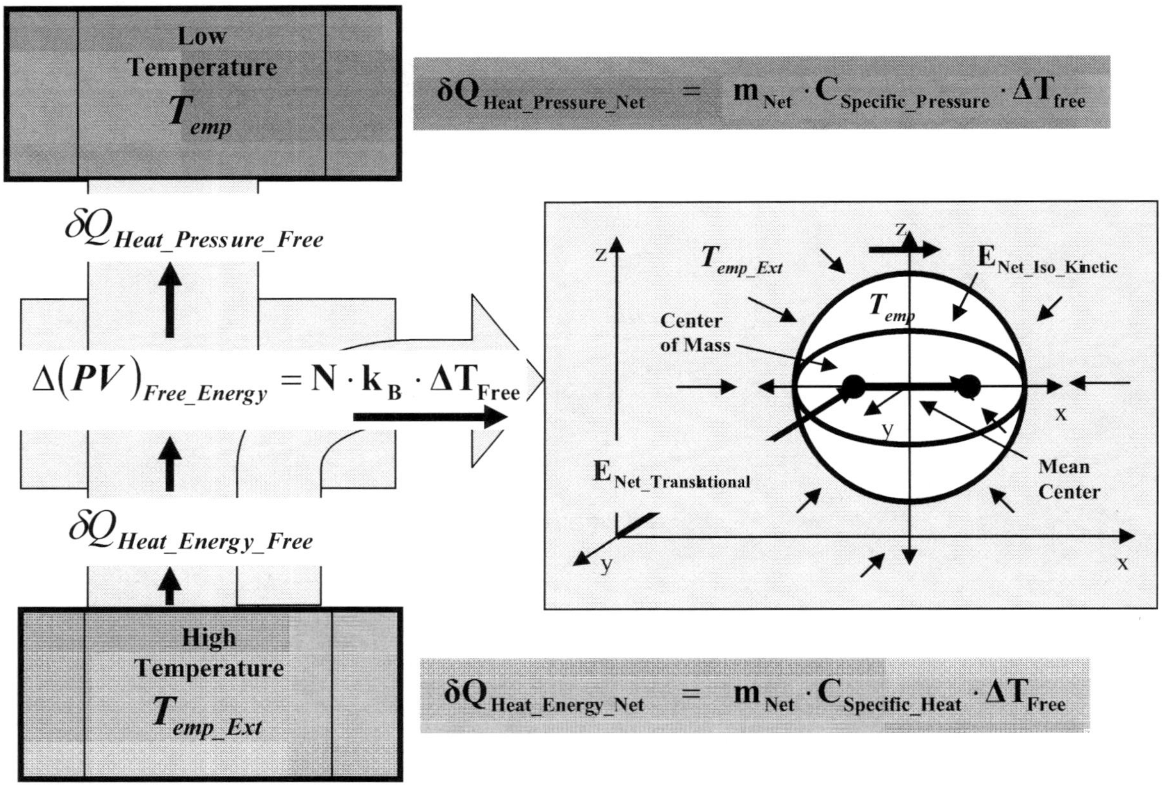

Figure 1.10: The total Aether Kinematic Work Energy transfer is a Heat Engine equal to the difference in the high "anisotropic" thermodynamic potential subtracted from the low "isotropic" thermodynamic potential.

Thus, when transforming thermal energy into mechanical energy, the thermal efficiency of a heat engine is the percentage of energy that is transformed into work *heating the body and cooling the environment*.

Aphorism 1.3: The **Carnot Thermal Engine "Isotropy" Efficiency Factor** ($\eta_{\text{Carnot\_Iso}}$) is the measure of the ***isotropy heat engine*** efficiency of an a isolated system mass body and its atmosphere environment doing work heating the mass body as it transfers heat energy from a higher temperature external anisotropic aerodynamic temperature (T_{emp_Ext}) reservoir to a lower temperature internal isotropic absolute temperature (T_{emp}) reservoir; and changes in inverse proportion to changes in the **Inertial Motion Foreshortening Factor** ($\phi_{\text{Loco\_Motion}}$).

$$\eta_{\text{Carnot\_Iso}} \propto \frac{1}{\phi_{\text{Loco\_Motion}}}$$

Heat Engines that operate on the Isotropic Carnot cycle can extract only a certain proportion of mechanical energy from the heat of the working fluid and this maximal amount is realized by the ideal Carnot heat engine.

If we allow the final temperature to be the Internal Absolute Temperature (T_{emp}) to be lower temperature heat sink of the system then the following condition holds,

1.35

$$\phi_{\text{Loco\_Motion}} \geq 1$$

$$T_{emp\_Ext} \geq T_{emp} \rightarrow K$$

$$\text{E}_{\text{Net\_Translational}} \geq \text{E}_{\text{Net\_Iso\_Kinetic}} \rightarrow kg \cdot m^2 / s^2$$

Starting with the Thermodynamic Entropy equation it can be demonstrated that the maximum efficiency possible by any sort of engine has a limit, defined by the following Carnot "isotropy" efficiency factor ($\eta_{\text{Carnot\_Iso}}$).

Since Heat flows between systems that are not in thermal equilibrium with each other in an attempt to reach equilibrium spontaneously flows from the areas of high temperature to areas of low temperature; in accordance with the Zeroth Law of Thermodynamics which states that a body in thermodynamic equilibrium with its environment will remain at that temperature unless another body comes in contact with it, which is of a different temperature than the first body.

The Entropy associated with the isotropy of the system is equally essential in predicting the extent of complex chemical reactions, for such applications the total entropy of a system must be incorporated in an expression that include both the system and its surroundings, as described below.

Thermodynamic Free Heat Energy Entropy

1.36

$$\Delta S_{Entropy\ Free\_Surroundings} = \Delta(PV)_{Free\_Energy} \cdot \left[\frac{1}{T_{emp\_Ext}} - \frac{1}{T_{emp}} \right]$$

$$\Delta S_{Entropy\ Free\_Surroundings} = \left(\frac{\Delta(PV)_{Free\_Energy}}{T_{emp}} \right) \cdot \left[\frac{1}{\left(\frac{T_{emp\_Ext}}{T_{emp}} \right)} - 1 \right]$$

$$\Delta S_{Entropy\ Free\_Surroundings} = \left(\frac{\Delta(PV)_{Free\_Energy}}{T_{emp}} \right) \cdot \left[\frac{T_{emp} - T_{emp\_Ext}}{T_{emp\_Ext}} \right]$$

$$\Delta S_{Entropy\ Free\_Surroundings} = \left(\frac{\Delta(PV)_{Free\_Energy}}{T_{emp}} \right) \cdot \left[\frac{\Delta T_{Free}}{T_{emp\_Ext}} \right] \rightarrow kg \cdot m^2 / s^2 \cdot K$$

It can be showed that the maximum efficiency possible by any sort of isotropic engine has a limit defined by the following efficiency factor:

Definition 1.4: The **Carnot Thermal Engine "Isotropy" Efficiency Factor** ($\eta_{\mathrm{Carnot\_Iso}}$) is the measure of the efficiency of a heat engine doing work heating the mass body as it transfers heat energy from a higher temperature anisotropic reservoir to a lower temperature isotropic reservoir of an isolated system mass body; and is equal to one subtracted from the ratio of the Final Low Temperature Reservoir divided by the Initial High Temperature Reservoir ($\left[\frac{1}{\left(\frac{T_{emp\_Ext}}{T_{emp}} \right)} - 1 \right]$).

Carnot Thermal Engine "Isotropy" Efficiency Factor

1.37

$$\eta_{\text{Carnot\_Iso}} = \left(\frac{T_{emp} \cdot \Delta S_{Syntropy}}{\Delta(PV)_{Free\_Energy}}\right) = \frac{\Delta S_{Syntropy}}{\left(\dfrac{\Delta(PV)_{Free\_Energy}}{T_{emp}}\right)} \rightarrow \text{Unit-less}$$

$$\eta_{\text{Carnot\_Iso}} = \left(\frac{\Delta T_{Free}}{T_{emp\_Ext}}\right) = \left[\frac{1}{\phi_{\text{Loco\_Motion}}} - 1\right]$$

$$\eta_{\text{Carnot\_Iso}} = \left(\frac{\Delta T_{Free}}{T_{emp\_Ext}}\right) = \left[\frac{1}{\left(\dfrac{T_{emp\_Ext}}{T_{emp}}\right)} - 1\right] = \left[\frac{1}{\dfrac{\gamma_{Heat}}{3} \cdot \left(\dfrac{\left|\overline{\text{v}}\right|^2_{\text{CM}}}{\text{c}^2_{\text{Sound}}}\right)} - 1\right]$$

$$\eta_{\text{Carnot\_Iso}} = \left[\frac{1}{\left(\dfrac{\left|\overline{\text{v}}\right|^2_{\text{CM}}}{\left|\text{v}^2\right|_{\text{Iso}}}\right)} - 1\right] = \left[\frac{1}{\left(\dfrac{\overline{\text{p}}^2_{\text{Inertial\_Net}}}{\overline{\text{p}}^2_{\text{Net\_Iso}}}\right)} - 1\right]$$

1.38

Scalar	
Magnitude	Units
Carnot Thermal Engine "Isotropy" Efficiency Factor	
$\eta_{\text{Carnot\_Iso}} = \left(\dfrac{\Delta T_{Free}}{T_{emp\_Ext}}\right) = \left[\dfrac{1}{\left(\dfrac{T_{emp\_Ext}}{T_{emp}}\right)} - 1\right]$	Unit-less

1.10 Kemp's Thermodynamic Efficiency Factor

Examples of everyday heat engines include: the steam engine, the diesel engine, and the gasoline (petrol) engine. All of these familiar heat engines are powered by the expansion of heated gases flowing towards cooler sinks. The general surroundings are the heat sink, providing relatively cool gases which, when heated, expand rapidly to drive the mechanical motion of an engine.

Nature provides a natural Aether Kinematic Heat Engine given by the Kemp Thermodynamic Efficiency Factor described below.

Definition 1.5: The **Kemp Thermal Engine Efficiency Factor** (η_{Kemp}) is the measure of the efficiency of a heat engine doing work as it transfers heat energy from a higher temperature reservoir to a lower temperature reservoir of an isolated system mass body; and is equal to one subtracted from the inverse of the **Specific Heat Capacity Adiabatic Index Ratio** ($\frac{1}{\gamma_{Heat}}$).

1.39

$$\eta_{\text{Kemp}} = \left[1 - \frac{1}{\gamma_{Heat}} \right]$$

Kemp's Thermodynamic Efficiency Factor

1.40

$$\eta_{\text{Kemp}} = \left[\frac{\Delta(PV)_{\text{Free\_Energy}}}{\delta Q_{\text{Heat\_Pressure\_Free}}} \right] = \left[\frac{\delta Q_{\text{Heat\_Pressure\_Free}} - \delta Q_{\text{Heat\_Energy\_Free}}}{\delta Q_{\text{Heat\_Pressure\_Free}}} \right]$$

$$\eta_{\text{Kemp}} = \left[1 - \left(\frac{\delta Q_{\text{Heat\_Energy\_Free}}}{\delta Q_{\text{Heat\_Pressure\_Free}}} \right) \right] = \left[1 - \left(\frac{C_{Specific\_Heat}}{C_{Specific\_Pressure}} \right) \right]$$

$$\eta_{\text{Kemp}} = \left[\frac{\Delta(PV)_{\text{Free\_Energy}}}{\delta Q_{\text{Heat\_Pressure\_Free}}} \right] = \left[\frac{\frac{2}{3} \cdot \left[\mathrm{E}_{\text{Net\_Iso\_Kinetic}} - \mathrm{E}_{\text{Net\_Translational}} \right]}{\left[\delta Q_{\text{Heat\_Energy\_Free}} + \frac{2}{3} \cdot \left[\mathrm{E}_{\text{Net\_Iso\_Kinetic}} - \mathrm{E}_{\text{Net\_Translational}} \right] \right]} \right]$$

Kemp's Thermodynamic Efficiency Factor

1.41

$$\eta_{\text{Kemp}} = \left[\frac{\Delta(PV)_{\text{Free\_Energy}}}{\delta Q_{\text{Heat\_Pressure\_Free}}}\right] = \left[\frac{\frac{1}{3}\cdot \mathrm{m}_{\text{Net}}\cdot\left[\left|\mathrm{v}^2\right|_{\text{Iso}} - \left|\overline{\mathrm{v}}\right|^2_{\text{CM}}\right]}{\left[\frac{\mathrm{m}_{\text{Net}}\cdot\left|\mathrm{v}^2\right|_{\text{Iso}}}{3\cdot(\gamma_{Heat} - 1)}\cdot\left[1 - \frac{\left|\overline{\mathrm{v}}\right|^2_{\text{CM}}}{\left|\mathrm{v}^2\right|_{\text{Iso}}}\right] + \frac{1}{3}\cdot \mathrm{m}_{\text{Net}}\cdot\left[\left|\mathrm{v}^2\right|_{\text{Iso}} - \left|\overline{\mathrm{v}}\right|^2_{\text{CM}}\right]\right]}\right]$$

$$\eta_{\text{Kemp}} = \left[\frac{\mathrm{N}\cdot \mathrm{k_B}}{\mathrm{m}_{\text{Net}}\cdot C_{Specific\_Pressure}}\right] = \left[1 - \frac{1}{\gamma_{Heat}}\right]$$

1.42

Scalar	
Magnitude	Units
Kemp Thermal Engine Efficiency Factor	
$\eta_{\text{Kemp}} = \left[\frac{\mathrm{N}\cdot \mathrm{k_B}}{\mathrm{m}_{\text{Net}}\cdot C_{Specific\_Pressure}}\right] = \left[1 - \left(\frac{C_{Specific\_Heat}}{C_{Specific\_Pressure}}\right)\right] = \left[1 - \frac{1}{\gamma_{Heat}}\right]$	Unit-less

1.43

Table of specific heat capacities at 25 °C unless otherwise specified

Substance	Molecule	(Cp) $C_{Specific\_Pressure}$ (kJ/kg K)	(Cv) $C_{Specific\_Heat}$ (kJ/kg K)	(C_p / C_v) γ_{Heat}	(cp - cv) $C_{Specific\_Temperature}$ (kJ/kg K)	Kemp Thermal Efficiency η_{Kemp}
Helium	He	5.19	3.12	1.667	2.08	40%
Neon		1.03	0.618	1.667	0.412	40%
Argon	Ar	0.52	0.312	1.667	0.208	40%
Blast furnace gas		1.03	0.73	1.41	0.3	29%
Hydrogen Chloride	HCl	0.8	0.57	1.41	0.23	29%
Hydrogen	H_2	14.32	10.16	1.405	4.12	29%
Nitrogen	N_2	1.04	0.743	1.4	0.297	29%
Carbon monoxide	CO	1.02	0.72	1.4	0.297	29%
Air		1.01	0.718	1.4	0.287	29%
Oxygen	O_2	0.919	0.659	1.395	0.26	28%
Nitric Oxide	NO	0.995	0.718	1.386	0.277	28%
Hydroxyl	OH	1.76	1.27	1.384	0.489	28%
Chlorine	Cl_2	0.48	0.36	1.34	0.12	25%
Water Vapor Steam 1 psia. 120 – 600 °F		1.93	1.46	1.32	0.462	24%
Ammonia	NH_3	2.19	1.66	1.31	0.53	24%
Steam 14.7 psia. 220 – 600 °F		1.97	1.5	1.31	0.46	24%
Methane	CH_4	2.22	1.7	1.304	0.518	23%
Sulfur dioxide (Sulphur dioxide)	SO_2	0.64	0.51	1.29	0.13	22%
Carbon dioxide	CO_2	0.844	0.655	1.289	0.189	22%
Steam 150 psia. 360 – 600 °F		2.26	1.76	1.28	0.5	22%
Bromine		0.25	0.2	1.28	0.05	22%
Natural Gas		2.34	1.85	1.27	0.5	21%
Nitrous oxide	N_2O	0.88	0.69	1.27	0.18	21%

Alcohol	CH_3OH	1.93	1.53	1.26	0.39	21%
Ethylene	C_2H_4	1.53	1.23	1.24	0.296	19%
Acetylene	C_2H_2	1.69	1.37	1.232	0.319	19%
Carbon disulphide		0.67	0.55	1.21	0.12	17%
Ethane	C_2H_6	1.75	1.48	1.187	0.276	16%
Propene (propylene)	C_3H_6	1.5	1.31	1.15	0.18	13%
Chloroform		0.63	0.55	1.15	0.08	13%
Alcohol	C_2H_5OH	1.88	1.67	1.13	0.22	12%
Propane	C_3H_8	1.67	1.48	1.127	0.189	11%
Benzene	C_6H_6	1.09	0.99	1.12	0.1	11%
Acetone		1.47	1.32	1.11	0.15	10%
Butane	C_4H_{10}	1.67	1.53	1.094	0.143	9%
Ether		2.01	1.95	1.03	0.06	3%
Nitrogen tetroxide		4.69	4.6	1.02	0.09	2%

Figure 1.11: Specific Heat Capacities – Table of various gases.

1.11 Variable Specific Heat Capacity Index

The Mach number ($M_{Mach} = \left(\frac{|\overline{v}|_{CM}}{c_{Sound}} \right)$) depends on the speed of sound in the gas, and the speed of sound depends on the type of gas substance, and the temperature of the gas. The speed of sound varies from planet to planet. On Earth, the atmosphere is composed of mostly diatomic nitrogen and oxygen, and the temperature depends on the altitude.

Square of the Mach number for Atomic Substance Medium

1.44

$$\beta^2_{Mach} = M^2_{Mach} = \left(\frac{|\overline{v}|^2_{CM}}{c^2_{Sound}} \right) = \frac{3}{\gamma_{Heat}} \cdot \left(\frac{|\overline{v}|^2_{CM}}{\left| v^2 \right|_{Iso}} \right) \rightarrow \textit{Unitless}$$

$$\beta^2_{Mach} = M^2_{Mach} = \frac{3 \cdot \phi_{Loco\_Motion}}{\gamma_{Heat}} = \frac{3 \cdot [1 - \eta_{Carnot}]}{\gamma_{Heat}}$$

$$\beta^2_{Mach} = M^2_{Mach} = \frac{3 \cdot \phi_{Loco\_Motion}}{\gamma_{Heat}} = \frac{3}{\gamma_{Heat}} \cdot \frac{1}{[1 + \eta_{Carnot\_Iso}]}$$

The heat capacity of most systems is not a constant, but depends on the state variables of the thermodynamic system under study. In particular it is dependent on temperature itself, as well as the pressure and on the volume of the system.

Different measurements of specific heat capacity may therefore be performed, most commonly at constant pressure and constant volume. Gases and liquids are typically also measured at constant volume.

Measurements under constant pressure produce larger values than those at constant volume because work must be performed in the former. This difference is particularly notable in gases where values under constant pressure are typically 30% to 66.7% greater than those at constant volume.

Specific Heat Capacity Adiabatic Index Ratio

1.45

$$\gamma_{Heat} = \left(\frac{C_{Specific\_Pressure}}{C_{Specific\_Heat}} \right) = \frac{3 \cdot (1 - \eta_{Carnot})}{\left(\frac{|\bar{v}|^2_{CM}}{c^2_{Sound}} \right)}$$

$$\gamma_{Heat} = \left(\frac{C_{Specific\_Pressure}}{C_{Specific\_Heat}} \right) = \frac{3}{\left(\frac{|\bar{v}|^2_{CM}}{c^2_{Sound}} \right)} \cdot \frac{1}{[1 + \eta_{Carnot\_Iso}]}$$

$$\gamma_{Heat} = \frac{3 \cdot \phi_{Loco\_Motion}}{\left(\frac{|\bar{v}|^2_{CM}}{c^2_{Sound}} \right)} = \frac{3}{\left(\frac{|v^2|_{Iso}}{c^2_{Sound}} \right)}$$

$$\gamma_{Heat} = \phi_{Loco\_Motion} \cdot \left(\frac{m_{Net} \cdot c^2_{Sound}}{N \cdot k_B \cdot T_{emp\_Ext}} \right) = \frac{m_{Net} \cdot c^2_{Sound}}{N \cdot k_B \cdot T_{emp}}$$

$$\gamma_{Heat} = \left(\frac{f_{Degrees\_Freedom} + 2}{f_{Degrees\_Freedom}} \right) \rightarrow Unitless$$

The negative Component of the Heat Capacity Index Ratio can typically be ignored. The positive Component of the Heat Capacity Index Ratio appears to be real.

Specific Heat Capacity Adiabatic Index Ratio (Negative Component)

1.46

$$\gamma(-)_{Heat} = \frac{3\cdot\left(1 - \eta(-)_{\text{Carnot}}\right)}{\left(\frac{|\overline{\mathrm{v}}|^2_{\text{CM}}}{\mathrm{c}^2_{\text{Sound}}}\right)} \rightarrow \textit{Unitless}$$

$$\gamma(-)_{Heat} = \frac{3}{\left(\frac{|\overline{\mathrm{v}}|^2_{\text{CM}}}{\mathrm{c}^2_{\text{Sound}}}\right)}\cdot\frac{1}{\left[1 + \eta(-)_{\text{Carnot\_Iso}}\right]}$$

$$\gamma(-)_{Heat} = \frac{3\cdot\phi(-)_{\text{Loco\_Motion}}}{\left(\frac{|\overline{\mathrm{v}}|^2_{\text{CM}}}{\mathrm{c}^2_{\text{Sound}}}\right)}$$

$$\gamma(-)_{Heat} = \frac{3\cdot\left(1 + \left[\begin{array}{l}\frac{\gamma_{Heat}}{6}\cdot\left(\frac{|\overline{\mathrm{v}}|^2_{\text{CM}}}{\mathrm{c}^2_{\text{Sound}}}\right) \\ + \sqrt{1 - \frac{\gamma_{Heat}}{3}\cdot\left(\frac{|\overline{\mathrm{v}}|^2_{\text{CM}}}{\mathrm{c}^2_{\text{Sound}}}\right) + \frac{\gamma^2_{Heat}}{12}\cdot\left(\frac{|\overline{\mathrm{v}}|^4_{\text{CM}}}{\mathrm{c}^4_{\text{Sound}}}\right)}\end{array}\right]\right)}{\left(\frac{|\overline{\mathrm{v}}|^2_{\text{CM}}}{\mathrm{c}^2_{\text{Sound}}}\right)}$$

Specific Heat Capacity Adiabatic Index Ratio (Positive Component)

1.47

$$\gamma(+)_{Heat} = \frac{3\cdot\left(1 - \eta(+)_{Carnot}\right)}{\left(\dfrac{\left|\overline{v}\right|^2_{CM}}{c^2_{Sound}}\right)} \rightarrow Unitless$$

$$\gamma(+)_{Heat} = \frac{3}{\left(\dfrac{\left|\overline{v}\right|^2_{CM}}{c^2_{Sound}}\right)}\cdot\frac{1}{\left[1 + \eta(+)_{Carnot\_Iso}\right]}$$

$$\gamma(+)_{Heat} = \frac{3\cdot\phi(+)_{Loco\_Motion}}{\left(\dfrac{\left|\overline{v}\right|^2_{CM}}{c^2_{Sound}}\right)}$$

$$\gamma(+)_{Heat} = \frac{3\cdot\left(1 + \left[\dfrac{\gamma_{Heat}}{6}\cdot\left(\dfrac{\left|\overline{v}\right|^2_{CM}}{c^2_{Sound}}\right) - \sqrt{1 - \dfrac{\gamma_{Heat}}{3}\cdot\left(\dfrac{\left|\overline{v}\right|^2_{CM}}{c^2_{Sound}}\right) + \dfrac{\gamma^2_{Heat}}{12}\cdot\left(\dfrac{\left|\overline{v}\right|^4_{CM}}{c^4_{Sound}}\right)}\right]\right)}{\left(\dfrac{\left|\overline{v}\right|^2_{CM}}{c^2_{Sound}}\right)}$$

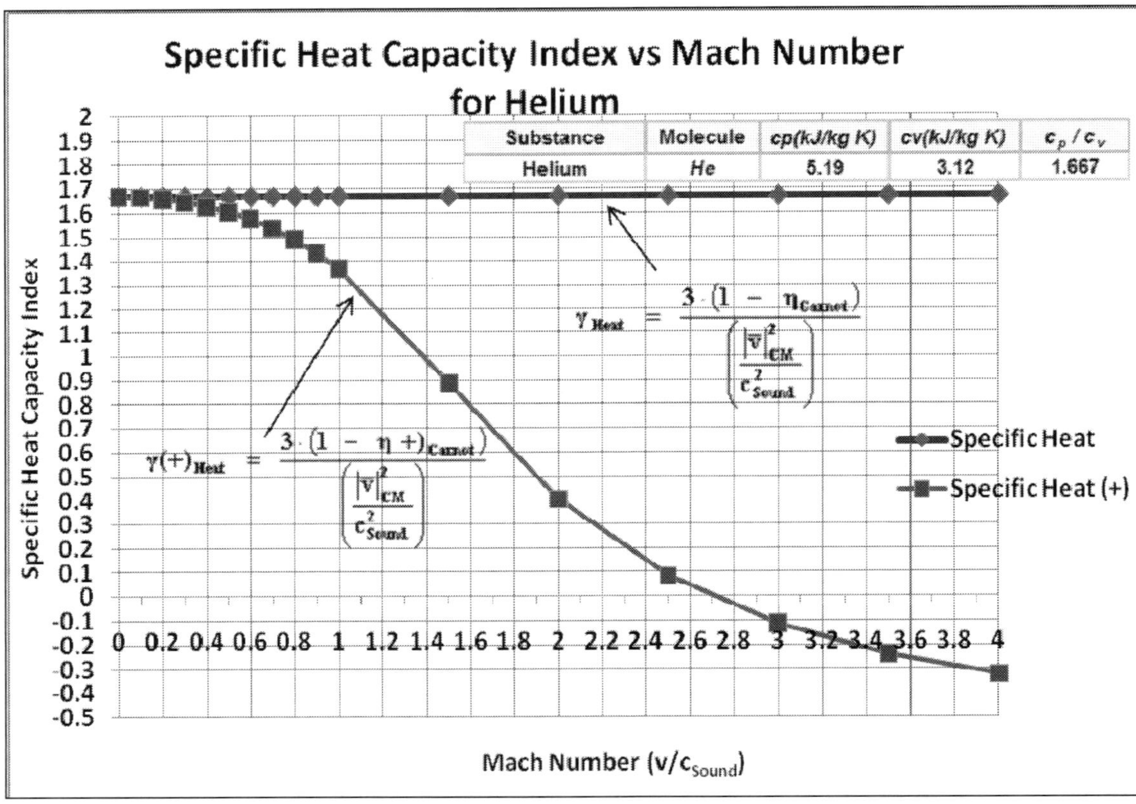

Figure 1.12: Specific Heat Capacity Index vs. Mach number for - Helium.

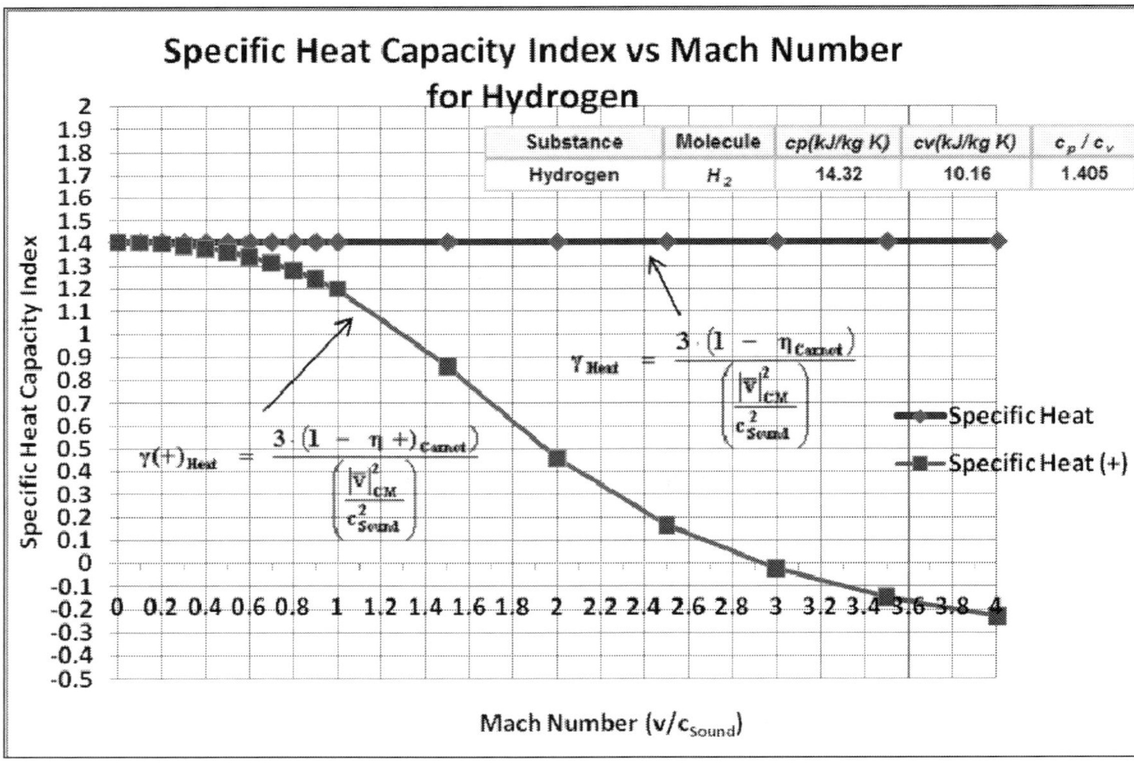

Figure 1.13: Specific Heat Capacity Index vs. Mach number for - Hydrogen.

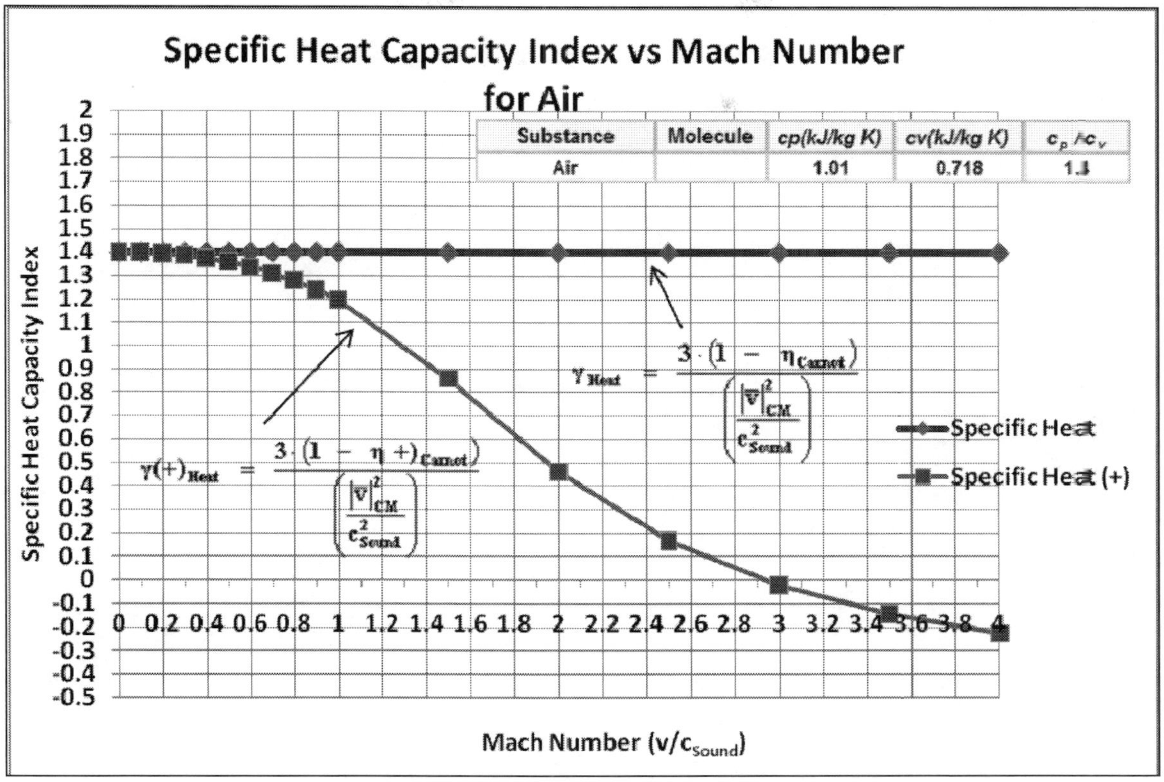

Figure 1.14: Specific Heat Capacity Index vs. Mach number for - Air.

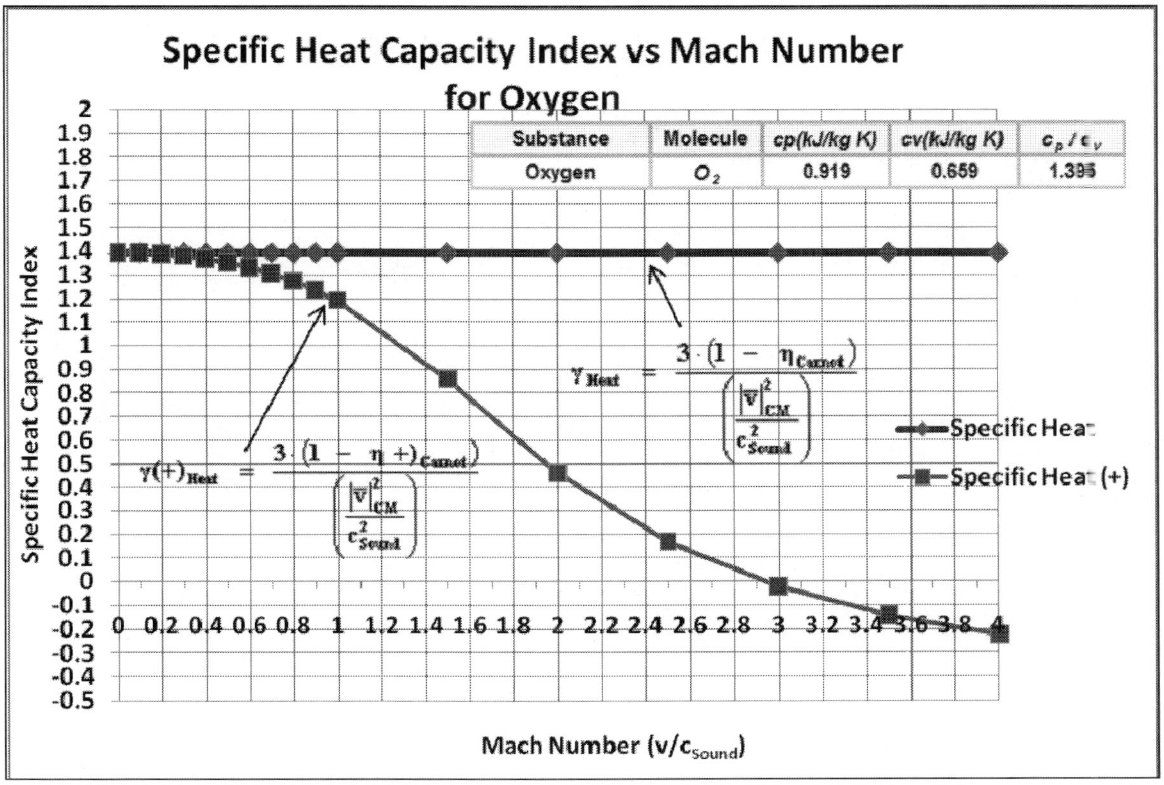

Figure 1.15: Specific Heat Capacity Index vs. Mach number for - Oxygen.

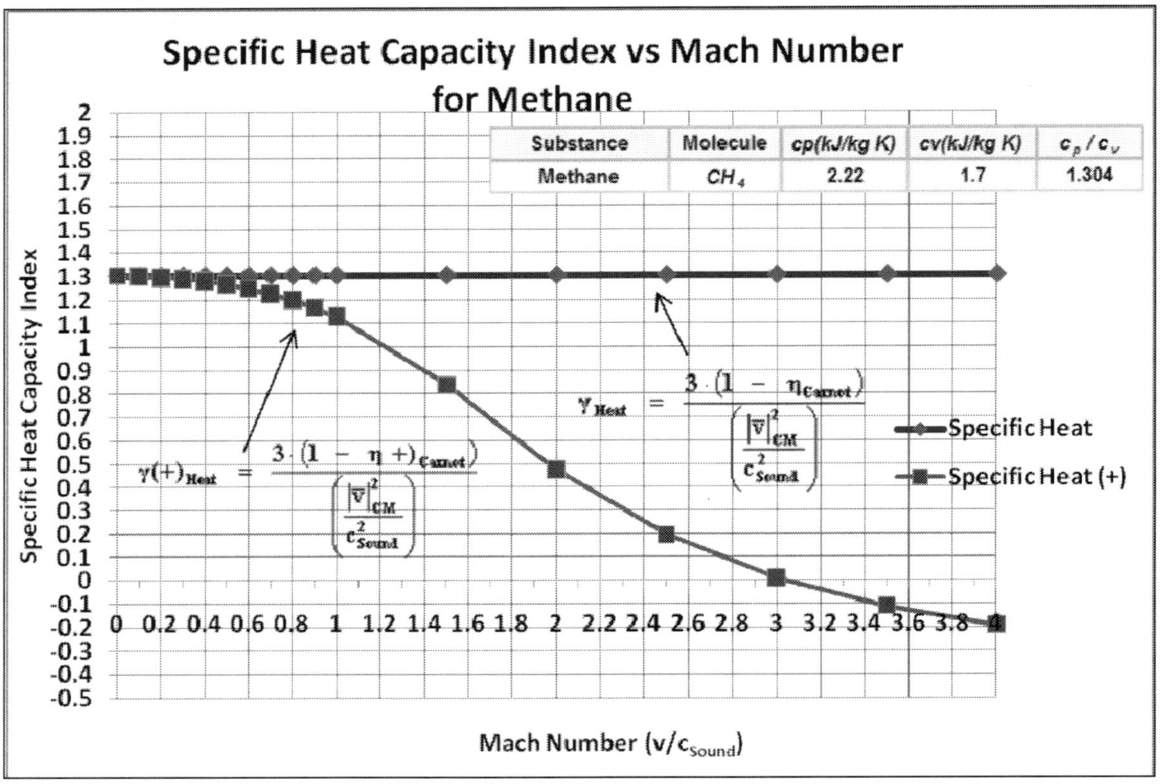

Figure 1.16: Specific Heat Capacity Index vs. Mach number for - Methane.

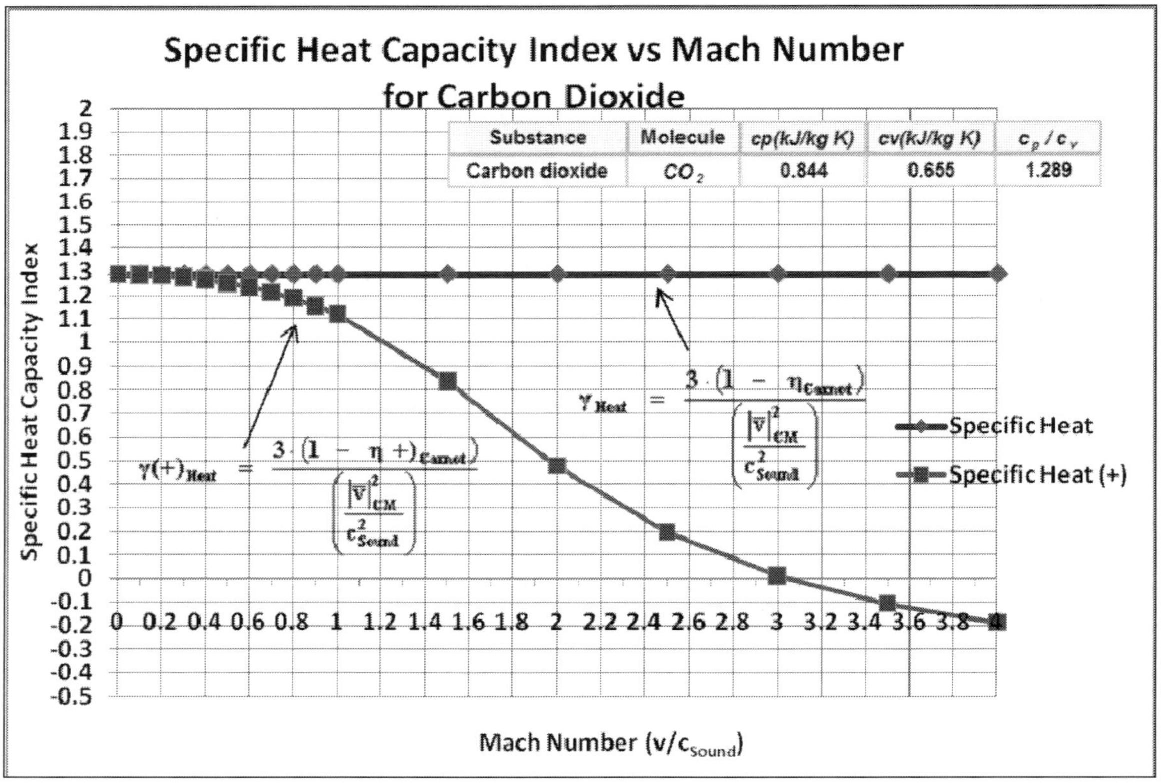

Figure 1.17: Specific Heat Capacity Index vs. Mach number for – Carbon Dioxide.

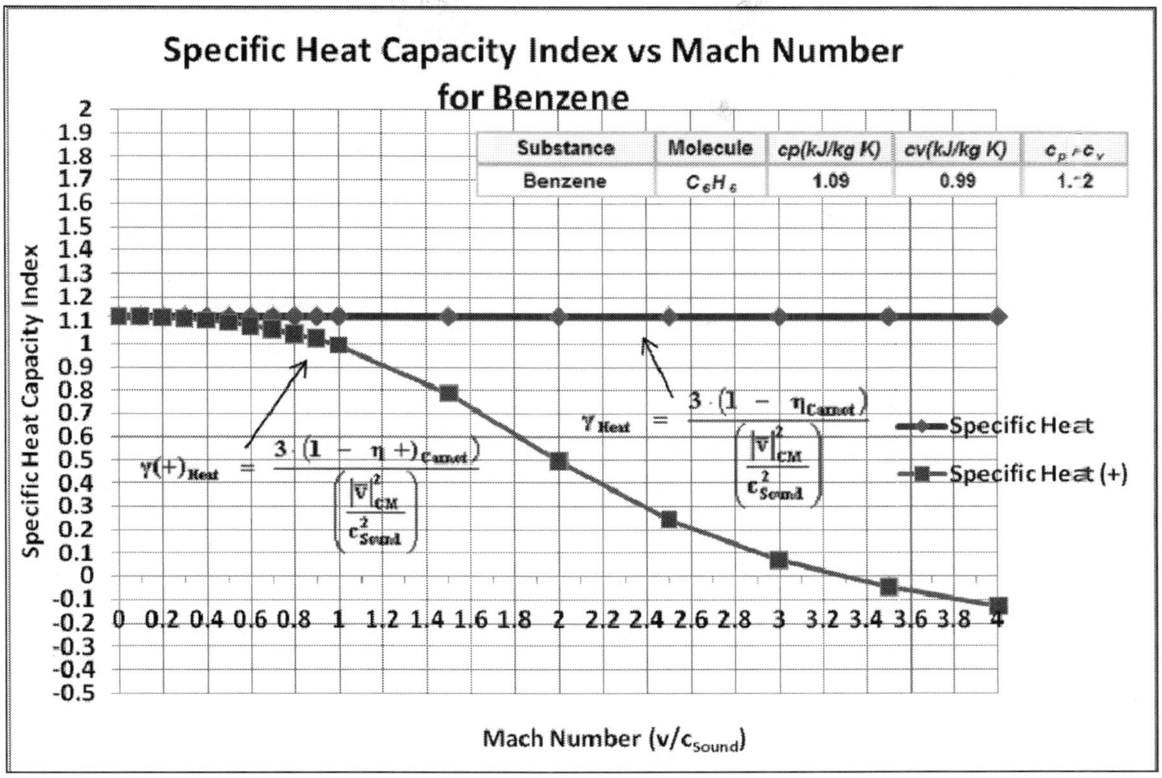

Figure 1.18: Specific Heat Capacity Index vs. Mach number for - Benzene.

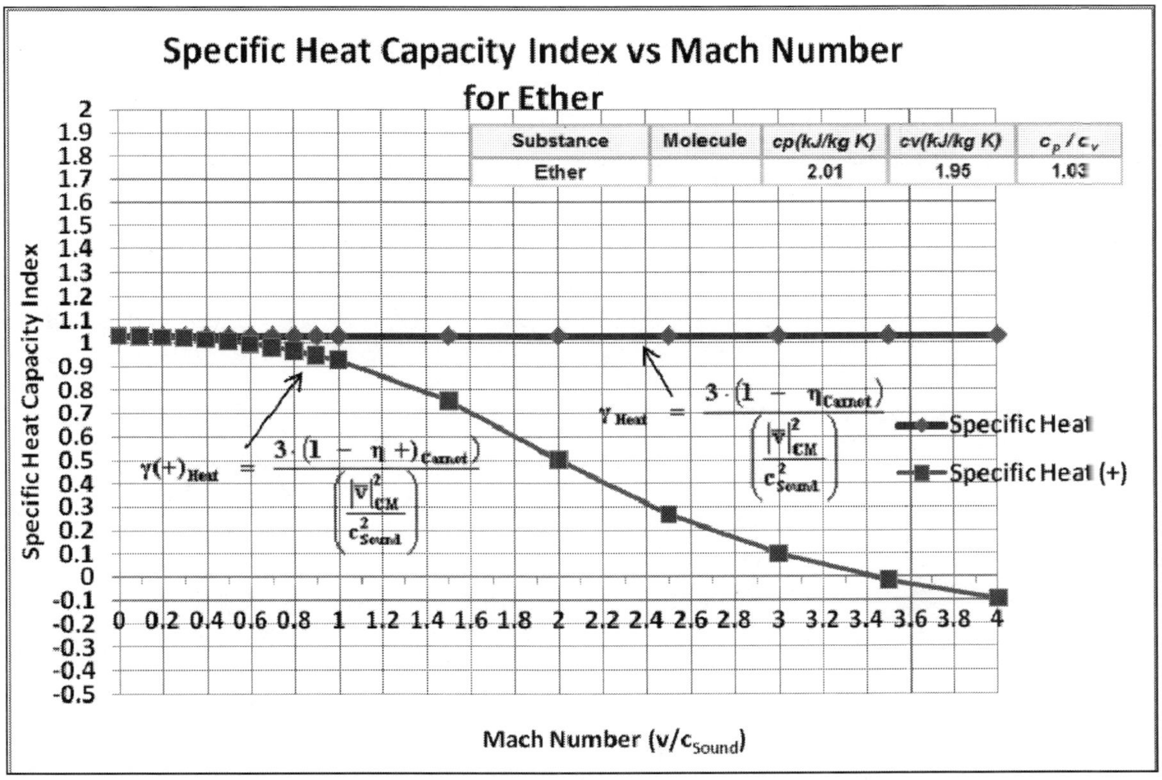

Figure 1.19: Specific Heat Capacity Index vs. Mach number for - Ether.

1.12 Derivation of Specific Heat Capacity Adiabatic Index Ratio

Specific Heat Capacity Equation

1.48

$$\left(\frac{c^2_{\text{Propagation\_Field}_{i_{Shell}}}}{T'_{emp\, i_{Shell}}}\right) = C_{Specific\_Heat} \cdot \left((\gamma_{Heat})^2 - \gamma_{Heat}\right)$$

$$0 = \left[C_{Specific\_Heat} \cdot (\gamma_{Heat})^2 - C_{Specific\_Heat} \cdot \gamma_{Heat} + \left(\frac{c^2_{\text{Propagation\_Field}_{i_{Shell}}}}{T'_{emp\, i_{Shell}}}\right)\right]$$

Obtaining the Quadratic Equation Coefficients

$$a = C_{Specific\_Heat}$$

$$b = -C_{Specific\_Heat}$$

$$c = \left(\frac{c^2_{\text{Propagation\_Field}_{i_{Shell}}}}{T'_{emp\, i_{Shell}}}\right)$$

$$0 = \left[a \cdot \gamma^2_{Heat} + b \cdot \gamma_{Heat} + c\right]$$

$$\gamma_{Heat} = \frac{-b \pm \sqrt{b^2 - 4 \cdot a \cdot c}}{2 \cdot a}$$

Specific Heat Capacity Index Ratio

1.49

$$\gamma_{Heat} = \frac{C_{Specific\_Heat} \pm \sqrt{C^2_{Specific\_Heat} - 4 \cdot C_{Specific\_Heat} \cdot \left(\frac{c^2_{\text{Propagation\_Field}_{i_{Shell}}}}{T'_{emp\, i_{Shell}}}\right)}}{2 \cdot C_{Specific\_Heat}}$$

Specific Heat Capacity Index Ratio

1.50

$$\gamma_{Heat} = \frac{1}{2} \pm \frac{1}{2} \cdot \sqrt{1 - \frac{4}{C_{Specific\_Heat}} \cdot \left(\frac{c^2_{Propagation\_Field_{i_{Shell}}}}{T'_{emp_{i_{Shell}}}} \right)}$$

Specific Heat Capacity Ratio

1.51

$$(2 \cdot \gamma_{Heat} - 1) = \pm \sqrt{1 - \frac{4}{C_{Specific\_Heat}} \cdot \left(\frac{c^2_{Propagation\_Field_{i_{Shell}}}}{T'_{emp_{i_{Shell}}}} \right)}$$

Specific Heat Capacity Ratio

1.52

$$\frac{1}{C_{Specific\_Heat}} \cdot \left(\frac{c^2_{Propagation\_Field_{i_{Shell}}}}{T'_{emp_{i_{Shell}}}} \right) = \frac{1}{4} \cdot \left[1 - (2 \cdot \gamma_{Heat} - 1)^2 \right]$$

Chapter 2

The Special Theory of Thermodynamics

(Thermodynamic Field)
(Irreversible "Free Aether" Heat Energy)

2.1 Irreversible “Aether” Heat Energy Temperature Difference

Heat flows between systems that are not in thermal equilibrium with each other in an attempt to reach equilibrium, which spontaneously flows from the areas of high temperature to areas of low temperature; in accordance with the Zeroth Law of Thermodynamics which states that a body in thermodynamic equilibrium with its environment will remain at that temperature unless another body comes in contact with it, which is of a different temperature than the first body.

For example, water is warmer than ice, which means that the molecules in water generally have more kinetic energy (and are thus moving faster) than the molecules in ice.

When we place an ice cube in a glass of water, the slow moving molecules in the ice are bombarded by the faster molecules in the water. In the resulting collisions, the molecules in the ice usually end up faster while the molecules in the water end up slower. Hence the ice gets warmer and the water gets colder! This transfer of energy from the warmer substance to the colder substance is called "heat."

In the early 1850s Rudolf Clausius (1822 – 1888) a German mathematician and physicist, set forth the concept of the thermodynamic system and postulated that in any irreversible process, a small amount of heat energy ($\delta Q_{\text{Heat\_Energy}}$) is incrementally dissipated across the system boundary.

In his 1854 memoir, Clausius develops the concepts of interior work, i.e. that "which the atoms of the body exert upon each other", and exterior work, i.e. that "which arise from foreign influences which the body may be exposed", which may act on a working body of fluid or gas, typically functioning to work a piston or to distribute energy across an expanding Thermodynamic Field.

In keeping with Clausius there are the three categories into which heat ($\delta Q_{\text{Heat\_Energy}}$) may be divided:

- Heat is Energy transferred between any two bodies resulting in increasing the heat in one body and reducing the heat in the other body.
- Heat is Energy transferred between any two bodies resulting in producing isotropic Omni-directional interior work.
- Heat is Energy transferred between any two bodies resulting in producing anisotropic directional exterior work.

When two bodies of different temperature come into thermal contact, they will exchange internal energy in the form of heat until their temperatures are equalized; that is, until they reach thermal equilibrium.

The adjective hot or cold is used as a relative term to compare any object's temperature to that of the surroundings. Likewise, the term heat is used to describe the flow of energy from a higher temperature "hot body" to a lower temperature "cold body" or that of the surroundings.

In the absence of work interactions, the heat that is transferred to a system mass body ends up getting stored in the object in the form of thermal, mass, kinetic, potential, rotational, and vibrational energy.

For a simple compressible system such as an ideal gas the changes in enthalpy and internal energy can be related to the heat capacity at constant temperature, pressure, and volume respectively.

For a reversible isothermal process, there is no transfer of heat energy and therefore the process is also adiabatic and the entropy is equal to zero. For an irreversible process, the entropy will always increase; as will be demonstrated mathematically in the following sections.

Hence removal of heat from the system (cooling) is necessary to maintain constant internal entropy for an irreversible process in order to make it isothermal. Thus an irreversible isentropic process is not adiabatic.

In the thermodynamics of this work, the term **Thermodynamic Free Energy** refers to the amount of Work Energy or Heat Energy that is lost from a system and escapes into the environment; very similar to the evaporation process.

In thermodynamics, a change in the thermodynamic state of a system is irreversible if the system cannot be restored to its former state by infinitesimal changes in some property of the system without expenditure of energy. An irreversible process increases the entropy of the system, which is a measure of the microscopic disorder of the system. Where a reversible process does not increase the entropy of the system, and is in fact equal to zero.

All complex natural processes are irreversible. The friction between a rolling ball and the ground generates heat and is a very common example of an irreversible process. The phenomenon of irreversibility results from the fact that if a thermodynamic system, which is any system of sufficient complexity, of interacting molecules is brought from one thermodynamic state to another, the configuration or arrangement of the atoms and molecules in the system will change in a way that is not easily predictable.

Thermodynamic Free Energy is an expendable energy of the Adiabatic Isothermal Thermodynamic Field at absolute temperature establishing equilibrium between the system body and its surroundings.

Thermodynamic Free Energy is work energy in that it can make things happen within finite amounts of time, and is subject to irreversible energy loss; and in the course of doing work is conserved, and may or may not be available for doing useful work.

For an Adiabatic Isentropic Thermodynamic Field each expanding volume of Ideal Gas potentials also represents a certain amount of "kinetic energy" at thermodynamic equilibrium that will be used as the molecules of the system do work on each other when they change from one state to another; changing pressures, densities, and decreasing temperatures distributed across the expanding volumes of the system mass body.

During this potential energy expansion, there will be a certain amount of heat energy loss or dissipation for each potential due to intermolecular friction and collisions as each volume potential establishes thermodynamic equilibrium between itself and the surroundings; the Thermodynamic Free Energy is repeatable however it may or may not be recoverable if the process is reversed.

In the traditional use, the term "free energy" was attached to Gibbs free energy, i.e., for systems at constant volume and temperature, and is the irreversible heat energy of constant volume work ($\mathrm{V_{ol}} \cdot dP_{Iso\_\mathrm{Pr}essure}$). Josiah Willard Gibbs (1839 – 1903) was an American theoretical physicist, chemist, and mathematician.

Likewise the term "free energy" is also attached to the Helmholtz free energy, i.e., for systems at constant Pressure and temperature, to mean available in the form of useful work ($P_{Iso\_\mathrm{Pr}essure} \cdot d\mathrm{V_{ol}}$). Hermann Helmholtz (1821 – 1894) was a German physician and physicist.

Because the Helmholtz function is equal to the negative of the work done by a system it can be used to describe reversible processes of a system; where its decrease is the maximum amount of work which can be done *by* a system, and it can increase at most by the same amount of work done *on* a system.

The Thermodynamic Free Heat Energy of a system is the upper limit for any isothermal (Constant Temperature), isochoric (Constant Volume), or isobaric (Constant Pressure) work that can be captured in the surroundings, or it may simply be dissipated, appearing as an increase in the entropy of the system and/or it's surrounding.

In thermodynamics, **Entropy** is the amount isotropic, random, Omni-directional disorder associated with any thermodynamic system.

- **Entropy** – a measure of a system's isotropy Omni-directional work energy going from order to disorder at the atomic, ionic, molecular, microscopic, macroscopic, and cosmological scales.
- **Entropy** – a measure of disorder in the universe; the higher the entropy the greater the disorder of the universe.
- **Entropy** – a measure of Isotropic Kinetic Energy in a reversible or irreversible thermodynamic process.

2.1

$$E_{Net\_Iso\_Kinetic} = \frac{1}{2}\cdot m_{Net}\cdot\left|v^2\right|_{Iso} = \frac{3}{2}\cdot N\cdot k_B\cdot T_{emp} = \begin{bmatrix} \frac{m_1\overline{v}_1^2}{2} & + & \frac{m_2\overline{v}_2^2}{2} \\ + \frac{m_3\overline{v}_3^2}{2} & +\cdots & \frac{m_N\overline{v}_N^2}{2} \end{bmatrix}$$

In thermodynamics, **Syntropy** is the amount of anisotropic, directional order associated with any thermodynamic system.

- **Syntropy** – a measure of a system's anisotropy directional work energy going from disorder to order at the atomic, ionic, molecular, microscopic, macroscopic, and cosmological scales.
- **Syntropy** – a measure of order in the universe; the higher the syntropy the greater the order of the universe.
- **Syntropy** – a measure of Anisotropic Translational Kinetic Energy in a reversible or irreversible thermodynamic process.

2.2

$$E_{Net\_Translational} = \frac{1}{2}m_{Net}\cdot\left|\overline{v}\right|_{CM}^2 = \frac{3}{2}\cdot N\cdot k_B\cdot T_{emp\_Ext} = \frac{\begin{pmatrix} m_1\overline{v}_1 + m_2\overline{v}_2 \\ + m_3\overline{v}_3 + \cdots m_N\overline{v}_N \end{pmatrix}^2}{2\cdot m_{Net}}$$

Consider a gas system body the value of the entropy of a distribution of atoms and molecules in a thermodynamic system is a measure of the disorder in the arrangements of its particles. And the value of the syntropy is a measure of the ordered arrangements of its particles.

Similarly, in a gas, where the atoms or molecules are arranging themselves in orderly patterns the system has large syntropy; or when there is more order in the system than disorder, the entropy is reduced when all the molecules are in one place. This local increase in order is, however, only possible at the expense of an entropy increase in the surroundings; here more disorder must be created.

Whereas when more of the gas is in a more disorderly state the syntropy is small and the measure of the entropy of the system has its largest value The disordering of the atoms and molecules becomes more random and chaotic with an increase in temperature.

In solids, for example, which are typically ordered on the molecular scale, usually have smaller entropy than liquids, and liquids have smaller entropy than gases and colder gases have smaller entropy than hotter gases.

Moreover, according to the third law of thermodynamics, at absolute zero of temperature, crystalline structures are approximated to have perfect "order" and zero entropy.

This correlation occurs because the numbers of different microscopic quantum energy states available to an ordered system are usually much smaller than the number of states available to a system that appears to be disordered.

According to the Austrian physicist Ludwig Boltzmann (1844 – 1906), increases in thermal motion, occur whenever heat is added to a working substance, the rest position of molecules will be pushed apart, the body will expand, and this will create more *disordered* distributions and arrangements of molecules. These disordered arrangements, contribute to an increase in the measure of entropy.

Rudolf Clausius (1822 – 1888) a German mathematician and physcist, introduces the measurement of entropy change ($\Delta S_{Entropy}$). Since heat always flow from the hotter to cooler spontaneously, Entropy change describes the direction and magnitude of changes in heat transfer between systems with different temperatures, pressures, and volumes.

Entropy is equally essential in predicting the extent of complex chemical reactions, for such applications the total entropy of a system must be incorporated in an expression that include both the system and its surroundings, as described below,

2.3

$$\Delta S_{\text{Entropy\_Universe}} = \Delta S_{\text{Entropy\_Surroundings}} + \Delta S_{\text{Entropy\_System}}$$

The Ideal Gas Potential Internal Energy ($\frac{2}{3}\cdot E_{\text{Iso\_Kinetic\_Energy}} = P_{Iso\_Pressure}\cdot V_{ol} = N\cdot k_B\cdot T_{emp}$) might be thought of as the energy required creating a system in the absence of changes in temperature or volume.

The Thermodynamic Aether Free Heat Energy, is expelled from the system mass body into the surroundings once equilibrium has been established at constant absolute isotropic temperature (T_{emp}) and will contribute an amount of heat energy ($Q_{\text{Free\_Heat\_Energy}} = T_{emp}\cdot \Delta S_{Entropy\_Free}$) into the surroundings, reducing the overall investment necessary for creating the system.

The internal energy ($E_{\text{Iso\_Kinetic\_Energy}} = U_{\text{Internal\_Energy}} = \frac{3}{2}\cdot N\cdot k_B\cdot T_{emp}$) might be thought of as the energy required to create a system in the absence of changes in temperature or volume. But if the process changes the volume, as in a chemical reaction which produces a gaseous product, then work must be done to produce the change in volume.

The Enthalpy ($H_{\text{Enthalpy\_Energy}} = \lfloor U_{\text{Internal\_Energy}} + P_{Iso\_Pressure}\cdot V_{ol}\rfloor$), of the system results in an additional amount of work ($P_{Iso\_Pressure}\cdot V_{ol}$) that must be done if the system is created from a very small volume in order to "create room" for the system.

For a constant pressure process the work you must do to produce a volume change (ΔV_{ol}) is ($P_{Iso\_Pressure}\cdot \Delta V_{ol}$) and can be interpreted as the work you must do to "create room" for the system if you presume it started at zero volume.

Free Aether Heat Energy Transfer ($Q_{Free\_Heat\_Energy}$) is a path function which occurs during constant Internal Energy ($U_{Internal\_Energy} = E_{Net\_Iso\_Kinetic}$), constant Volume ($V_{ol}$), constant Pressure ($P_{Total\_Pressure}$), and constant absolute temperature (T_{Temp}) and requires a process such as a change in the static isotropic and dynamic anisotropic temperature ($\Delta T_{Free} = T_{Temp} - T_{Temp\_Ext}$) of an isolated system body equal to the **Isotropic Internal Temperature** (T_{Temp}) subtracted from the **Anisotropic External Temperature** (T_{Temp_Ext}).

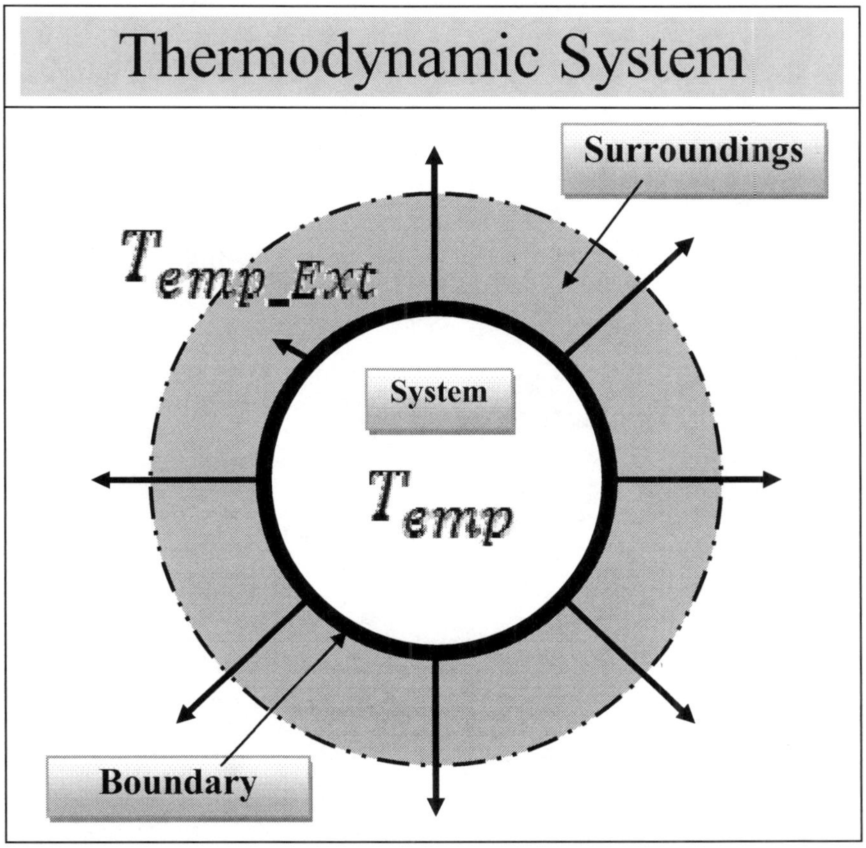

Figure 2.1: A Thermodynamic System and its Surroundings exchanges irreversible heat energy during a state of thermodynamic equilibrium or uniform motion.

The **Aether Free Energy Temperature Difference** is a measure of the temperature difference of an irreversible heat energy exchange between a system mass body and an external surroundings or environment during equilibrium and constant absolute Isotropic Temperature (T_{Temp}) of which the system body is bathed.

2.4

$$\Delta T_{Free} = -\Delta T_{Free\_Surrounding}$$

$$\lfloor \Delta T_{Free} + \Delta T_{Free\_Surrounding} \rfloor = 0$$

The Thermodynamic Aether Free Heat Energy is the energy needed to "make space for additional molecules" produced by various processes. The Aether Free Energy escapes the system mass body spontaneously as irreversible heat “lost” to the surrounding, such that heat energy is “gained” by the surroundings and is therefore conserved.

The **Aether Free Energy Temperature Difference** is a fixed quantity measure of the irreversible loss in heat energy during an isothermal process between the Order and Disorder, Entropy and Syntropy, or Isotropy and Anisotropy of a system mass body at constant absolute temperature.

The **Aether Free Energy Temperature Difference** is a fixed quantity measure during an isothermal and constant volume process in which a temperature difference still exists between the entropy and syntropy of a system, although no external work is being done to the system.

The **Aether Free Energy Temperature Difference** is the thermodynamic temperature measure of difference between order and disorder in an isolated system. The disorder of a system given by **Isotropic Internal Temperature** (T_{Temp}) and is always greater than or equal to the order of a system given by the **Anisotropic External Temperature** (T_{Temp_Ext}).

2.5

$$(Disorder \quad \geq \quad Order)$$

$$(T_{Temp} \quad \geq \quad T_{Temp\_Ext})$$

The upper limit of the **Aether Free Energy Temperature Difference** is a value that approaches the Internal Isotropic Absolute Temperature (T_{Temp}) and the closer the temperature difference to the Isotropic temperature the more disorder is in the system relative to the order.

2.6

$$(\Delta T_{Free} \quad \cong \quad T_{Temp} \quad \equiv \quad \text{Perfect Disorder})$$

The lower limit of the **Aether Free Energy Temperature Difference** is a value that approaches an infinitesimal value (close to zero), and the closer the temperature difference to zero the more order there is in the system relative to the disorder.

2.7

$$(\Delta T_{Free} \quad \cong \quad 0 \quad \equiv \quad \text{Perfect Order})$$

Definition 2.1: The **Aether Free Energy Temperature Difference** (ΔT_{Free}) of an isolated system mass body exist during an isothermal process which establishes thermodynamic equilibrium with the surroundings, and is equal the **Isotropic Internal Absolute Temperature** (T_{Temp}) subtracted from the **Anisotropic External Temperature** (T_{Temp_Ext}); and likewise defines a fixed amount of heat energy lost/gained by a system and gained/lost by its surrounding, and is a measure of the difference in the order (syntropy) subtracted from the disorder (entropy).

Aether Free Energy Temperature Difference

2.8

$$\Delta T_{Free} = -\Delta T_{Free\_to\_Surrounding} \rightarrow K$$

$$\Delta T_{Free} = \lfloor T_{Temp} - T_{Temp\_Ext} \rfloor$$

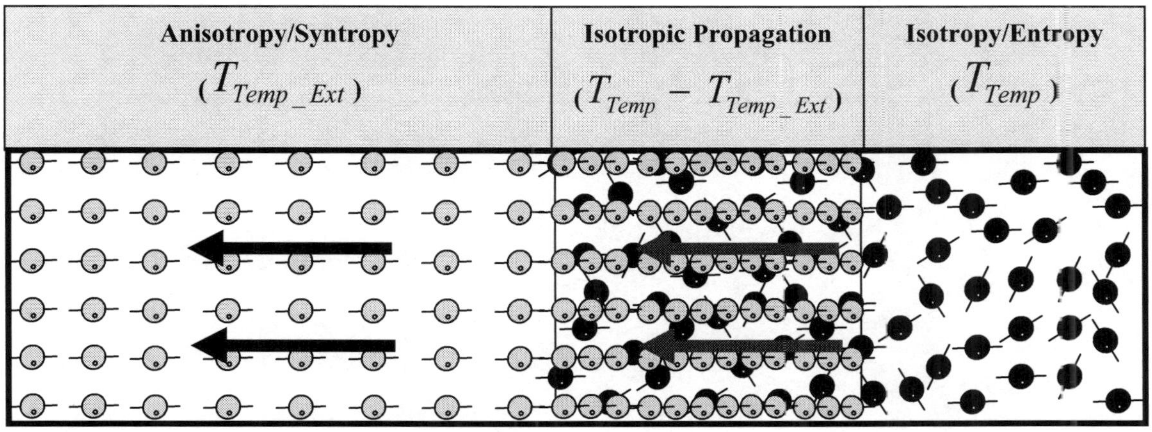

Figure 2.2: Aether Free Energy Temperature Difference is the measure of the irreversible heat energy that is lost from a system and is gained by the surroundings.

Aether Free Energy Temperature Difference

2.9

$$\Delta T_{Free} = -\Delta T_{Freee\_to\_Surrounding} \rightarrow K$$

$$\Delta T_{Free} = \lfloor T_{Temp} - T_{Temp\_Ext} \rfloor$$

$$\Delta T_{Free} = T_{Temp} \cdot \left[1 - \left(\frac{T_{Temp\_Ext}}{T_{Temp}} \right) \right]$$

$$\Delta T_{Free} = \frac{\Delta(PV)_{\mathrm{Free\_Energy}}}{\mathrm{N} \cdot \mathrm{k_B}} \cdot \left[1 - \left(\frac{T_{Temp\_Ext}}{T_{Temp}} \right) \right]$$

$$\Delta T_{Free} = T_{Temp} \cdot \left[1 - \frac{1}{\mathrm{m_{Net}}} \cdot \left[\frac{\left(\sum_{i=1}^{N} (\mathrm{m_i \bar{v}_i}) \right)^2}{\sum_{i=1}^{N} (\mathrm{m_i} \cdot \mathrm{\bar{v}_i^2})} \right] \right]$$

$$\Delta T_{Free} = T_{Temp} \cdot \left[1 - \frac{1}{\begin{pmatrix} \mathrm{m_1} + \mathrm{m_2} \\ + \mathrm{m_3} + \cdots \mathrm{m_N} \end{pmatrix}} \cdot \left[\frac{\begin{pmatrix} \mathrm{m_1 \bar{v}_1} + \mathrm{m_2 \bar{v}_2} \\ + \mathrm{m_3 \bar{v}_3} + \cdots \mathrm{m_N \bar{v}_N} \end{pmatrix}^2}{\begin{bmatrix} \mathrm{m_1 \bar{v}_1^2} + \mathrm{m_2 \bar{v}_2^2} \\ + \mathrm{m_3 \bar{v}_3^2} + \cdots \mathrm{m_N \bar{v}_N^2} \end{bmatrix}} \right] \right]$$

Aether Free Energy Temperature Difference

2.10

$$\Delta T_{Free} = \left(\frac{2}{3 \cdot \mathrm{N} \cdot \mathrm{k_B}}\right) \cdot \left[\mathrm{E}_{\mathrm{Net\_Iso\_Kinetic}} - \mathrm{E}_{\mathrm{Net\_Translational}}\right]$$

$$\Delta T_{Free} = \left(\frac{\mathrm{V_{ol}}}{\mathrm{N} \cdot \mathrm{k_B}}\right) \cdot \left[P_{Iso\_Pressure} - q_{Dynamic\_Pressure}\right]$$

$$\Delta T_{Free} = \left(\frac{\mathrm{m_{Net}}}{3 \cdot \mathrm{N} \cdot \mathrm{k_B}}\right) \cdot \left[\left.\mathrm{v}^2\right|_{\mathrm{Iso}} - \left|\overline{\mathrm{v}}\right|^2_{\mathrm{CM}}\right]$$

$$\Delta T_{Free} = \left(\frac{\mathrm{m_{Net}} \cdot \left.\mathrm{v}^2\right|_{\mathrm{Iso}}}{3 \cdot \mathrm{N} \cdot \mathrm{k_B}}\right) \cdot \left[1 - \frac{\left|\overline{\mathrm{v}}\right|^2_{\mathrm{CM}}}{\left.\mathrm{v}^2\right|_{\mathrm{Iso}}}\right]$$

$$\Delta T_{Free} = \left(\frac{2}{3 \cdot \mathrm{N} \cdot \mathrm{k_B}}\right) \cdot \left[\left[\frac{\mathrm{m_1} \cdot \overline{\mathrm{v}}_1^2}{2} + \frac{\mathrm{m_2} \cdot \overline{\mathrm{v}}_2^2}{2} + \frac{\mathrm{m_3} \cdot \overline{\mathrm{v}}_3^2}{2} + \cdots \frac{\mathrm{m_N} \cdot \overline{\mathrm{v}}_N^2}{2}\right] - \left[\frac{\left(\mathrm{m_1}\overline{\mathrm{v}}_1 + \mathrm{m_2}\overline{\mathrm{v}}_2 + \mathrm{m_3}\overline{\mathrm{v}}_3 + \cdots \mathrm{m_N}\overline{\mathrm{v}}_N\right)^2}{2 \cdot \mathrm{m_{Net}}}\right]\right]$$

2.11

Scalar	
Magnitude	Units
Aether Free Energy Temperature Difference	
$\Delta T_{Free} = \left[T_{Temp} - T_{Temp\_Ext}\right] = \left(\frac{\Delta(PV)_{\mathrm{Free\_Energy}}}{\mathrm{N} \cdot \mathrm{k_B}}\right) \cdot \left[1 - \left(\frac{T_{Temp\_Ext}}{T_{Temp}}\right)\right]$	K

2.2 Thermodynamic Field Irreversible Heat "Free Aether" Work Energy Transfer

The theory of Heat or Mechanical Theory of Heat was postulated predominantly in 1824 by the French physicist Sadi Carnot (1796 – 1832), in which he states that heat and mechanical work are equivalent. However, it is more complete to state that heat, mechanical work, and the aether are equivalent.

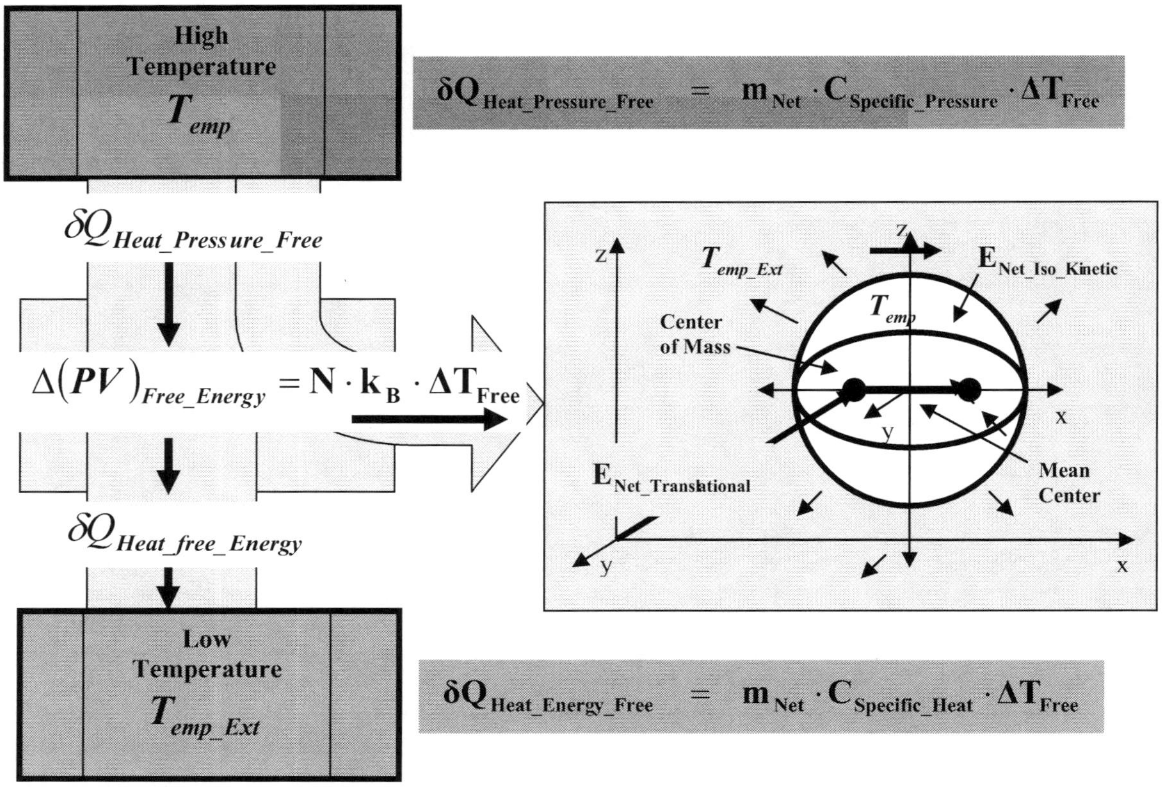

Figure 2.3: Aether Free Work Energy Difference is the measure of the irreversible heat energy that is lost from a system and is gained by the surroundings.

Energy transformation process in which the resultant condition lacks the driving potential needed to reverse the process; the measure of this loss is expressed by the entropy increase of the system

We have seen from our study of entropy that the universe inexorably moves towards a state of higher and higher entropy or disorder, and must always increase.

However, the entropy can only be a constant if the system is in the highest possible state of disorder, such as a gas that always was, and always will be, uniformly spread out in its container.

The existence of a thermodynamic arrow of time implies that the system is highly ordered in one time direction, which would by definition be the "past".

Can we simply define physical time to be the flow of time in a direction for which entropy always increases?

The answer is no, and in fact many physicists think this is so, and consider entropy to be the quantity which defines what is called the ``arrow of time".

The fact that the flow of the time is irreversible is said to originate in the irreversible processes which increase entropy. Experience shows that time can never flow backwards; events which have occurred in the “past” can never be undone.

It is the Free Heat energy that is transferred between bodies which determines the amount and flow from disorder to order in the system.

Quantum physics introduces a different kind of irreversibility into physics. Once a measurement is performed on a quantum system, the effects of the measurement are irreversible, in that the changes resulting from measurement cannot be undone.

The thermodynamic arrow is also often linked to the cosmological arrow of time, because it is ultimately about the boundary conditions of the early universe.

According to the Big Bang theory, the Universe was initially very hot with energy distributed uniformly. For a system in which gravity is important, such as the universe, this is a low-entropy state (compared to a high-entropy state of having all matter collapsed into black holes, a state to which the system may eventually evolve).

As the Universe grows, its temperature drops, which leave less energy available to perform useful, work in the future than was available in the past. Additionally, perturbations in the energy density grow (eventually forming galaxies and stars).

Thus the Universe itself has a well-defined thermodynamic arrow of time. But this doesn't address the question of why the initial state of the universe was that of low entropy.

If cosmic expansion were to halt and reverse due to gravity, the temperature of the Universe would once again increase, but its entropy would continue to increase due to the continued growth of perturbations and eventually black hole formation.

Given the fairly ordered state of the Universe at the present state, with billions of galaxies, stars burning, new stars being formed, life existing on earth, and so on, most are led to conclude that the Universe must have started with a beginning whose origin is a Big Bang.

This Big Bang origin must have existed early on in a very low entropy state; that is, in a highly ordered state. Ever since this Big Bang the universe has been expanding, racing towards a state with greater and greater disorder.

Why was the initial state of the universe so highly ordered? Let us state that the large Entropy and more disorder of the system is the result of the increase of more stuff being added into the system or universe.

More stuff could mean more particles, more mass, more energy, more volume potential, etc...

Assuming that the flow of Time itself originated with the Big Bang, which is not necessarily true, a “heat death of the universe” theory was postulated. However, Time could have existed before the Big Bang.

The hypothesis of a universal heat death originates in the 1850s with William Thomson, 1st Baron Kelvin. The idea of time and heat death became widespread in the nineteenth century when the idea of entropy was first understood.

Since the universe must move in a direction of greater and greater entropy, it was assumed that ultimately the universe would reach a state of equilibrium for which entropy would be a maximum.

When the universe reaches maximum entropy, time would effectively cease to exist since no change would take place. This ultimate equilibrium state was said to indicate that the universe was heading towards a final ``heat death." Once again this assumes that time originated with the big bang.

Another type of Free Aether Heat Energy is observed in Neutrino production in the nucleus of the atom.

Neutrinos are elementary particles that often travel close to the speed of light, lack an electric charge, are able to pass through ordinary matter almost undisturbed and are thus extremely difficult to detect. Neutrinos have a minuscule, but nonzero mass.

Neutrinos are created as a result of certain types of radioactive decay or nuclear reactions such as those that take place in the Sun, in nuclear reactors, or when cosmic rays hit atoms. Most neutrinos passing through the Earth emanate from the Sun, and more than 50 trillion solar electron neutrinos pass through the human body every second.

Energy and mass are conserved within all interacting systems, and Free Heat Energy which is energy transfer (in time) is irreversible (takes place in one direction only) from higher to lower potential, which then results in continuous generation (increase) of energy-displacement, called entropy generation, which is a fundamental measure of irreversibility, or permanent changes to a system mass body.

There are not two different types of entropy, the entropy of a system for a given state is the same, regardless whether it is reached by reversible heat transfer or irreversible heat or irreversible work transfer.

In accordance with the First Law of Motion a body at rest or in uniform motion, the energy of Internal Isotropy is transforms into the energy of External Anisotropy which represents heat flow from the internal (high temperature) to the external (lower temperature) of a system body.

Definition 2.2: The **Thermodynamic Field Irreversible Heat "Free Aether" Work Energy Transfer** ($\Delta(PV)_{\mathrm{Free\_Energy}}$) at constant **Temperature** is the heat energy gained or lost by an inertial mass substance equal to the **Net Inertial Mass** of the substance ($\mathrm{m}_{\mathrm{Net}}$) multiplied by its **Specific Heat Capacity** ($C_{Specific\_Temperature}$) at constant Temperature multiplied by the **Aether Free Energy Temperature Difference** (ΔT_{Free}) which is equal the **Isotropic Internal Absolute Temperature** (T_{Temp}) subtracted from the **Anisotropic External Temperature** ($\Delta T_{Free} = \lfloor T_{Temp} - T_{Temp\_Ext} \rfloor$) of the substance system mass body.

Definition 2.3: The **Thermodynamic Field Irreversible Heat “Free Aether” Work Energy Transfer** ($\Delta(PV)_{\mathrm{Free\_Energy}}$) is a thermodynamic measurement of the heat energy gained or lost equal to an amount of work done by an isolated system mass body establishing thermodynamic equilibrium between a system and its surroundings, and is equal to the product of the **Number of mass bodies** (N) multiplied by the, **Microscopic Boltzmann Gas Energy-Temperature Constant** ($\mathrm{k_B}$) and multiplied by the **Aether Free Energy Temperature Difference** (ΔT_{Free}) which is equal the **Isotropic Internal Absolute Temperature** (T_{Temp}) subtracted from the **Anisotropic External Temperature**.

Finally, the term Thermodynamic Field Irreversible Heat “Free Aether” Work Energy Transfer ($\Delta(PV)_{\mathrm{Free\_Energy}}$) for an isolated system body assuming no external heating from the outside environment, is the measure of a fixed amount of Internal Work done to the external environment system as the result of the Static Isotropic Pressure transforming into the Dynamic Pressure at constant volume.

Definition 2.4: The amount of **Thermodynamic Field Irreversible Heat “Free Aether” Work Energy Transfer** ($\Delta(PV)_{\mathrm{Free\_Energy}}$) at constant **Temperature** and a Fixed Compressive Volume Potential lost by a substance is equal to the **Isotropic Internal Omni-directional Static Pressure** subtracted from the **Translational Anisotropic External Dynamic Pressure**, multiplied by the Fixed Compressive Volume Potential ($\lfloor P_{Iso\_\Pr essure} - q_{Dynamic\_\Pr essure} \rfloor \cdot \mathrm{V_{ol}}$) of a substance system mass body.

Thermodynamic Field Irreversible Heat “Free Aether” Work Energy Transfer

2.12

$$\Delta(PV)_{\text{Free\_Energy}} = \text{N}\cdot\text{k}_{\text{B}}\cdot\left[\int_{T_{Initial}}^{T'_{Final}} dT_{Free}\right] = \text{N}\cdot\text{k}_{\text{B}}\cdot\left[T_{Temp} - T_{Temp\_Ext}\right]$$

$$\Delta(PV)_{\text{Free\_Energy}} = \int_{Q_{\text{Heat\_Energy\_Free}}}^{Q_{\text{Heat\_Pressure\_Free}}} d(PV)_{\text{Free\_Energy}}$$

$$\Delta(PV)_{\text{Free\_Energy}} = \text{m}_{\text{Net}}\cdot C_{Specific\_Temperature}\cdot\Delta T_{Free}$$

$$\Delta(PV)_{\text{Free\_Energy}} = \left[\delta Q_{\text{Heat\_Pressure\_Free}} - \delta Q_{\text{Heat\_Energy\_Free}}\right]$$

$$\Delta(PV)_{\text{Free\_Energy}} = T_{emp}\cdot\Delta S_{\text{Free\_Energy}} = \text{N}\cdot\text{k}_{\text{B}}\cdot\Delta T_{Free}$$

$$\Delta(PV)_{\text{Free\_Energy}} = \text{m}_{\text{Net}}\cdot\left[C_{Specific\_Pressure} - C_{Specific\_Hect}\right]\cdot\Delta T_{Free}$$

$$\Delta(PV)_{\text{Free\_Energy}} = -\,\text{m}_{\text{Net}}\cdot C_{Specific\_Heat}\cdot\left[1 - \gamma_{Heat}\right]\cdot\Delta T_{Free}$$

$$\Delta(PV)_{\text{Free\_Energy}} = -\,\text{m}_{\text{Net}}\cdot C_{Specific\_Pressure}\cdot\left[\frac{1 - \gamma_{Heat}}{\gamma_{Heat}}\right]\cdot\Delta T_{Free}$$

$$\Delta(PV)_{Free\_Energy} = \left(\frac{\text{m}_{\text{Net}}\cdot\left|\text{v}^2\right|_{\text{Iso}}}{3}\right)\cdot\left(\frac{\Delta T_{Free}}{T_{emp}}\right) \rightarrow kg\cdot m^2/s^2$$

Thermodynamic Field Irreversible Heat “Free Aether” Work Energy Transfer

2.13

$$\Delta(PV)_{\text{Free\_Energy}} = \text{V}_{\text{ol}} \cdot \left[\int_{P_{Iso\_Pressure}}^{q_{Dynamic\_Pressure}} d\left(P_{Iso\_Pressure}\right) \right] \rightarrow kg \cdot m^2 / s^2$$

$$\Delta(PV)_{\text{Free\_Energy}} = \left[P_{Iso\_Pressure} - q_{Dynamic\_Pressure} \right] \cdot \text{V}_{\text{ol}}$$

$$\Delta(PV)_{\text{Free\_Energy}} = \frac{2}{3} \cdot \left[\text{E}_{\text{Net\_Iso\_Kinetic}} - \text{E}_{\text{Net\_Translational}} \right]$$

$$\Delta(PV)_{\text{Free\_Energy}} = \left(\frac{\text{m}_{\text{Net}}}{3} \right) \cdot \left[\text{v}^2 \Big|_{\text{Iso}} - \left| \overline{\text{v}} \right|_{\text{CM}}^2 \right]$$

$$\Delta(PV)_{\text{Free\_Energy}} = \left(\frac{\text{m}_{\text{Net}} \cdot \left| \text{v}^2 \right|_{\text{Iso}}}{3} \right) \cdot \left[1 - \frac{\left| \overline{\text{v}} \right|_{\text{CM}}^2}{\left| \text{v}^2 \right|_{\text{Iso}}} \right]$$

$$\Delta(PV)_{\text{Free\_Energy}} = \left(\frac{\text{m}_{\text{Net}} \cdot \text{c}_{\text{Propagation}}^2}{\gamma_{Heat}} \right) \cdot \left[1 - \frac{\left| \overline{\text{v}} \right|_{\text{CM}}^2}{\left| \text{v}^2 \right|_{\text{Iso}}} \right]$$

$$\Delta(PV)_{\text{Free\_Energy}} = \left(\frac{\text{m}_{\text{Net}} \cdot \text{c}_{\text{Propagation}}^2}{\gamma_{Heat}} \right) \cdot \left[1 - \frac{\gamma_{Heat}}{3} \cdot \left(\frac{\left| \overline{\text{v}} \right|_{\text{CM}}^2}{\text{c}_{\text{Propagation}}^2} \right) \right]$$

Thermodynamic Field Irreversible Heat "Free Aether" Work Energy Transfer

2.14

$$\Delta(PV)_{\text{Free\_Energy}} = T_{emp} \cdot \Delta S_{\text{Free\_Energy}} = \mathrm{N} \cdot \mathrm{k}_{\mathrm{B}} \cdot \Delta T_{Free} \rightarrow kg \cdot m^2 / s^2$$

$$\Delta(PV)_{\text{Free\_Energy}} = \mathrm{m}_{\text{Net}} \cdot C_{Specific\_Temperature} \cdot \Delta T_{Free}$$

$$\Delta(PV)_{\text{Free\_Energy}} = \left[\frac{\Delta(PV)_{\text{Field\_Energy}}}{ln\left(\dfrac{T'_{emp\ Final}}{T_{emp\ Initial}} \right)} \right] \cdot \left(\frac{\Delta T_{Free}}{T_{emp\ Initial}} \right)$$

$$\Delta(PV)_{\text{Free\_Energy}} = -\frac{1}{(\gamma_{Heat} - 1)} \cdot \left[\frac{\Delta(PV)_{\text{Field\_Energy}}}{ln\left(\dfrac{\mathrm{V}'_{\text{ol}\,Final}}{\mathrm{V}_{\text{ol}\,Initial}} \right)} \right] \cdot \left[1 - \frac{\left|\overline{\mathrm{v}}\right|^2_{\text{CM}}}{\left|\mathrm{v}^2\right|_{\text{Iso}}} \right]$$

$$\Delta(PV)_{\text{Free\_Energy}} = \frac{\gamma_{Heat}}{(\gamma_{Heat} - 1)} \cdot \left[\frac{\Delta(PV)_{\text{Field\_Energy}}}{ln\left(\dfrac{P'_{Iso\_\text{Pr}essure\ Final}}{P_{Iso\_\text{Pr}essure\ Initial}} \right)} \right] \cdot \left[1 - \left(\frac{T_{Temp\_Ext}}{T_{Temp}} \right) \right]$$

Thermodynamic Field Irreversible Heat "Free Aether" Work Energy Transfer

2.15

$$\Delta(PV)_{\text{Free\_Energy}} = \left[\begin{array}{l} \dfrac{1}{\gamma_{Heat}}\cdot\left(\text{N}\cdot\text{k}_\text{B}\cdot T_{emp\ Initial}\right) \\ \quad + \left[\dfrac{\Delta(PV)_{\text{Field\_Energy}}}{ln\left(\dfrac{P'_{Iso\_Pressure\ Final}}{P_{Iso\_Pressure\ Initial}}\right)}\right]\end{array}\right]\cdot\left[1 - \dfrac{\left|\overline{\text{v}}\right|^2_{\text{CM}}}{\left|\text{v}^2\right|_{\text{Iso}}}\right] \rightarrow kg\cdot m^2/s^2$$

$$\Delta(PV)_{\text{Free\_Energy}} = \frac{1}{\gamma_{Heat}}\cdot\left[\begin{array}{l}\text{N}\cdot\text{k}_\text{B}\cdot T_{emp\ Initial} \\ \quad - \dfrac{\Delta(PV)_{\text{Field\_Energy}}}{ln\left(\dfrac{\text{V}'_{\text{ol}\,Final}}{\text{V}_{\text{ol}\,Initial}}\right)}\end{array}\right]\cdot\left[1 - \frac{\gamma_{Heat}}{3}\cdot\left(\frac{\left|\overline{\text{v}}\right|^2_{\text{CM}}}{\text{c}^2_{\text{Propagation}}}\right)\right]$$

$$\Delta(PV)_{\text{Free\_Energy}} = \frac{1}{\gamma_{Heat}}\cdot\left[\begin{array}{l}\text{N}\cdot\text{k}_\text{B}\cdot T_{emp\ Initial} \\ \quad + \left(\gamma_{Heat} - 1\right)\cdot\dfrac{\Delta(PV)_{\text{Field\_Energy}}}{ln\left(\dfrac{T'_{emp\ Final}}{T_{emp\ Initial}}\right)}\end{array}\right]\cdot\left[1 - \left(\frac{T_{Temp\_Ext}}{T_{Temp}}\right)\right]$$

Thermodynamic Field Irreversible Heat "Free Aether" Work Energy Transfer

2.16

$$\Delta(PV)_{\text{Free\_Energy}} = \text{N}\cdot\text{k}_\text{B}\cdot\Delta T_{Free} = \text{N}\cdot\text{k}_\text{B}\cdot\left\lfloor T_{emp} - T_{emp\_Ext}\right\rfloor \rightarrow kg\cdot m^2/s^2$$

$$\Delta(PV)_{\text{Free\_Energy}} = \text{n}_{\text{Moles}}\cdot\text{R}\cdot\left\lfloor T_{emp} - T_{emp\_Ext}\right\rfloor$$

$$\Delta(PV)_{\text{Free\_Energy}} = \left\lfloor P_{Iso\_Pressure} - q_{Dynamic\_Pressure}\right\rfloor\cdot\text{V}_{\text{ol}}$$

Thermodynamic Field Irreversible Heat “Free Aether” Work Energy Transfer

2.17

$$\Delta(PV)_{Free\_Energy} = \frac{1}{3}\cdot\left(m_{Net} - m_{Inertia\_Mass}\right)\cdot\left|v^2\right|_{Iso} \rightarrow kg\cdot m^2/s^2$$

$$\Delta(PV)_{Free\_Energy} = \frac{1}{3}\cdot\left(\left(\frac{m^2_{Net}}{m_{Inertia\_Mass}}\right) - m_{Net}\right)\cdot\left|\overline{v}\right|^2_{CM}$$

$$\Delta(PV)_{Free\_Energy} = \left(\frac{1}{m_{Inertia\_Mass}} - \frac{1}{m_{Net}}\right)\cdot\frac{\overline{p}^2_{Inertial\_Net}}{3}$$

$$\Delta(PV)_{Free\_Energy} = \left(\frac{1}{m_{Net}} - \frac{m_{Inertia\_Mass}}{m^2_{Net}}\right)\cdot\frac{\overline{p}^2_{Iso\_Net}}{3}$$

$$\Delta(PV)_{Free\_Energy} = \frac{1}{\gamma_{Heat}}\cdot\left(m_{Net} - m_{Inertia\_Mass}\right)\cdot c^2_{Propagation}$$

$$\Delta(PV)_{Free\_Energy} = \left(m_{Net} - m_{Inertia\_Mass}\right)\cdot\frac{v^2_{Out\_Flow\_Velocity}}{\left[\gamma_{Heat} - 1\right]}$$

$$\Delta(PV)_{Free\_Energy} = -\left(m_{Net} - m_{Inertia\_Mass}\right)\cdot\frac{v^2_{In\_Flow\_Velocity}}{\left[\gamma_{Heat} - 1\right]}$$

Thermodynamic Field Irreversible Heat "Free Aether" Work Energy Transfer — "Before State"

2.18

$$\Delta(PV)_{\text{Free\_Energy}} = \frac{2}{3}\cdot\left[E_{\text{Net\_Iso\_Kinetic}} - E_{\text{Net\_Translational}}\right] \rightarrow kg\cdot m^2/s^2$$

$$\Delta(PV)_{\text{Free\_Energy}} = \frac{2}{3}\cdot\left[\sum_{i=1}^{N}\left(\frac{m_i\cdot\overline{v}_i^2}{2}\right) - \frac{1}{2\cdot m_{\text{Net}}}\cdot\left(\sum_{i=1}^{N}(m_i\overline{v}_i)\right)^2\right]$$

$$\Delta(PV)_{\text{Free\_Energy}} = \frac{2}{3}\cdot\left[\left[\frac{m_1\overline{v}_1^2}{2} + \frac{m_2\overline{v}_2^2}{2} + \frac{m_3\overline{v}_3^2}{2} + \cdots\frac{m_N\overline{v}_N^2}{2}\right] - \frac{(m_1\overline{v}_1 + m_2\overline{v}_2 + m_3\overline{v}_3 + \cdots m_N\overline{v}_N)^2}{2\cdot m_{\text{Net}}}\right]$$

$$\Delta(PV)_{\text{Free\_Energy}} = \frac{2}{3}\cdot\left[\sum_{i=1}^{N}\left(\frac{m_i\cdot\overline{v}_{Ai}^2}{2}\right) - \frac{1}{2\cdot m_{\text{Net}}}\cdot\left(\sum_{i=1}^{N}(m_i\overline{v}_i)\right)^2\right]$$

$$\Delta(PV)_{\text{Free\_Energy}} = \frac{2}{3}\cdot\left[\left[\frac{m_1\overline{v}_{A1}^2}{2} + \frac{m_2\overline{v}_{A2}^2}{2} + \frac{m_3\overline{v}_{A3}^2}{2} + \cdots\frac{m_N\overline{v}_{AN}^2}{2}\right] - \frac{(m_1\overline{v}_1 + m_2\overline{v}_2 + m_3\overline{v}_3 + \cdots m_N\overline{v}_N)^2}{2\cdot m_{\text{Net}}}\right]$$

Thermodynamic Field Irreversible Heat "Free Aether" Work Energy Transfer — "After State"

2.19

$$\Delta(PV)_{\text{Free\_Energy}} = \frac{1}{3}\cdot m_{\text{Net}}\cdot\left[\left|v^2\right|_{\text{Iso}} - \left|\overline{v}\right|^2_{\text{CM}}\right] \rightarrow kg\cdot m^2/s^2$$

$$\Delta(PV)_{\text{Free\_Energy}} = \frac{m_{\text{Net}}}{6\cdot N}\cdot\left[\begin{array}{l}\left[\overline{v}'^2_{\text{A\_Iso\_Total}} + \overline{v}^2_{\text{Iso\_Total}}\right] \\ \qquad + 2\cdot N\cdot\left|\overline{v}\right|^2_{\text{CM}} - 4\cdot\left|\overline{v}\right|_{\text{CM}}\cdot\overline{v}_{\text{Inertial\_Net}}\end{array}\right]$$

$$\Delta(PV)_{\text{Free\_Energy}} = \frac{m_{\text{Net}}}{6\cdot N}\cdot\left[\begin{array}{l}\left[2\cdot N\cdot\left|v^2\right|_{\text{Iso}} - \left(\overline{v}'^2_{\text{Iso\_Total}} - \overline{v}^2_{\text{Iso\_Total}}\right)\right] \\ \qquad + 2\cdot N\cdot\left|\overline{v}\right|^2_{\text{CM}} - 4\cdot\left|\overline{v}\right|_{\text{CM}}\cdot\overline{v}_{\text{Inertial\_Net}}\end{array}\right]$$

Thermodynamic Field Irreversible Heat "Free Aether" Work Energy Transfer — "Before State"

2.20

$$\Delta(PV)_{\text{Free\_Energy}} = \frac{1}{3}\cdot m_{\text{Net}}\cdot\left[\left|v^2\right|_{\text{Iso}} - \left|\overline{v}\right|^2_{\text{CM}}\right] \rightarrow kg\cdot m^2/s^2$$

$$\Delta(PV)_{\text{Free\_Energy}} = \frac{m_{\text{Net}}}{6\cdot N}\cdot\left[\begin{array}{l}\left[\overline{v}^2_{\text{A\_Iso\_Total}} + \overline{v}'^2_{\text{Iso\_Total}}\right] \\ \qquad + 2\cdot N\cdot\left|\overline{v}\right|^2_{\text{CM}} - 4\cdot\left|\overline{v}\right|_{\text{CM}}\cdot\overline{v}'_{\text{Inertial\_Net}}\end{array}\right]$$

$$\Delta(PV)_{\text{Free\_Energy}} = \frac{m_{\text{Net}}}{6\cdot N}\cdot\left[\begin{array}{l}\left[2\cdot N\cdot\left|v^2\right|_{\text{Iso}} + \left(\overline{v}'^2_{\text{Iso\_Total}} - \overline{v}^2_{\text{Iso\_Total}}\right)\right] \\ \qquad + 2\cdot N\cdot\left|\overline{v}\right|^2_{\text{CM}} - 4\cdot\left|\overline{v}\right|_{\text{CM}}\cdot\overline{v}'_{\text{Inertial\_Net}}\end{array}\right]$$

Thermodynamic Field Irreversible Heat “Free Aether” Work Energy Transfer

2.21

$$\Delta(PV)_{\text{Free\_Energy}} = \mathrm{N} \cdot \mathrm{k}_{\mathrm{B}} \cdot T_{emp} \cdot \left[1 - \left(\frac{T_{emp\_Ext}}{T_{emp}}\right)\right] \rightarrow kg \cdot m^2 / s^2$$

$$\Delta(PV)_{\text{Free\_Energy}} = \frac{1}{3} \cdot \mathrm{m}_{\text{Net}} \cdot \left|\mathrm{v}^2\right|_{\text{Iso}} \cdot \left[1 - \left(\frac{\left|\overline{\mathrm{v}}\right|_{\text{CM}}^2}{\left|\mathrm{v}^2\right|_{\text{Iso}}}\right)\right]$$

$$\Delta(PV)_{\text{Free\_Energy}} = \frac{1}{3} \cdot \mathrm{m}_{\text{Net}} \cdot \left|\mathrm{v}^2\right|_{\text{Iso}} \cdot \left[1 - \left(\frac{\mathrm{m}_{\text{Inertia\_Mass}}}{\mathrm{m}_{\text{Net}}}\right)\right]$$

$$\Delta(PV)_{\text{Free\_Energy}} = \frac{1}{3} \cdot \left(\frac{\mathrm{m}_{\text{Net}}^2}{\mathrm{m}_{\text{Inertia\_Mass}}}\right) \cdot \left|\overline{\mathrm{v}}\right|_{\text{CM}}^2 \cdot \left[1 - \left(\frac{\mathrm{m}_{\text{Inertia\_Mass}}}{\mathrm{m}_{\text{Net}}}\right)\right]$$

$$\Delta(PV)_{\text{Free\_Energy}} = \left(\frac{1}{3} \cdot \mathrm{m}_{\text{Net}} \cdot \left|\overline{\mathrm{v}}\right|_{\text{CM}}^2\right) \cdot \left(\frac{\mathrm{m}_{\text{Net}}}{\mathrm{m}_{\text{Inertia\_Mass}}}\right) \cdot \left[1 - \left(\frac{\mathrm{m}_{\text{Inertia\_Mass}}}{\mathrm{m}_{\text{Net}}}\right)\right]$$

$$\Delta(PV)_{\text{Free\_Energy}} = \mathrm{N} \cdot \mathrm{k}_{\mathrm{B}} \cdot T_{emp} \cdot \left[1 - \left(\frac{1}{\mathrm{m}_{\text{Net}}}\right) \cdot \left[\frac{\left(\sum_{\mathrm{i}=1}^{\mathrm{N}} (\mathrm{m}_{\mathrm{i}} \overline{\mathrm{v}}_{\mathrm{i}})\right)^2}{\sum_{\mathrm{i}=1}^{\mathrm{N}} \left(\mathrm{m}_{\mathrm{i}} \cdot \overline{\mathrm{v}}_{\mathrm{i}}^2\right)}\right]\right]$$

$$\Delta(PV)_{\text{Free\_Energy}} = \mathrm{N} \cdot \mathrm{k}_{\mathrm{B}} \cdot T_{emp} \cdot \left[1 - \left(\frac{1}{\mathrm{m}_{\text{Net}}}\right) \cdot \left[\frac{\left(\sum_{\mathrm{i}=1}^{\mathrm{N}} (\mathrm{m}_{\mathrm{i}} \overline{\mathrm{v}}_{\mathrm{i}})\right)^2}{\sum_{\mathrm{i}=1}^{\mathrm{N}} \left(\mathrm{m}_{\mathrm{i}} \cdot \overline{\mathrm{v}}_{\mathrm{Ai}}^2\right)}\right]\right]$$

Thermodynamic Field Irreversible Heat “Free Aether” Work Energy Transfer — “Before State”

2.22

$$\Delta(PV)_{\text{Free\_Energy}} = \mathrm{N}\cdot \mathrm{k}_{\mathrm{B}}\cdot T_{emp}\cdot\left[1-\left(\frac{1}{\mathrm{m}_{\mathrm{Net}}}\right)\cdot\left[\frac{\left(\begin{array}{c}\mathrm{m}_1\cdot\overline{\mathrm{v}}_1 + \mathrm{m}_2\cdot\overline{\mathrm{v}}_2\\ + \mathrm{m}_3\cdot\overline{\mathrm{v}}_3 + \cdots \mathrm{m}_{\mathrm{N}}\cdot\overline{\mathrm{v}}_{\mathrm{N}}\end{array}\right)^2}{\left[\begin{array}{c}\mathrm{m}_1\cdot\overline{\mathrm{v}}_1^2 + \mathrm{m}_2\cdot\overline{\mathrm{v}}_2^2\\ + \mathrm{m}_3\cdot\overline{\mathrm{v}}_3^2 + \cdots \mathrm{m}_{\mathrm{N}}\cdot\overline{\mathrm{v}}_{\mathrm{N}}^2\end{array}\right]}\right]\right]$$

$$\Delta(PV)_{\text{Free\_Energy}} = \frac{1}{3}\cdot\left[\begin{array}{l}\mathrm{m}_1\cdot\overline{\mathrm{v}}_1^2\\ + \mathrm{m}_2\cdot\overline{\mathrm{v}}_2^2\\ + \mathrm{m}_3\cdot\overline{\mathrm{v}}_3^2\\ +\cdots \mathrm{m}_{\mathrm{N}}\cdot\overline{\mathrm{v}}_{\mathrm{N}}^2\end{array}\right]\cdot\left[1-\left(\frac{1}{\mathrm{m}_{\mathrm{Net}}}\right)\cdot\left[\frac{\left(\begin{array}{l}\mathrm{m}_1\cdot\overline{\mathrm{v}}_1\\ + \mathrm{m}_2\cdot\overline{\mathrm{v}}_2\\ + \mathrm{m}_3\cdot\overline{\mathrm{v}}_3\\ + \cdots \mathrm{m}_{\mathrm{N}}\cdot\overline{\mathrm{v}}_{\mathrm{N}}\end{array}\right)^2}{\left[\begin{array}{l}\mathrm{m}_1\cdot\overline{\mathrm{v}}_1^2\\ + \mathrm{m}_2\cdot\overline{\mathrm{v}}_2^2\\ + \mathrm{m}_3\cdot\overline{\mathrm{v}}_3^2\\ +\cdots \mathrm{m}_{\mathrm{N}}\cdot\overline{\mathrm{v}}_{\mathrm{N}}^2\end{array}\right]}\right]\right]$$

$$\Delta(PV)_{\text{Free\_Energy}} = \frac{1}{3}\cdot \mathrm{m}_{\mathrm{Net}}\cdot\left|\mathrm{v}^2\right|_{\mathrm{Iso}}\cdot\left[1-\left(\frac{1}{\mathrm{m}_{\mathrm{Net}}}\right)\cdot\left[\frac{\left(\begin{array}{c}\mathrm{m}_1\cdot\overline{\mathrm{v}}_1 + \mathrm{m}_2\cdot\overline{\mathrm{v}}_2\\ + \mathrm{m}_3\cdot\overline{\mathrm{v}}_3 + \cdots \mathrm{m}_{\mathrm{N}}\cdot\overline{\mathrm{v}}_{\mathrm{N}}\end{array}\right)^2}{\left[\begin{array}{c}\mathrm{m}_1\cdot\overline{\mathrm{v}}_{\mathrm{A1}}^2 + \mathrm{m}_2\cdot\overline{\mathrm{v}}_{\mathrm{A2}}^2\\ + \mathrm{m}_3\cdot\overline{\mathrm{v}}_{\mathrm{A3}}^2 + \cdots \mathrm{m}_{\mathrm{N}}\cdot\overline{\mathrm{v}}_{\mathrm{AN}}^2\end{array}\right]}\right]\right]$$

Thermodynamic Field Irreversible Heat “Free Aether” Work Energy Transfer

2.23

$$\Delta(PV)_{\text{Free\_Energy}} = \frac{1}{3}\cdot \mathrm{m}_{\text{Net}} \cdot \left[\left|\mathrm{v}^2\right|_{\text{Iso}} - \left|\overline{\mathrm{v}}\right|^2_{\text{CM}}\right]$$

$$\Delta(PV)_{\text{Free\_Energy}} = \frac{\overline{\mathrm{p}}^2_{\text{Inertial\_Net}}}{3\cdot\left(\dfrac{\mathrm{m}_{\text{Inertia\_Mass}}\cdot \mathrm{m}_{\text{Net}}}{\mathrm{m}_{\text{Net}} - \mathrm{m}_{\text{Inertia\_Mass}}}\right)}$$

$$\Delta(PV)_{\text{Free\_Energy}} = \frac{\mathrm{m}^2_{\text{Net}}\cdot\left|\overline{\mathrm{v}}\right|^2_{\text{CM}}}{3\cdot\left(\dfrac{\mathrm{m}_{\text{Inertia\_Mass}}\cdot \mathrm{m}_{\text{Net}}}{\mathrm{m}_{\text{Net}} - \mathrm{m}_{\text{Inertia\_Mass}}}\right)}$$

$$\Delta(PV)_{\text{Free\_Energy}} = \left(\frac{\overline{\mathrm{p}}^2_{\text{Inertial\_Net}}}{3\cdot \mathrm{m}_{\text{Net}}}\right)\cdot\left[\frac{\left(1 - \left(\dfrac{\left|\overline{\mathrm{v}}\right|^2_{\text{CM}}}{\left|\mathrm{v}^2\right|_{\text{Iso}}}\right)\right)}{\left(\dfrac{\left|\overline{\mathrm{v}}\right|^2_{\text{CM}}}{\left|\mathrm{v}^2\right|_{\text{Iso}}}\right)}\right]$$

Thermodynamic Field Irreversible Heat “Free Aether” Work Energy Transfer — “Before State”

2.24

$$\Delta(PV)_{\text{Free\_Energy}} = \Delta p_{\textit{Thermdynamic\_Impulse}}\cdot \mathrm{c}_{\text{Propagation}} \rightarrow kg\cdot m^2/s^2$$

$$\Delta(PV)_{\text{Free\_Energy}} = \left[\int_0^{\mathrm{d}_{i_{Shell}}} F_{\textit{Thermo\_Free\_Force}}\cdot d\left(\mathrm{d}_{i_{Shell}}\right)\right]$$

$$\Delta(PV)_{\text{Free\_Energy}} = F_{\textit{Thermo\_Free\_Force}}\cdot \mathrm{d}_{i_{Shell}}$$

The **Thermodynamic Field Irreversible Heat "Free Aether" Work Energy Transfer,** in three dimensional Cartesian coordinates x, y, and z, are given by the following,

2.25

$$\Delta(PV)_{\text{Free\_Energy}\,x} = T_{emp} \cdot \Delta S_{Entropy\_Free\,x} = \frac{1}{3} \cdot \mathrm{m}_{\text{Net}} \cdot \left[\left| \mathrm{v}^2 \right|_{\text{Iso}\,x} - \left| \overline{\mathrm{v}} \right|^2_{\text{CM}\,x} \right]$$

$$\Delta(PV)_{\text{Free\_Energy}\,y} = T_{emp} \cdot \Delta S_{Entropy\_Free\,y} = \frac{1}{3} \cdot \mathrm{m}_{\text{Net}} \cdot \left[\left| \mathrm{v}^2 \right|_{\text{Iso}\,y} - \left| \overline{\mathrm{v}} \right|^2_{\text{CM}\,y} \right]$$

$$\Delta(PV)_{\text{Free\_Energy}\,z} = T_{emp} \cdot \Delta S_{Entropy\_Free\,z} = \frac{1}{3} \cdot \mathrm{m}_{\text{Net}} \cdot \left[\left| \mathrm{v}^2 \right|_{\text{Iso}\,z} - \left| \overline{\mathrm{v}} \right|^2_{\text{CM}\,z} \right]$$

2.26

Scalar/Tensor		
Magnitude	Magnitude/Tensor	Units
Thermodynamic Field Irreversible Heat "Free Aether" Work Energy Transfer		
$\Delta(PV)_{\text{Free\_Energy}} = \begin{bmatrix} \Delta(PV)_{\text{Free\_Energy}\,x} \\ + \Delta(PV)_{\text{Free\_Energy}\,y} \\ + \Delta(PV)_{\text{Free\_Energy}\,z} \end{bmatrix}$	$\Delta(PV)_{\text{Free\_Energy}} = \begin{bmatrix} \Delta(PV)_{\text{Free\_Energy}\,x} \\ \Delta(PV)_{\text{Free\_Energy}\,y} \\ \Delta(PV)_{\text{Free\_Energy}\,z} \end{bmatrix}$	$\frac{\text{kg} \cdot \text{m}^2}{\text{s}^2}$

2.3 Thermodynamic Field Irreversible Heat "Free Aether" Pressure Potential Energy Transfer

An isolated mass system body constrained to have constant pressure in a Thermodynamic Field Potential is the net or total amount of **Thermodynamic Field Irreversible Heat "Free Aether" Pressure Potential Energy Transfer** between a system and its surroundings.

The Thermodynamic Field Irreversible Heat "Free Aether" **Pressure Potential** Energy Transfer is at a ***higher potential*** in any system than the Thermodynamic Field Irreversible Heat "Free Aether" **Volume Potential** Energy Transfer.

Definition 2.5: The total amount of The Thermodynamic Field Irreversible Heat "Free Aether" **Pressure Potential** Energy Transfer ($\delta Q_{\mathrm{Heat\_Pressure\_Free}}$) is the heat energy lost/gained by an inertial mass substance at constant Pressure and gained/lost by the surroundings and is equal to the **Net Inertial Mass** of the substance ($\mathrm{m}_{\mathrm{Net}}$) multiplied by its **Specific Heat Capacity of an Atomic Substance** at constant **Pressure** **(**$C_{Specific\_Pressure}$**)** multiplied by the **Aether Free Energy Temperature Difference** (ΔT_{Free}) which is equal the **Isotropic Internal Absolute Temperature** (T_{Temp}) subtracted from the **Anisotropic External Temperature** ($\Delta T_{Free} = \lfloor T_{Temp} - T_{Temp\_Ext} \rfloor$) of the substance system mass body.

Thermodynamic Field Irreversible Heat "Free Aether" Pressure Potential Energy Transfer

2.27

$$\delta Q_{\text{Heat\_Pressure\_Free}} = \mathrm{m}_{\text{Net}} \cdot C_{Specific\_Pressure} \cdot \left[\int_{T_{Initial}}^{T'_{Final}} dT_{Free} \right] \rightarrow kg \cdot m^2 / s^2$$

$$\delta Q_{\text{Heat\_Pressure\_Free}} = \mathrm{m}_{\text{Net}} \cdot C_{Specific\_Pressure} \cdot \Delta T_{Free}$$

$$\delta Q_{\text{Heat\_Pressure\_Free}} = \mathrm{m}_{\text{Net}} \cdot C_{Specific\_Pressure} \cdot \left[T_{Temp} - T_{Temp\_Ext} \right]$$

$$\delta Q_{\text{Heat\_Pressure\_Free}} = \left[\delta Q_{\text{Heat\_Energy\_Free}} + \Delta(PV)_{\text{Free\_Energy}} \right]$$

$$\delta Q_{\text{Heat\_Pressure\_Free}} = \left[\mathrm{m}_{\text{Net}} \cdot C_{Specific\_Heat} + \mathrm{N} \cdot \mathrm{k_B} \right] \cdot \Delta T_{Free}$$

$$\delta Q_{\text{Heat\_Pressure\_Free}} = \left[\mathrm{m}_{\text{Net}} \cdot C_{Specific\_Heat} + \mathrm{N} \cdot \mathrm{k_B} \right] \cdot \left[T_{Temp} - T_{Temp\_Ext} \right]$$

$$\delta Q_{\text{Heat\_Pressure\_Free}} = \frac{\mathrm{m}_{\text{Net}} \cdot \mathrm{c}^2_{\text{Propagation}}}{(\gamma_{Heat} - 1)} \cdot \left(\frac{\Delta T_{Free}}{T_{emp\ Initial}} \right)$$

$$\delta Q_{\text{Heat\_Pressure\_Free}} = \frac{\mathrm{m}_{\text{Net}} \cdot \mathrm{c}^2_{\text{Propagation}}}{(\gamma_{Heat} - 1)} \cdot \left[1 - \left(\frac{T_{Temp\_Ext}}{T_{Temp}} \right) \right]$$

Thermodynamic Field Irreversible Heat "Free Aether" Pressure Potential Energy Transfer

2.28

$$\delta Q_{\text{Heat\_Pressure\_Free}} = \left\lfloor \delta Q_{\text{Heat\_Energy\_Free}} + \Delta(PV)_{\text{Free\_Energy}} \right\rfloor \rightarrow kg \cdot m^2 / s^2$$

$$\delta Q_{\text{Heat\_Pressure\_Free}} = \left[\delta Q_{\text{Heat\_Energy\_Free}} + \frac{2}{3} \cdot \left[\mathrm{E}_{\text{Net\_Iso\_Kinetic}} - \mathrm{E}_{\text{Net\_Translational}} \right] \right]$$

$$\delta Q_{\text{Heat\_Pressure\_Free}} = \left[\delta Q_{\text{Heat\_Energy\_Free}} + \frac{1}{3} \cdot \mathrm{m}_{\text{Net}} \cdot \left[\left| \mathrm{v}^2 \right|_{\text{Iso}} - \left| \overline{\mathrm{v}} \right|^2_{\text{CM}} \right] \right]$$

$$\delta Q_{\text{Heat\_Pressure\_Free}} = \frac{\mathrm{m}_{\text{Net}} \cdot \left| \mathrm{v}^2 \right|_{\text{Iso}}}{3 \cdot \left(1 - \dfrac{1}{\gamma_{Heat}} \right)} \cdot \left(\frac{\Delta T_{Free}}{T_{emp\ Initial}} \right)$$

$$\delta Q_{\text{Heat\_Pressure\_Free}} = \frac{\mathrm{m}_{\text{Net}} \cdot \left[\dfrac{\left| \mathrm{v}^2 \right|_{\text{Iso}}}{3} + \mathrm{v}^2_{\text{Out\_Flow\_Velocity}} \right]}{\left(\gamma_{Heat} - 1 \right)} \cdot \left[1 - \frac{T_{Temp\_Ext}}{T_{Temp}} \right]$$

$$\delta Q_{\text{Heat\_Pressure\_Free}} = \frac{\mathrm{m}_{\text{Net}} \cdot \left| \mathrm{v}^2 \right|_{\text{Iso}}}{3 \cdot \left(1 - \dfrac{1}{\gamma_{Heat}} \right)} \cdot \left[1 - \frac{\left| \overline{\mathrm{v}} \right|^2_{\text{CM}}}{\left| \mathrm{v}^2 \right|_{\text{Iso}}} \right]$$

$$\delta Q_{\text{Heat\_Pressure\_Free}} = \frac{\mathrm{m}_{\text{Net}} \cdot \mathrm{c}^2_{\text{Propagation}}}{\left(\gamma_{Heat} - 1 \right)} \cdot \left[1 - \frac{\gamma_{Heat}}{3} \cdot \left(\frac{\left| \overline{\mathrm{v}} \right|^2_{\text{CM}}}{\mathrm{c}^2_{\text{Propagation}}} \right) \right]$$

Thermodynamic Field Irreversible Heat "Free Aether" Pressure Potential Energy Transfer

2.29

$$\delta Q_{\text{Heat\_Pressure\_Free}} = \frac{\left[\dfrac{m_{\text{Net}} \cdot \left|v^2\right|_{\text{Iso}}}{3} - \dfrac{\Delta(PV)_{\text{Field\_Energy}}}{ln\left(\dfrac{V'_{\text{ol}\,Final}}{V_{\text{ol}\,Initial}} \right)} \right]}{\left(\gamma_{Heat} - 1\right)} \cdot \left[1 - \frac{\left|\bar{v}\right|^2_{\text{CM}}}{\left|v^2\right|_{\text{Iso}}} \right]$$

$$\delta Q_{\text{Heat\_Pressure\_Free}} = \frac{\left[\dfrac{m_{\text{Net}} \cdot c^2_{\text{Propagation}}}{\gamma_{Heat}} + \gamma_{Heat} \cdot \left[\dfrac{\Delta(PV)_{\text{Field\_Energy}}}{ln\left(\dfrac{P'_{Iso\_Pressure\,Final}}{P_{Iso\_Pressure\,Initial}} \right)} \right] \right]}{\left(\gamma_{Heat} - 1\right)} \cdot \left[1 - \frac{\gamma_{Heat}}{3} \cdot \left(\frac{\left|\bar{v}\right|^2_{\text{CM}}}{c^2_{\text{Propagation}}} \right) \right]$$

$$\delta Q_{\text{Heat\_Pressure\_Free}} = \frac{\left[N \cdot k_B \cdot T_{emp\,Initial} + \left(\gamma_{Heat} - 1\right) \cdot \left[\dfrac{\Delta(PV)_{\text{Field\_Energy}}}{ln\left(\dfrac{T'_{emp\,Final}}{T_{emp\,Initial}} \right)} \right] \right]}{\left(\gamma_{Heat.} - 1\right)} \cdot \left[1 - \left(\frac{T_{Temp\_Ext}}{T_{emp\,Initial}} \right) \right]$$

The **Thermodynamic Field Irreversible Heat "Free Aether" Pressure Potential Energy Transfer,** in three dimensional Cartesian coordinates x, y, and z, are given by the following,

2.30

$$\delta Q_{Heat\_Pressure\_Free_x} = \frac{m_{Net} \cdot \left|v^2\right|_{Iso\,x}}{3 \cdot \left(1 - \dfrac{1}{\gamma_{Heat}}\right)} \cdot \left[1 - \frac{\left|\overline{v}\right|^2_{CM\,x}}{\left|v^2\right|_{Iso\,x}}\right]$$

$$\delta Q_{Heat\_Pressure\_Free_y} = \frac{m_{Net} \cdot \left|v^2\right|_{Iso\,y}}{3 \cdot \left(1 - \dfrac{1}{\gamma_{Heat}}\right)} \cdot \left[1 - \frac{\left|\overline{v}\right|^2_{CM\,y}}{\left|v^2\right|_{Iso\,y}}\right]$$

$$\delta Q_{Heat\_Pressure\_Free_z} = \frac{m_{Net} \cdot \left|v^2\right|_{Iso\,z}}{3 \cdot \left(1 - \dfrac{1}{\gamma_{Heat}}\right)} \cdot \left[1 - \frac{\left|\overline{v}\right|^2_{CM\,z}}{\left|v^2\right|_{Iso\,z}}\right]$$

2.31

Scalar/Tensor		
Magnitude	Magnitude/Tensor	Units
Thermodynamic Field Irreversible Heat "Free Aether" Pressure Potential Energy Transfer		
$\delta Q_{Heat\_Pressure\_Free} = \begin{bmatrix} \delta Q_{Heat\_Pressure\_Free_x} \\ + \delta Q_{Heat\_Pressure\_Free_y} \\ + \delta Q_{Heat\_Pressure\_Free_z} \end{bmatrix}$	$\delta Q_{Heat\_Pressure\_Free} = \begin{bmatrix} \delta Q_{Heat\_Pressure\_Free_x} \\ \delta Q_{Heat\_Pressure\_Free_y} \\ \delta Q_{Heat\_Pressure\_Free_z} \end{bmatrix}$	$\frac{kg \cdot m^2}{s^2}$

2.4 Thermodynamic Field Irreversible Heat "Free Aether" Volume Potential Energy Transfer

An isolated mass system body constrained to have constant volume in a Thermodynamic Field Potential is the amount of **Thermodynamic Field Irreversible Heat "Free Aether" Volume Potential Energy Transfer** between a system and its surroundings.

The Thermodynamic Field Irreversible Heat "Free Aether" **Volume Potential** Energy Transfer is at a ***lower potential*** in any system than the Thermodynamic Field Irreversible Heat "Free Aether" **Pressure Potential** Energy Transfer.

Definition 2.6: The total amount of The Thermodynamic Field Irreversible Heat "Free Aether" **Volume Potential** Energy Transfer ($\delta Q_{\mathrm{Heat\_Energy\_Free}}$) is the heat energy lost/gained by an inertial mass substance at constant Pressure and gained/lost by the surroundings and is equal to the **Net Inertial Mass** of the substance ($\mathrm{m}_{\mathrm{Net}}$) multiplied by its **Specific Heat Capacity of an Atomic Substance** at constant **Volume** ($C_{Specific\_Heat}$) multiplied by the **Aether Free Energy Temperature Difference** (ΔT_{Free}) which is equal the **Isotropic Internal Absolute Temperature** (T_{Temp}) subtracted from the **Anisotropic External Temperature** ($\Delta T_{Free} = \lfloor T_{Temp} - T_{Temp\_Ext} \rfloor$) of the substance system mass body.

Thermodynamic Field Irreversible Heat "Free Aether" Volume Potential Energy Transfer

2.32

$$\delta Q_{\mathrm{Heat\_Energy\_Free}} = \mathrm{m}_{\mathrm{Net}} \cdot C_{Specific\_Heat} \cdot \left[\int_{T_{Initial}}^{T'_{Final}} dT_{Free} \right] \rightarrow kg \cdot m^2 / s^2$$

$$\delta Q_{\mathrm{Heat\_Energy\_Free}} = \mathrm{m}_{\mathrm{Net}} \cdot C_{Specific\_Heat} \cdot \Delta T_{Free}$$

$$\delta Q_{\mathrm{Heat\_Energy\_Free}} = \mathrm{m}_{\mathrm{Net}} \cdot C_{Specific\_Heat} \cdot \left[T_{Temp} - T_{Temp\_Ext} \right]$$

$$\delta Q_{\mathrm{Heat\_Energy\_Free}} = \left[\delta Q_{\mathrm{Heat\_Pressure\_Free}} - \Delta(PV)_{\mathrm{Free\_Energy}} \right]$$

$$\delta Q_{\mathrm{Heat\_Energy\_Free}} = \left[\mathrm{m}_{\mathrm{Net}} \cdot C_{Specific\_Pressure} - \mathrm{N} \cdot \mathrm{k_B} \right] \cdot \Delta T_{Free}$$

$$\delta Q_{\mathrm{Heat\_Energy\_Free}} = \left[\mathrm{m}_{\mathrm{Net}} \cdot C_{Specific\_Pressure} - \mathrm{N} \cdot \mathrm{k_B} \right] \cdot \left[T_{Temp} - T_{Temp\_Ext} \right]$$

$$\delta Q_{\mathrm{Heat\_Energy\_Free}} = \frac{\mathrm{m}_{\mathrm{Net}} \cdot \mathrm{c}^2_{\mathrm{Propagation}}}{\left((\gamma_{Heat})^2 - \gamma_{Heat} \right)} \cdot \left(\frac{\Delta T_{Free}}{T_{emp\ Initial}} \right)$$

$$\delta Q_{\mathrm{Heat\_Energy\_Free}} = \frac{\mathrm{m}_{\mathrm{Net}} \cdot \mathrm{c}^2_{\mathrm{Propagation}}}{\left((\gamma_{Heat})^2 - \gamma_{Heat} \right)} \cdot \left[1 - \frac{T_{Temp\_Ext}}{T_{Temp}} \right]$$

Thermodynamic Field Irreversible Heat “Free Aether” Volume Potential Energy Transfer

2.33

$$\delta Q_{\text{Heat\_Energy\_Free}} = \left[\delta Q_{\text{Heat\_Pressure\_Free}} - \Delta(PV)_{\text{Free\_Energy}}\right] \rightarrow kg \cdot m^2 / s^2$$

$$\delta Q_{\text{Heat\_Energy\_Free}} = \left[\delta Q_{\text{Heat\_Pressure\_Free}} - \frac{2}{3} \cdot \left[E_{\text{Net\_Iso\_Kinetic}} - E_{\text{Net\_Translational}}\right]\right]$$

$$\delta Q_{\text{Heat\_Energy\_Free}} = \left[\delta Q_{\text{Heat\_Pressure\_Free}} - \frac{1}{3} \cdot m_{\text{Net}} \cdot \left[\left|v^2\right|_{\text{Iso}} - \left|\overline{v}\right|^2_{\text{CM}}\right]\right]$$

$$\delta Q_{\text{Heat\_Energy\_Free}} = \frac{m_{\text{Net}} \cdot \left|v^2\right|_{\text{Iso}}}{3 \cdot (\gamma_{Heat} - 1)} \cdot \left(\frac{\Delta T_{Free}}{T_{emp\ Initial}}\right)$$

$$\delta Q_{\text{Heat\_Energy\_Free}} = \frac{m_{\text{Net}} \cdot \left[\frac{\left|v^2\right|_{\text{Iso}}}{3} + v^2_{\text{Out\_Flow\_Velocity}}\right]}{\left((\gamma_{Heat})^2 - \gamma_{Heat}\right)} \cdot \left[1 - \frac{T_{Temp\_Ext}}{T_{Temp}}\right]$$

$$\delta Q_{\text{Heat\_Energy\_Free}} = \frac{m_{\text{Net}} \cdot \left|v^2\right|_{\text{Iso}}}{3 \cdot (\gamma_{Heat} - 1)} \cdot \left[1 - \frac{\left|\overline{v}\right|^2_{\text{CM}}}{\left|v^2\right|_{\text{Iso}}}\right]$$

$$\delta Q_{\text{Heat\_Energy\_Free}} = \frac{m_{\text{Net}} \cdot c^2_{\text{Propagation}}}{\left((\gamma_{Heat})^2 - \gamma_{Heat}\right)} \cdot \left[1 - \frac{\gamma_{Heat}}{3} \cdot \left(\frac{\left|\overline{v}\right|^2_{\text{CM}}}{c^2_{\text{Propagation}}}\right)\right]$$

Thermodynamic Field Irreversible Heat "Free Aether" Volume Pressure Potential Energy Transfer

2.34

$$\delta Q_{\text{Heat\_Energy\_Free}} = \frac{\left[\dfrac{m_{\text{Net}} \cdot \left|v^2\right|_{\text{Iso}}}{3} - \dfrac{\Delta(PV)_{\text{Field\_Energy}}}{ln\left(\dfrac{V'_{ol\,Final}}{V_{ol\,Initial}}\right)}\right]}{\left((\gamma_{Heat})^2 - \gamma_{Heat}\right)} \cdot \left[1 - \frac{\left|\overline{v}\right|^2_{\text{CM}}}{\left|v^2\right|_{\text{Iso}}}\right]$$

$$\delta Q_{\text{Heat\_Energy\_Free}} = \frac{\left[\dfrac{m_{\text{Net}} \cdot c^2_{\text{Propagation}}}{\gamma_{Heat}} + \gamma_{Heat} \cdot \left[\dfrac{\Delta(PV)_{\text{Field\_Energy}}}{ln\left(\dfrac{P'_{Iso\_Pressure\,Final}}{P_{Iso\_Pressure\,Initial}}\right)}\right]\right]}{\left((\gamma_{Heat})^2 - \gamma_{Heat}\right)} \cdot \left[1 - \frac{\gamma_{Heat}}{3} \cdot \left(\frac{\left|\overline{v}\right|^2_{\text{CM}}}{c^2_{\text{Propagation}}}\right)\right]$$

$$\delta Q_{\text{Heat\_Energy\_Free}} = \frac{\left[N \cdot k_B \cdot T_{emp\,Initial} + (\gamma_{Heat} - 1) \cdot \left[\dfrac{\Delta(PV)_{\text{Field\_Energy}}}{ln\left(\dfrac{T'_{emp\,Final}}{T_{emp\,Initial}}\right)}\right]\right]}{\left((\gamma_{Heat})^2 - \gamma_{Heat}\right)} \cdot \left[1 - \left(\frac{T_{Temp\_Ext}}{T_{Temp}}\right)\right]$$

The **Thermodynamic Field Irreversible Heat "Free Aether" Potential Energy Transfer,** in three dimensional Cartesian coordinates x, y, and z, are given by the following,

2.35

$$\delta Q_{\text{Heat\_Energy\_Free}_x} = \frac{m_{\text{Net}} \cdot \left|v^2\right|_{\text{Iso x}}}{3 \cdot \left(\gamma_{Heat} - 1\right)} \cdot \left[1 - \frac{\left|\bar{v}\right|^2_{\text{CM x}}}{\left|v^2\right|_{\text{Iso x}}}\right]$$

$$\delta Q_{\text{Heat\_Energy\_Free}_y} = \frac{m_{\text{Net}} \cdot \left|v^2\right|_{\text{Iso y}}}{3 \cdot \left(\gamma_{Heat} - 1\right)} \cdot \left[1 - \frac{\left|\bar{v}\right|^2_{\text{CM y}}}{\left|v^2\right|_{\text{Iso y}}}\right]$$

$$\delta Q_{\text{Heat\_Energy\_Free}_z} = \frac{m_{\text{Net}} \cdot \left|v^2\right|_{\text{Iso z}}}{3 \cdot \left(\gamma_{Heat} - 1\right)} \cdot \left[1 - \frac{\left|\bar{v}\right|^2_{\text{CM z}}}{\left|v^2\right|_{\text{Iso z}}}\right]$$

2.36

Scalar/Tensor		
Magnitude	Magnitude/Tensor	Units
Thermodynamic Field Irreversible Heat "Free Aether" Volume Potential Energy Transfer		
$\delta Q_{\text{Heat\_Energy\_Free}} = \begin{bmatrix} \delta Q_{\text{Heat\_Energy\_Free}_x} \\ + \delta Q_{\text{Heat\_Energy\_Free}_y} \\ + \delta Q_{\text{Heat\_Energy\_Free}_z} \end{bmatrix}$	$\delta Q_{\text{Heat\_Energy\_Free}} = \begin{bmatrix} \delta Q_{\text{Heat\_Energy\_Free}_x} \\ \delta Q_{\text{Heat\_Energy\_Free}_y} \\ \delta Q_{\text{Heat\_Energy\_Free}_z} \end{bmatrix}$	$\frac{kg \cdot m^2}{s^2}$

2.5 Adiabatic Isentropic Thermodynamic Field Free Aether Entropy

In a thermodynamic isolated system, a "Universe" can be described by a system and its surroundings. An isolated thermodynamic system mass body is made up of quantities of matter, pressure differences, density differences, and temperature differences which all tend to equalize over time.

Thus, when we image an enclosed room and a cup of ice water as an isolated dynamic system it can be considered a “Universe” in which the cup and the room are initially at different temperatures. But after some finite time the universe will have reached thermodynamic “temperature” equilibrium.

When the enclosed universe reaches thermodynamic equilibrium the entropy change varies from some initial minimal entropy state to a final maximum entropy state. The entropy of the thermodynamic system is thus, a measure of how far the equalization has progressed.

In any isolated system, the overall change in entropy is always positive, that is, it will always increase. Where the energy is always conserved (the first law of thermodynamics states that energy cannot be created or destroyed), the same cannot be said about entropy. Entropy always increases in an isolated system that reaches thermodynamic equilibrium.

The second law of thermodynamics says that whenever energy is exchanged or converted from one form to another, the potential for energy to do work gets less.

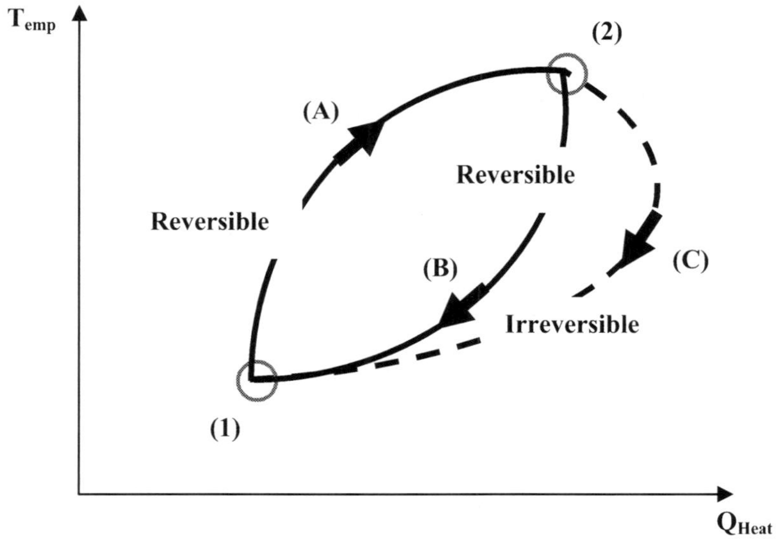

Figure 2.4: The Entropy or Heat Flow between two different processes, reversible and irreversible.

Thus, entropy is a measure of the lack of potential or quality of energy; and once that energy has been exchanged or converted, it cannot revert back to a higher state, without any additional input of heat energy.

Consider a cycle to be composed of two thermodynamic processes, one reversible and the other irreversible, as shown in the above diagram. Suppose that a reversible path is traced going from position (1) to position (2) along path (A) and (B).

Thus for any reversible process the Entropy or Entropy Change is equal to zero.

2.37

$$\underset{\text{Along A}}{\int_1^2 \frac{\delta Q_{Heat}}{T_{emp}}} + \underset{\text{Along B}}{\int_2^1 \frac{\delta Q_{Heat}}{T_{emp}}} = 0$$

Next, consider the irreversible path is traced going from position (1) to position (2) along path (A) and (C). Thus for any irreversible process the Entropy or Entropy Change is negative or less than zero.

2.38

$$\underset{\text{Along A}}{\int_1^2 \frac{\delta Q_{Heat}}{T_{emp}}} + \underset{\text{Along C}}{\int_2^1 \frac{\delta Q_{Heat}}{T_{emp}}} < 0$$

This implies that the entropy change along an irreversible path is less than the entropy change along a reversible path.

2.39

$$\underset{\text{Along B}}{\int_2^1 \frac{\delta Q_{Heat}}{T_{emp}}} > \underset{\text{Along C}}{\int_2^1 \frac{\delta Q_{Heat}}{T_{emp}}}$$

The equality holds true for a system undergoing a reversible process, and an irreversible process. Thus, the entropy production is zero for a reversible process and positive for an irreversible process or any real process.

2.40

$$\left(\frac{\delta Q_{Heat}}{T_{emp}}\right)_{\text{Production}} = \left(\frac{\delta Q_{Heat}}{T_{emp}}\right)_{\text{System}} + \left(\frac{\delta Q_{Heat}}{T_{emp}}\right)_{\text{Surroundings}} \geq 0$$

Furthermore for any real process the entropy of an isolated system must increase or have a positive value. Considering a complete system to be composed of any system mass body plus the surrounding, the heat and work transferred to the system under consideration are negative of the heat and work transferred from the surroundings.

This combination of the system and the surroundings is often referred to as the universe.

The natural state of motion for all processes in the Universe is to end up at the same temperature, so the entropy of the Universe is always increasing. When heat is added to a system at high temperature, the increase in entropy is small. When heat is added to a system at low temperature, the increase in entropy is great.

In Isothermal systems held at constant temperature, the change in entropy, *ΔS*, is given by the equation.

2.41

$$T_{emp} = T_{emp\_Ext}$$

$$\Delta S_{Entropy\ Free\_Surroundings} = \Delta(PV)_{\text{Free\_Energy}} \cdot \left[\frac{1}{T_{emp\_Ext}} - \frac{1}{T_{emp}} \right] = 0$$

$$\Delta S_{Entropy\ Free\_Surroundings} = \Delta S_{Entropy\ Free\_Energy} = \frac{\Delta(PV)_{\text{Free\_Energy}}}{T_{emp}}$$

$$\Delta S_{Entropy\ Free\_Surroundings} = \Delta S_{Entropy\ Free\_Energy} = \frac{\Delta(PV)_{\text{Free\_Energy}}}{T_{emp\_Ext}}$$

Where ($\Delta(PV)_{\text{Free\_Energy}}$) is the amount of heat gained/lost by the system in an isothermal and reversible process in which the system goes from one state to another, and *T* is the absolute temperature at which the process is occurring.

If the temperature of the system is not constant, then the relationship becomes a differential equation and the entropy of the system is increased as described by the following

2.42

$$T_{emp} \neq T_{emp\_Ext}$$

$$\Delta S_{Entropy\ Free\_Energy} = \left(\frac{\mathrm{V}_{\mathrm{ol}}}{T_{emp}}\right) \cdot \left[\int_{P_{Iso\_Pressure}}^{q_{Dynamic\_Pressure}} d\left(P_{Iso\_Pressure}\right) \right]$$

$$\Delta S_{Entropy\ Free\_Energy} = \frac{\Delta(PV)_{\mathrm{Free\_Energy}}}{T_{emp}}$$

Then the total change in entropy for the surroundings is given by the following

2.43

$$\Delta S_{Entropy\ Free\_Surroundings} = \Delta(PV)_{\mathrm{Free\_Energy}} \cdot \left[\frac{1}{T_{emp\_Ext}} - \frac{1}{T_{emp}} \right]$$

$$\Delta S_{Entropy\ Free\_Surroundings} = \mathrm{N} \cdot \mathrm{k_B} \cdot \left[\frac{\left[T_{emp} - T_{emp\_Ext}\right]}{T_{emp\_Ext}} - \frac{\left[T_{emp} - T_{emp\_Ext}\right]}{T_{emp}} \right]$$

Then the production of entropy is given by the following

2.44

$$\Delta S_{Entropy\ Free\_Production} = \left[\frac{\Delta(PV)_{\mathrm{Free\_Energy}}}{T_{emp\_Ext}} \right] = \left[\begin{array}{l} \Delta S_{Entropy\ Free\_Energy} \\ + \ \Delta S_{Entropy\ Free\_Surroundings} \end{array} \right]$$

Adiabatic Isentropic Thermodynamic Field Free Aether Entropy – System
2.45

$$\Delta S_{\text{Free\_Entropy}} = \frac{\Delta(PV)_{\text{Free\_Energy}}}{T_{emp}} = \mathrm{N}\cdot \mathrm{k_B}\cdot\left(\frac{\Delta T_{Free}}{T_{emp}}\right) \rightarrow kg\cdot m^2 \big/ s^2\cdot K$$

$$\Delta S_{Entropy\ Free\_Energy} = \left(\frac{\mathrm{V_{ol}}}{T_{emp}}\right)\cdot\left[\int_{P_{Iso\_Pressure}}^{q_{Dynamic\_Pressure}} d\left(P_{Iso\_Pressure}\right)\right]$$

$$\Delta S_{Entropy\ Free\_Energy} = \frac{\Delta(PV)_{Free\_Energy}}{T_{emp}} = \left[\frac{\delta Q_{\text{Heat\_Pressure\_Free}} - \delta Q_{\text{Heat\_Energy\_Free}}}{T_{emp}}\right]$$

$$\Delta S_{Entropy\ Free\_Energy} = \frac{\Delta(PV)_{Free\_Energy}}{T_{emp}} = \left[\frac{P_{Iso\_Pressure} - q_{Dynamic\_Pressure}}{T_{emp}}\right]\cdot \mathrm{V_{ol}}$$

$$\Delta S_{Entropy\ Free\_Energy} = \frac{\Delta(PV)_{Free\_Energy}}{T_{emp}} = \mathrm{N}\cdot \mathrm{k_B}\cdot\left[\frac{T_{emp} - T_{emp\_Ext}}{T_{emp}}\right]$$

$$\Delta S_{Entropy\ Free\_Energy} = \frac{\Delta(PV)_{Free\_Energy}}{T_{emp}} = \frac{2}{3}\cdot\left[\frac{\mathrm{E_{Net\_Iso\_Kinetic}} - \mathrm{E_{Net\_Translational}}}{T_{emp}}\right]$$

$$\Delta S_{Entropy\ Free\_Energy} = \frac{\Delta(PV)_{Free\_Energy}}{T_{emp}} = \left(\frac{\mathrm{m_{Net}}\cdot\left|\mathrm{v}^2\right|_{\mathrm{Iso}}}{3}\right)\cdot\left(\frac{\Delta T_{Free}}{T_{emp}^2}\right)$$

Adiabatic Isentropic Thermodynamic Field Free Aether Entropy – System

2.46

$$\Delta S_{Entropy\ Free\_Energy} = \mathrm{N} \cdot \mathrm{k_B} \cdot \left[\frac{T_{emp} - T_{emp\_Ext}}{T_{emp}} \right] \rightarrow kg \cdot m^2 / s^2 \cdot K$$

$$\Delta S_{Entropy\ Free\_Energy} = \mathrm{n_{Moles}} \cdot \mathrm{R} \cdot \left[1 - \left(\frac{T_{emp\_Ext}}{T_{emp}} \right) \right]$$

$$\Delta S_{Entropy\ Free\_Energy} = \mathrm{N} \cdot \mathrm{k_B} \cdot \left[1 - \frac{\left|\overline{\mathrm{v}}\right|^2_{\mathrm{CM}}}{\left|\mathrm{v}^2\right|_{\mathrm{Iso}}} \right]$$

$$\Delta S_{Entropy\ Free\_Energy} = \mathrm{N} \cdot \mathrm{k_B} \cdot \left[1 - \frac{\gamma_{Heat}}{3} \cdot \left(\frac{\left|\overline{\mathrm{v}}\right|^2_{\mathrm{CM}}}{\mathrm{c}^2_{\mathrm{Propagation}}} \right) \right]$$

$$\Delta S_{Entropy\ Free\_Energy} = \frac{\Delta(PV)_{Free\_Energy}}{T_{emp}} = \left(\frac{\mathrm{m_{Net}} \cdot \left|\mathrm{v}^2\right|_{\mathrm{Iso}}}{3} \right) \cdot \left[\frac{T_{emp\_Ext} - T_{emp}}{T^2_{emp}} \right]$$

The **Adiabatic Isentropic Thermodynamic Field Free Aether Entropy – System,** in three dimensional Cartesian coordinates x, y, and z, are given by the following,

2.47

$$\Delta S_{Entropy\ Free\_Energy\ x} = \frac{\Delta(PV)_{Free\_Energy\ x}}{T_{emp}} = \left(\frac{m_{Net} \cdot \left|v^2\right|_{Iso\ x}}{3}\right) \cdot \left[\frac{T_{emp\_Ext} - T_{emp}}{T_{emp}^2}\right]$$

$$\Delta S_{Entropy\ Free\_Energy\ y} = \frac{\Delta(PV)_{Free\_Energy\ y}}{T_{emp}} = \left(\frac{m_{Net} \cdot \left|v^2\right|_{Iso\ y}}{3}\right) \cdot \left[\frac{T_{emp\_Ext} - T_{emp}}{T_{emp}^2}\right]$$

$$\Delta S_{Entropy\ Free\_Energy\ z} = \frac{\Delta(PV)_{Free\_Energy\ z}}{T_{emp}} = \left(\frac{m_{Net} \cdot \left|v^2\right|_{Iso\ z}}{3}\right) \cdot \left[\frac{T_{emp\_Ext} - T_{emp}}{T_{emp}^2}\right]$$

2.48

Scalar/Tensor		
Magnitude	Magnitude/Tensor	Units
Adiabatic Isentropic Thermodynamic Field Free Aether Entropy – System		
$\Delta S_{Entropy\ Free\_Energy} = \begin{bmatrix} \Delta S_{Entropy\ Free\_Energy\ z} \\ + \Delta S_{Entropy\ Free\_Energy\ y} \\ + \Delta S_{Entropy\ Free\_Energy\ z} \end{bmatrix}$	$\Delta S_{Entropy\ Free\_Energy} = \begin{bmatrix} \Delta S_{Entropy\ Free\_Energy\ z} \\ \Delta S_{Entropy\ Free\_Energy\ y} \\ \Delta S_{Entropy\ Free\_Energy\ z} \end{bmatrix}$	$\frac{kg \cdot m^2}{s^2 \cdot K}$

Adiabatic Isentropic Thermodynamic Field Entropy – Surroundings

2.49

$$\Delta S_{Entropy\ Free\_Surroundings} = \Delta(PV)_{\text{Free\_Energy}} \cdot \left[\frac{1}{T_{emp\_Ext}} - \frac{1}{T_{emp}}\right] \rightarrow kg \cdot m^2 / s^2 \cdot K$$

$$\Delta S_{Entropy\ Free\_Surroundings} = \left(\mathrm{N} \cdot \mathrm{k_B} \cdot \Delta T_{Free}\right) \cdot \left[\frac{1}{T_{emp\_Ext}} - \frac{1}{T_{emp}}\right]$$

$$\Delta S_{Entropy\ Free\_Surroundings} = \mathrm{N} \cdot \mathrm{k_B} \cdot \left[\frac{\left[T_{emp} - T_{emp\_Ext}\right]}{T_{emp\_Ext}} - \frac{\left[T_{emp} - T_{emp\_Ext}\right]}{T_{emp}}\right]$$

$$\Delta S_{Entropy\ Free\_Surroundings} = \mathrm{N} \cdot \mathrm{k_B} \cdot \left[\left(\frac{T_{emp}}{T_{emp\_Ext}}\right) + \left(\frac{T_{emp\_Ext}}{T_{emp}}\right) - 2\right]$$

$$\Delta S_{Entropy\ Free\_Surroundings} = \mathrm{N} \cdot \mathrm{k_B} \cdot \left[\left(\frac{\left|\mathrm{v}^2\right|_{\mathrm{Iso}}}{\left|\overline{\mathrm{v}}\right|^2_{\mathrm{CM}}}\right) + \left(\frac{\left|\overline{\mathrm{v}}\right|^2_{\mathrm{CM}}}{\left|\mathrm{v}^2\right|_{\mathrm{Iso}}}\right) - 2\right]$$

$$\Delta S_{Entropy\ Free\_Surroundings} = \Delta(PV)_{\text{Free\_Energy}} \cdot \left[\frac{T_{emp} - T_{emp\_Ext}}{T_{emp\_Ext} \cdot T_{emp}}\right]$$

$$\Delta S_{Entropy\ Free\_Surroundings} = \mathrm{N} \cdot \mathrm{k_B} \cdot \left[\frac{\left(\Delta T_{Free}\right)^2}{T_{emp\_Ext} \cdot T_{emp}}\right]$$

Thermodynamic Field Entropy – Surroundings

2.50

$$\Delta S_{Entropy\ Free\_Surroundings} = \mathrm{N}\cdot \mathrm{k_B}\cdot\left[\frac{\left(1 - \left(\dfrac{T_{emp\_Ext}}{T_{emp}}\right)\right)^2}{\left(\dfrac{T_{emp\_Ext}}{T_{emp}}\right)}\right] \rightarrow kg\cdot m^2 / s^2\cdot K$$

$$\Delta S_{Entropy\ Free\_Surroundings} = \mathrm{N}\cdot \mathrm{k_B}\cdot\left[\frac{\left(1 - \left(\dfrac{\left|\overline{\mathrm{v}}\right|^2_{\mathrm{CM}}}{\left|\mathrm{v}^2\right|_{\mathrm{Iso}}}\right)\right)^2}{\left(\dfrac{\left|\overline{\mathrm{v}}\right|^2_{\mathrm{CM}}}{\left|\mathrm{v}^2\right|_{\mathrm{Iso}}}\right)}\right]$$

$$\Delta S_{Entropy\ Free\_Surroundings} = \frac{\Delta(PV)_{\mathrm{Free\_Energy}}}{T_{emp\_Ext}}\cdot\left[1 - \left(\frac{T_{emp\_Ext}}{T_{emp}}\right)\right]$$

$$\Delta S_{Entropy\ Free\_Surroundings} = \frac{\Delta(PV)_{\mathrm{Free\_Energy}}}{T_{emp\_Ext}}\cdot\left[1 - \left(\frac{\left|\overline{\mathrm{v}}\right|^2_{\mathrm{CM}}}{\left|\mathrm{v}^2\right|_{\mathrm{Iso}}}\right)\right]$$

$$\Delta S_{Entropy\ Free\_Surroundings} = \frac{\Delta(PV)_{\mathrm{Free\_Energy}}}{T_{emp}}\cdot\left[\frac{\left(1 - \left(\dfrac{\left|\overline{\mathrm{v}}\right|^2_{\mathrm{CM}}}{\left|\mathrm{v}^2\right|_{\mathrm{Iso}}}\right)\right)}{\left(\dfrac{\left|\overline{\mathrm{v}}\right|^2_{\mathrm{CM}}}{\left|\mathrm{v}^2\right|_{\mathrm{Iso}}}\right)}\right]$$

The **Thermodynamic Field Free Aether Entropy – Surroundings,** in three dimensional Cartesian coordinates x, y, and z, are given by the following,

2.51

$$\Delta S_{Entropy\ Free\_Surroundings\ x} = \frac{m_{Net} \cdot \left|v^2\right|_{Iso\ x}}{3} \cdot \left[\left(\frac{1}{T_{emp\_Ext}} \right) + \left(\frac{T_{emp\_Ext}}{T_{emp}^2} \right) - \frac{2}{T_{emp}} \right]$$

$$\Delta S_{Entropy\ Free\_Surroundings\ y} = \frac{m_{Net} \cdot \left|v^2\right|_{Iso\ y}}{3} \cdot \left[\left(\frac{1}{T_{emp\_Ext}} \right) + \left(\frac{T_{emp\_Ext}}{T_{emp}^2} \right) - \frac{2}{T_{emp}} \right]$$

$$\Delta S_{Entropy\ Free\_Surroundings\ z} = \frac{m_{Net} \cdot \left|v^2\right|_{Iso\ z}}{3} \cdot \left[\left(\frac{1}{T_{emp\_Ext}} \right) + \left(\frac{T_{emp\_Ext}}{T_{emp}^2} \right) - \frac{2}{T_{emp}} \right]$$

2.52

Scalar/Tensor		
Magnitude	Magnitude/Tensor	Units
Thermodynamic Field Free Aether Entropy – Surroundings		
$\Delta S_{Entropy\ Free\_Surroundings} = \begin{bmatrix} \Delta S_{Entropy\ Free\_Surroundings\ x} \\ + \Delta S_{Entropy\ Free\_Surroundings\ y} \\ + \Delta S_{Entropy\ Free\_Surroundings\ z} \end{bmatrix}$	$\Delta S_{Entropy\ Free\_Surroundings} = \begin{bmatrix} \Delta S_{Entropy\ Free\_Surroundings\ x} \\ \Delta S_{Entropy\ Free\_Surroundings\ y} \\ \Delta S_{Entropy\ Free\_Surroundings\ z} \end{bmatrix}$	$\frac{kg \cdot m^2}{s^2\ K}$

Adiabatic Isentropic Thermodynamic Field Entropy – Production

2.53

$$\Delta S_{Entropy\ Free\_Production} = \left[\frac{\Delta(PV)_{Free\_Energy}}{T_{emp\_Ext}}\right] \rightarrow kg \cdot m^2 / s^2 \cdot K$$

$$\Delta S_{Entropy\ Free\_Production} = N \cdot k_B \cdot \left[\frac{\Delta T_{Free}}{T_{emp\_Ext}}\right]$$

$$\Delta S_{Entropy\ Free\_Production} = N \cdot k_B \cdot \left[\left(\frac{T_{emp}}{T_{emp\_Ext}}\right) - 1\right]$$

$$\Delta S_{Entropy\ Free\_Production} = N \cdot k_B \cdot \left[\frac{\left(1 - \left(\frac{|\overline{v}|^2_{CM}}{|v^2|_{Iso}}\right)\right)}{\left(\frac{|\overline{v}|^2_{CM}}{|v^2|_{Iso}}\right)}\right]$$

$$\Delta S_{Entropy\ Free\_Production} = \left\lfloor \Delta S_{Entropy\ Free\_Energy} + \Delta S_{Entropy\ Free\_Surroundings} \right\rfloor$$

2.54

Scalar/Tensor		
Magnitude	Magnitude/Tensor	Units
Adiabatic Isentropic Thermodynamic Field Free Aether Entropy – Production		
$\Delta S_{Entropy\ Free\_Production} = \begin{bmatrix} \Delta S_{Entropy\ Free\_Production_x} \\ + \Delta S_{Entropy\ Free\_Production_y} \\ + \Delta S_{Entropy\ Free\_Production_z} \end{bmatrix}$	$\Delta S_{Entropy\ Free\_Production} = \begin{bmatrix} \Delta S_{Entropy\ Free\_Production_x} \\ \Delta S_{Entropy\ Free\_Production_y} \\ \Delta S_{Entropy\ Free\_Production_z} \end{bmatrix}$	$\frac{kg \cdot m^2}{s^2 \cdot K}$

Chapter 3

The Special Theory of Thermodynamics

(Hydrodynamic Fluid Resistance) & (Compressibility Effects)

3.1 Hydrodynamic Compressibility and the Mach Velocity Ratio

As an object moves through the earth's gas atmosphere, the gas molecules of the atmosphere near the object are disturbed and move around the object. Aerodynamic forces are generated between the gas and the object. The magnitude of these forces depends on: the shape of the object, the speed of the object, the mass of the gas going by the object, the viscosity, or stickiness, of the gas, and the compressibility, or springiness, of the gas.

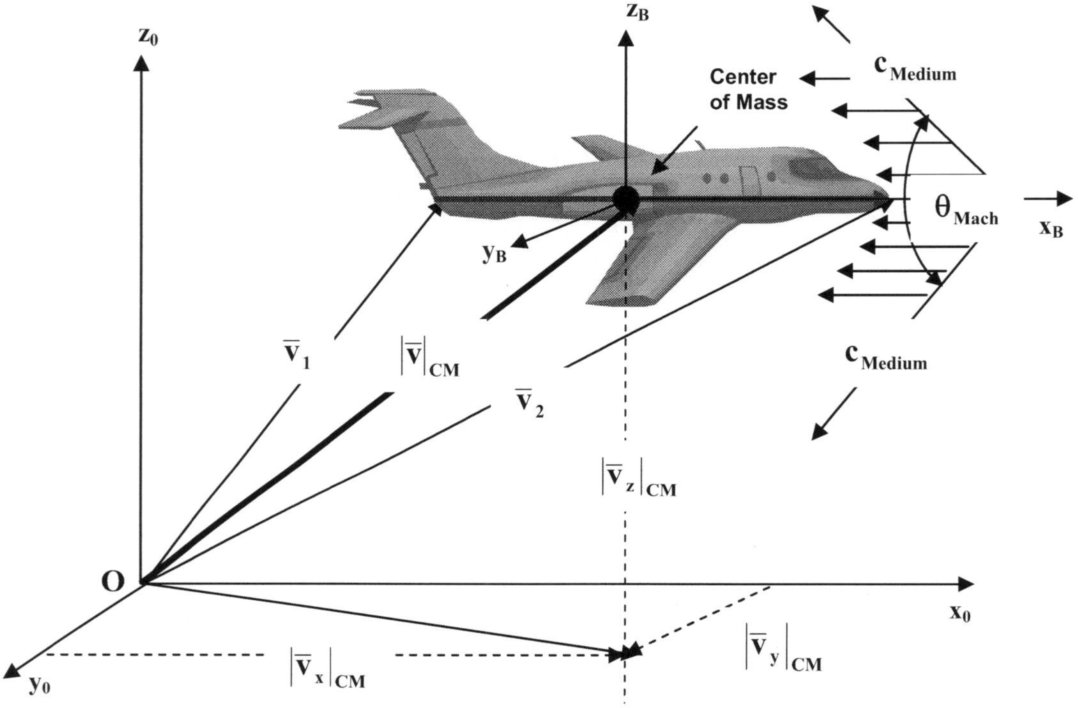

Figure 3.1: Center of Mass Velocity for a moving body

As an object moves through a gas, the gas molecules stick to the surface. This creates a layer of air near the surface, called a boundary layer, which, in effect, changes the shape of the object. The flow of gas reacts to the edge of the boundary layer as if it was the physical surface of the object.

While the body is in motion the boundary layer may separate from the body and create an effective shape much different from the physical shape. And to make it even more confusing, the flow conditions in and near the boundary layer are often unsteady (changing in time).

Understanding the boundary layer is very important in determining the drag force on an object. To determine and predict these boundary conditions, aerodynamicists rely on wind tunnel testing and computer analysis.

The nature of Aerodynamic forces on moving bodies also depends on the compressibility of the gas medium though which it is moving. When a mass body object moves through a gas medium, the molecules of the medium move around and relative to the object itself. If the object passes through the gas at a low velocity (typically less than 200 mph) the density of the fluid remains constant.

But for high velocities where the speed of the mass body approaches the Propagation speed of the medium, some of the energy of the object goes into compressing the fluid and changing the density, which alters the amount of resulting force on the object, and can produce an increasing mass effect. This effect becomes more important as speed increases.

When a system mass body is moving with a velocity that is 99.995% the Propagation speed of the medium a sharp disturbance generates a shock wave in the medium that requires and additional force to keep moving uniformly because the boundary layer adds and additional mass to the system body.

Thus, when the speed of an aircraft is near, or equal to the speed of sound in air (about 330 m/s or 700 mph on earth) a sharp disturbance in the air medium is generated, which produces a shock wave[1] that affects both the lift and drag of an aircraft.

Figure 3.2: A Shock is produced by a mass body traveling at a speed equal to or greater than the propagation speed of the medium through which it is traveling

[1] U.S. Navy photo by Ensign John Gay, U.S. Navy McDonnell Douglas F/A-18 Hornet 7 July 1999(1999-07-07) http://www.news.navy.mil/management/photodb/photos/990707-N-6483G-001.jpg

Unfortunately the math equations predict infinities when the velocity of the system mass body is equal to the shock Propagation speed of the medium; however this does not prevent us from exceeding the propagation speed of the medium as demonstrated in the image above.

Shock waves are also produced in particle accelerators. As electrons or protons are accelerated within the accelerators to speeds close to the speed of light in a vacuum, resistance occurs that increase the mass of the electron or proton.

This additional quantity of hydrodynamic resistance motion that is produced as a result of a mass body traveling at a speed very close to the average dissipation speed of the medium is called the Relativistic Mass Increase.

In honor of Czech/Austrian physicist and philosopher Ernst Mach (1838 – 1916) the Mach number ($M = \beta_{\text{Mach}}$) is the ratio of the speed of the Center of Mass Velocity to the dissipation speed of Propagation in the gas determines the magnitude of many of the compressibility effects.

Because of the importance of this speed ratio, aerodynamicists have designated it with a special parameter called, the Mach number ($M = \beta_{\text{Mach}}$) which allows us to define flight regimes in which compressibility effects vary.

Definition 3.1: Relative to an inertial frame, the Mean Center, and the Center of Mass of an isolated system body, the **Mach Number in Air** ($\beta_{\text{Mach}} = \left(\frac{|\overline{\mathbf{v}}|_{\mathbf{CM}}}{\mathbf{c}_{\mathbf{Sound}}}\right)$) is defined as the ratio of the **Center of Mass Velocity** divided by the **Speed of Sound in an Atomic Substance Medium**.

Mach number for Speed of Sound in an Atomic Substance Medium

3.1

$$\beta_{\text{Mach}} = M_{\text{Mach}} = \left(\frac{|\overline{\mathbf{v}}|_{\mathbf{CM}}}{\mathbf{c}_{\mathbf{Sound}}}\right) \rightarrow \textit{Unitless}$$

Mach Compressibility Conditions for Air:

1. **Subsonic Flight** conditions occur for Mach numbers that are less than 0.3, $\mathbf{M} = \beta_{\text{Mach}} < \mathbf{0.3}$. For the lowest subsonic conditions, compressibility can be ignored.

2. **Sonic Flight** conditions occur for Mach numbers that are greater than 0.3 and less than 0.8, $\mathbf{0.3} < \mathbf{M} = \beta_{\text{Mach}} < \mathbf{0.8}$. For the highest subsonic conditions, compressibility occurs on the body.

3. **Transonic Flight** conditions occur as the speed of the object approaches the speed of sound, the flight Mach number is nearly equal to one, $\mathbf{0.8} < \mathbf{M} = \beta_{\text{Mach}} < \mathbf{1}$, and the flow is said to be transonic. At some places on the object, the local speed exceeds the speed of sound. Because of the high drag associated with compressibility effects, aircraft do not cruise near Mach 1.

4. **Supersonic Flight** conditions occur for Mach numbers greater than one, $\mathbf{1} < \mathbf{M} = \beta_{\text{Mach}} < \mathbf{3}$. Compressibility effects are important for supersonic aircraft, and shock waves are generated by the surface of the object. For high supersonic speeds, $\mathbf{3} < \mathbf{M} = \beta_{\text{Mach}} < \mathbf{5}$, aerodynamic heating also becomes very important for aircraft design.

5. **Hypersonic Flight** conditions occur for speeds greater than five times the speed of sound, $\mathbf{M} = \beta_{\text{Mach}} > \mathbf{5}$, the flow is said to be hypersonic. The Space Shuttle travels at M = 25 At these speeds, some of the energy of the object now goes into exciting the chemical bonds which hold together the nitrogen and oxygen molecules of the air. At hypersonic speeds, the chemistry of the air must be considered when determining forces on the object. Under these conditions, the heated air becomes ionized plasma of gas and the spacecraft must be insulated from the high temperatures

The drag divergence Mach number is the Mach number at which the aerodynamic drag on an airfoil or airframe begins to increase rapidly as the Mach number continues to increase. This increase can cause the drag coefficient to rise to more than ten times its low speed value.

The value of the drag divergence Mach number is typically greater than (β_{Sonic_Mach} = M = 0.6); therefore it is a transonic effect. Generally, the drag coefficient peaks at Mach 1.0 and begins to decrease again after the transition into the supersonic regime above approximately (β_{Sonic_Mach} = M = 1.2).

The large increase in drag is caused by the formation of a shock wave on the upper surface of the airfoil, which can induce flow separation and adverse pressure gradients on the aft portion of the wing. To compensate for this effect aircraft that are intended to fly at supersonic speeds must have the capability to generate large amounts of thrust.

In early development of transonic and supersonic aircraft, a steep dive was often used to provide extra acceleration through the high drag region around (β_{Sonic_Mach} = M = 1.0).

In the early days of aviation, this steep increase in drag gave rise to the popular false notion of an unbreakable sound barrier, because it seemed that no aircraft technology would have enough propulsive force or control authority to overcome the hypothetical barrier.

One of the popular analytical methods for calculating drag at high speeds, is the Prandtl-Glauert rule, which predicts an infinite amount of drag at (β_{Sonic_Mach} = M = 1.0). We now, know that this hypothetical sound barrier can actually be overcome, and fifth generation fighter aircraft reach Mach numbers above (β_{Sonic_Mach} = M = 1.0), all of the time.

The "Wave Drag" is an aerodynamics term that refers to a sudden and very powerful form of drag that appears on aircraft and blade tips moving at high-subsonic and supersonic speeds. Wave drag is caused by the formation of shock waves around the aircraft.

Shock waves radiate away a considerable amount of energy; energy that is experienced by the aircraft as drag. Although shock waves are typically associated with supersonic flow, they can form at much lower speeds at areas on the aircraft where local airflow accelerates to supersonic speeds.

Shock waves form when the speed of a gas changes by more than the speed of sound in that medium. A shock wave may be described as the furthest point upstream of a moving object which "knows" about the approach of the object. The shock wave position is defined as the boundary between the zone having no information about the shock-driving event, and the zone aware of the shock-driving event, analogous with the light cone described in the theory of special relativity.

At the region where the shock wave occurs the sound waves traveling against the flow reach a point where they cannot travel any further upstream and the pressure progressively builds in that region, and a high pressure shock wave rapidly forms.

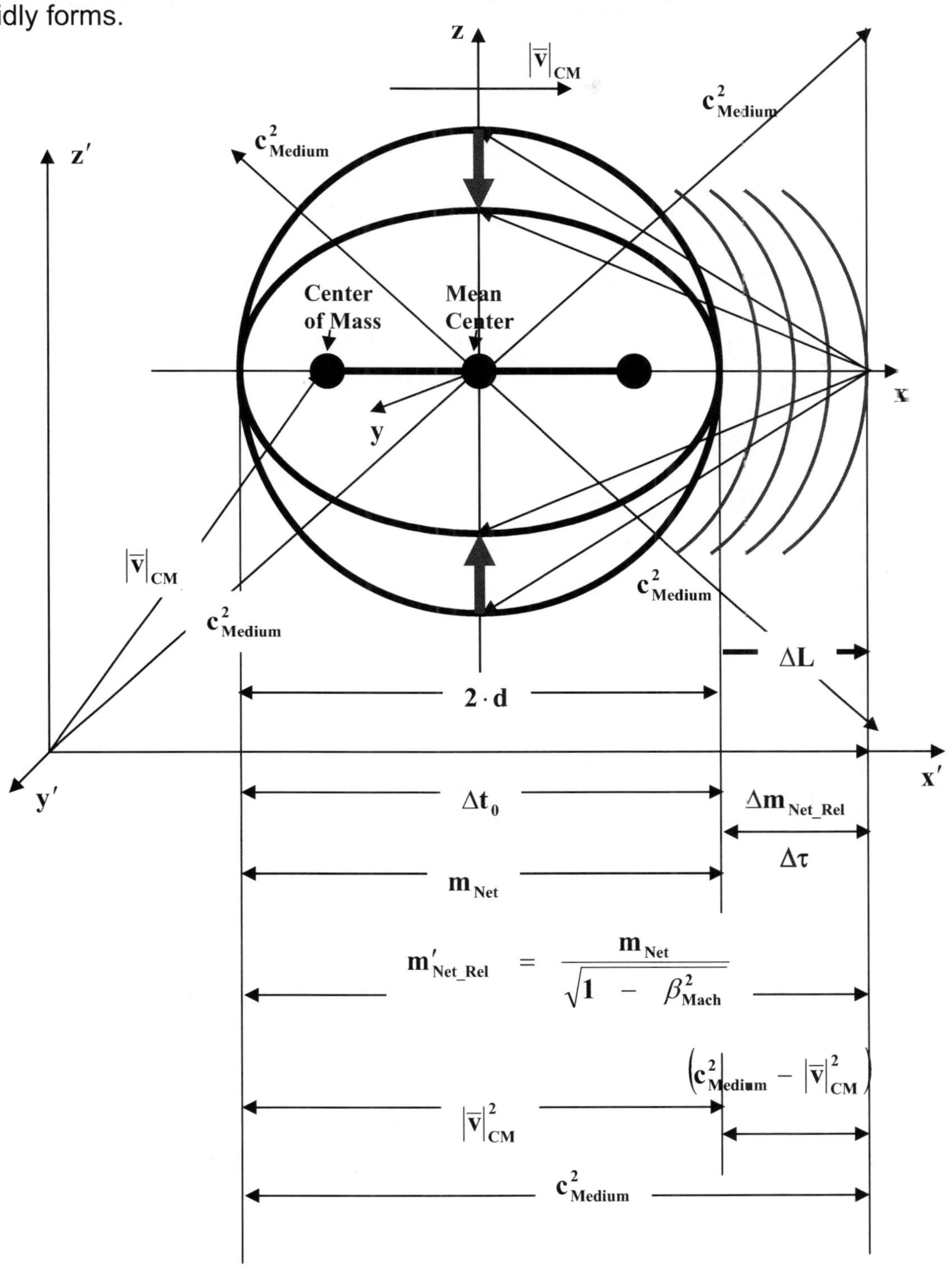

Figure 3.3: The total relativistic mass (of a body or system of bodies) includes a contribution from the kinetic energy of motion; and the contribution from the kinetic energy is larger the faster the body moves.

Aphorism 3.1: A Shock Wave occurs when a system body moves through an isotropic medium faster than the information about it can be propagated away from the body into the surrounding fluid, the fluid near the surface of the body and the instantaneous disturbance cannot react or "get out of the way" before the body surface of the disturbance arrives.

In a shock wave the properties of the fluid (temperature, velocity, density, and pressure) change almost instantaneously. Since shock waves are made of compressed air molecules, measurements of the thickness of shock waves have resulted in values approximately one order of magnitude greater than the mean free path of the gas investigated.

A shock wave can carry energy and propagate through a medium (solid, liquid or gas).

A shock wave can carry energy in the Vacuum of SpaceTime via a photon, or electromagnetic field. Shock waves are characterized by an abrupt, nearly discontinuous change in the characteristics of the medium.

The energy of any shock wave dissipates relatively quickly with distance depending on the medium. When a shock wave passes through matter, the total energy is preserved but the energy which can be extracted as work decreases and entropy increases.

In air wave fronts of the air waves heat the medium near each neighboring pressure front, due to adiabatic compression of the air itself, so that high pressure fronts outrun the corresponding pressure troughs; therefore the expansion wave approaches and eventually merges with the shock wave, partially cancelling it out.

The effect is typically seen at transonic speeds above about Mach 0.8, but it is possible to notice the problem at any speed over that of the critical Mach of that aircraft's wing. The magnitude of the rise in drag typically peaks at about four times the normal subsonic drag.

This shock near critical Mach is so powerful that it was thought for some time that engines would not be able to provide enough power to easily overcome the effect, which led to the concept of a "sound barrier".

To overcome this shock effect a supercritical airfoil is an airfoil typically designed, primarily, to delay the onset of wave drag in the transonic speed range. Supercritical airfoils are characterized by their flattened upper surface, highly cambered (curved) aft section, and greater leading edge radius as compared to traditional airfoil shapes.

Research aircraft of the 1950s and 60s found it difficult to break the sound barrier, or even reach Mach 0.9, with conventional airfoils. However today we build aircraft which easily exceed Mach 2.0.

Supersonic airflow over the upper surface of the traditional airfoil induced excessive wave drag and a form of stability loss called Mach Tuck. Due to the airfoil shape used, supercritical wings experience these problems less severely and at much higher speeds, thus allowing the wing to maintain high performance at speeds closer to Mach 1.

Techniques learned from studies of the original supercritical airfoil sections are used to design airfoils for high-speed subsonic and transonic aircraft from the AV-8B Harrier II to the Boeing 777.

Supercritical airfoils have four main benefits:

- higher Mach Number
- develop shock waves farther aft of the airframe body
- reduce shock-induced boundary layer separation,
- Supercritical airfoil geometry allows for more efficient wing design (e.g., a thicker wing and/or reduced wing sweep, each of which may allow for a lighter wing).

At a particular speed for a given airfoil section, the Mach Number, flow over the upper surface of an airfoil can become locally supersonic, but slow down to match the pressure at the trailing edge of the lower surface without a shock.

However, at a certain higher speed and high Mach Numbers, a shock is required to recover enough pressure to match the pressures at the trailing edge. This shock causes transonic wave drag, and induces flow separation behind it; both have negative effects on the airfoil's performance.

At a certain point along the airfoil, a shock is generated, which increases the pressure coefficient to a critical value where the local flow velocity will be ($\beta_{\text{Sonic\_Mach}} = M = 1.0$).

The position of this shockwave is determined by the geometry of the airfoil; a supercritical foil is more efficient because the shockwave is minimized and is created as far aft as possible thus reducing drag.

Compared to a typical airfoil section, the supercritical airfoil creates more of its lift at the aft end, due to its more even pressure distribution over the upper surface.

In addition to improved transonic performance, a supercritical wing's enlarged leading edge gives it excellent high-lift characteristics. As a result, aircraft utilizing a supercritical wing have superior takeoff and landing performance. This makes the supercritical wing a favorite for designers of cargo transport aircraft.

Mach tuck is another aerodynamic effect, whereby the nose of an aircraft tends to pitch downwards as the airflow around the wing reaches supersonic speeds.

Shock wave on upper surface of wing moves rearwards as aircraft mach increases. Initially as airspeed is increased past the Mach 1.0 number, the wing develops an increasing amount of lift therefore causing a significantly higher nose-down force and requiring a nose-up input or trim to maintain level flight.

With increased speed, and the aft movement of the shock wave the wing's center of pressure also moves aft, causing the start of a nose-down tendency or "tuck." If allowed to progress unchecked, in an aircraft not designed for supersonic flight, Mach tuck may occur.

Although Mach tuck develops gradually, if it is allowed to progress significantly, the center of pressure can move so far rearward that there is no longer enough elevator authority available to counteract it, and the airplane could enter a steep, sometimes unrecoverable dive.

In addition as the shockwave goes towards the rear, it can impinge upon the elevator control surfaces and this can greatly exacerbate the nose down tendency.

Historically, recovery from a mach tuck has not always been possible. In some cases as the aircraft descends the air density increases and the extra drag will slow the aircraft and control will return.

For aircraft such as supersonic bombers or supersonic transports that spend long periods in supersonic, Mach tuck is often compensated for by moving fuel between tanks in the fuselage to change the position of the Center of Mass. This minimizes the amount of trim required and significantly reduces aerodynamic drag.

Aphorism 3.2: **Hydrodynamic Motion Law (1)** – Relative to an external observer frame, the Mean Center, and the Center of Mass of an isolated system mass body, the Net Inertial "Rest" Mass moving relative to the atomic substance gaseous fluid medium, experiences fluid resistance due to any motion in the gaseous fluid with a velocity greater than zero; and the faster the velocity of the Net Inertial "Rest" Mass body relative to the medium the more hydrodynamic resistance that body will experience eventually increasing the mass, momentum, kinetic energy, and hydrodynamic Force of the moving mass body.

Aphorism 3.3: **Hydrodynamic Motion Law (2)** – Relative to an external observer frame, the Mean Center, and the Center of Mass, of an isolated system mass body, the average speed of sound ($\mathbf{c}_{\mathbf{Sound}}$) and in an atomic substance medium has the same value for all observers measuring the same temperature and mass, regardless of their relative motion, or of the motion of the source.

$$c^2_{\text{Sound}} = \gamma_{Heat} \cdot \left(\frac{\text{N} \cdot \text{k}_{\text{B}} \cdot T_{emp\ Initial}}{\text{m}_{\text{Net}}} \right) = \gamma_{Heat} \cdot \left(\frac{\left| \text{v}^2 \right|_{\text{Iso}}}{3} \right)$$

Aphorism 3.4: **Hydrodynamic Motion Law (3)** – Relative to an external observer frame, the Mean Center, and the Center of Mass, of an isolated system, as a Net Inertial "Rest" Mass body travels through an atomic substance gaseous fluid medium at a speed that is very close to the average speed of propagation of the medium, a lagging boundary layer is added to the rest mass and the total mass of the system is an effective Total Relativistic Mass being the sum of the inertial mass and added condensed/compressed fluid motion mass.

3.2 Lorentz/Einstein – Vacuum of Spacetime Compressibility Factor

Describing the Mass Increase in terms of resistance through a medium makes claims for real relative motion between an inertial system mass body, an external observer, and the Aether in the Vacuum of Spacetime.

The **Lorentz Spacetime Compressibility Factor** ($\gamma_{Lorentz} = \frac{1}{\sqrt{\sigma_{\text{CM\_Propagation}}}}$) named after the Dutch physicist Hendrik Lorentz (1853 – 1928) appears in several equations in special relativity, including time dilation, length contraction, and the relativistic mass increase formula.

3.2

$$\gamma_{Lorentz} = \frac{1}{\sqrt{\sigma_{\text{CM\_Propagation}}}} = \frac{1}{\sqrt{1 - \left(\frac{\left|\overline{\mathrm{V}}\right|^2_{\mathrm{CM}}}{\mathrm{c}^2_{\mathrm{Light}}}\right)}} \rightarrow \mathit{Unitless}$$

The Lorentz Factor represents the ratio of the increase of the fluid resistance as the Center of Mass Velocity of the system body moves relative to the Vacuum of Spacetime medium.

Albert Einstein in 1905 concluded that the Vacuum of SpaceTime is empty and the only possible explanation for the requirement of an excess force is that the inertial mass of the particle must increase as its speed approaches the speed of light.

Like others before, Jules Henri Poincaré (1854 –1912) French mathematician, theoretical physicist, and a philosopher of science around (1900) discovered a relation between mass and electromagnetic energy.

While studying the conflict between the action/reaction principle and Lorentz ether theory, Poincaré tried to determine whether the center of gravity still moves with a uniform velocity when electromagnetic fields are included. He noticed that the action/reaction principle does not hold for matter alone, but that the electromagnetic field has its own momentum. Poincaré concluded that the electromagnetic field energy of an electromagnetic wave behaves like a fictitious fluid with a mass ($\mathrm{m}_{\mathrm{Net}} = \frac{\mathrm{E}_{\text{Rest\_Energy}}}{\mathrm{c}^2_{\mathrm{Light}}}$) and mass density of ($\rho_{\mathrm{Net}} = \frac{1}{\mathrm{V}_{\mathrm{ol}}} \cdot \left(\frac{\mathrm{E}_{\text{Rest\_Energy}}}{\mathrm{c}^2_{\mathrm{Light}}}\right)$).

According to Poincaré, if the center of mass frame is defined by both the mass of matter and the mass of the fictitious "aether" fluid, and if the fictitious "aether" fluid is indestructible — it's neither created nor destroyed — then the motion of the center of mass frame remains uniform.

However, according to Albert Einstein electromagnetic energy can be converted into other forms of energy. So Poincaré assumed that there is a non-electric energy fluid at each point of space, into which electromagnetic energy can be transformed and which also carries a mass proportional to the energy. In this way, the motion of the center of mass remains uniform.

Steven Rado's aether theory debuted in 1994, postulated that as a mass body reaches a speed that is close to the average dissipation speed of the medium, an additional hydrodynamic force is required to overcome the resistance and density increase established by a mass body and the medium through which it is travelling.

Thus, as the system mass body Center of Mass Velocity approaches the spontaneous Propagation velocity of the medium a density disturbance due to the motion relative to the medium results in an increase in the mass, momentum, pressure, kinetic energy, and hydrodynamic force.

The **Lorentz Spacetime Compressibility Factor** as a measurable quantity is a unit-less ratio which measures the compressibility of the medium relative to internal and external observers.

According to Lorentz, Poincaré, Rado and their respective aether theories, any elastic unit in nature, whether it is macroscopic matter, a living cell, an atom, or sub atomic matter, when an external directional force exerted upon it, it must undergo some distortion of its original undisturbed shape even before any translational displacement of the body occurs.

Thus, for the translational motion of a moving mass, the relativistic mass (of a body or system of bodies) includes a contribution from the "net" kinetic energy of motion of the center of mass of the body; and is larger the faster the body moves.

The Lorentz compressibility factor describes a limiting compressibility of the medium which is the Vacuum of SpaceTime which produces an increasing magnitude of Aether resistance related to the ratio between the speed of the System mass body and the speed of light in a vacuum.

The Lorentz Compressibility Factor appears in several equations in special relativity, including time dilation, length contraction, and the relativistic mass and energy formula, as described in the equation below.

Definition 3.2: **Lorentz Compressibility Factor** ($\gamma_{Lorentz} = \dfrac{1}{\sqrt{1 - \dfrac{|\overline{v}|^2_{CM}}{c^2_{Light}}}}$) is a unit-less ratio defined as the hydrodynamic fluid motion resistance of the Aether and Matter relative to the anisotropic and isotropic density of the inertial mass body and the atmosphere surrounding Vacuum of Spacetime medium, and represents a density increase due to the increasing Center of Mass Velocity of the system as a whole; and is equal to the inverse square root of one subtracted from the square of the **Mach Light Velocity Ratio** ($\dfrac{|\overline{v}|^2_{CM}}{c^2_{Light}}$), or one subtracted from the ratio of the square of the **Center of Mass Velocity** ($|\overline{v}|^2_{CM}$) divided by the square of the **Speed of Propagation of the medium** (c^2_{Light}).

Lorentz Compressibility Factor for the Vacuum of SpaceTime

3.3

$$\gamma_{Lorentz} = \frac{1}{\sqrt{1 - \dfrac{|\overline{v}|^2_{CM}}{c^2_{Light}}}} = \frac{1}{\sqrt{1 - \beta^2_{Mach}}} \rightarrow \textit{Unitless}$$

$$\gamma_{Lorentz} = \frac{m'_{Net\_Rel}}{m_{Net}}$$

$$\gamma_{Lorentz} = \frac{\overline{p}'_{Inertial\_Rel}}{\overline{p}_{Inertial\_Net}} = \frac{E'_{Relativistic\_Energy}}{E_{Rest\_Energy}}$$

$$\gamma_{Lorentz} = \frac{4 \cdot E'_{Translational}}{4 \cdot E_{Translational}} = \frac{E'_{Momentum\_Energy}}{E_{Momentum\_Energy}}$$

3.3 Prandtl-Glauert – Atomic Substance - Spacetime Compressibility Factor

Describing the Mass Increase in terms of resistance through a medium makes claims for real relative motion between an inertial system mass body, an external observer, and the Aether in an atomic substance.

The Prandtl-Glauert **Atomic Substance - Spacetime Compressibility Factor** ($\gamma_{Prandtl} = \dfrac{1}{\sqrt{\left\|1 - \dfrac{|\overline{v}|^2_{CM}}{c^2_{Sound}}\right\|}}$) named after the German scientist **Ludwig Prandtl** (1875 –1953) and the British aerodynamicist **Hermann Glauert** (1892 – 1934) appears in several for subsonic airflow to describe the compressbility effects of air at high speeds, and various Mach Numbers.

The Prandtl-Glauert equation is found by taking the ratio of the compressibility to incompressibility of the fluid flow. It was discovered that the pressures in such a flow were equal to those found from incompressible flow theory multiplied by a correction factor. This **Atomic Substance - Spacetime Compressibility Factor** correction is given below.

Definition 3.3: **Prandtl-Glauert Atomic Substance - Spacetime Compressibility Factor** ($\gamma_{Prandtl} = \dfrac{1}{\sqrt{\left\|1 - \dfrac{|\overline{v}|^2_{CM}}{c^2_{Sound}}\right\|}}$) is a unit-less ratio defined as the hydrodynamic fluid motion resistance of the Aether and Matter relative to the anisotropic and isotropic density of the inertial mass body and the atomic substance atmosphere surrounding medium, and represents a density increase due to the increasing Center of Mass Velocity of the system as a whole; and is equal to the inverse square root of one subtracted from the square of the **Mach Velocity Ratio** ($\dfrac{|\overline{v}|^2_{CM}}{c^2_{Sound}}$), or one subtracted from the ratio of the square of the **Center of Mass Velocity** ($|\overline{v}|^2_{CM}$) divided by the square of the **Speed of Propagation of the medium** (c^2_{Sound}).

Prandtl-Glauert Atomic Substance - Spacetime Compressibility Factor

3.4

$$\gamma_{\Pr andtl} = \frac{C'_{\text{Compression}}}{C_{\text{Incompression}}} = \frac{1}{\sqrt{\left\| 1 - \frac{\left|\overline{v}\right|^2_{\text{CM}}}{c^2_{\text{Medium}}} \right\|}} = \frac{1}{\sqrt{\left\| 1 - \beta^2_{\text{Mach}} \right\|}} \rightarrow \textit{Unitless}$$

At lower altitudes, air has a higher density and is considered incompressible for theoretical and experimental purposes. Notice in the above equation that inside the square root symbol there is the absolute value of one minus the Mach Number ($\left\| 1 - \beta^2_{\text{Mach}} \right\|$).

This Prandtl-Glauert Atomic Substance - Spacetime Compressibility correction factor works well for all Mach numbers $0.3 < \beta^2_{\text{Sonic\_Mach}} < 0.7$.

The Prandtl-Glauert Atomic Substance - Spacetime Compressibility Factor aso represents the ratio of the increase of the fluid resistance as the Center of Mass Velocity of the system body moves relative to the atomic substance medium.

Prandtl-Glauert and Mach's Theory of the 1920s placed a limit on the compressibility of a medium which produces an increasing resistance to motion, related to the ratio between the speed of the system mass body and the speed of Propagation of the atomic substance medium.

In subsonic flow the compressibility of the atomic substance fluid (often air) becomes more and more influential with increasing velocity.

Near the sonic speed ($\beta_{\text{Sonic\_Mach}}$ = M = 1.0) the discussed equation features a singularity, although this point is not within the area of validity. The singularity is also called the Prandtl–Glauert singularity, and the flow resistance is calculated to approach infinity.

In reality aerodynamic and thermodynamic perturbations get amplified strongly near the sonic speed; and some sort of singularity "barrier" resistance to motion does not occur.

An explanation for this is that the Prandtl-Glauert transformation is an approximation of compressible, in-viscid potential flow. As the flow approaches sonic ($\beta_{\text{Sonic\_Mach}}$ = M = 1.0), the nonlinear phenomena dominate within the flow, which this transformation completely ignores for the sake of simplicity.

This hypothetical incompressibility mainly means that in an isotropic fluid, like that of air, or the vacuum of spacetime, all locally produced density fluctuations around a moving mass body are continuously dissipated in the form of waves with the Propagation speed of the medium.

Therefore, as long as the speed of the object is much smaller than that of the medium in which it is moving, there are no bulk density changes accumulating in front of the moving object.

If the speed of the object approaches speeds that are much closer to the Propagation speed of the medium, the medium is unable to dissipate the accumulating density in front of the moving solid and therefore cannot preserve its isotropic density.

Due to incapability of the medium to disperse fluid mass energy in front of the moving body; and this result in accumulation of a local density increase in front of the motion of the body and therefore an increase in the loca air-resistance applied to the body.

Both the Lorentz compressibility factor for a vacuum, and the Prandtl Factor for atomic substance mediums (i.e. airflow) both predict infinities or a singularity ($\gamma = \frac{1}{0} \rightarrow \infty$), when the Mach Number is equal to one ($\beta_{\text{Mach}} = \left(\frac{|\bar{v}|_{\text{CM}}}{c_{\text{Medium}}} \right) = 1$); However for Air Mach Numbers greater than one are very easily achieved by today's modern aircraft.

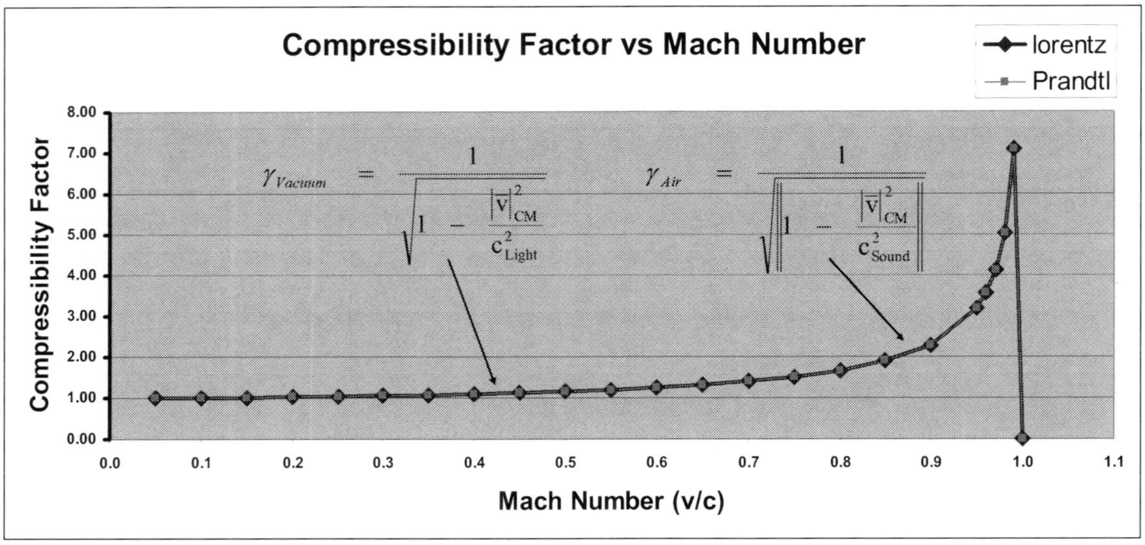

Figure 3.4: The Lorentz Vacuum of Spacetime Compressibility factor predicts similar physics to the Prandtl Compressibility factor as demonstrated in the graph above.

The image above demonstrates that the motion through the Vacuum of Spacetime and the motion though an atomic substance produce similar mass density increase effects.

Square of the Prandtl-Glauert Atomic Substance - Spacetime Compressibility Factor

3.5

$$\gamma^2_{Prandtl} = \frac{1}{\left[1 - \frac{3}{\gamma_{Heat}} \cdot \left[1 - \eta_{Carnot}\right]\right]} = \frac{1}{\left(1 - \left(\frac{|\overline{v}|^2_{CM}}{c^2_{Sound}}\right)\right)} \rightarrow \textit{Unitless}$$

$$\gamma^2_{Prandtl} = \frac{1}{\left[1 - \frac{3}{\gamma_{Heat}} \cdot \left[1 - \frac{1}{\gamma^2_{Inertial}}\right]\right]} = \frac{1}{\left(1 - \left(\frac{|\overline{v}|^2_{CM}}{c^2_{Sound}}\right)\right)}$$

$$\gamma^2_{Prandtl} = \frac{1}{\left[1 - \frac{3}{\gamma_{Heat}} \cdot \frac{1}{\left[1 + \frac{1}{\gamma^2_{Inertial\_Iso}}\right]}\right]} = \frac{1}{\left(1 - \left(\frac{|\overline{v}|^2_{CM}}{c^2_{Sound}}\right)\right)}$$

$$\gamma^2_{Prandtl} = \frac{1}{\left[1 - \frac{3}{\gamma_{Heat}} \cdot \frac{1}{\left[1 + \eta_{Carnot\_Iso}\right]}\right]} = \frac{1}{\left(1 - \left(\frac{|\overline{v}|^2_{CM}}{c^2_{Sound}}\right)\right)}$$

Prandtl-Glauert Atomic Substance - Spacetime Compressibility Factor

3.6

$$\gamma_{Prandtl} = \frac{1}{\sqrt{\left[1 - \frac{3}{\gamma_{Heat}} \cdot \left[1 - \eta_{Carnot}\right]\right]}} = \frac{1}{\sqrt{1 - \left(\frac{|\overline{v}|^2_{CM}}{c^2_{Sound}}\right)}} \rightarrow Unitless$$

$$\gamma_{Prandtl} = \frac{1}{\sqrt{\left[1 - \frac{3}{\gamma_{Heat}} \cdot \left[1 - \frac{1}{\gamma^2_{Inertial}}\right]\right]}} = \frac{1}{\sqrt{1 - \left(\frac{|\overline{v}|^2_{CM}}{c^2_{Sound}}\right)}}$$

$$\gamma_{Prandtl} = \frac{1}{\sqrt{\left[1 - \frac{3}{\gamma_{Heat}} \cdot \frac{1}{\left[1 + \frac{1}{\gamma^2_{Inertial\_Iso}}\right]}\right]}} = \frac{1}{\sqrt{1 - \left(\frac{|\overline{v}|^2_{CM}}{c^2_{Sound}}\right)}}$$

$$\gamma_{Prandtl} = \frac{1}{\sqrt{\left[1 - \frac{3}{\gamma_{Heat}} \cdot \frac{1}{\left[1 + \eta_{Carnot\_Iso}\right]}\right]}} = \frac{1}{\sqrt{1 - \left(\frac{|\overline{v}|^2_{CM}}{c^2_{Sound}}\right)}}$$

3.4 Mach Number for a Specific Atomic Substance

Ernst Mach (1838 – 1916) was an Austrian physicist and philosopher, remembered for his contributions to physics such as the Mach number and the study of shock waves.

The Mach number ($M_{Mach} = \left(\frac{|\overline{v}|_{CM}}{c_{Sound}} \right)$) is a ratio that depends on the speed of the center of mass of a body and the speed of sound in an atomic substance medium. And the speed of sound depends on the type of atomic substance, and the temperature of that atomic substance.

The speed of sound varies from planet to planet. On Earth, the atmosphere is composed of mostly diatomic nitrogen and oxygen, and the temperature depends on the altitude.

Square of the Mach number for Atomic Substance Medium – Anisotropic Engine

3.7

$$\beta^2_{Mach} = M^2_{Mach} = \left(\frac{|\overline{v}|^2_{CM}}{c^2_{Sound}} \right) = \frac{3}{\gamma_{Heat}} \cdot \left[1 - \frac{1}{\gamma^2_{Inertial}} \right]$$

$$\beta^2_{Mach} = M^2_{Mach} = \frac{3 \cdot \phi_{Loco\_Motion}}{\gamma_{Heat}} = \frac{3 \cdot [1 - \eta_{Carnot}]}{\gamma_{Heat}} \rightarrow Unitless$$

Square of the Mach number for Atomic Substance Medium – Isotropic Engine

3.8

$$\beta^2_{Mach} = M^2_{Mach} = \left(\frac{|\overline{v}|^2_{CM}}{c^2_{Sound}} \right) = \frac{3}{\gamma_{Heat}} \cdot \frac{1}{\left[1 + \frac{1}{\gamma^2_{Inertial\_Iso}} \right]}$$

$$\beta^2_{Mach} = M^2_{Mach} = \frac{3 \cdot \phi_{Loco\_Motion}}{\gamma_{Heat}} = \frac{3}{\gamma_{Heat}} \cdot \frac{1}{[1 + \eta_{Carnot\_Iso}]} \rightarrow Unitless$$

Mach number for Atomic Substance Medium – Anisotropic Engine

3.9

$$\beta_{\text{Mach}} = \text{M}_{\text{Mach}} = \left(\frac{|\bar{v}|_{\text{CM}}}{c_{\text{Sound}}} \right) = \sqrt{\frac{3 \cdot \left[1 - \dfrac{1}{\gamma^2_{Inertial}} \right]}{\gamma_{Heat}}}$$

$$\beta_{\text{Mach}} = \text{M}_{\text{Mach}} = \sqrt{\frac{3 \cdot \phi_{\text{Loco\_Motion}}}{\gamma_{Heat}}} = \sqrt{\frac{3 \cdot [1 - \eta_{\text{Carnot}}]}{\gamma_{Heat}}} \rightarrow \textit{Unitless}$$

Square of the Mach number for Atomic Substance Medium – Isotropic Engine

3.10

$$\beta_{\text{Mach}} = \text{M}_{\text{Mach}} = \left(\frac{|\bar{v}|_{\text{CM}}}{c_{\text{Sound}}} \right) = \sqrt{\frac{3}{\gamma_{Heat}} \cdot \frac{1}{\left[1 + \dfrac{1}{\gamma^2_{Inertial\_Iso}} \right]}}$$

$$\beta_{\text{Mach}} = \text{M}_{\text{Mach}} = \sqrt{\frac{3 \cdot \phi_{\text{Loco\_Motion}}}{\gamma_{Heat}}} = \sqrt{\frac{3}{\gamma_{Heat}} \cdot \frac{1}{[1 + \eta_{\text{Carnot\_Iso}}]}} \rightarrow \textit{Unitless}$$

3.11

Scalar	
Magnitude	Units
Speed of Light Propagation in Vacuum	
$$\mathbf{c}_{\mathbf{Light}} = 2.99792480 \times 10^{8}\ {m}/{s}$$	$\frac{m}{s}$
Speed of Sound Propagation in Air	
The approximate speed of sound in dry (0% humidity) air, in meters per second, at temperatures near 0 °C, can be calculated using the following equation: $$\mathbf{c}_{\mathbf{Sound}} = \left(331.3\ {m}/{s}\right)\cdot\sqrt{1 + \frac{T^{\circ}_{\text{Temp\_Celsius}}}{273.15^{\circ}C}}$$	$\frac{m}{s}$

3.12

SpaceTime Warp Factors	
Ratio	Units
Mach Number for Vacuum	
$$\beta_{\text{Mach\_Light}} = \left(\frac{\lvert\overline{\mathbf{v}}\rvert_{\mathbf{CM}}}{\mathbf{c}_{\mathbf{Light}}}\right)$$	Unit-less
Lorentz Compressibility Factor for Vacuum	
$$\gamma_{\text{Lorentz}} = \frac{1}{\sqrt{1 - \frac{\lvert\overline{v}\rvert^{2}_{CM}}{c^{2}_{Light}}}} = \frac{1}{\sqrt{1 - \beta^{2}_{\text{Mach\_Light}}}}$$	Unit-less

SpaceTime Warp Factors	
Ratio	Units
Mach Number for Atomic Substance Medium	
$\beta_{\text{Mach}} = \text{M}_{\text{Mach}} = \left(\dfrac{\lvert \overline{\mathbf{v}} \rvert_{\mathbf{CM}}}{\mathbf{c}_{\mathbf{Sound}}} \right)$	Unit-less
Prandtl Compressibility Factor for Atomic Substance Medium	
$\gamma_{\text{Pr}\,andtl} = \dfrac{1}{\sqrt{\left\lVert 1 - \dfrac{\lvert \overline{\text{v}} \rvert^2_{\text{CM}}}{\text{c}^2_{\text{Sound}}} \right\rVert}} = \dfrac{1}{\sqrt{\left\lVert 1 - \beta^2_{\text{Mach}} \right\rVert}}$	Unit-less

Chapter 4

The Special Theory of Thermodynamics

(Aether Working Fluid Ratio)
(Locomotion Foreshortening Factor)

4.1 Universal Geometric Mean Ratio – Locomotion Foreshortening Factor

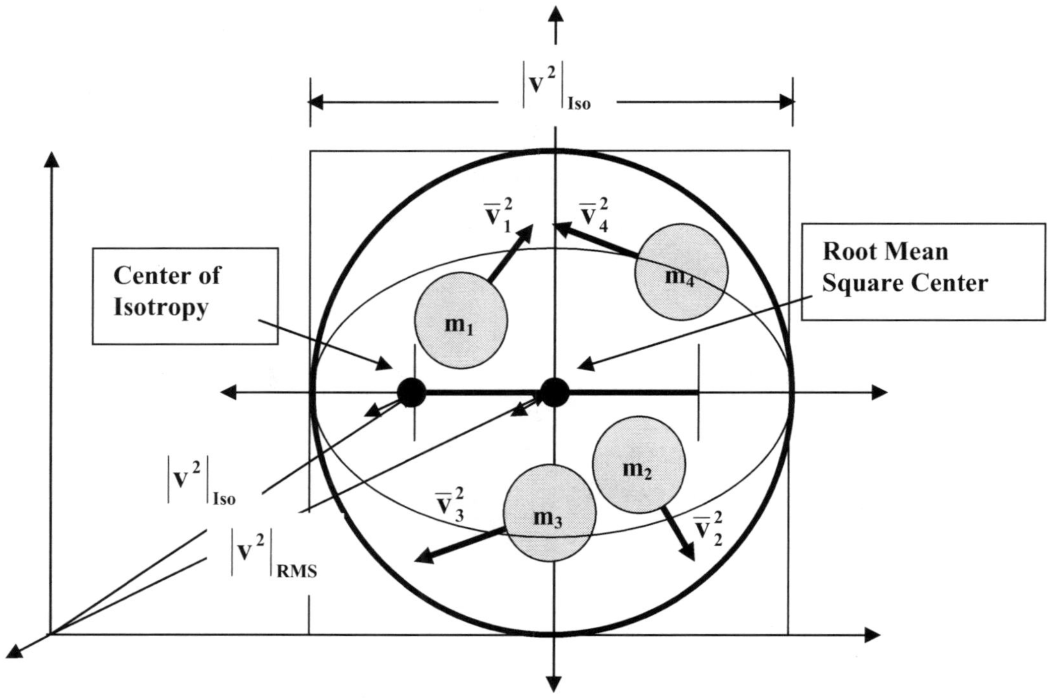

Figure 4.1: Isotropy and the Center of Isotropy.

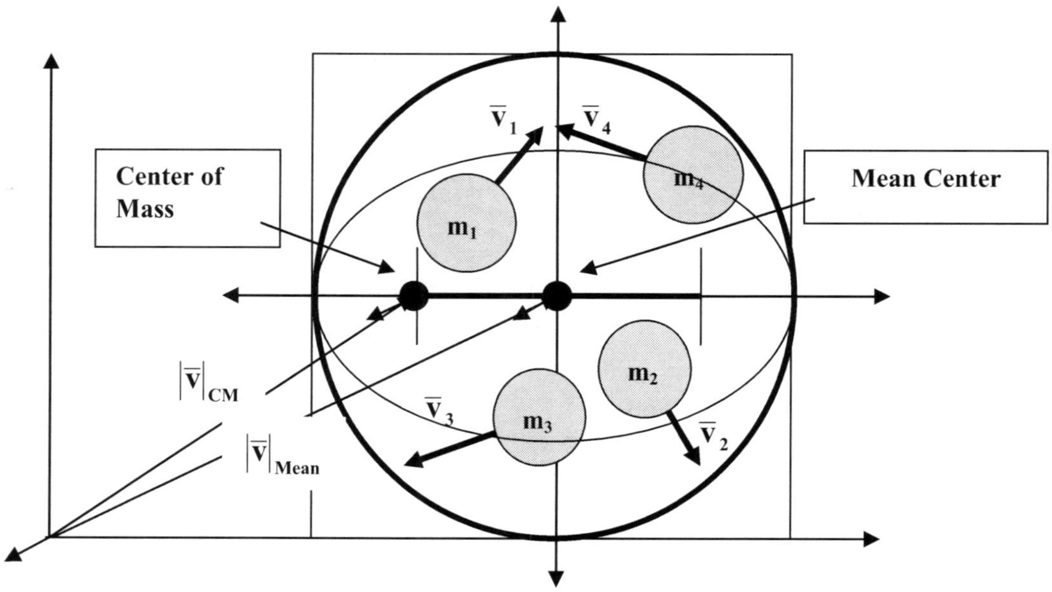

Figure 4.2: Anisotropy and the Center of Mass.

Starting with the Anisotropy Carnot Efficiency Factor

$$\eta_{Carnot} = \frac{1}{\gamma^2_{Inertial}} = \left[\frac{W_{Free\_Work\_Energy}}{E_{Net\_Iso\_Kinetic}}\right] = \left(1 - \frac{\left|\overline{v}\right|^2_{CM}}{\left|v^2\right|_{Iso}}\right)$$

$$\gamma^2_{Inertial} = \frac{1}{\eta_{Carnot}} = \frac{1}{1 - \dfrac{\left|\overline{v}\right|^2_{CM}}{\left|v^2\right|_{Iso}}}$$

$$1 - \frac{1}{\gamma^2_{Inertial}} = \frac{\left|\overline{v}\right|^2_{CM}}{\left|v^2\right|_{Iso}}$$

Inertial Locomotion Foreshortening Factor — Carnot Anisotropy Engine

4.1

$$\phi_{Loco\_Motion} = \left(\frac{\left|\overline{v}\right|^2_{CM}}{\left|v^2\right|_{Iso}}\right) \rightarrow \textit{Unitless}$$

$$\phi_{Loco\_Motion} = \frac{\left|\overline{v}\right|^2_{CM}}{\left|v^2\right|_{Iso}} = \left[1 - \frac{1}{\gamma^2_{Inertial}}\right] = \left[1 - \eta_{Carnot}\right]$$

Starting with the Isotropy Carnot Efficiency Factor

$$\eta_{\text{Carnot\_Iso}} = \frac{1}{\gamma^2_{Inertial\_Iso}} = \left[\frac{W_{\text{Free\_Work\_Energy}}}{E_{\text{Net\_Translational}}}\right] = \left(\frac{1}{\dfrac{\left|\overline{v}\right|^2_{\text{CM}}}{\left|v^2\right|_{\text{Iso}}}} - 1\right)$$

$$\gamma^2_{Inertial\_Iso} = \frac{1}{\eta_{\text{Carnot\_Iso}}} = \frac{1}{\left(\dfrac{1}{\dfrac{\left|\overline{v}\right|^2_{\text{CM}}}{\left|v^2\right|_{\text{Iso}}}} - 1\right)}$$

$$\frac{\left|\overline{v}\right|^2_{\text{CM}}}{\left|v^2\right|_{\text{Iso}}} = \left(\frac{1}{\left[\dfrac{1}{\gamma^2_{Inertial\_Iso}} + 1\right]}\right)$$

Inertial Locomotion Foreshortening Factor – Carnot Isotropy Engine

4.2

$$\phi_{\text{Loco\_Motion}} = \left(\frac{\left|\overline{v}\right|^2_{\text{CM}}}{\left|v^2\right|_{\text{Iso}}}\right) \rightarrow \textit{Unitless}$$

$$\phi_{\text{Loco\_Motion}} = \left(\frac{\left|\overline{v}\right|^2_{\text{CM}}}{\left|v^2\right|_{\text{Iso}}}\right) = \left(\frac{1}{\left[\dfrac{1}{\gamma^2_{Inertial\_Iso}} + 1\right]}\right) = \left(\frac{1}{\left[\eta_{\text{Carnot\_Iso}} + 1\right]}\right)$$

Inertial Mass Locomotion Foreshortening Factor

Definition 4.1: The **Inertial Locomotion Foreshortening Factor** ($\phi_{\text{Loco\_Motion}}$) is Carnot Heat Engine mechanism defined as the Anisotropy to Isotropy ratio of an isolated dynamic mass system, and is the measure of the isotropic to anisotropic pressure difference, kinetic energy difference, and the isotropic temperature difference between an isolated system mass body and its atmosphere environment medium.

Universal Geometric Mean Ratio

4.3

$$\frac{(Min)}{(Mean)} = \frac{(Mean)}{(Max)} = \sqrt{\frac{(Min)}{(Max)}} = \left(\frac{(Mean) - (Min)}{(Max) - (Mean)}\right)$$

Inertial Locomotion Foreshortening Factor

$$\phi_{\text{Loco\_Motion}} = \left(\frac{E_{\text{Net\_Translational}}}{E_{\text{Net\_Iso\_Kinetic}}}\right) = \left(\frac{\text{External Anisotropic Tranlational Kinetic Energy}}{\text{Internal Isotropic Omni - directional Kinetic Energy}}\right)$$

$$\phi_{\text{Loco\_Motion}} = \left(\frac{\overline{p}^{\,2}_{\text{Inertial\_Net}}}{\overline{p}^{\,2}_{\text{Net\_Iso}}}\right) = \left(\frac{\left|\overline{v}\right|^2_{\text{CM}}}{\left|v^2\right|_{\text{Iso}}}\right) = \left(\frac{\begin{matrix}\text{External Anisotropic}\\ \text{Center of Mass Squared Velocity Inertia}\end{matrix}}{\begin{matrix}\text{Internal Isotropic Omni - directional}\\ \text{Center of Isotropy Squared Velocity Inertia}\end{matrix}}\right)$$

$$\phi_{\text{Loco\_Motion}} = \left(\frac{T_{emp\_Ext}}{T_{emp}}\right) = \left(\frac{\begin{matrix}\text{External Anisotropic Tranlational}\\ \text{Aerodynamic Temperature}\end{matrix}}{\begin{matrix}\text{Internal Isotropic Omni - directional}\\ \text{Absolute Temperature}\end{matrix}}\right)$$

Aphorism 4.1: The **Inertial Locomotion Foreshortening Factor** ($\phi_{\mathrm{Loco\_Motion}}$) is the Carnot Heat Engine Anisotropy to Isotropy ratio of an isolated dynamic mass system, and is directly proportional to the product of the atomic substance Specific Heat Capacity Index (γ_{Heat}) and the square of the Mach number ($\left(\frac{\left|\overline{\mathrm{v}}\right|_{\mathrm{CM}}^{2}}{\mathrm{c}_{\mathrm{Sound}}^{2}}\right)$).

$$\phi_{\mathrm{Loco\_Motion}} = \left(\frac{\left|\overline{\mathrm{v}}\right|_{\mathrm{CM}}^{2}}{\left|\mathrm{v}^{2}\right|_{\mathrm{Iso}}}\right) = \frac{\gamma_{Heat}}{3}\cdot\left(\frac{\left|\overline{\mathrm{v}}\right|_{\mathrm{CM}}^{2}}{\mathrm{c}_{\mathrm{Sound}}^{2}}\right) = \frac{\gamma_{Heat}}{3}\cdot\beta_{Mach}^{2}$$

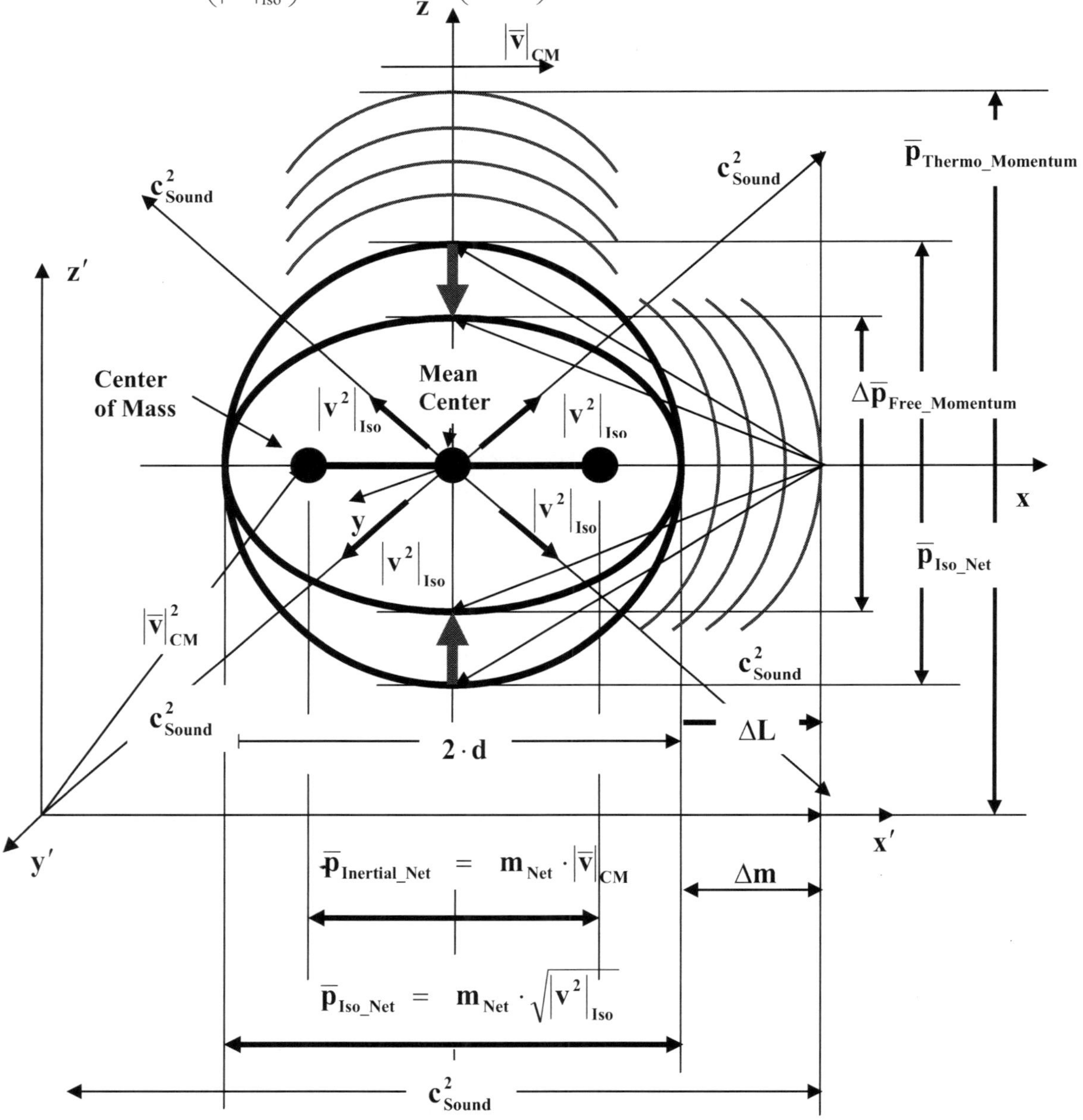

Figure 4.3: Isolated Dynamic System Mass Body.

Universal Geometric Mean Ratio – Inertial Locomotion Foreshortening Factor

4.4

$$\phi_{Loco\_Motion} = \frac{m_{Inertia\_Mass}}{m_{Net}} = \frac{m_{Net}}{m_{Dark\_Matter}} = \sqrt{\frac{m_{Inertia\_Mass}}{m_{Dark\_Matter}}}$$

$$\phi_{Loco\_Motion} = \frac{m_{Inertia\_Mass}}{m_{Net}} = \frac{1}{\left(\sum_{i=1}^{N} m_i\right)} \cdot \left[\frac{\left(\sum_{i=1}^{N}(m_i \bar{v}_i)\right)^2}{\sum_{i=1}^{N}(m_i \cdot \bar{v}_i^2)}\right]$$

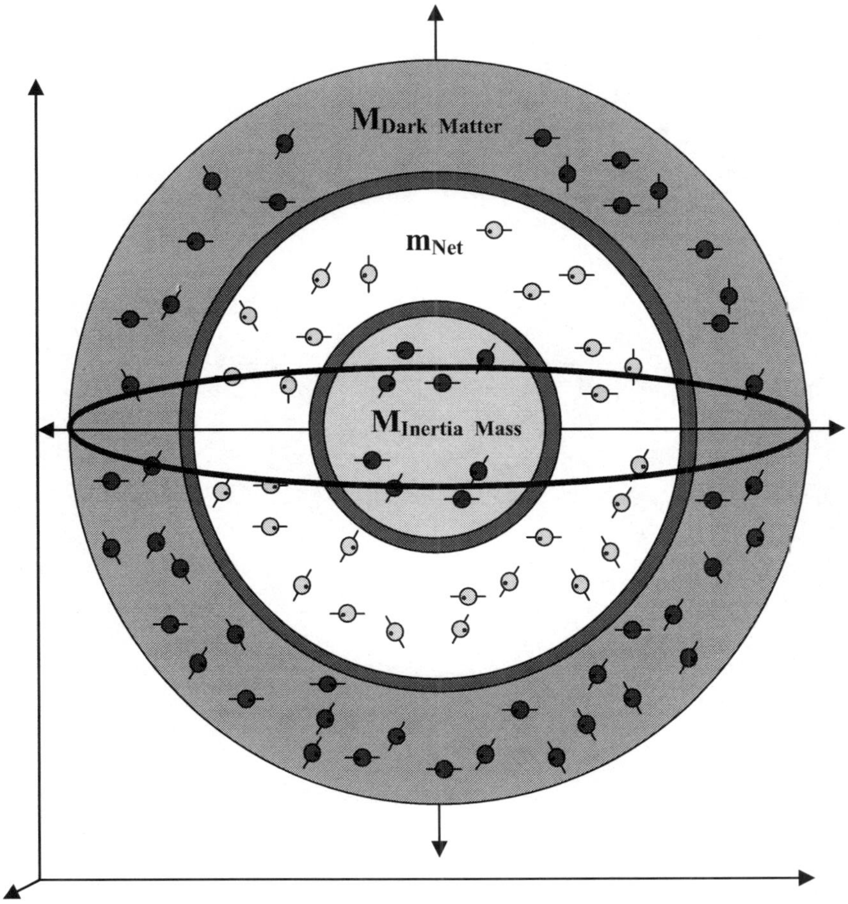

Figure 4.4: The Inertial Locomotion Foreshortening Factor is the Universal Geometric Mean of the inertia mass for an isolated mass system body, which consist of the Rest Mass, Dark Matter, and the Inherent Inertia Mass.

Inertia Geometric Mean Ratio – Inertial Locomotion Foreshortening Factor

4.5

$$\phi_{\text{Loco\_Motion}} = \frac{1}{\left(\begin{matrix} m_1 + m_2 \\ + m_3 + \cdots m_N \end{matrix}\right)} \cdot \left[\frac{\left(m_1\bar{v}_1 + m_2\bar{v}_2 + m_3\bar{v}_3 + \cdots m_N\bar{v}_N\right)^2}{\left[m_1\bar{v}_1^2 + m_2\bar{v}_2^2 + m_3\bar{v}_3^2 + \cdots m_N\bar{v}_N^2\right]} \right]$$

$$\phi_{\text{Loco\_Motion}} = \frac{m_{\text{Inertia\_Mass}}}{m_{\text{Net}}} = \frac{m_{\text{Net}}}{m_{\text{Dark\_Matter}}} = \sqrt{\frac{m_{\text{Inertia\_Mass}}}{m_{\text{Dark\_Matter}}}}$$

$$\phi_{\text{Loco\_Motion}} = \frac{\left(m_{\text{Net}} - m_{\text{Inertia\_Mass}}\right)}{\left(m_{\text{Dark\_Matter}} - m_{\text{Net}}\right)}$$

$$\phi_{\text{Loco\_Motion}} = \left(\frac{m_{\text{Inertia\_Mass}}}{m_{\text{Net}}}\right) = \left[1 - \left(\frac{\Delta(PV)_{\text{Free\_Energy}}}{\left(1 - \frac{1}{\gamma_{Heat}}\right) \cdot m_{\text{Net}} \cdot C_{Specific\_Pressure} \cdot T_{emp}}\right)\right]$$

$$\phi_{\text{Loco\_Motion}} = \left(\frac{m_{\text{Inertia\_Mass}}}{m_{\text{Net}}}\right) = \left[1 - \left(\frac{\Delta(PV)_{\text{Free\_Energy}}}{N \cdot k_B \cdot T_{emp}}\right)\right]$$

$$\phi_{\text{Loco\_Motion}} = \left(\frac{m_{\text{Net}}}{m_{\text{Dark\_Matter}}}\right) = \frac{1}{\left[1 + \left(\frac{\Delta(PV)_{\text{Free\_Energy}}}{\left(1 - \frac{1}{\gamma_{Heat}}\right) \cdot m_{\text{Net}} \cdot C_{Specific\_Pressure} \cdot T_{emp\_Ext}}\right)\right]}$$

$$\phi_{\text{Loco\_Motion}} = \left(\frac{m_{\text{Net}}}{m_{\text{Dark\_Matter}}}\right) = \frac{1}{\left[1 + \left(\frac{\Delta(PV)_{\text{Free\_Energy}}}{N \cdot k_B \cdot T_{emp\_Ext}}\right)\right]} \rightarrow \textit{Unitless}$$

Inertia Geometric Mean Ratio – Inertial Locomotion Foreshortening Factor

4.6

$$\phi_{\text{Loco\_Motion}} = \left(\frac{m_{\text{Inertia\_Mass}}}{m_{\text{Net}}}\right) = \left(\frac{m_{\text{Net}}}{m_{\text{Dark\_Matter}}}\right) = \sqrt{\frac{m_{\text{Inertia\_Mass}}}{m_{\text{Dark\_Matter}}}} \rightarrow \textit{Unitless}$$

$$\phi_{\text{Loco\_Motion}} = \left(\frac{m_{\text{Inertia\_Mass}}}{m_{\text{Net}}}\right) = \left(\frac{E_{\text{Net\_Translational}}}{E_{\text{Net\_Iso\_Kinetic}}}\right) = \left(\frac{q_{Dynamic\_Pressure}}{P_{Iso\_Pressure}}\right)$$

$$\phi_{\text{Loco\_Motion}} = \left(\frac{m_{\text{Inertia\_Mass}}}{m_{\text{Net}}}\right) = \left(\frac{T_{emp\_Ext}}{T_{emp}}\right) = \frac{\left[m_{\text{Net}} + m_{\text{Inertia\_Mass}}\right]}{\left[m_{\text{Dark\_Matter}} + m_{\text{Net}}\right]}$$

$$\phi_{\text{Loco\_Motion}} = \left(\frac{m_{\text{Net}}}{m_{\text{Dark\_Matter}}}\right) = \left(\frac{\left|\overline{v}\right|^2_{\text{CM}}}{\left|v^2\right|_{\text{Iso}}}\right) = \left(\frac{\overline{p}^2_{\text{Inertial\_Net}}}{\overline{p}^2_{\text{Net\_Iso}}}\right)$$

$$\phi_{\text{Loco\_Motion}} = \left(\frac{m_{\text{Net}}}{m_{\text{Dark\_Matter}}}\right) = \left(\frac{\left|\overline{v}\right|^2_{\text{CM}}}{\left|v^2\right|_{\text{Iso}}}\right) = \frac{1}{m_{\text{Net}}} \cdot \left[\frac{\left(\sum_{i=1}^{N}\left(m_i \overline{v}_i\right)\right)^2}{\sum_{i=1}^{N}\left(m_i \cdot \overline{v}_i^2\right)}\right]$$

$$\phi_{\text{Loco\_Motion}} = \left(\frac{\left|\overline{v}\right|^2_{\text{CM}}}{\left|v^2\right|_{\text{Iso}}}\right) = \frac{1}{\left(m_1 + m_2 + m_3 + \cdots m_N\right)} \cdot \left[\frac{\left(m_1\overline{v}_1 + m_2\overline{v}_2 + m_3\overline{v}_3 + \cdots m_N\overline{v}_N\right)^2}{\left[m_1\overline{v}_1^2 + m_2\overline{v}_2^2 + m_3\overline{v}_3^2 + \cdots m_N\overline{v}_N^2\right]}\right]$$

Inertia Geometric Mean Ratio – Inertial Locomotion Foreshortening Factor

4.7

$$\phi_{\text{Loco\_Motion}} = \left(\frac{\left|\bar{v}\right|^2_{\text{CM}}}{\left|v^2\right|_{\text{Iso}}} \right) = \frac{\gamma_{Heat}}{3} \cdot \left(\frac{\left|\bar{v}\right|^2_{\text{CM}}}{c^2_{\text{Sound}}} \right) = \frac{\gamma_{Heat}}{3} \cdot \beta^2_{Mach}$$

$$\phi_{\text{Loco\_Motion}} = \left(\frac{m_{\text{Inertia\_Mass}}}{m_{\text{Net}}} \right) = \left(\frac{\left|\bar{v}\right|^2_{\text{CM}}}{\left|v^2\right|_{\text{Iso}}} \right) = \frac{1}{m_{\text{Net}}} \cdot \left[\frac{\left(\sum_{i=1}^{N} \left(m_i \bar{v}_i \right) \right)^2}{\sum_{i=1}^{N} \left(m_i \cdot \bar{v}^2_{Ai} \right)} \right]$$

$$\phi_{\text{Loco\_Motion}} = \left(\frac{\left|\bar{v}\right|^2_{\text{CM}}}{\left|v^2\right|_{\text{Iso}}} \right) = \frac{1}{\left(m_1 + m_2 + m_3 + \cdots m_N \right)} \cdot \left[\frac{\left(m_1 \cdot \bar{v}_1 + m_2 \cdot \bar{v}_2 + m_3 \cdot \bar{v}_3 + \cdots m_N \cdot \bar{v}_N \right)^2}{\left[m_1 \cdot \bar{v}^2_{A1} + m_2 \cdot \bar{v}^2_{A2} + m_3 \cdot \bar{v}^2_{A3} + \cdots m_N \cdot \bar{v}^2_{AN} \right]} \right]$$

$$\phi_{\text{Loco\_Motion}} = \frac{\left[\left|\bar{v}\right|^2_{\text{CM x}} + \left|\bar{v}\right|^2_{\text{CM y}} + \left|\bar{v}\right|^2_{\text{CM z}} \right]}{\left[\left|v^2_x\right|_{\text{Iso}} + \left|v^2_y\right|_{\text{Iso}} + \left|v^2_z\right|_{\text{Iso}} \right]} \rightarrow \textit{Unitless}$$

Two Body – Inertia Geometric Mean Ratio – Inertial Locomotion Foreshortening Factor

4.8

$$\phi_{\text{Loco\_Motion}} = \frac{m_{\text{Inertia\_Mass}}}{m_{\text{Net}}} = \frac{m_{\text{Net}}}{m_{\text{Dark\_Matter}}} = \sqrt{\frac{m_{\text{Inertia\_Mass}}}{m_{\text{Dark\_Matter}}}}$$

$$\phi_{\text{Loco\_Motion}} = \frac{1}{\left(m_1 + m_2 \right)} \cdot \left[\frac{\left(m_1 \bar{v}_1 + m_2 \bar{v}_2 \right)^2}{m_1 \bar{v}^2_1 + m_2 \bar{v}^2_2} \right] \rightarrow \textit{Unitless}$$

Inertial Locomotion Foreshortening Factor

4.9

$$\phi_{\text{Loco\_Motion}} = \left(\frac{\left|\overline{\text{v}}\right|^2_{\text{CM}}}{\left|\text{v}^2\right|_{\text{Iso}}} \right) = \left[1 - \frac{1}{\gamma^2_{Inertial}} \right] = \left[1 - \eta_{\text{Carnot}} \right]$$

$$\phi_{\text{Loco\_Motion}} = \frac{1}{\left[1 + \frac{1}{\gamma^2_{Inertial\_Iso}} \right]} = \frac{1}{\left[1 + \eta_{\text{Carnot\_Iso}} \right]}$$

$$\phi_{\text{Loco\_Motion}} = \left(\frac{\left|\overline{\text{v}}\right|^2_{\text{CM}}}{\left|\text{v}^2\right|_{\text{Iso}}} \right) = \frac{\gamma_{Heat}}{3} \cdot \left(\frac{\left|\overline{\text{v}}\right|^2_{\text{CM}}}{\text{c}^2_{\text{Sound}}} \right)$$

$$\phi_{\text{Loco\_Motion}} = -\left[\eta(+)_{\text{Carnot}} + \eta(-)_{\text{Carnot}} \right] \rightarrow Unitless$$

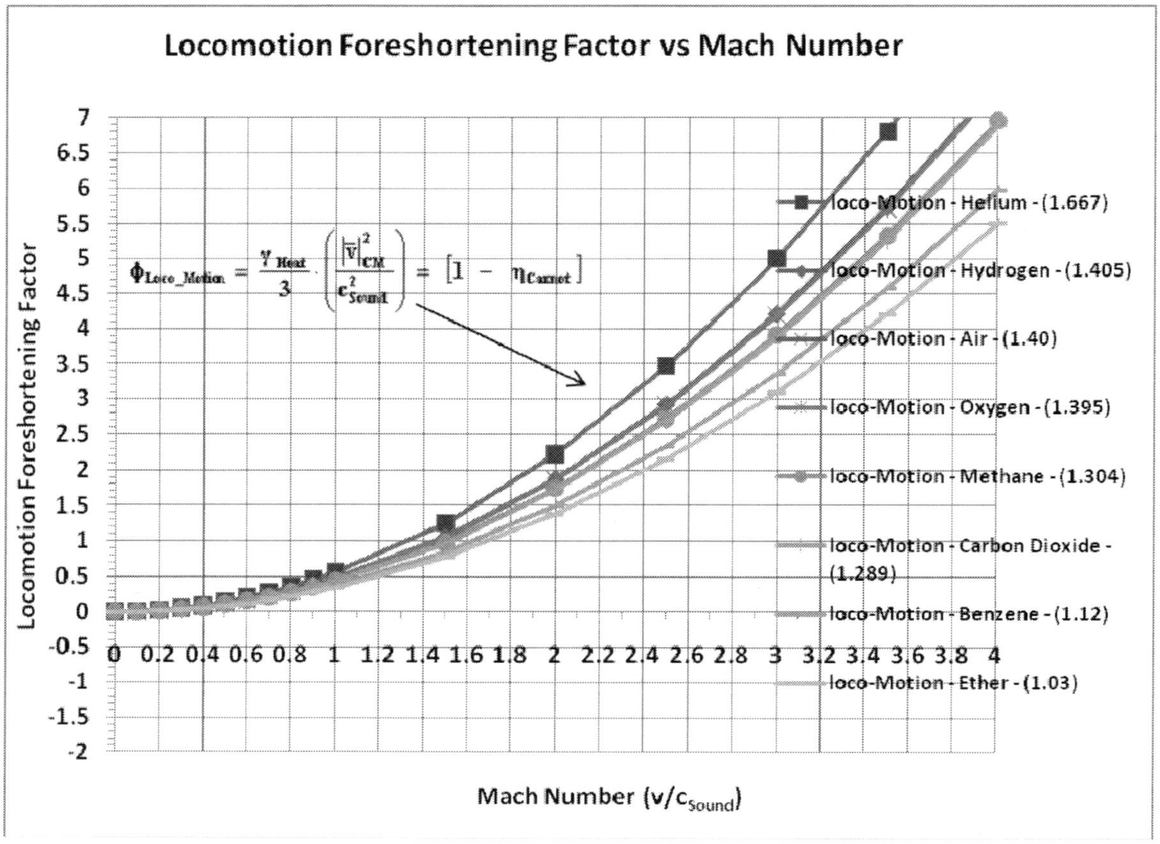

Figure 4.5: Locomotion Foreshortening Factor vs. Mach number for – various atomic elements.

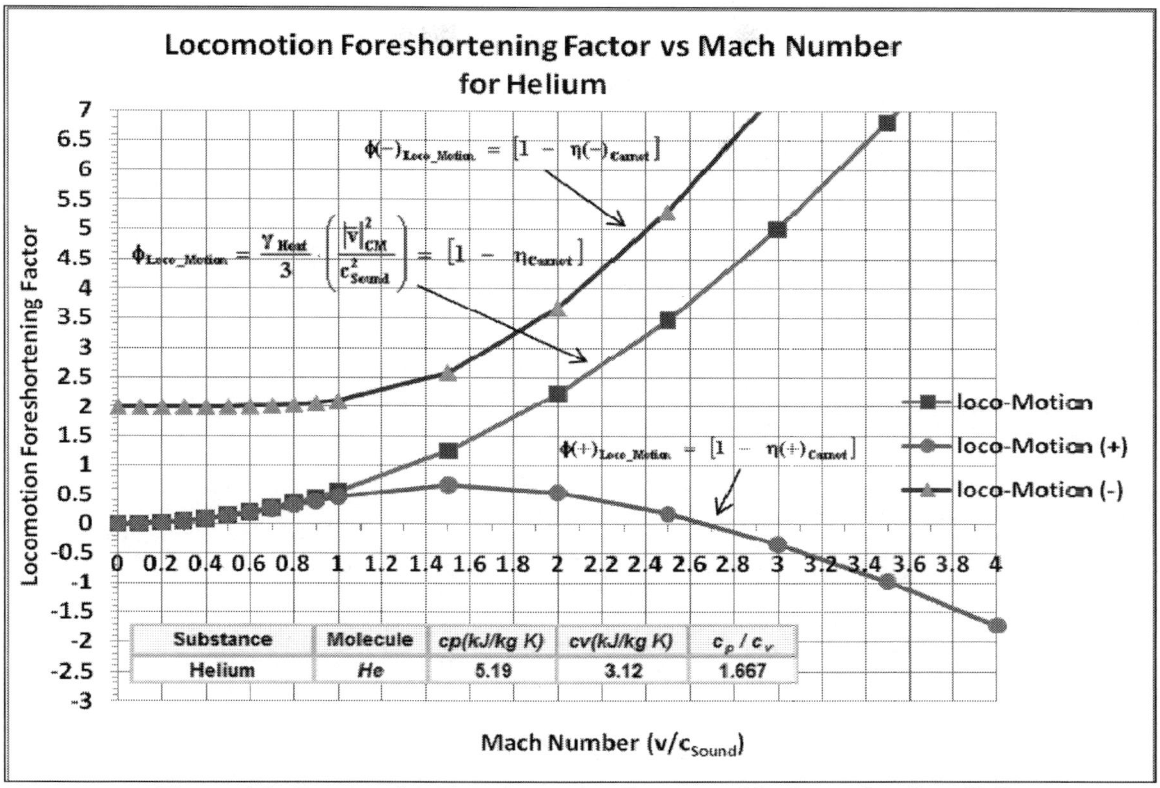

Figure 4.6: Locomotion Foreshortening Factor vs. Mach number for - Helium.

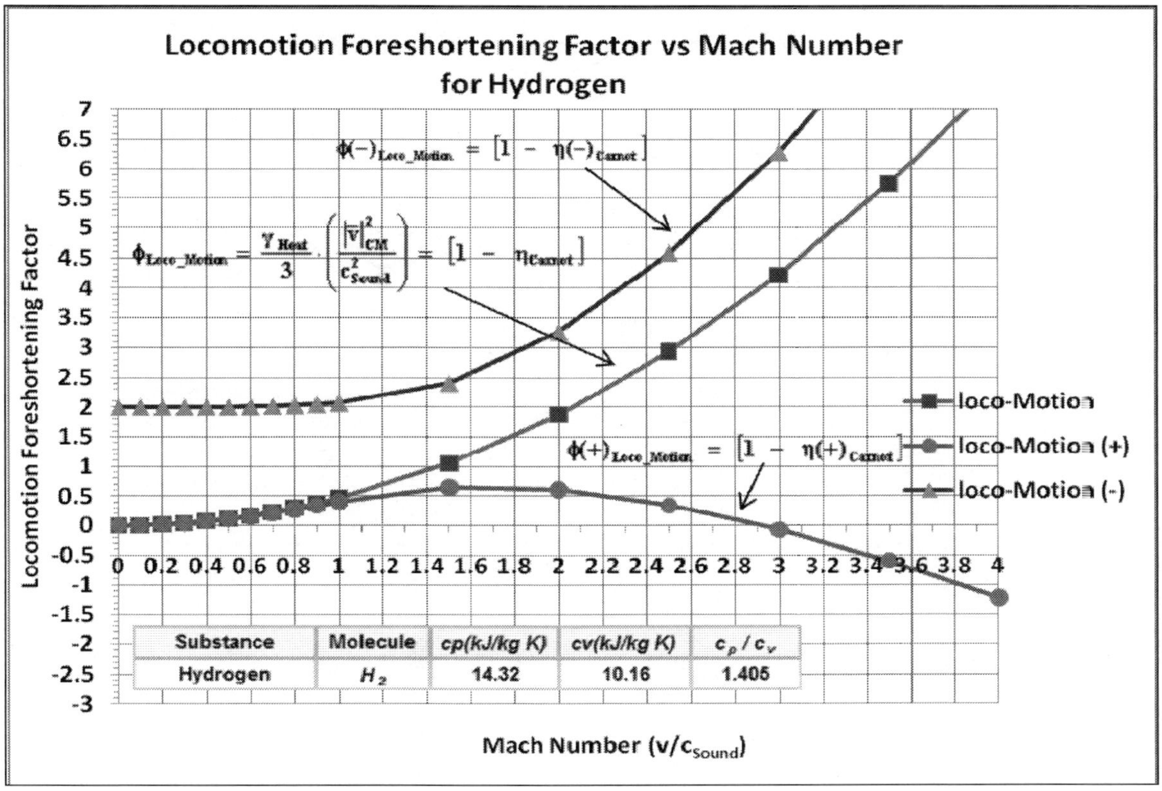

Figure 4.7: Locomotion Foreshortening Factor vs. Mach number for - Hydrogen.

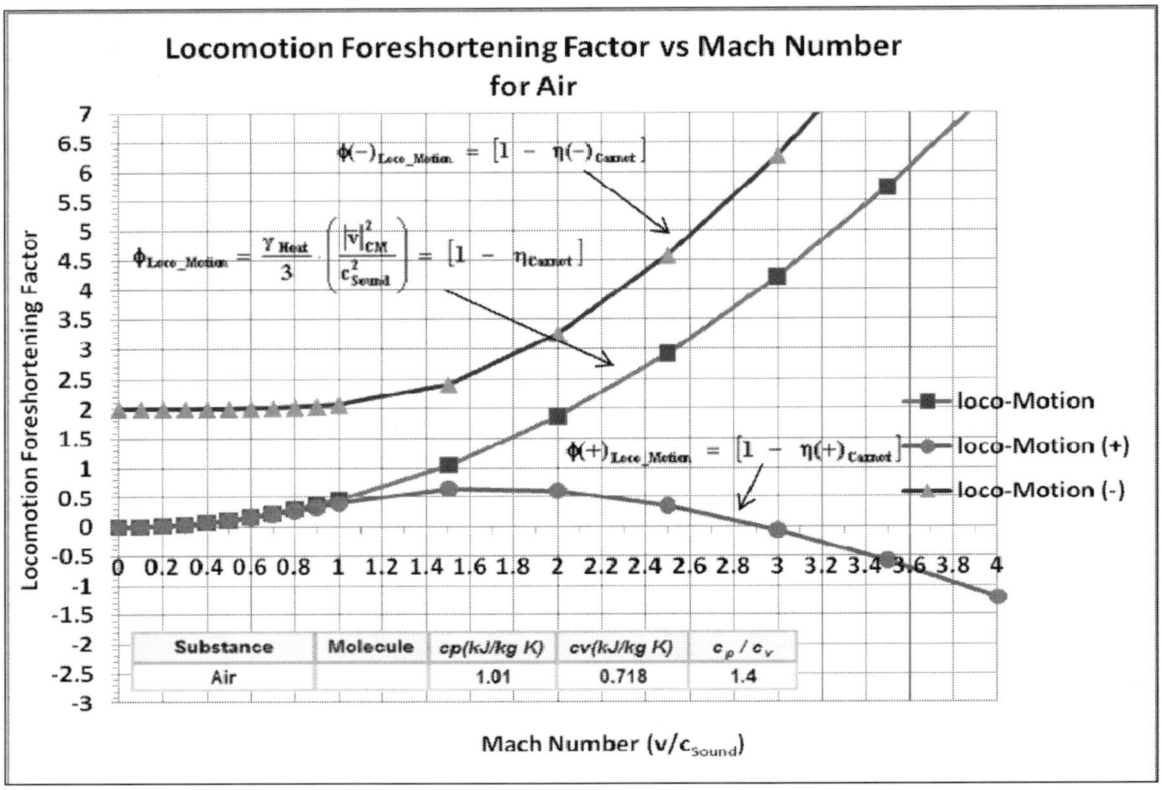

Figure 4.8: Locomotion Foreshortening Factor vs. Mach number for - Air.

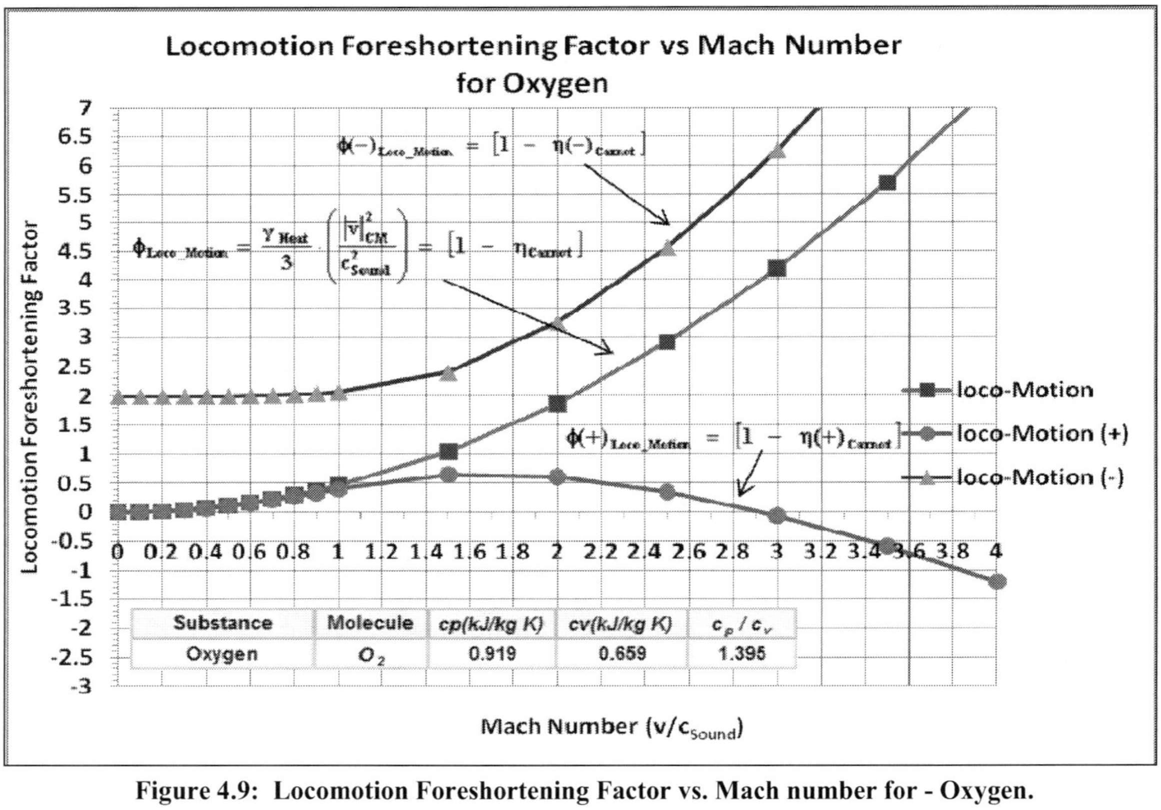

Figure 4.9: Locomotion Foreshortening Factor vs. Mach number for - Oxygen.

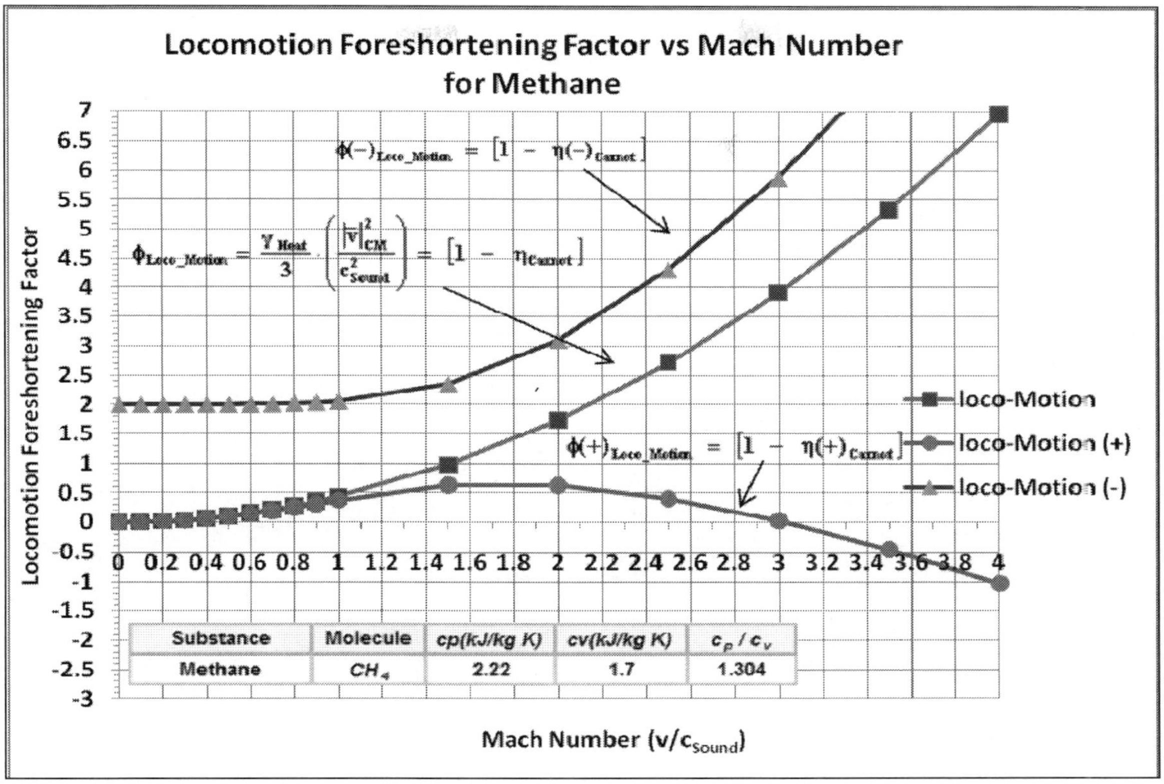

Figure 4.10: Locomotion Foreshortening Factor vs. Mach number for - Methane.

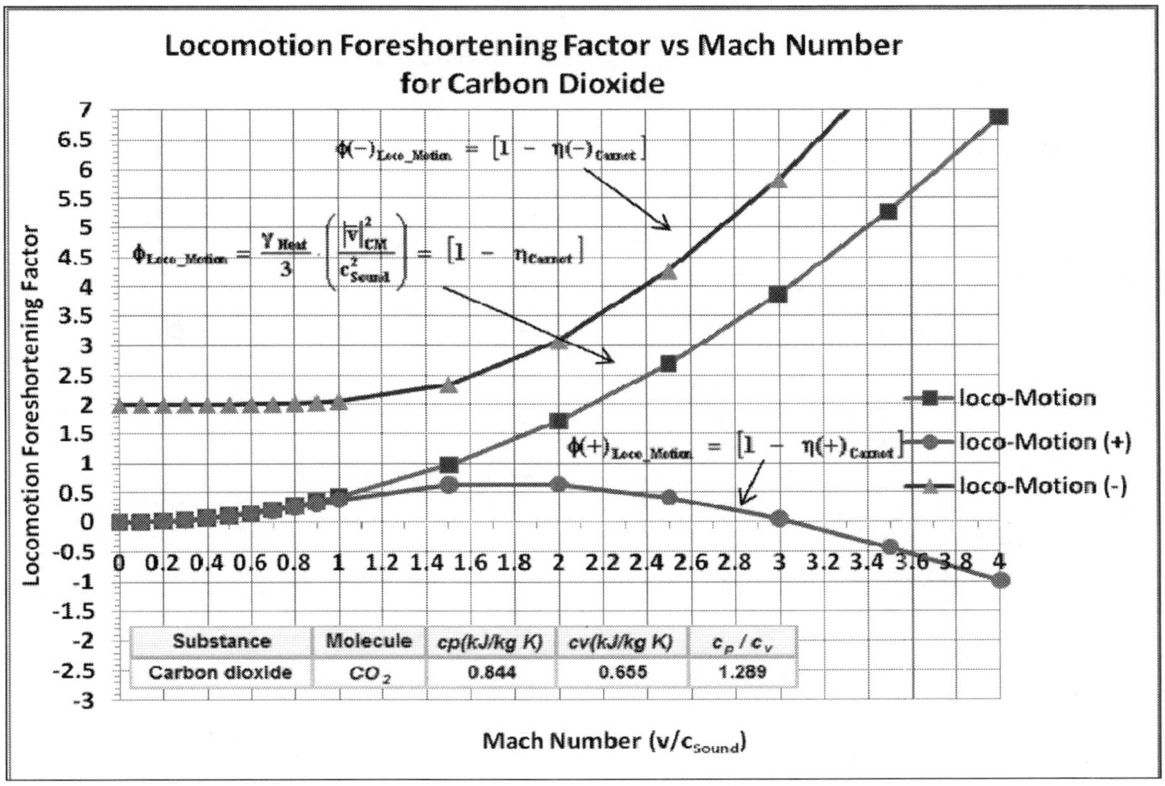

Figure 4.11: Locomotion Foreshortening Factor vs. Mach number for – Carbon Dioxide.

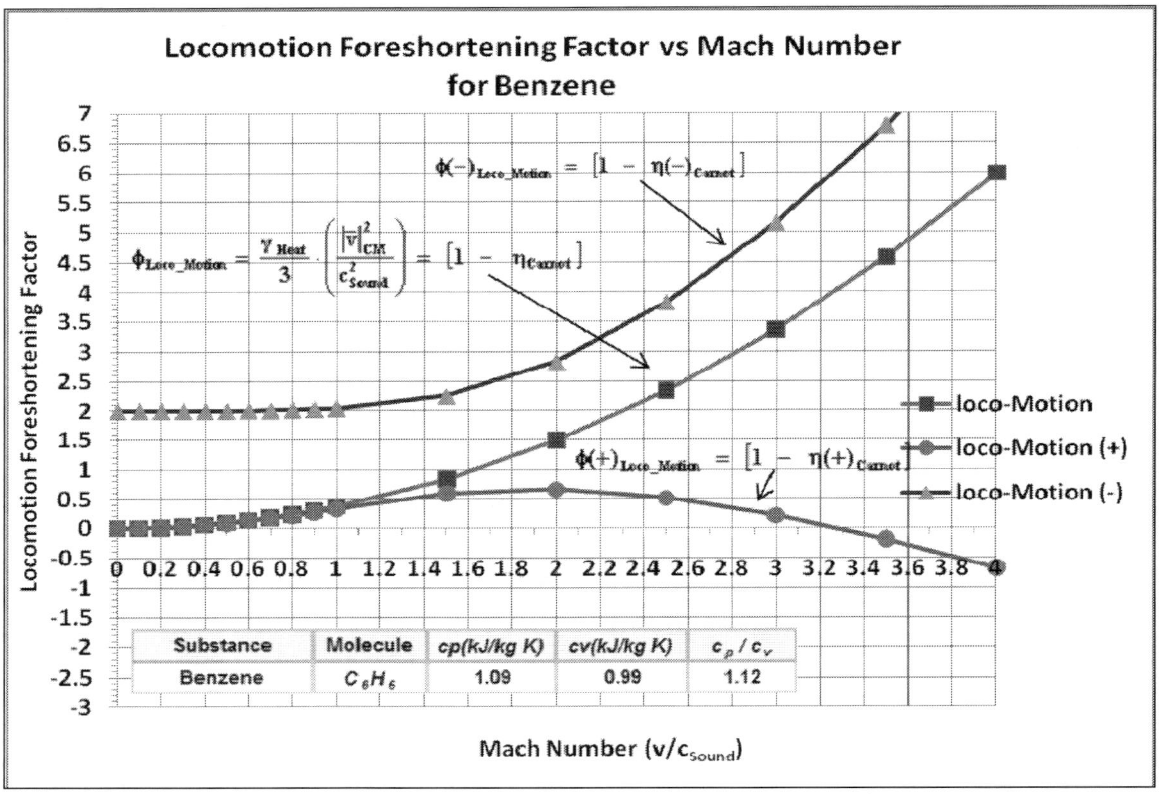

Figure 4.12: Locomotion Foreshortening Factor vs. Mach number for - Benzene.

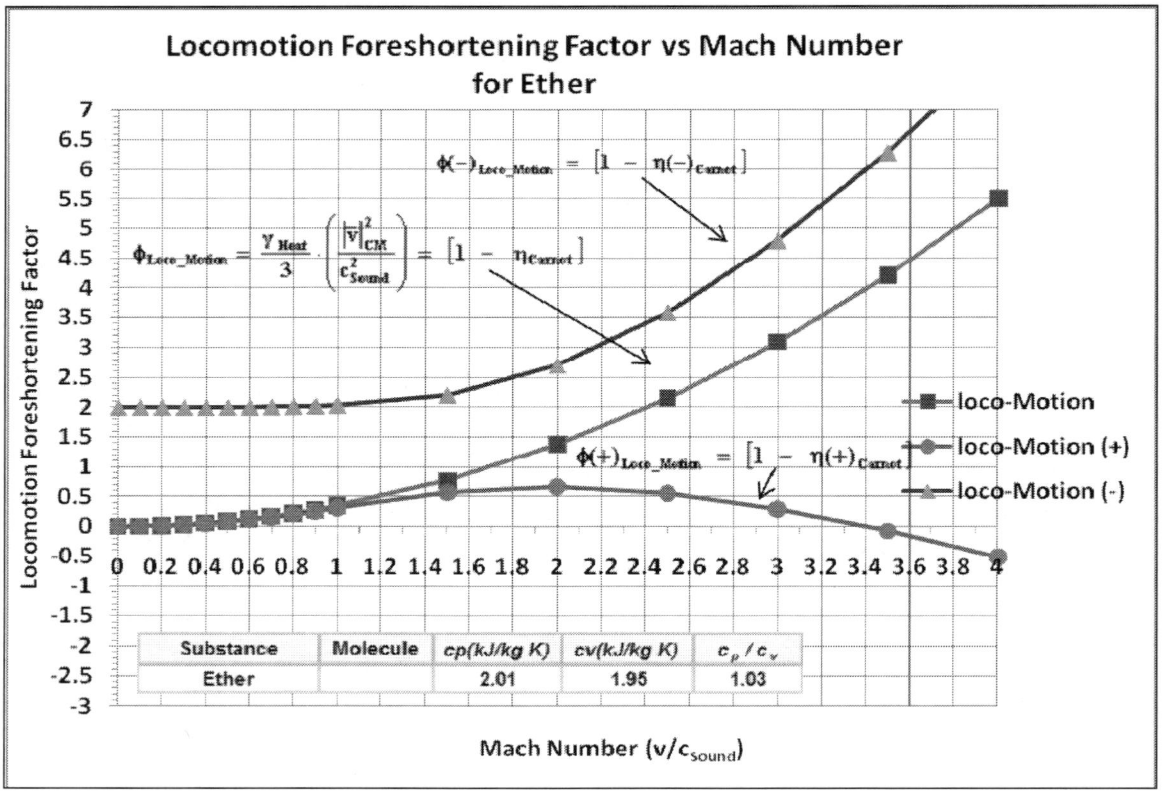

Figure 4.13: Locomotion Foreshortening Factor vs. Mach number for - Ether.

4.2 The Speed of Sound in an Atomic Substance Medium

Square of the Speed of Sound in an Atomic Material

4.10

$$c^2_{\text{Sound}} = \gamma_{Heat} \cdot \left(\frac{N \cdot k_B \cdot T_{emp}}{m_{\text{Net}}} \right) = \gamma_{Heat} \cdot \left(\frac{\left| v^2 \right|_{\text{Iso}}}{3} \right) \rightarrow m^2/s^2$$

$$c^2_{\text{Sound}} = \frac{\gamma_{Heat} \cdot \left| \bar{v} \right|^2_{\text{CM}}}{3 \cdot \left[1 - \frac{1}{\gamma^2_{Inertial}} \right]} = \frac{\gamma_{Heat}}{\left[1 - \eta_{\text{Carnot}} \right]} \cdot \left(\frac{\left| \bar{v} \right|^2_{\text{CM}}}{3} \right)$$

Speed of Sound in an Atomic Material

4.11

$$c_{\text{Sound}} = \left| \bar{v} \right|_{\text{CM}} \cdot \sqrt{\frac{\gamma_{Heat}}{3 \cdot \left[1 - \frac{1}{\gamma^2_{Inertial}} \right]}} = \left| \bar{v} \right|_{\text{CM}} \cdot \sqrt{\frac{\gamma_{Heat}}{3 \cdot \left[1 - \eta_{\text{Carnot}} \right]}}$$

$$c_{\text{Sound}} = \sqrt{\gamma_{Heat} \cdot \left(\frac{N \cdot k_B \cdot T_{emp}}{m_{\text{Net}}} \right)} = \sqrt{\gamma_{Heat} \cdot \left(\frac{\left| v^2 \right|_{\text{Iso}}}{3} \right)} \rightarrow m/s$$

4.12

Vector	
Magnitude	Units
Speed of Sound Propagation in Air	
The approximate speed of sound in dry (0% humidity) air, in meters per second, at temperatures near 0 °C, and a specific heat capacity index ($\gamma_{Heat} = 1.4$) can be calculated using the following equation: $$\mathbf{c_{Sound}} = \left(331.3\ \mathrm{m/s} \right) \cdot \sqrt{1 + \frac{T^{\circ}_{\text{Temp\_Celsius}}}{273.15^{\circ}\,\mathrm{C}}}$$	$\frac{\mathrm{m}}{\mathrm{s}}$

4.3 Aether Working Fluid Mass Ratio & per Temperature Change

Aether Kinematic Working Fluid Ratio

4.13

$$\frac{\left[\left(T_{emp} - T_{emp\_Ext}\right)\left(T_{emp\_Ext} + T_{emp}\right)\right]}{T_{emp} \cdot T_{emp\_Ext}} = \left[\frac{\left[m_{\text{Dark\_Matter}} - m_{\text{Inertia\_Mass}}\right]}{m_{\text{Net}}}\right]$$

$$\frac{\left[\left(\left|v^2\right|_{\text{Iso}} - \left|\overline{v}\right|^2_{\text{CM}}\right)\left(\left|\overline{v}\right|^2_{\text{CM}} + \left|v^2\right|_{\text{Iso}}\right)\right]}{\left|v^2\right|_{\text{Iso}} \cdot \left|\overline{v}\right|^2_{\text{CM}}} = \left[\frac{m_{\text{Dark\_Matter}}}{m_{\text{Net}}} - \frac{m_{\text{Inertia\_Mass}}}{m_{\text{Net}}}\right]$$

$$\frac{\left[\left(\left|v^4\right|_{\text{Iso}} - \left|\overline{v}\right|^4_{\text{CM}}\right)\right]}{\left|v^2\right|_{\text{Iso}} \cdot \left|\overline{v}\right|^2_{\text{CM}}} = \left[\frac{1}{\phi_{\text{Loco\_Motion}}} - \phi_{\text{Loco\_Motion}}\right]$$

$$\frac{\left[\left(T^2_{emp} - T^2_{emp\_Ext}\right)\right]}{T_{emp} \cdot T_{emp\_Ext}} = \left[\frac{1 - \phi^2_{\text{Loco\_Motion}}}{\phi_{\text{Loco\_Motion}}}\right] \rightarrow Unitless$$

$$\frac{\left[\left(T^2_{emp} - T^2_{emp\_Ext}\right)\right]}{T_{emp} \cdot T_{emp\_Ext}} = m_{\text{Net}} \cdot \left[\left(\frac{1}{m_{\text{Inertia\_Mass}}}\right) - \left(\frac{1}{m_{\text{Dark\_Matter}}}\right)\right]$$

$$\frac{\left[\begin{pmatrix} P_{Iso\_Pressure} \\ - q_{Dynamic\_Pressure} \end{pmatrix}\begin{pmatrix} q_{Dynamic\_Pressure} \\ + P_{Iso\_Pressure} \end{pmatrix}\right]}{P_{Iso\_Pressure} \cdot q_{Dynamic\_Pressure}} = m_{\text{Net}} \cdot \left(\frac{m_{\text{Dark\_Matter}} - m_{\text{Inertia\_Mass}}}{m_{\text{Inertia\_Mass}} \cdot m_{\text{Dark\_Matter}}}\right)$$

Aether Kinematic Working Fluid Ratio

4.14

$$\frac{\left[\left(E^2_{Net\_Iso\_Kinetic} - E^2_{Net\_Translational}\right)\right]}{E_{Net\_Iso\_Kinetic} \cdot E_{Net\_Translational}} = \left[\frac{1}{\phi_{Loco\_Motion}} - \phi_{Loco\_Motion}\right] \rightarrow \textit{Unitless}$$

$$\frac{\left[\begin{pmatrix} E_{Net\_Iso\_Kinetic} \\ - E_{Net\_Translational} \end{pmatrix} \cdot \begin{pmatrix} E_{Net\_Iso\_Kinetic} \\ + E_{Net\_Translational} \end{pmatrix}\right]}{E_{Net\_Iso\_Kinetic} \cdot E_{Net\_Translational}} = \left[\frac{\left[m_{Dark\_Matter} - m_{Inertia\_Mass}\right]}{m_{Net}}\right]$$

Starting with the **Aether Kinematic Working Fluid Ratio** equation derive the Inertial Locomotion Foreshortening Factor

4.15

$$\left[\phi^2_{\text{Loco\_Motion}} + \phi_{\text{Loco\_Motion}} \cdot \left[\frac{\left[\left(T^2_{emp} - T^2_{emp\_Ext}\right)\right]}{T_{emp} \cdot T_{emp\_Ext}}\right] - 1\right] = 0$$

Obtaining the Quadratic Equation Coefficients

$$a = 1$$

$$b = \left[\frac{\left[\left(T^2_{emp} - T^2_{emp\_Ext}\right)\right]}{T_{emp} \cdot T_{emp\_Ext}}\right] = \left[\frac{\left[\left(\left|\text{v}^4\right|_{\text{Iso}} - \left|\overline{\text{v}}\right|^4_{\text{CM}}\right)\right]}{\left|\text{v}^2\right|_{\text{Iso}} \cdot \left|\overline{\text{v}}\right|^2_{\text{CM}}}\right]$$

$$c = -1$$

$$0 = \left[a \cdot \phi^2_{\text{Loco\_Motion}} + b \cdot \phi_{\text{Loco\_Motion}} + c\right]$$

$$\phi_{\text{Loco\_Motion}} = \frac{-b \pm \sqrt{b^2 - 4 \cdot a \cdot c}}{2 \cdot a}$$

Net Inertial Mass of an Isolated System

4.16

$$\phi_{\text{Loco\_Motion}} = \left[\begin{array}{l} -\frac{1}{2}\left[\frac{\left[\left(T^2_{emp} - T^2_{emp\_Ext}\right)\right]}{T_{emp} \cdot T_{emp\_Ext}}\right] \\ \pm \frac{1}{2} \cdot \sqrt{\left[\frac{\left[\left(T^2_{emp} - T^2_{emp\_Ext}\right)\right]}{T_{emp} \cdot T_{emp\_Ext}}\right]^2 + 4} \end{array}\right]$$

$$\phi_{\text{Loco\_Motion}} = \left[\begin{array}{l} -\frac{1}{2}\left[\frac{\left[\left(\left|\text{v}^4\right|_{\text{Iso}} - \left|\overline{\text{v}}\right|^4_{\text{CM}}\right)\right]}{\left|\text{v}^2\right|_{\text{Iso}} \cdot \left|\overline{\text{v}}\right|^2_{\text{CM}}}\right] \\ \pm \frac{1}{2} \cdot \sqrt{\left[\frac{\left[\left(\left|\text{v}^4\right|_{\text{Iso}} - \left|\overline{\text{v}}\right|^4_{\text{CM}}\right)\right]}{\left|\text{v}^2\right|_{\text{Iso}} \cdot \left|\overline{\text{v}}\right|^2_{\text{CM}}}\right]^2 + 4} \end{array}\right]$$

$$\phi_{\text{Loco\_Motion}} = \left[-\frac{1}{2}A \pm \frac{1}{2} \cdot \sqrt{A^2 + 4}\right] \rightarrow \textit{Unitless}$$

Aether Kinematic Working Fluid Mass per Temperature Change

4.17

$$\mathrm{m}_{\mathrm{Net}} \cdot \left[\frac{T_{emp\_Ext} + T_{emp}}{T_{emp} \cdot T_{emp\_Ext}} \right] = \left[\frac{\mathrm{m}_{\mathrm{Dark\_Matter}} - \mathrm{m}_{\mathrm{Inertia\_Mass}}}{T_{emp} - T_{emp\_Ext}} \right]$$

$$\mathrm{m}_{\mathrm{Net}} \cdot \left[\frac{T_{emp\_Ext} + T_{emp}}{T_{emp} \cdot T_{emp\_Ext}} \right] = \left[\frac{\Delta \mathrm{m}_{\mathrm{Aniso\_WFM}}}{T_{emp\_Ext}} \right]$$

$$\mathrm{m}_{\mathrm{Net}} \cdot \left[\frac{T_{emp\_Ext} + T_{emp}}{T_{emp} \cdot T_{emp\_Ext}} \right] = \left[\frac{\Delta \mathrm{m}_{\mathrm{Iso\_WFM}}}{T_{emp}} \right] \rightarrow {kg}/{K}$$

Anisotropic Aether Kinematic Working Fluid Mass

4.18

$$\Delta \mathrm{m}_{\mathrm{Aniso\_WFM}} = \left[\frac{\left[\mathrm{m}_{\mathrm{Dark\_Matter}} - \mathrm{m}_{\mathrm{Inertia\_Mass}} \right]}{\left(T_{emp\_Ext} + T_{emp} \right)} \right] \cdot T_{emp\_Ext} \rightarrow kg$$

Isotropic Aether Kinematic Working Fluid Mass

4.19

$$\Delta \mathrm{m}_{\mathrm{Iso\_WFM}} = \left[\frac{\left[\mathrm{m}_{\mathrm{Dark\_Matter}} - \mathrm{m}_{\mathrm{Inertia\_Mass}} \right]}{\left(T_{emp\_Ext} + T_{emp} \right)} \right] \cdot T_{emp} \rightarrow kg$$

Chapter 5

The Special Theory of Thermodynamics

(Carnot's Thermal "Anisotropy") (Engine Efficiency)

5.1 Carnot's Thermal Engine "Anisotropy" Efficiency Factor

The Carnot's Thermal Engine "Anisotropy" Efficiency Factor is a measure of an isolated system mass body that is used to determine the cooling of a dynamic system mass body and a heating of the atmosphere surroundings medium environment.

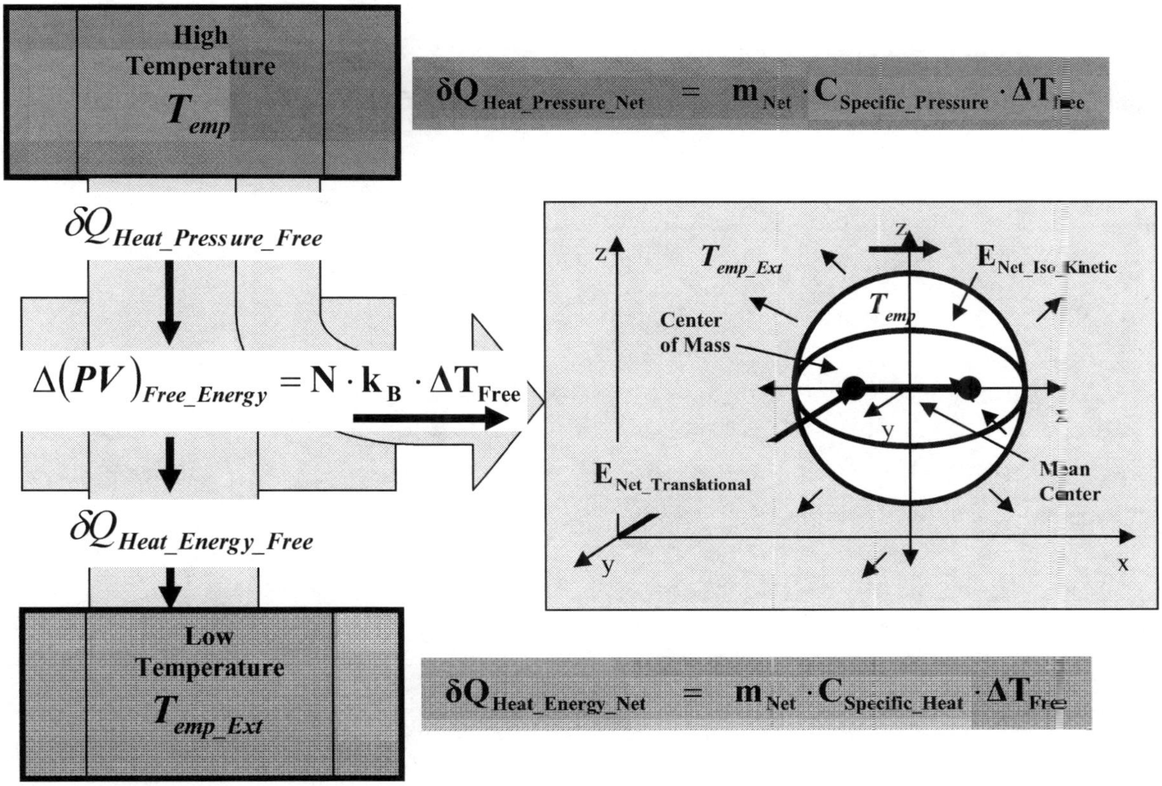

Figure 5.1: The Carnot "Anisotropy Engine is total Aether Kinematic Work Energy transfer which is equal to the difference in the high thermodynamic potential subtracted from the low thermodynamic potential.

Carnot's theorem sets essential limitations on the yield of a cyclic heat engine such as steam engines, internal combustion engines, atoms, and matter motion, which operate on the **Anisotropic Carnot Cycle**; given by the following **Inertial Locomotion Foreshortening Factor** condition.

5.1

$$\phi_{Loco\_Motion} = \left(\frac{\left|\overline{V}\right|^2_{CM}}{\left|V^2\right|_{Iso}} \right) \leq 1$$

Heat Engines that operate on the Carnot cycle can extract only a certain proportion of mechanical energy from the heat of the working fluid, and this maximal amount is realized by the ***Ideal Anisotropic Carnot Heat Engine***.

The Ideal Anisotropic Carnot Heat Engine of an isolated system mass body is divided into two heat energy transfer reservoirs, an anisotropy reservoir and an isotropy reservoir which can be considered initial and final states of the system. The two reservoirs also represent higher potential and lower potential, in which energy only flows from high potential to low potential.

For the ***Ideal Anisotropic Carnot Heat Engine*** the initial high potential state is the Internal Isotropy in which the Internal Absolute Temperature (T_{emp}), Internal Isotropic Omni-directional Kinetic Energy ($\mathrm{E}_{\mathrm{Net\_Iso\_Kinetic}}$), Square of the Center of Isotropy Rectilinear Momentum ($\overline{\mathrm{p}}^2_{\mathrm{Net\_Iso}}$), and the Center of Mass Isotropy Velocity Inertia ($\left|\mathrm{v}^2\right|_{\mathrm{Iso}}$) are all considered to be the higher potential and temperature heat source of the system.

For the ***Ideal Anisotropic Carnot Heat Engine*** the final low potential state is the External Anisotropy in which the is the Aerodynamic Temperature (T_{emp_Ext}), External Translational Kinetic Energy ($\mathrm{E}_{\mathrm{Net\_Translational}}$), Square of the Center of Mass Rectilinear Momentum ($\overline{\mathrm{p}}^2_{\mathrm{Inertial\_Net}}$), and the Center of Mass Squared Velocity Inertia ($\left|\overline{\mathrm{v}}\right|^2_{\mathrm{CM}}$), are all considered be lower potential and temperature heat sink of the system.

Thus the following conditions hold for the Anisotropic Carnot Heat Engine,

5.2

$$T_{emp\_Ext} \leq T_{emp}$$

$$\overline{\mathrm{p}}^2_{\mathrm{Inertial\_Net}} \leq \overline{\mathrm{p}}^2_{\mathrm{Net\_Iso}}$$

$$\mathrm{E}_{\mathrm{Net\_Translational}} \leq \mathrm{E}_{\mathrm{Net\_Iso\_Kinetic}}$$

$$\left|\overline{\mathrm{v}}\right|^2_{\mathrm{CM}} \leq \left|\mathrm{v}^2\right|_{\mathrm{Iso}}$$

$$\phi_{\mathrm{Loco\_Motion}} = \left(\frac{\left|\overline{\mathrm{v}}\right|^2_{\mathrm{CM}}}{\left|\mathrm{v}^2\right|_{\mathrm{Iso}}} \right) \leq 1$$

The **Carnot Thermal Engine "Anisotropy" Efficiency Factor** ($\eta_{Carnot} = \frac{1}{\gamma^2_{Inertial}}$) is an observer invariant (covariant) unit-less ratio which can be described as the percentage of the anisotropy energy transfer.

This anisotropy energy transfer is a measure of the quantity of Internal Isotropic Omni-directional center of isotropy motion that is converted into the External Anisotropic Translational center of mass motion.

The Carnot Thermal Engine "Anisotropy" Efficiency Factor being a ratio of the Isotropy divided by the Anisotropy of an isolated system is also a dimension invariant ratio. The Anisotropy represents the rectilinear translational first order motions of the closed system, and the Isotropy represents the Omni-directional second order motions of a closed system.

What do I mean when I state that the Anisotropy is a first order motion? The anisotropy naturally defines one dimensional linear space and time through the **net anisotropic translational rectilinear momentum conservation**.

5.3

$$E_{Net\_Translational} = \frac{\bar{p}^2_{Inertial\_Net}}{2 \cdot m_{Net}}$$

$$E_{Net\_Translational} = \frac{1}{2} m_{Net} \cdot \left|\bar{v}\right|^2_{CM} = \frac{3}{2} \cdot N \cdot k_B \cdot T_{emp\_Ext} = \frac{\begin{pmatrix} m_1\bar{v}_1 + m_2\bar{v}_2 \\ + m_3\bar{v}_3 + \cdots m_N\bar{v}_N \end{pmatrix}^2}{2 \cdot m_{Net}}$$

What do I mean when I state that the Isotropy is a second order motion? The isotropy naturally defines area or two dimensional space and squared time through the **net Isotropic Omni-directional kinetic energy conservation**.

5.4

$$E_{Net\_Iso\_Kinetic} = \frac{\bar{p}^2_{Net\_Iso}}{2 \cdot m_{Net}}$$

$$E_{Net\_Iso\_Kinetic} = \frac{1}{2} \cdot m_{Net} \cdot \left|v^2\right|_{Iso} = \frac{3}{2} \cdot N \cdot k_B \cdot T_{emp} = \begin{bmatrix} \frac{m_1\bar{v}_1^2}{2} + \frac{m_2\bar{v}_2^2}{2} \\ + \frac{m_3\bar{v}_3^2}{2} + \cdots \frac{m_N\bar{v}_N^2}{2} \end{bmatrix}$$

Definition 5.1: The **Carnot Thermal Engine "Anisotropy" Efficiency Factor** ($\eta_{\text{Carnot}} = \left[1 - \left(\frac{T_{emp\_Ext}}{T_{emp}}\right)\right]$) is the measure of the efficiency of a heat engine doing work as it transfers heat energy from a higher temperature Isotropy reservoir to a lower temperature Anisotropy reservoir of an isolated system mass body; and is equal to one subtracted from the ratio of the Final **Low Temperature Reservoir – External Anisotropic Aerodynamic Temperature** (T_{emp_Ext}) divided by the Initial **High Temperature Reservoir – Internal Isotropic Absolute Temperature** (T_{emp}).

Definition 5.2: The **Carnot Thermal Engine "Anisotropy" Efficiency Factor** ($\eta_{\text{Carnot}} = \left[1 - \left(\frac{\text{E}_{\text{Net\_Translational}}}{\text{E}_{\text{Net\_Iso\_Kinetic}}}\right)\right] = \left[1 - \left(\frac{\left|\overline{\text{v}}\right|^2_{\text{CM}}}{\left|\text{v}^2\right|_{\text{Iso}}}\right)\right]$) is the measure of the efficiency of a heat engine doing work as it transfers heat energy from a higher temperature Isotropy reservoir to a lower temperature Anisotropy reservoir of an isolated system mass body; and is equal to one subtracted from the ratio of the "final" **External Anisotropic Translational Kinetic Energy** ($\text{E}_{\text{Net\_Translational}}$) divided by the "initial" **Internal Isotropic Omni-directional Kinetic Energy** ($\text{E}_{\text{Net\_Iso\_Kinetic}}$).

Definition 5.3: **Carnot Thermal Engine "Anisotropy" Efficiency Factor** ($\eta_{\text{Carnot}} = \left[\frac{\text{E}_{\text{Net\_Iso\_Kinetic}} - \text{E}_{\text{Net\_Translational}}}{\text{E}_{\text{Net\_Iso\_Kinetic}}}\right]$) is an dimension invariant unit-less ratio that is the measure of the percentage of the energy of Isotropy that is being converted into the energy of Anisotropy; and when defined in the Center of mass/Momentum frame or the Center of Isotropy frame, is a ratio equal to **Internal Isotropic Omni-directional Kinetic Energy** ($\text{E}_{\text{Net\_Iso\_Kinetic}}$) subtracted from the **External Anisotropic Translational Kinetic Energy** ($\text{E}_{\text{Net\_Translational}}$) divided by the **Internal Isotropic Omni-directional Kinetic Energy** of an isolated system body.

Carnot Thermal Engine "Anisotropy" Efficiency Factor

5.5

$$\eta_{\text{Carnot}} = \frac{1}{\gamma^2_{Inertial}} = \left[\frac{W_{\text{Free\_Work\_Energy}}}{E_{\text{Net\_Iso\_Kinetic}}}\right] = \left(1 - \frac{|\overline{v}|^2_{\text{CM}}}{|v^2|_{\text{Iso}}}\right) \rightarrow Unitless$$

$$\eta_{\text{Carnot}} = \left[\frac{E_{\text{Net\_Iso\_Kinetic}} - E_{\text{Net\_Translational}}}{E_{\text{Net\_Iso\_Kinetic}}}\right] = \frac{\left[|v^2|_{\text{Iso}} - |\overline{v}|^2_{\text{CM}}\right]}{|v^2|_{\text{Iso}}}$$

$$\eta_{\text{Carnot}} = \frac{\Delta\overline{p}^2_{\text{Free\_Momentum}}}{\overline{p}^2_{\text{Net\_Iso}}} = \left(1 - \frac{|\overline{v}|^2_{\text{CM}}}{|v^2|_{\text{Iso}}}\right)$$

$$\eta_{\text{Carnot}} = \frac{\left[\overline{p}^2_{\text{Net\_Iso}} - \overline{p}^2_{\text{Inertial\_Net}}\right]}{\overline{p}^2_{\text{Net\_Iso}}} = \left(1 - \frac{\gamma_{Heat}}{3}\cdot\left(\frac{|\overline{v}|^2_{\text{CM}}}{c^2_{\text{Sound}}}\right)\right)$$

$$\eta_{\text{Carnot}} = \frac{\left[P_{Iso\_Pressure} - q_{Dynamic\_Pressure}\right]}{P_{Iso\_Pressure}} = \left(1 - \phi_{\text{Loco\_Motion}}\right)$$

$$\eta_{\text{Carnot}} = \frac{1}{\gamma^2_{Inertial}} = \left(1 - \frac{|\overline{v}|^2_{\text{CM}}}{|v^2|_{\text{Iso}}}\right) = \left(1 - \phi_{\text{Loco\_Motion}}\right)$$

Carnot Thermal Engine "Anisotropy" Efficiency Factor

5.6

$$\eta_{\text{Carnot}} = \frac{1}{\gamma^2_{Inertial}} = \left[\frac{W_{\text{Free\_Work\_Energy}}}{E_{\text{Net\_Iso\_Kinetic}}}\right] = \left(1 - \frac{|\bar{v}|^2_{\text{CM}}}{|v^2|_{\text{Iso}}}\right) \rightarrow \textit{Unitless}$$

$$\eta_{\text{Carnot}} = \left(\frac{\Delta T_{Free}}{T_{emp}}\right) = \left(\frac{T_{emp} - T_{emp\_Ext}}{T_{emp}}\right) = \left[1 - \phi_{\text{Loco\_Motion}}\right]$$

$$\eta_{\text{Carnot}} = \left(\frac{\Delta T_{Free}}{T_{emp}}\right) = \left[1 - \left(\frac{T_{emp\_Ext}}{T_{emp}}\right)\right]$$

$$\eta_{\text{Carnot}} = \frac{1}{\gamma^2_{Inertial}} = \left(1 - \frac{\gamma_{Heat}}{3} \cdot \left(\frac{|\bar{v}|^2_{\text{CM}}}{c^2_{\text{Sound}}}\right)\right)$$

$$\eta_{\text{Carnot}} = \frac{1}{\gamma^2_{Inertial}} = \left(1 - \frac{m_{\text{Net}}|\bar{v}|^2_{\text{CM}}}{3 \cdot \left(N \cdot k_B \cdot T_{emp}\right)}\right)$$

$$\eta_{\text{Carnot}} = \left[1 - \left(\frac{m_{\text{Net}}}{m_{\text{Dark\_Matter}}}\right)\right] = \left[1 - \left(\frac{\bar{p}^2_{\text{Inertial\_Net}}}{\bar{p}^2_{\text{Net\_Iso}}}\right)\right]$$

$$\eta_{\text{Carnot}} = \frac{1}{\gamma^2_{Inertial}} = \eta_{\text{Carnot\_Iso}} \cdot \left(\frac{\bar{p}^2_{\text{Inertial\_Net}}}{\bar{p}^2_{\text{Net\_Iso}}}\right) = \frac{1}{\gamma^2_{Inertial\_Iso}} \cdot \left(\frac{\bar{p}^2_{\text{Inertial\_Net}}}{\bar{p}^2_{\text{Net\_Iso}}}\right)$$

Carnot Thermal Engine "Anisotropy" Efficiency Factor

5.7

$$\eta_{\text{Carnot}} = \left(\frac{\Delta m_{\text{Aniso\_WFM}}}{m_{\text{Net}}}\right) = \left[1 - \phi_{\text{Loco\_Motion}}\right] \rightarrow \textit{Unitless}$$

$$\eta_{\text{Carnot}} = \left(\frac{\Delta m_{\text{Aniso\_WFM}}}{m_{\text{Net}}}\right) = \left[1 - \phi_{\text{Loco\_Motion}}\right] = \left(\frac{\Delta(PV)_{\text{Free\_Energy}}}{\text{N} \cdot \text{k}_{\text{B}} \cdot T_{emp}}\right)$$

$$\eta_{\text{Carnot}} = \left[1 - \phi_{\text{Loco\_Motion}}\right] = \left[1 - \left(\frac{m_{\text{Inertia\_Mass}}}{m_{\text{Net}}}\right)\right]$$

$$\eta_{\text{Carnot}} = \left[1 - \phi_{\text{Loco\_Motion}}\right] = \left(\frac{\Delta(PV)_{\text{Free\_Energy}}}{\left(1 - \dfrac{1}{\gamma_{Heat}}\right) \cdot m_{\text{Net}} \cdot C_{Specific\_Pressure} \cdot T_{emp}}\right)$$

$$\eta_{\text{Carnot}} = \left[1 - \phi_{\text{Loco\_Motion}}\right] = \left[1 - \left(\frac{m_{\text{Net}}}{m_{\text{Dark\_Matter}}}\right)\right]$$

$$\eta_{\text{Carnot}} = \left[1 - \phi_{\text{Loco\_Motion}}\right] = \left[1 - \frac{1}{\left[1 + \left(\dfrac{\Delta m_{\text{Iso\_WFM}}}{m_{\text{Net}}}\right)\right]}\right]$$

Carnot Thermal Engine "Anisotropy" Efficiency Factor

5.8

$$\eta_{\text{Carnot}} = \left[1 - \phi_{\text{Loco\_Motion}}\right] = \left[1 - \frac{1}{\left[1 + \left(\frac{\Delta(PV)_{\text{Free\_Energy}}}{\mathrm{N}\cdot\mathrm{k}_{\mathrm{B}}\cdot T_{emp\_Ext}}\right)\right]}\right]$$

$$\eta_{\text{Carnot}} = \left[1 - \frac{1}{\left[1 + \left(\frac{\Delta(PV)_{\text{Free\_Energy}}}{\left(1 - \frac{1}{\gamma_{Heat}}\right)\cdot\mathrm{m}_{\text{Net}}\cdot C_{Specific\_Pressure}\cdot T_{emp\_Ext}}\right)\right]}\right]$$

$$\eta_{\text{Carnot}} = \frac{1}{\gamma^2_{Inertial}} = \eta_{\text{Carnot\_Iso}}\cdot\left(\frac{\overline{\mathrm{p}}^2_{\text{Inertial\_Net}}}{\overline{\mathrm{p}}^2_{\text{Net\_Iso}}}\right) = \frac{1}{\gamma^2_{Inertial\_Iso}}\cdot\left(\frac{\overline{\mathrm{p}}^2_{\text{Inertial\_Net}}}{\overline{\mathrm{p}}^2_{\text{Net\_Iso}}}\right)$$

$$\eta_{\text{Carnot}} = \frac{\eta_{\text{Carnot\_Iso}}}{\left[1 + \eta_{\text{Carnot\_Iso}}\right]} \rightarrow \textit{Unitless}$$

Carnot Thermal Engine "Anisotropy" Efficiency Factor

5.9

$$\eta_{\text{Carnot}} = \left[1 - \phi_{\text{Loco\_Motion}}\right] = \left[1 - \left(\frac{\left|\overline{\text{v}}\right|^2_{\text{CM}}}{\left|\text{v}^2\right|_{\text{Iso}}}\right)\right] \rightarrow Unitless$$

$$\eta_{\text{Carnot}} = \left[1 - \phi_{\text{Loco\_Motion}}\right] = \left[1 - \frac{\gamma_{Heat}}{3} \cdot \left(\frac{\left|\overline{\text{v}}\right|^2_{\text{CM}}}{\text{c}^2_{\text{Sound}}}\right)\right]$$

$$\eta_{\text{Carnot}} = \left[1 - \phi_{\text{Loco\_Motion}}\right] = \left[1 - \left(\frac{T_{emp\_Ext}}{T_{emp}}\right)\right]$$

$$\eta_{\text{Carnot}} = \left[1 - \phi_{\text{Loco\_Motion}}\right] = \left[1 - \left(\frac{\text{E}_{\text{Net\_Translational}}}{\text{E}_{\text{Net\_Iso\_Kinetic}}}\right)\right]$$

$$\eta_{\text{Carnot}} = \left[1 - \phi_{\text{Loco\_Motion}}\right] = \left[1 - \left(\frac{q_{Dynamic\_\text{Pr}essure}}{P_{Iso\_\text{Pr}essure}}\right)\right]$$

$$\eta_{\text{Carnot}} = \left[1 - \phi_{\text{Loco\_Motion}}\right] = \left[1 - \left(\frac{\overline{\text{p}}^2_{\text{Inertial\_Net}}}{\overline{\text{p}}^2_{\text{Net\_Iso}}}\right)\right]$$

Carnot Thermal Engine "Anisotropy" Efficiency Factor

5.10

$$\eta_{\text{Carnot}} = \frac{1}{\gamma^2_{Inertial}} = \left[\frac{W_{\text{Free\_Work\_Energy}}}{E_{\text{Net\_Iso\_Kinetic}}}\right] \rightarrow \textit{Unitless}$$

$$\eta_{\text{Carnot}} = \left[\frac{\left[E_{\text{Net\_Iso\_Kinetic}} - \left[\frac{E'_{\text{Relativistic\_Energy}}}{2}\cdot\left(\frac{m'_{\text{Net\_Rel}}}{m_{\text{Net}}}\right) - \frac{E_{\text{Rest\_Energy}}}{2}\right]\cdot\left(1 - \frac{|\overline{v}|^2_{\text{CM}}}{c^2_{\text{Light}}}\right)\right]}{E_{\text{Net\_Iso\_Kinetic}}}\right]$$

$$\eta_{\text{Carnot}} = \left[1 - \left[\frac{E'_{\text{Relativistic\_Energy}}}{2\cdot E_{\text{Net\_Iso\_Kinetic}}}\cdot\left(\frac{m'_{\text{Net\_Rel}}}{m_{\text{Net}}}\right) - \frac{E_{\text{Rest\_Energy}}}{2\cdot E_{\text{Net\_Iso\_Kinetic}}}\right]\cdot\left(1 - \frac{|\overline{v}|^2_{\text{CM}}}{c^2_{\text{Light}}}\right)\right]$$

$$\eta_{\text{Carnot}} = \left[1 - \left[\frac{\left(\frac{c^2_{\text{Light}}}{\left|v^2\right|_{\text{Iso}}}\right)}{\left(1 - \frac{|\overline{v}|^2_{\text{CM}}}{c^2_{\text{Light}}}\right)} - \left(\frac{c^2_{\text{Light}}}{\left|v^2\right|_{\text{Iso}}}\right)\right]\cdot\left(1 - \frac{|\overline{v}|^2_{\text{CM}}}{c^2_{\text{Light}}}\right)\right]$$

$$\eta_{\text{Carnot}} = \left[1 - \left(\frac{c^2_{\text{Light}}}{\left|v^2\right|_{\text{Iso}}}\right) + \left(\frac{c^2_{\text{Light}}}{\left|v^2\right|_{\text{Iso}}}\right)\cdot\left(1 - \frac{|\overline{v}|^2_{\text{CM}}}{c^2_{\text{Light}}}\right)\right]$$

Carnot Thermal Engine "Anisotropy" Efficiency Factor

5.11

$$\eta_{Carnot} = \frac{1}{\gamma^2_{Inertial}} = \left[\frac{W_{Free\_Work\_Energy}}{E_{Net\_Iso\_Kinetic}}\right] \rightarrow \textit{Unitless}$$

$$\eta_{Carnot} = \left(1 - \frac{|\overline{v}|^2_{CM}}{|v^2|_{Iso}}\right) = \left(1 - \phi_{Loco\_Motion}\right)$$

$$\eta_{Carnot} = \frac{1}{\gamma^2_{Inertial}} = \left(1 - \frac{|\overline{v}|^2_{CM}}{|v^2|_{Iso}}\right) = \left(1 - \frac{\gamma_{Heat}}{3}\cdot\left(\frac{|\overline{v}|^2_{CM}}{c^2_{Sound}}\right)\right)$$

$$\eta_{Carnot} = \left[-\frac{1}{2}\cdot\left(\frac{E_{Net\_Translational}}{E_{Net\_Iso\_Kinetic}}\right) \pm \sqrt{1 - \left(\frac{E_{Net\_Translational}}{E_{Net\_Iso\_Kinetic}}\right) + \frac{1}{4}\cdot\left(\frac{E_{Net\_Translational}}{E_{Net\_Iso\_Kinetic}}\right)^2}\right]$$

$$\eta_{Carnot} = \left[-\frac{1}{2}\cdot\left(\frac{|\overline{v}|^2_{CM}}{|v^2|_{Iso}}\right) \pm \sqrt{1 - \left(\frac{|\overline{v}|^2_{CM}}{|v^2|_{Iso}}\right) + \frac{1}{4}\cdot\left(\frac{|\overline{v}|^4_{CM}}{|v^4|_{Iso}}\right)}\right]$$

$$\eta_{Carnot} = \left[-\frac{\gamma_{Heat}}{6}\cdot\left(\frac{|\overline{v}|^2_{CM}}{c^2_{Sound}}\right) \pm \sqrt{1 - \frac{\gamma_{Heat}}{3}\cdot\left(\frac{|\overline{v}|^2_{CM}}{c^2_{Sound}}\right) + \frac{\gamma^2_{Heat}}{12}\cdot\left(\frac{|\overline{v}|^4_{CM}}{c^4_{Sound}}\right)}\right]$$

5.2 Carnot Thermal Engine "Anisotropy Efficiency Factor – Positive and Negative Components

Carnot Thermal Engine "Anisotropy" Efficiency Factor

5.12

$$\eta_{\text{Carnot}} = \frac{1}{\gamma^2_{Inertial}} = \left(1 - \frac{\left|\overline{\text{v}}\right|^2_{\text{CM}}}{\left|\text{v}^2\right|_{\text{Iso}}}\right) = \left(1 - \frac{\gamma_{Heat}}{3}\cdot\left(\frac{\left|\overline{\text{v}}\right|^2_{\text{CM}}}{\text{c}^2_{\text{Sound}}}\right)\right)$$

$$\eta_{\text{Carnot}} = \left[-\frac{\gamma_{Heat}}{6}\cdot\left(\frac{\left|\overline{\text{v}}\right|^2_{\text{CM}}}{\text{c}^2_{\text{Sound}}}\right) \pm \sqrt{1 - \frac{\gamma_{Heat}}{3}\cdot\left(\frac{\left|\overline{\text{v}}\right|^2_{\text{CM}}}{\text{c}^2_{\text{Sound}}}\right) + \frac{\gamma^2_{Heat}}{12}\cdot\left(\frac{\left|\overline{\text{v}}\right|^4_{\text{CM}}}{\text{c}^4_{\text{Sound}}}\right)}\right]$$

$$\eta_{\text{Carnot}} = \left(1 - \phi_{\text{Loco\_Motion}}\right)$$

$$\eta_{\text{Carnot}} = \left[\eta(+)_{\text{Carnot}} + \eta(-)_{\text{Carnot}} + 1\right] \rightarrow \textit{Unitless}$$

Carnot Thermal Engine "Anisotropy" Efficiency Factor (Positive Component)

5.13

$$\eta(+)_{\text{Carnot}} = \left[-\frac{\gamma_{Heat}}{6}\cdot\left(\frac{\left|\overline{\text{v}}\right|^2_{\text{CM}}}{\text{c}^2_{\text{Sound}}}\right) + \sqrt{1 - \frac{\gamma_{Heat}}{3}\cdot\left(\frac{\left|\overline{\text{v}}\right|^2_{\text{CM}}}{\text{c}^2_{\text{Sound}}}\right) + \frac{\gamma^2_{Heat}}{12}\cdot\left(\frac{\left|\overline{\text{v}}\right|^4_{\text{CM}}}{\text{c}^4_{\text{Sound}}}\right)}\right]$$

Carnot Thermal Engine "Anisotropy" Efficiency Factor (Negative Component)

5.14

$$\eta(-)_{\text{Carnot}} = \left[-\frac{\gamma_{Heat}}{6}\cdot\left(\frac{\left|\overline{\text{v}}\right|^2_{\text{CM}}}{\text{c}^2_{\text{Sound}}}\right) - \sqrt{1 - \frac{\gamma_{Heat}}{3}\cdot\left(\frac{\left|\overline{\text{v}}\right|^2_{\text{CM}}}{\text{c}^2_{\text{Sound}}}\right) + \frac{\gamma^2_{Heat}}{12}\cdot\left(\frac{\left|\overline{\text{v}}\right|^4_{\text{CM}}}{\text{c}^4_{\text{Sound}}}\right)}\right]$$

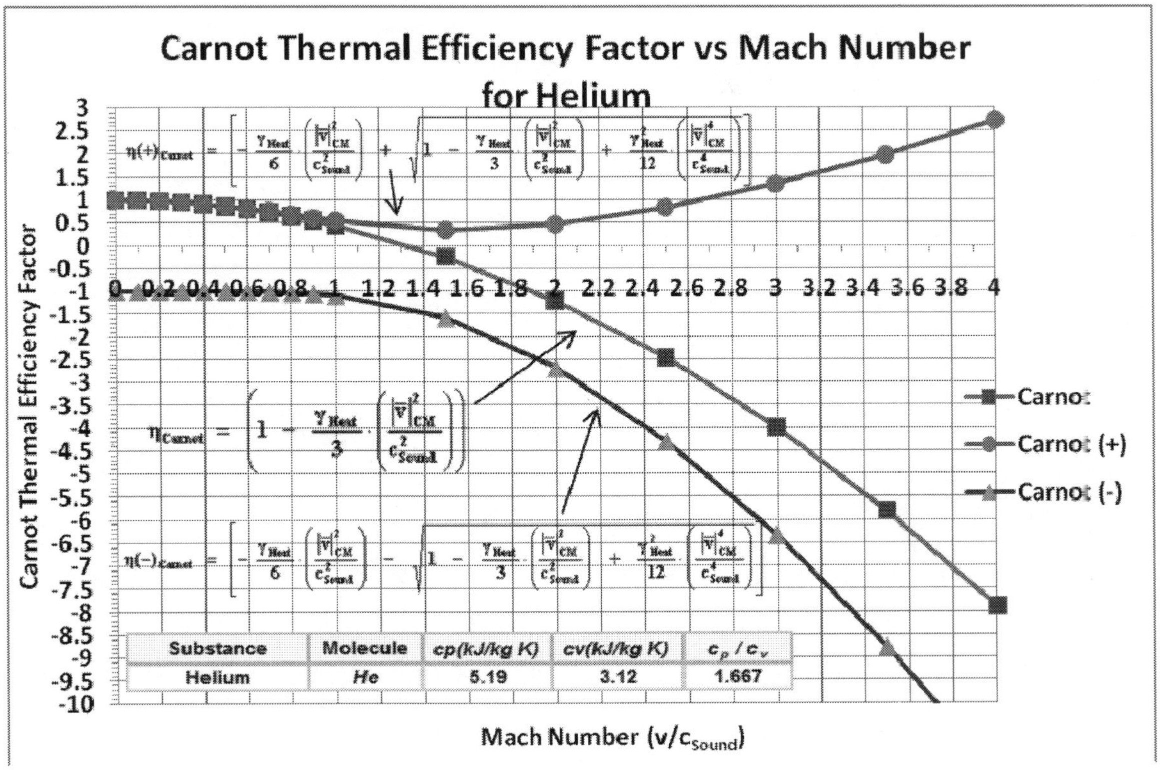

Figure 5.2: Carnot Thermal "Anisotropy" Efficiency Factor vs. Mach number for - Helium.

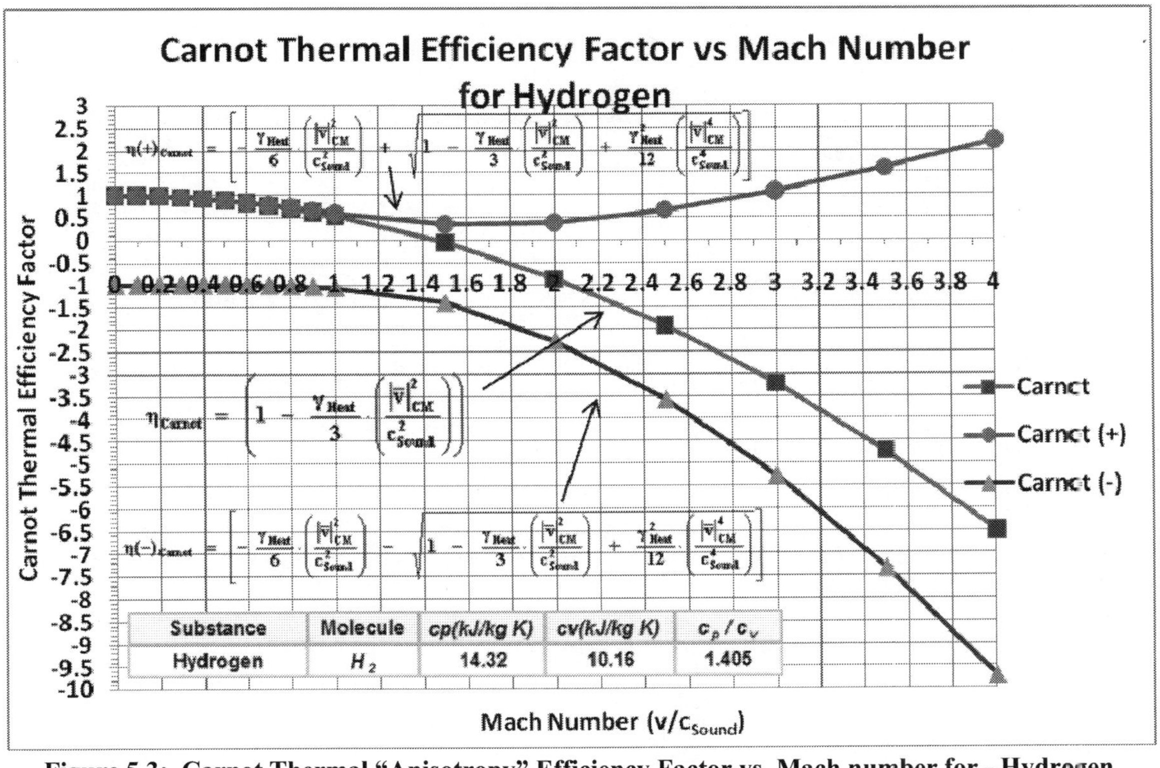

Figure 5.3: Carnot Thermal "Anisotropy" Efficiency Factor vs. Mach number for - Hydrogen.

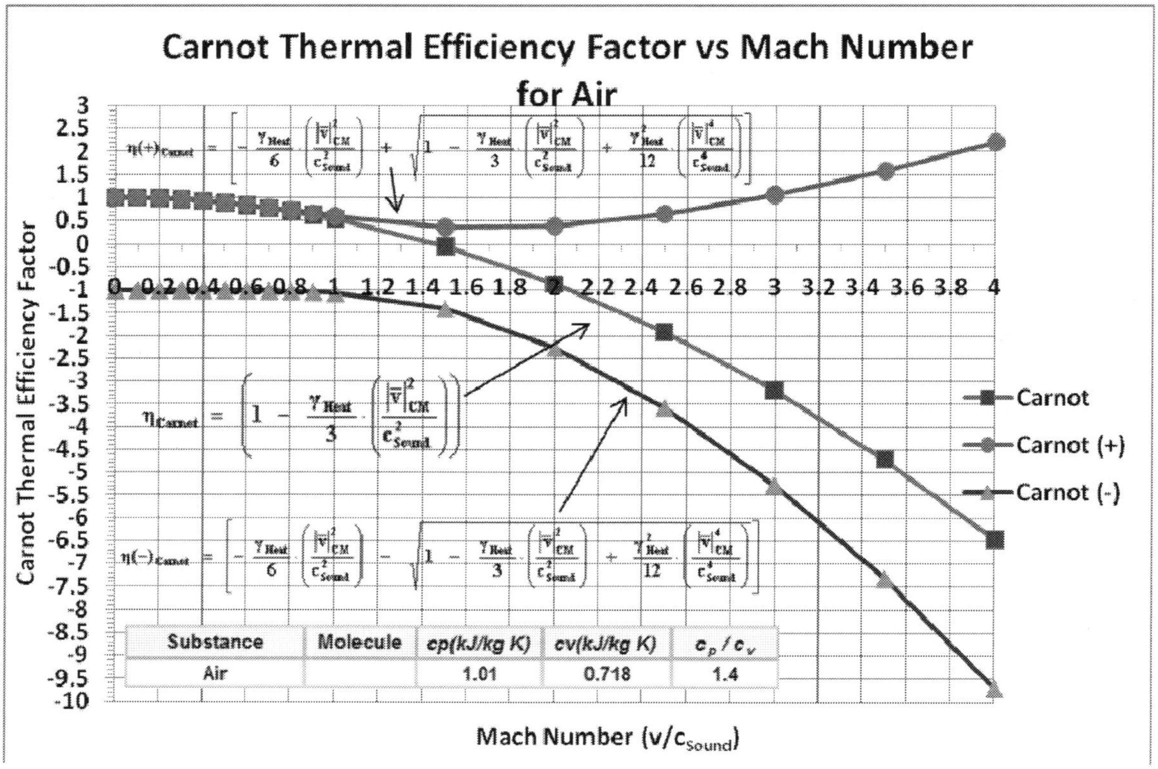

Figure 5.4: Carnot Thermal "Anisotropy" Efficiency Factor vs. Mach number for - Air.

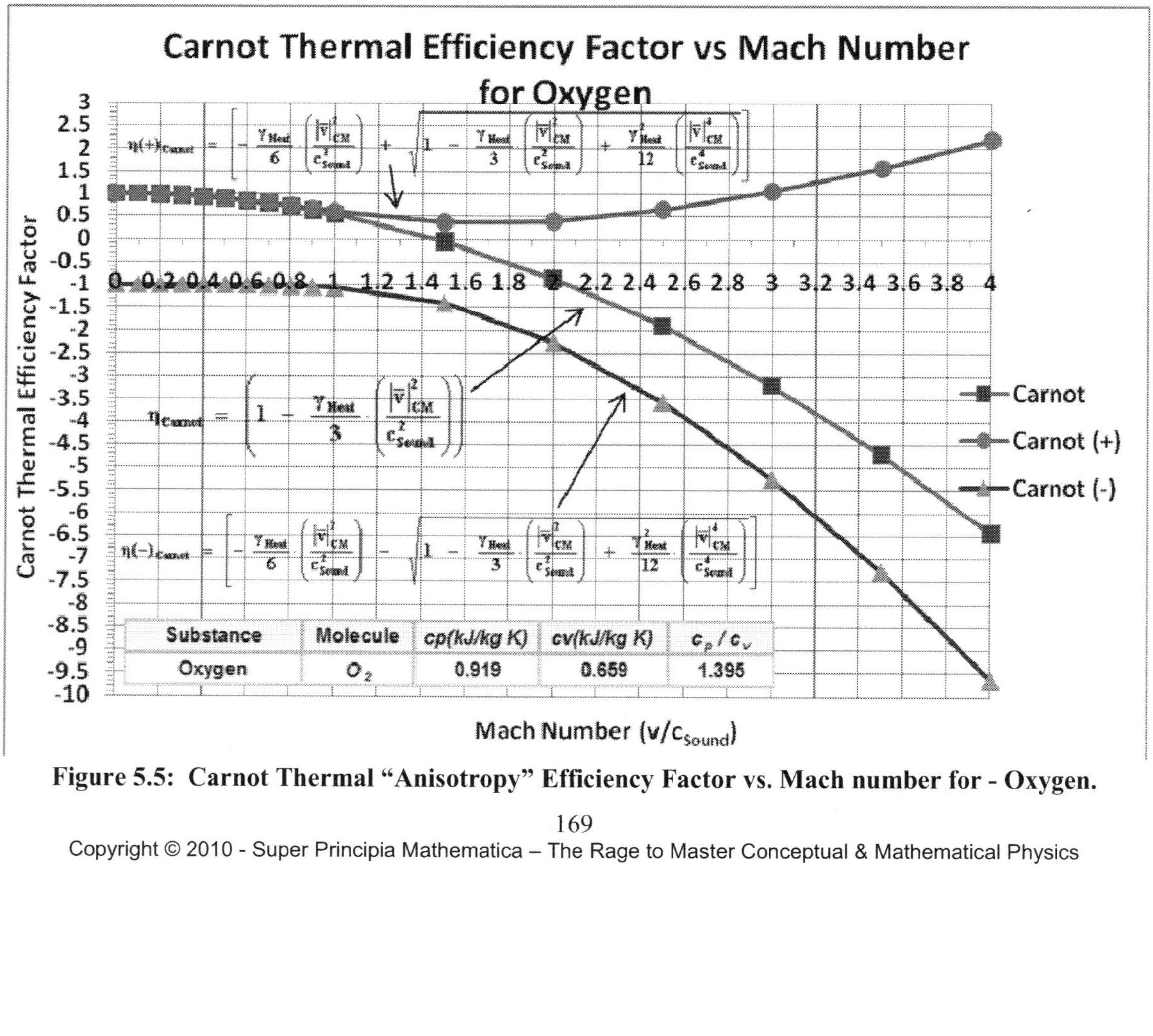

Figure 5.5: Carnot Thermal "Anisotropy" Efficiency Factor vs. Mach number for - Oxygen.

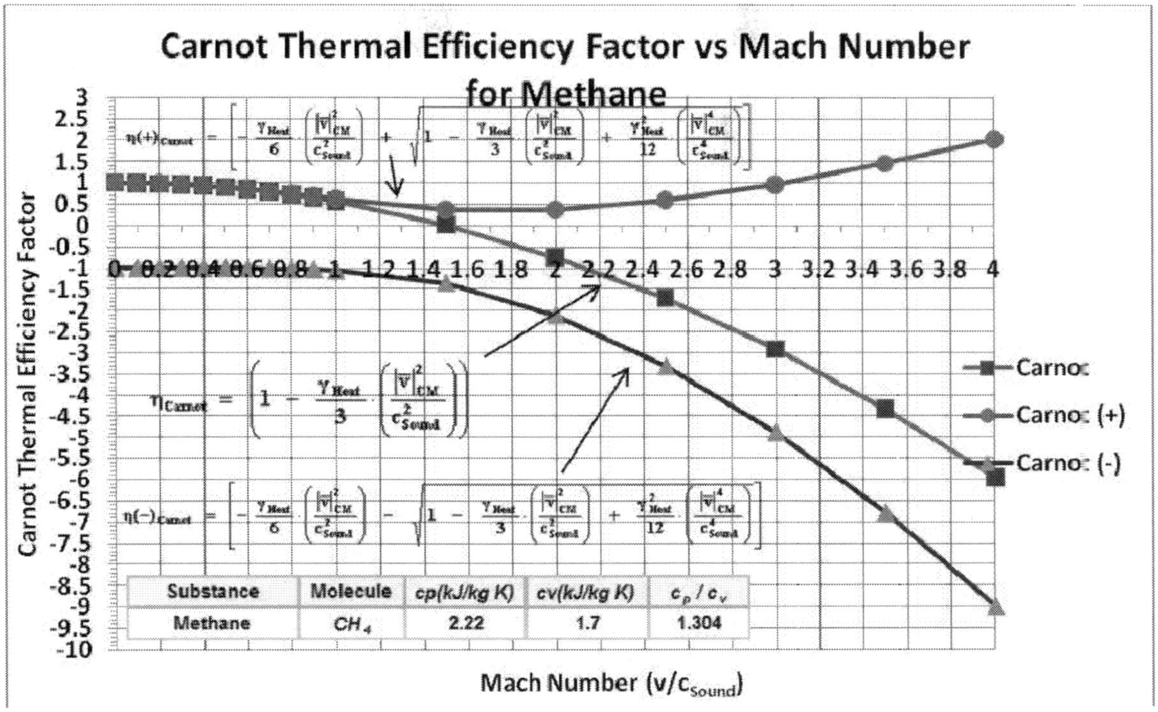

Figure 5.6: Carnot Thermal "Anisotropy" Efficiency Factor vs. Mach number for - Methane.

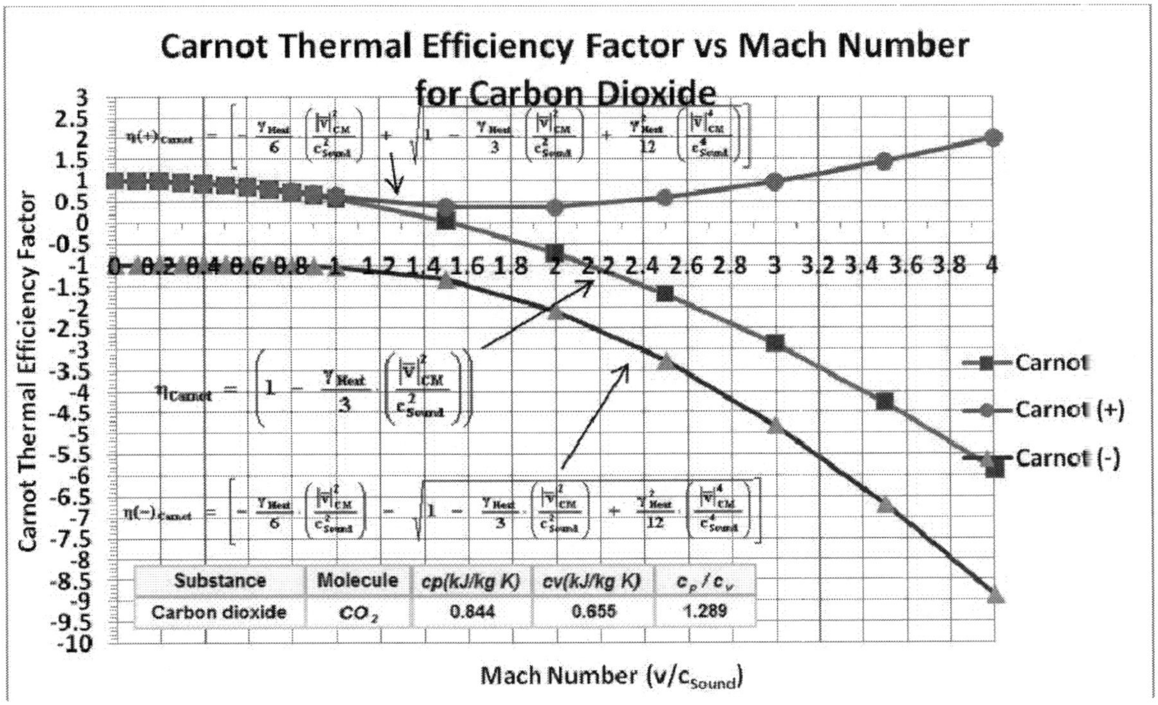

Figure 5.7: Carnot Thermal "Anisotropy" Efficiency Factor vs. Mach number for – Carbon Dioxide.

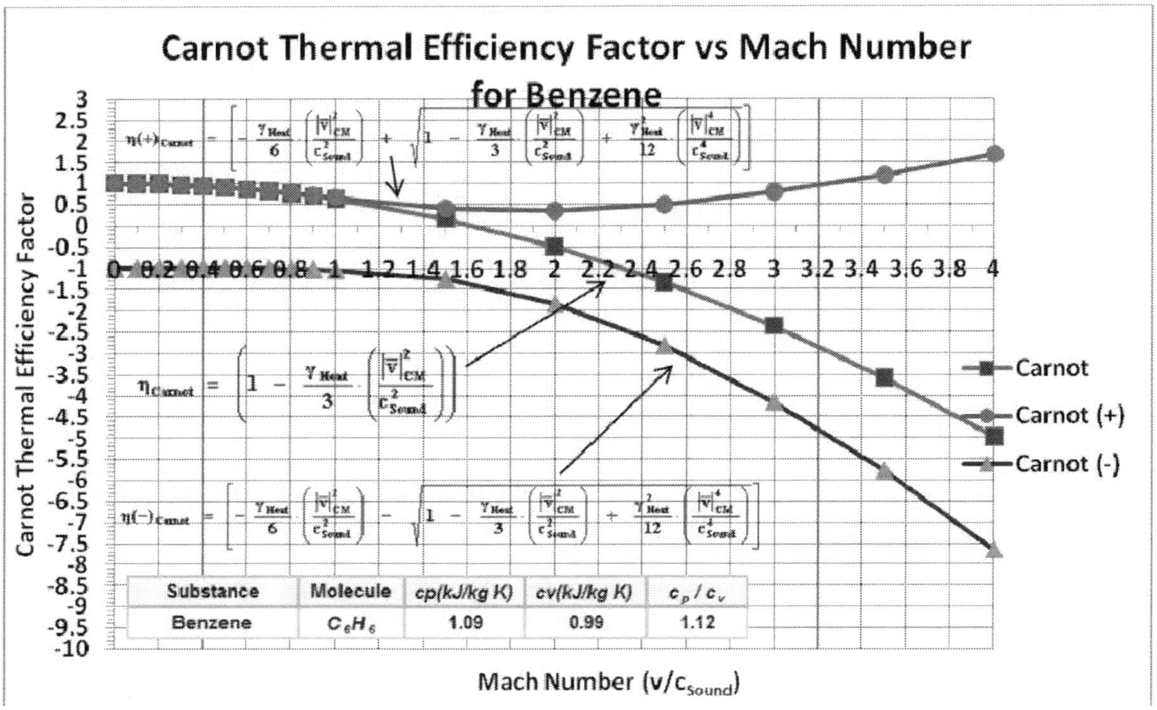

Figure 5.8: Carnot Thermal "Anisotropy" Efficiency Factor vs. Mach number for - Benzene.

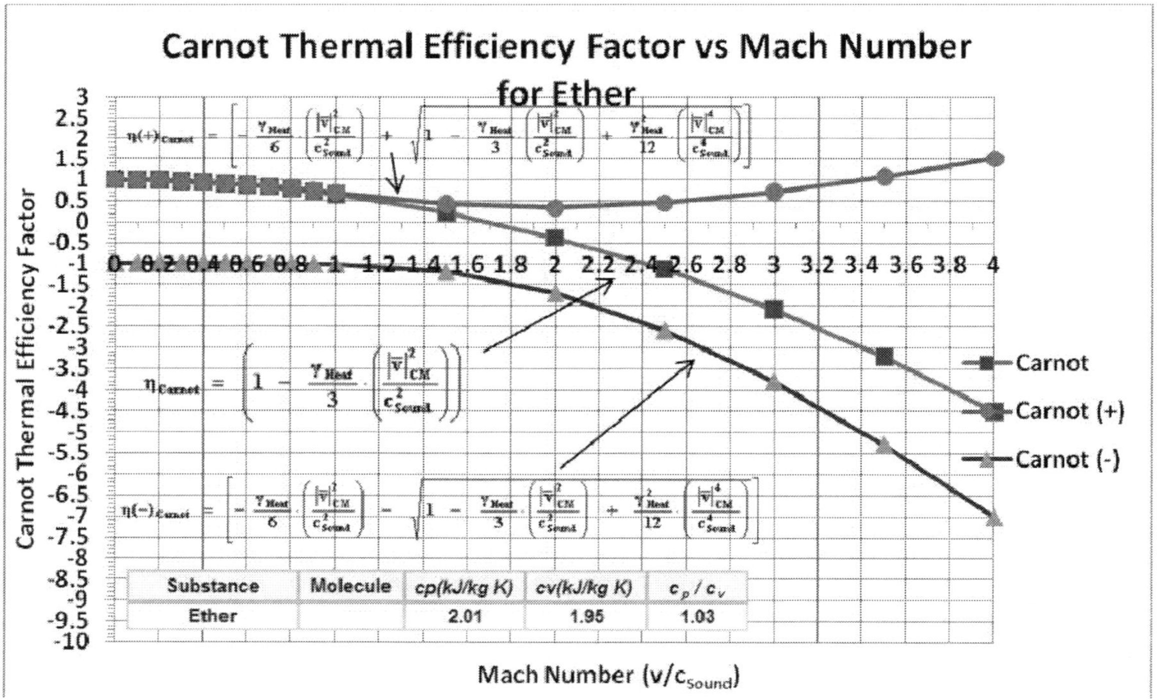

Figure 5.9: Carnot Thermal "Anisotropy" Efficiency Factor vs. Mach number for - Ether.

Chapter 6

The Special Theory of Thermodynamics

(Carnot's Thermal "Isotropy") (Engine Efficiency)

6.1 Carnot's Thermal Engine "Isotropy" Efficiency Factor

The Carnot's Thermal Engine "Isotropy" Efficiency Factor is a measure of an isolated system mass body that is used to determine the heating of a dynamic system mass body and a cooling of the atmosphere surroundings medium environment.

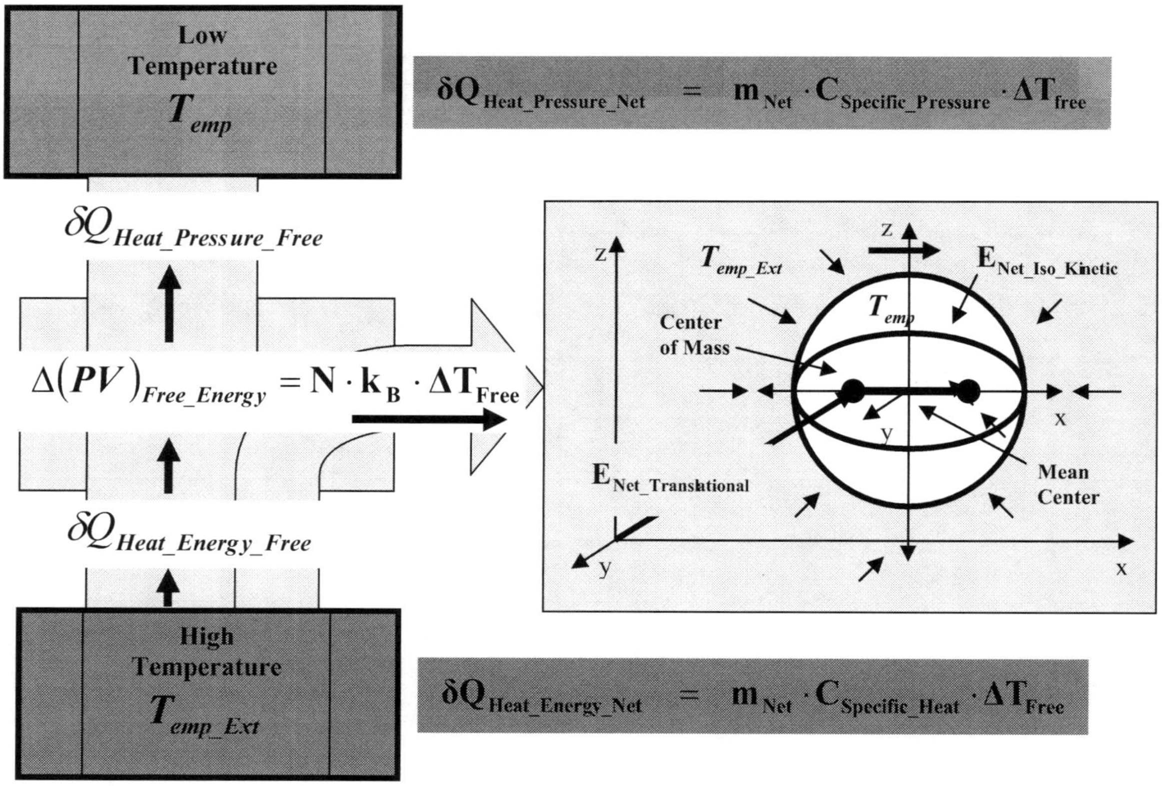

Figure 6.1: The Carnot "Isotropy" Engine is the total Aether Kinematic Work Energy transfer which is equal to the difference in the high thermodynamic potential subtracted from the low thermodynamic potential.

Carnot's theorem sets essential limitations on the yield of a cyclic heat engine such as steam engines, internal combustion engines, atoms, and matter motion, which operate on the **Isotropic Carnot Cycle**; given by the following **Inertial Locomotion Foreshortening Factor** condition.

6.1

$$\phi_{\text{Loco\_Motion}} = \left(\frac{\left|\overline{V}\right|^2_{CM}}{\left|V^2\right|_{Iso}} \right) \geq 1$$

Heat Engines that operate on the Carnot cycle can extract only a certain proportion of mechanical energy from the heat of the working fluid, and this maximal amount is realized by the ***Ideal Isotropic Carnot Heat Engine***.

The Ideal Isotropic Carnot Heat Engine of an isolated system mass body is divided into two heat energy transfer reservoirs, an anisotropy reservoir and an isotropy reservoir which can be considered initial and final states of the system. The two reservoirs also represent higher potential and lower potential, in which energy only flows from high potential to low potential.

For the ***Ideal Isotropic Carnot Heat Engine*** the initial high potential state is the External Anisotropy in which the Aerodynamic Temperature (T_{emp_Ext}), External Translational Kinetic Energy ($\mathrm{E}_{\mathrm{Net\_Translational}}$), Square of the Center of Mass Rectilinear Momentum ($\overline{\mathrm{p}}^2_{\mathrm{Inertial\_Net}}$), and the Center of Mass Squared Velocity Inertia ($|\overline{\mathrm{v}}|^2_{\mathrm{CM}}$), are all considered be higher potential and temperature heat source of the system

For the ***Ideal Isotropic Carnot Heat Engin***e the final low potential state is the Internal Isotropy in which the Internal Absolute Temperature (T_{emp}), Internal Isotropic Omni-directional Kinetic Energy ($\mathrm{E}_{\mathrm{Net\_Iso\_Kinetic}}$), Square of the Center of Isotropy Rectilinear Momentum ($\overline{\mathrm{p}}^2_{\mathrm{Net\_Iso}}$), and the Center of Mass Isotropy Velocity Inertia ($|\mathrm{v}^2|_{\mathrm{Iso}}$), are all considered to be the lower potential and temperature heat sink of the system.

Thus the following conditions hold for the Isotropic Carnot Heat Engine,

6.2

$$T_{emp\_Ext} \geq T_{emp}$$

$$\overline{\mathrm{p}}^2_{\mathrm{Inertial\_Net}} \geq \overline{\mathrm{p}}^2_{\mathrm{Net\_Iso}}$$

$$\mathrm{E}_{\mathrm{Net\_Translational}} \geq \mathrm{E}_{\mathrm{Net\_Iso\_Kinetic}}$$

$$|\overline{\mathrm{v}}|^2_{\mathrm{CM}} \geq |\mathrm{v}^2|_{\mathrm{Iso}}$$

$$\phi_{\text{Loco\_Motion}} = \left(\frac{|\overline{v}|^2_{\text{CM}}}{|v^2|_{\text{Iso}}} \right) \geq 1$$

The **Carnot Thermal Engine "Isotropy" Efficiency Factor** ($\eta_{\text{Carnot\_Iso}} = \frac{1}{\gamma^2_{Inertial\_Iso}}$) is an observer invariant (covariant) unit-less ratio which can be described as the percentage of the isotropy energy transfer.

This isotropy energy transfer is a measure of the quantity of External Anisotropic Translational center of mass motion that is converted into the Internal Isotropic Omni-directional center of isotropy motion.

The Carnot Thermal Engine "Isotropy" Efficiency Factor being a ratio of the Anisotropy divided by the Isotropy of an isolated system is also a dimension invariant ratio. The Anisotropy represents the rectilinear translational first order motions of the closed system, and the Isotropy represents the Omni-directional second order motions of a closed system.

What do I mean when I state that the Anisotropy is a first order motion? The anisotropy naturally defines one dimensional linear space and time through the **net anisotropic translational rectilinear momentum conservation**.

6.3

$$E_{\text{Net\_Translational}} = \frac{\overline{p}^2_{\text{Inertial\_Net}}}{2 \cdot m_{\text{Net}}}$$

$$E_{\text{Net\_Translational}} = \frac{1}{2} m_{\text{Net}} \cdot |\overline{v}|^2_{\text{CM}} = \frac{3}{2} \cdot N \cdot k_B \cdot T_{emp\_Ext} = \frac{\left(\begin{array}{c} m_1\overline{v}_1 + m_2\overline{v}_2 \\ + m_3\overline{v}_3 + \cdots m_N\overline{v}_N \end{array} \right)^2}{2 \cdot m_{\text{Net}}}$$

What do I mean when I state that the Isotropy is a second order motion? The isotropy naturally defines area or two dimensional space and squared time through the **net Isotropic Omni-directional kinetic energy conservation**.

6.4

$$E_{\text{Net\_Iso\_Kinetic}} = \frac{\overline{p}^2_{\text{Net\_Iso}}}{2 \cdot m_{\text{Net}}}$$

$$E_{\text{Net\_Iso\_Kinetic}} = \frac{1}{2} \cdot m_{\text{Net}} \cdot |v^2|_{\text{Iso}} = \frac{3}{2} \cdot N \cdot k_B \cdot T_{emp} = \left[\begin{array}{cc} \frac{m_1\overline{v}_1^2}{2} + & \frac{m_2\overline{v}_2^2}{2} \\ + \frac{m_3\overline{v}_3^2}{2} & + \cdots \frac{m_N\overline{v}_N^2}{2} \end{array} \right]$$

Definition 6.1: The **Carnot Thermal Engine "Isotropy" Efficiency Factor** ($\eta_{\text{Carnot\_Iso}} = \left[\dfrac{1}{\left(\dfrac{T_{emp\_Ext}}{T_{emp}} \right)} - 1 \right]$) is the measure of the effic ency of a heat engine doing work as it transfers heat energy from a higher temperature Anisotropy reservoir to a lower temperature Isotropy reservoir of an isolated system mass body; and is equal to the inverse ratio of the Final **Low Temperature Reservoir – External Anisotropic Aerodynamic Temperature** (T_{emp_Ext}) divided by the nitial **High Temperature Reservoir – Internal Isotropic Absolute Temperature** (T_{emp}) subtracted from one.

Definition 6.2: The **Carnot Thermal Engine "Isotropy" Efficiency Factor** ($\eta_{\text{Carnot\_Iso}} = \left[\dfrac{1}{\left(\dfrac{E_{\text{Net\_Translational}}}{E_{\text{Net\_Iso\_Kinetic}}} \right)} - 1 \right]$) is the measure of the efficiency of a heat engine doing work as it transfers heat energy from a higher temperature Anisotropy reservoir to a lower temperature Isotropy reservoir of an isolated system mass body; and is equal to the ratio of the "final" **External Anisotropic Translational Kinetic Energy** ($E_{\text{Net\_Translational}}$) divided by the "initial" **Internal Isotropic Omni-directional Kinetic Energy** ($E_{\text{Net\_Iso\_Kinetic}}$) subtracted from one.

Definition 6.3: **Carnot Thermal Engine "Isotropy" Efficiency Factor** ($\eta_{\text{Carnot\_Iso}} = \left[\dfrac{E_{\text{Net\_Iso\_Kinetic}} - E_{\text{Net\_Translational}}}{E_{\text{Net\_Translational}}} \right]$) is an dimension invariant unit-less ratio that is the measure of the percentage of the energy of Anisotropy that is being converted into the energy of Isotropy; and when defined in the Center of mass/Momentum frame or the Center of Isotropy frame is a ratio equal to **Internal Isotropic Omni-directional Kinetic Energy** ($E_{\text{Net\_Iso\_Kinetic}}$) subtracted from the **External Anisotropic Translational Kinetic Energy** ($E_{\text{Net\_Translational}}$) divided by the **Anisotropic Translational Kinetic Energy** of an isolated system body.

Carnot Thermal Engine "Isotropy" Efficiency Factor

6.5

$$\eta_{\text{Carnot\_Iso}} = \frac{1}{\gamma^2_{Inertial\_Iso}} = \left[\frac{\mathrm{W}_{\text{Free\_Work\_Energy}}}{\mathrm{E}_{\text{Net\_Translational}}}\right] = \left(\frac{1}{\dfrac{\left|\overline{\mathrm{v}}\right|^2_{\text{CM}}}{\left|\mathrm{v}^2\right|_{\text{Iso}}}} - 1\right) \rightarrow \textit{Unitless}$$

$$\eta_{\text{Carnot\_Iso}} = \left[\frac{\mathrm{E}_{\text{Net\_Iso\_Kinetic}} - \mathrm{E}_{\text{Net\_Translational}}}{\mathrm{E}_{\text{Net\_Translational}}}\right] = \frac{\left[\left|\mathrm{v}^2\right|_{\text{Iso}} - \left|\overline{\mathrm{v}}\right|^2_{\text{CM}}\right]}{\left|\overline{\mathrm{v}}\right|^2_{\text{CM}}}$$

$$\eta_{\text{Carnot\_Iso}} = \frac{\Delta\overline{\mathrm{p}}^2_{\text{Free\_Momentum}}}{\overline{\mathrm{p}}^2_{\text{Inertial\_Net}}} = \left(\frac{1}{\dfrac{\left|\overline{\mathrm{v}}\right|^2_{\text{CM}}}{\left|\mathrm{v}^2\right|_{\text{Iso}}}} - 1\right)$$

$$\eta_{\text{Carnot\_Iso}} = \frac{\left[\overline{\mathrm{p}}^2_{\text{Net\_Iso}} - \overline{\mathrm{p}}^2_{\text{Inertial\_Net}}\right]}{\overline{\mathrm{p}}^2_{\text{Inertial\_Net}}} = \left(\frac{1}{\dfrac{\gamma_{Heat}}{3}\cdot\left(\dfrac{\left|\overline{\mathrm{v}}\right|^2_{\text{CM}}}{\mathrm{c}^2_{\text{Sound}}}\right)} - 1\right)$$

$$\eta_{\text{Carnot\_Iso}} = \frac{\left[P_{Iso\_Pressure} - q_{Dynamic\_Pressure}\right]}{q_{Dynamic\_Pressure}} = \left(\frac{1}{\phi_{\text{Loco\_Motion}}} - 1\right)$$

$$\eta_{\text{Carnot\_Iso}} = \frac{1}{\gamma^2_{Inertial\_Iso}} = \left(\frac{\left|\mathrm{v}^2\right|_{\text{Iso}}}{\left|\overline{\mathrm{v}}\right|^2_{\text{CM}}} - 1\right) = \left(\frac{1}{\phi_{\text{Loco\_Motion}}} - 1\right)$$

Carnot Thermal Engine "Isotropy" Efficiency Factor

6.6

$$\eta_{\text{Carnot\_Iso}} = \frac{1}{\gamma^2_{Inertial\_Iso}} = \left[\frac{\mathrm{W}_{\text{Free\_Work\_Energy}}}{\mathrm{E}_{\text{Net\_Translational}}}\right] = \left(\frac{1}{\dfrac{\left|\overline{\mathrm{v}}\right|^2_{\text{CM}}}{\left|\mathrm{v}^2\right|_{\text{Iso}}}} - 1\right)$$

$$\eta_{\text{Carnot\_Iso}} = \left(\frac{\Delta T_{Free}}{T_{emp\_Ext}}\right) = \left(\frac{T_{emp} - T_{emp\_Ext}}{T_{emp\_Ext}}\right) = \left[\frac{1}{\phi_{\text{Loco\_Motion}}} - 1\right]$$

$$\eta_{\text{Carnot\_Iso}} = \left(\frac{\Delta T_{Free}}{T_{emp\_Ext}}\right) = \left[\frac{1}{\left(\dfrac{T_{emp\_Ext}}{T_{emp}}\right)} - 1\right]$$

$$\eta_{\text{Carnot\_Iso}} = \frac{1}{\gamma^2_{Inertial\_Iso}} = \left(\frac{1}{\dfrac{\gamma_{Heat}}{3}\cdot\left(\dfrac{\left|\overline{\mathrm{v}}\right|^2_{\text{CM}}}{\mathrm{c}^2_{\text{Sound}}}\right)} - 1\right)$$

$$\eta_{\text{Carnot\_Iso}} = \frac{1}{\gamma^2_{Inertial\_Iso}} = \left(\frac{1}{\dfrac{\mathrm{m}_{\text{Net}}\left|\overline{\mathrm{v}}\right|^2_{\text{CM}}}{3\cdot\left(\mathrm{N}\cdot\mathrm{k}_{\text{B}}\cdot T_{emp}\right)}} - 1\right) \rightarrow \textit{Unitless}$$

Carnot Thermal Engine "Isotropy" Efficiency Factor

6.7

$$\eta_{Carnot\_Iso} = \left[\frac{1}{\left(\dfrac{\overline{p}^{2}_{Inertial\_Net}}{\overline{p}^{2}_{Net\_Iso}} \right)} - 1 \right] = \left(\frac{1}{\dfrac{\gamma_{Heat}}{3} \cdot \left(\dfrac{\left|\overline{v}\right|^{2}_{CM}}{c^{2}_{Sound}} \right)} - 1 \right) \rightarrow Unitless$$

$$\eta_{Carnot\_Iso} = \left[\frac{1}{\left(\dfrac{m_{Inertia\_Mass}}{m_{Net}} \right)} - 1 \right] = \left[\frac{1}{\left(\dfrac{m_{Net}}{m_{Dark\_Matter}} \right)} - 1 \right]$$

$$\eta_{Carnot\_Iso} = \left[\frac{1}{\phi_{Loco\_Motion}} - 1 \right] = \left[\frac{1}{\left(\dfrac{E_{Net\_Translational}}{E_{Net\_Iso\_Kinetic}} \right)} - 1 \right]$$

$$\eta_{Carnot\_Iso} = \left[\frac{1}{\phi_{Loco\_Motion}} - 1 \right] = \left[\frac{1}{\left(\dfrac{q_{Dynamic\_Pressure}}{P_{Iso\_Pressure}} \right)} - 1 \right]$$

Carnot Thermal Engine "Isotropy" Efficiency Factor

6.8

$$\eta_{\text{Carnot}} = \left(\frac{\Delta \text{m}_{\text{Iso\_WFM}}}{\text{m}_{\text{Net}}} \right) = \left[\frac{1}{\phi_{\text{Loco\_Motion}}} - 1 \right] \rightarrow \textit{Unitless}$$

$$\eta_{\text{Carnot}} = \left(\frac{\Delta \text{m}_{\text{Iso\_WFM}}}{\text{m}_{\text{Net}}} \right) = \left[\frac{1}{\phi_{\text{Loco\_Motion}}} - 1 \right] = \left(\frac{\Delta (PV)_{\text{Free\_Energy}}}{\text{N} \cdot \text{k}_{\text{B}} \cdot T_{emp\_Ext}} \right)$$

$$\eta_{\text{Carnot\_Iso}} = \left[\frac{1}{\phi_{\text{Loco\_Motion}}} - 1 \right] = \left[\frac{1}{\left(\frac{\text{m}_{\text{Inertia\_Mass}}}{\text{m}_{\text{Net}}} \right)} - 1 \right]$$

$$\eta_{\text{Carnot\_Iso}} = \left[\frac{1}{\phi_{\text{Loco\_Motion}}} - 1 \right] = \left[\frac{1}{\left(\frac{\text{m}_{\text{Net}}}{\text{m}_{\text{Dark\_Matter}}} \right)} - 1 \right]$$

$$\eta_{\text{Carnot\_Iso}} = \left[\frac{1}{\phi_{\text{Loco\_Motion}}} - 1 \right] = \left(\frac{\Delta (PV)_{\text{Free\_Energy}}}{\left(1 - \frac{1}{\gamma_{Heat}} \right) \cdot \text{m}_{\text{Net}} \cdot C_{Specific\_Pressure} \cdot T_{emp\_Ext}} \right)$$

$$\eta_{\text{Carnot}} = \left[\frac{1}{\phi_{\text{Loco\_Motion}}} - 1 \right] = \left[1 - \frac{1}{\left[1 + \left(\frac{\Delta \text{m}_{\text{Iso\_WFM}}}{\text{m}_{\text{Net}}} \right) \right]} \right]$$

Carnot Thermal Engine "Isotropy" Efficiency Factor

6.9

$$\eta_{\mathrm{Carnot\_Iso}} = \left[\frac{1}{\phi_{\mathrm{Loco\_Motion}}} - 1 \right] = \left(\frac{\Delta(PV)_{\mathrm{Free\_Energy}}}{\mathrm{N} \cdot \mathrm{k_B} \cdot T_{emp\_Ext}} \right)$$

$$\eta_{\mathrm{Carnot}} = \left(\frac{\Delta(PV)_{\mathrm{Free\_Energy}}}{\left(1 - \dfrac{1}{\gamma_{Heat}} \right) \cdot \mathrm{m}_{\mathrm{Net}} \cdot C_{Specific\_Pressure} \cdot T_{emp\_Ext}} \right)$$

$$\eta_{\mathrm{Carnot\_Iso}} = \frac{1}{\gamma^2_{Inertial\_Iso}} = \frac{\eta_{\mathrm{Carnot}}}{\left(\dfrac{\overline{\mathrm{p}}^2_{\mathrm{Inertial\_Net}}}{\overline{\mathrm{p}}^2_{\mathrm{Net\_Iso}}} \right)} = \frac{1}{\gamma^2_{Inertial} \cdot \left(\dfrac{\overline{\mathrm{p}}^2_{\mathrm{Inertial\_Net}}}{\overline{\mathrm{p}}^2_{\mathrm{Net\_Iso}}} \right)}$$

$$\eta_{\mathrm{Carnot\_Iso}} = \frac{\eta_{\mathrm{Carnot}}}{\left[1 - \eta_{\mathrm{Carnot}} \right]} \rightarrow Unitless$$

Carnot Thermal Engine "Isotropy" Efficiency Factor

6.10

$$\eta_{\text{Carnot\_Iso}} = \frac{1}{\gamma^2_{Inertial\_Iso}} = \left[\frac{W_{\text{Free\_Work\_Energy}}}{E_{\text{Net\_Translational}}}\right] = \left(\frac{1}{\dfrac{|\overline{v}|^2_{\text{CM}}}{|v^2|_{\text{Iso}}}} - 1\right) \rightarrow \textit{Unitless}$$

$$\eta_{\text{Carnot\_Iso}} = \left[\frac{\left[E_{\text{Net\_Iso\_Kinetic}} - \left[\dfrac{E'_{\text{Relativistic\_Energy}}}{2}\cdot\left(\dfrac{m'_{\text{Net\_Rel}}}{m_{\text{Net}}}\right) - \dfrac{E_{\text{Rest\_Energy}}}{2}\right]\cdot\left(1 - \dfrac{|\overline{v}|^2_{\text{CM}}}{c^2_{\text{Light}}}\right)\right]}{E_{\text{Net\_Translational}}}\right]$$

$$\eta_{\text{Carnot\_Iso}} = \left[\frac{E_{\text{Net\_Iso\_Kinetic}}}{E_{\text{Net\_Translational}}} - \left[\frac{E'_{\text{Relativistic\_Energy}}}{2\cdot E_{\text{Net\_Translational}}}\cdot\left(\frac{m'_{\text{Net\_Rel}}}{m_{\text{Net}}}\right) - \frac{E_{\text{Rest\_Energy}}}{2\cdot E_{\text{Net\_Translational}}}\right]\cdot\left(1 - \frac{|\overline{v}|^2_{\text{CM}}}{c^2_{\text{Light}}}\right)\right]$$

$$\eta_{\text{Carnot\_Iso}} = \left[\left(\frac{|v^2|_{\text{Iso}}}{|\overline{v}|^2_{\text{CM}}}\right) - \left[\frac{\left(\dfrac{c^2_{\text{Light}}}{|\overline{v}|^2_{\text{CM}}}\right)}{\left(1 - \dfrac{|\overline{v}|^2_{\text{CM}}}{c^2_{\text{Light}}}\right)} - \left(\frac{c^2_{\text{Light}}}{|\overline{v}|^2_{\text{CM}}}\right)\right]\cdot\left(1 - \frac{|\overline{v}|^2_{\text{CM}}}{c^2_{\text{Light}}}\right)\right]$$

Carnot Thermal Engine "Isotropy" Efficiency Factor

6.11

$$\eta_{\mathrm{Carnot\_Iso}} = \frac{1}{\gamma^2_{Inertial\_Iso}} = \left[\frac{W_{\mathrm{Free\_Work\_Energy}}}{E_{\mathrm{Net\_Translational}}}\right] \rightarrow \textit{Unitless}$$

$$\eta_{\mathrm{Carnot\_Iso}} = \left(\frac{1}{\dfrac{\left|\overline{v}\right|^2_{\mathrm{CM}}}{\left|v^2\right|_{\mathrm{Iso}}}} - 1\right) = \left(\frac{1}{\phi_{\mathrm{Loco\_Motion}}} - 1\right) = \left(\frac{1}{\dfrac{\gamma_{Heat}}{3}\cdot\left(\dfrac{\left|\overline{v}\right|^2_{\mathrm{CM}}}{c^2_{\mathrm{Sound}}}\right)} - 1\right)$$

$$\eta_{\mathrm{Carnot}} = \left[\frac{1}{2}\cdot\left(\frac{E_{\mathrm{Net\_Iso\_Kinetic}}}{E_{\mathrm{Net\_Translational}}}\right) \pm \frac{1}{2}\cdot\sqrt{4 - \left(\frac{E_{\mathrm{Net\_Iso\_Kinetic}}}{E_{\mathrm{Net\_Translational}}}\right) + \frac{1}{4}\cdot\left(\frac{E_{\mathrm{Net\_Iso\_Kinetic}}}{E_{\mathrm{Net\_Translational}}}\right)^2}\right]$$

$$\eta_{\mathrm{Carnot\_Iso}} = \left[\frac{1}{2}\cdot\frac{1}{\left(\dfrac{\left|\overline{v}\right|^2_{\mathrm{CM}}}{\left|v^2\right|_{\mathrm{Iso}}}\right)} \pm \frac{1}{2}\cdot\sqrt{4 - \frac{1}{\left(\dfrac{\left|\overline{v}\right|^2_{\mathrm{CM}}}{\left|v^2\right|_{\mathrm{Iso}}}\right)} + \frac{1}{4}\cdot\frac{1}{\left(\dfrac{\left|\overline{v}\right|^4_{\mathrm{CM}}}{\left|v^4\right|_{\mathrm{Iso}}}\right)}}\right]$$

$$\eta_{\mathrm{Carnot\_Iso}} = \left[\frac{3}{2\cdot\gamma_{Heat}}\cdot\left(\frac{c^2_{\mathrm{Sound}}}{\left|\overline{v}\right|^2_{\mathrm{CM}}}\right) \pm \frac{1}{2}\cdot\sqrt{4 - \frac{3}{\gamma_{Heat}}\cdot\left(\frac{c^2_{\mathrm{Sound}}}{\left|\overline{v}\right|^2_{\mathrm{CM}}}\right) + \frac{3}{4\cdot\gamma^2_{Heat}}\cdot\left(\frac{c^4_{\mathrm{Sound}}}{\left|\overline{v}\right|^4_{\mathrm{CM}}}\right)}\right]$$

6.2 Carnot Thermal Engine "Isotropy" Efficiency Factor – Positive and Negative Components

Carnot Thermal Engine "Isotropy" Efficiency Factor

6.12

$$\eta_{\text{Carnot\_Iso}} = \frac{1}{\gamma^2_{Inertial\_Iso}} = \left(\frac{1}{\dfrac{|\overline{\text{v}}|^2_{\text{CM}}}{|\text{v}^2|_{\text{Iso}}}} - \right) = \left(\frac{1}{\dfrac{\gamma_{Heat}}{3} \cdot \left(\dfrac{|\overline{\text{v}}|^2_{\text{CM}}}{\text{c}^2_{\text{Sound}}} \right)} - 1 \right)$$

$$\eta_{\text{Carnot\_Iso}} = \left[\frac{3}{2 \cdot \gamma_{Heat}} \cdot \left(\frac{\text{c}^2_{\text{Sound}}}{|\overline{\text{v}}|^2_{\text{CM}}} \right) \pm \frac{1}{2} \cdot \sqrt{4 - \frac{3}{\gamma_{Heat}} \cdot \left(\frac{\text{c}^2_{\text{Sound}}}{|\overline{\text{v}}|^2_{\text{CM}}} \right) + \frac{3}{4 \cdot \gamma^2_{Heat}} \cdot \left(\frac{\text{c}^4_{\text{Sound}}}{|\overline{\text{v}}|^4_{\text{CM}}} \right)} \right]$$

$$\eta_{\text{Carnot\_Iso}} = \left(\frac{1}{\phi_{\text{Loco\_Motion}}} - 1 \right)$$

$$\eta_{\text{Carnot\_Iso}} = \lfloor \eta(+)_{\text{Carnot\_Iso}} + \eta(-)_{\text{Carnot\_Iso}} - 1 \rfloor \rightarrow \textit{Unitless}$$

Carnot Thermal Engine "Isotropy" Efficiency Factor (Positive Component)

6.13

$$\eta(+)_{\text{Carnot\_Iso}} = \left[\frac{3}{2\cdot\gamma_{Heat}}\cdot\left(\frac{c^2_{\text{Sound}}}{\left|\overline{\text{v}}\right|^2_{\text{CM}}}\right) + \frac{1}{2}\cdot\sqrt{4 - \frac{3}{\gamma_{Heat}}\cdot\left(\frac{c^2_{\text{Sound}}}{\left|\overline{\text{v}}\right|^2_{\text{CM}}}\right) + \frac{3}{4\cdot\gamma^2_{Heat}}\cdot\left(\frac{c^4_{\text{Sound}}}{\left|\overline{\text{v}}\right|^4_{\text{CM}}}\right)} \right]$$

Carnot Thermal Engine "Isotropy" Efficiency Factor (Negative Component)

6.14

$$\eta(-)_{\text{Carnot\_Iso}} = \left[\frac{3}{2\cdot\gamma_{Heat}}\cdot\left(\frac{c^2_{\text{Sound}}}{\left|\overline{\text{v}}\right|^2_{\text{CM}}}\right) - \frac{1}{2}\cdot\sqrt{4 - \frac{3}{\gamma_{Heat}}\cdot\left(\frac{c^2_{\text{Sound}}}{\left|\overline{\text{v}}\right|^2_{\text{CM}}}\right) + \frac{3}{4\cdot\gamma^2_{Heat}}\cdot\left(\frac{c^4_{\text{Sound}}}{\left|\overline{\text{v}}\right|^4_{\text{CM}}}\right)} \right]$$

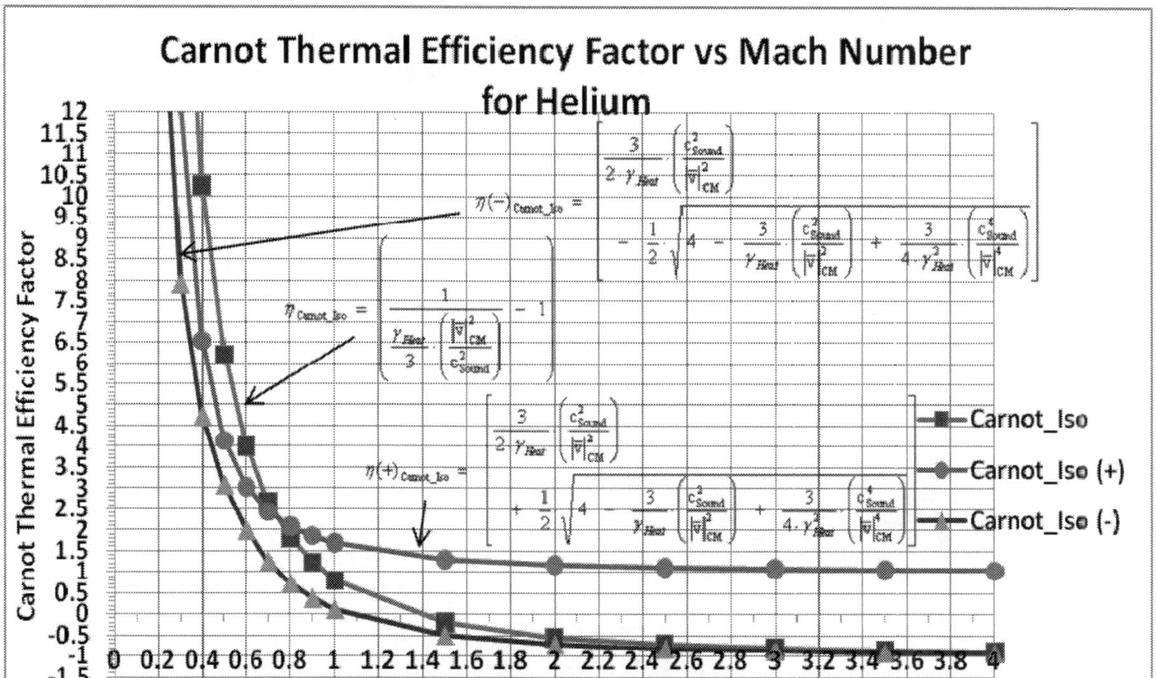

Substance	Molecule	cp(kJ/kg K)	cv(kJ/kg K)	c_p / c_v
Helium	He	5.19	3.12	1.667

Figure 6.2: Carnot Thermal "Isotropy" Efficiency Factor vs. Mach number for - Helium.

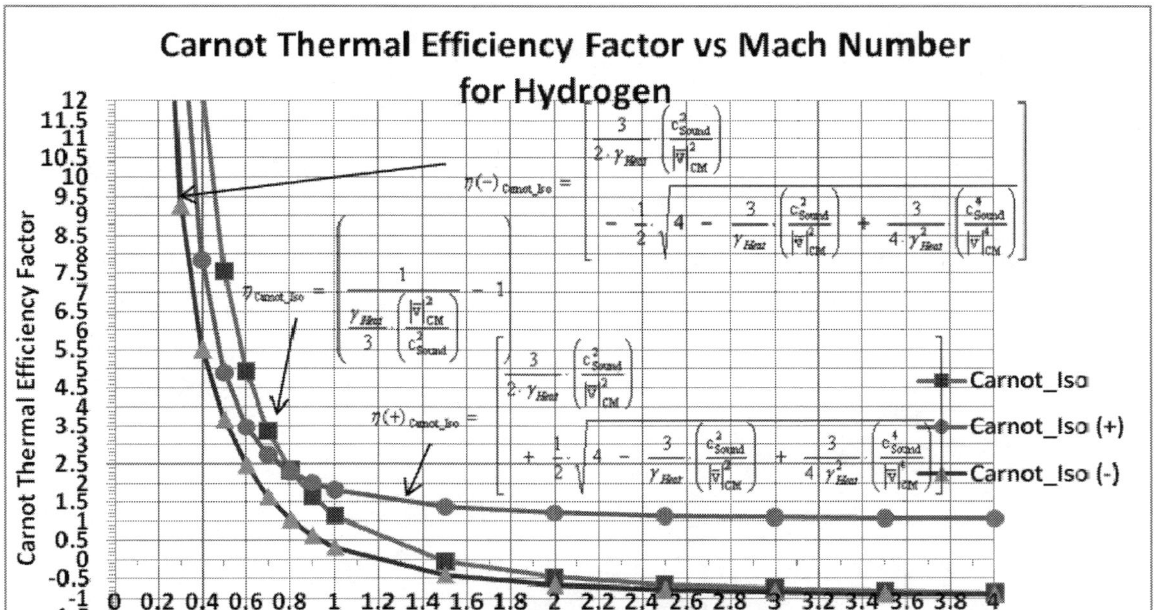

Substance	Molecule	cp(kJ/kg K)	cv(kJ/kg K)	c_p / c_v
Hydrogen	H_2	14.32	10.16	1.405

Figure 6.3: Carnot Thermal "Isotropy" Efficiency Factor vs. Mach number for - Hydrogen.

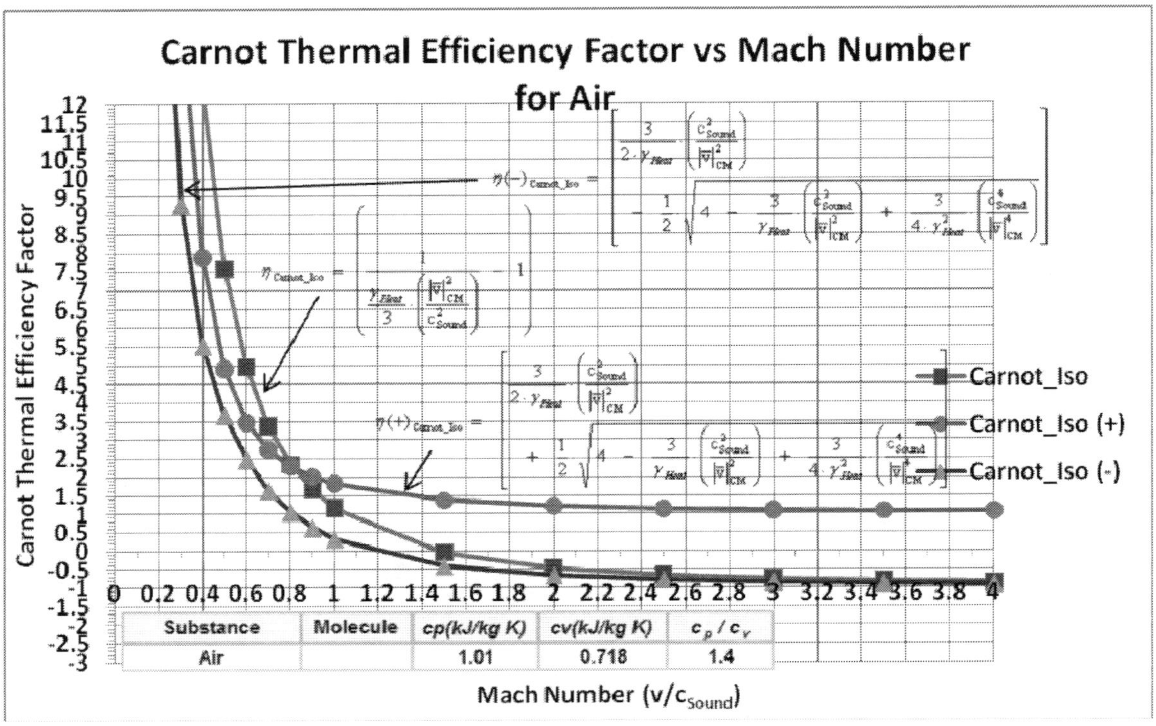

Figure 6.4: Carnot Thermal "Isotropy" Efficiency Factor vs. Mach number for - Air.

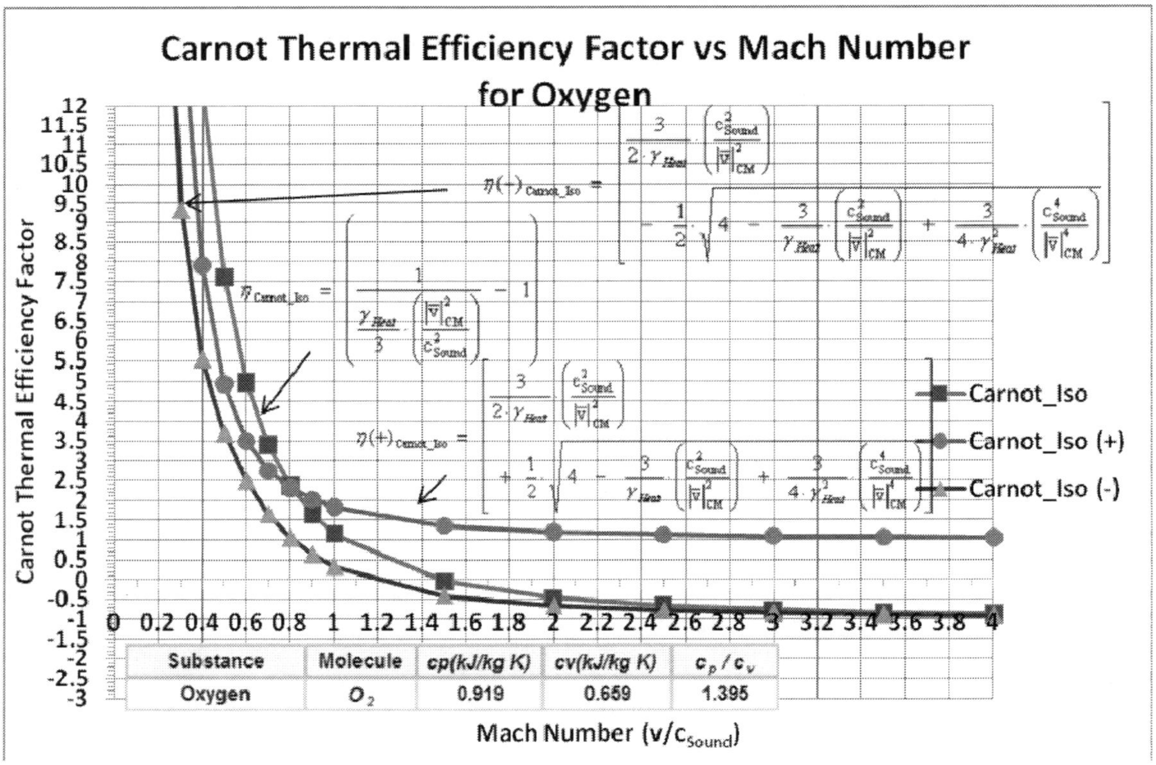

Figure 6.5: Carnot Thermal "Isotropy" Efficiency Factor vs. Mach number for - Oxygen.

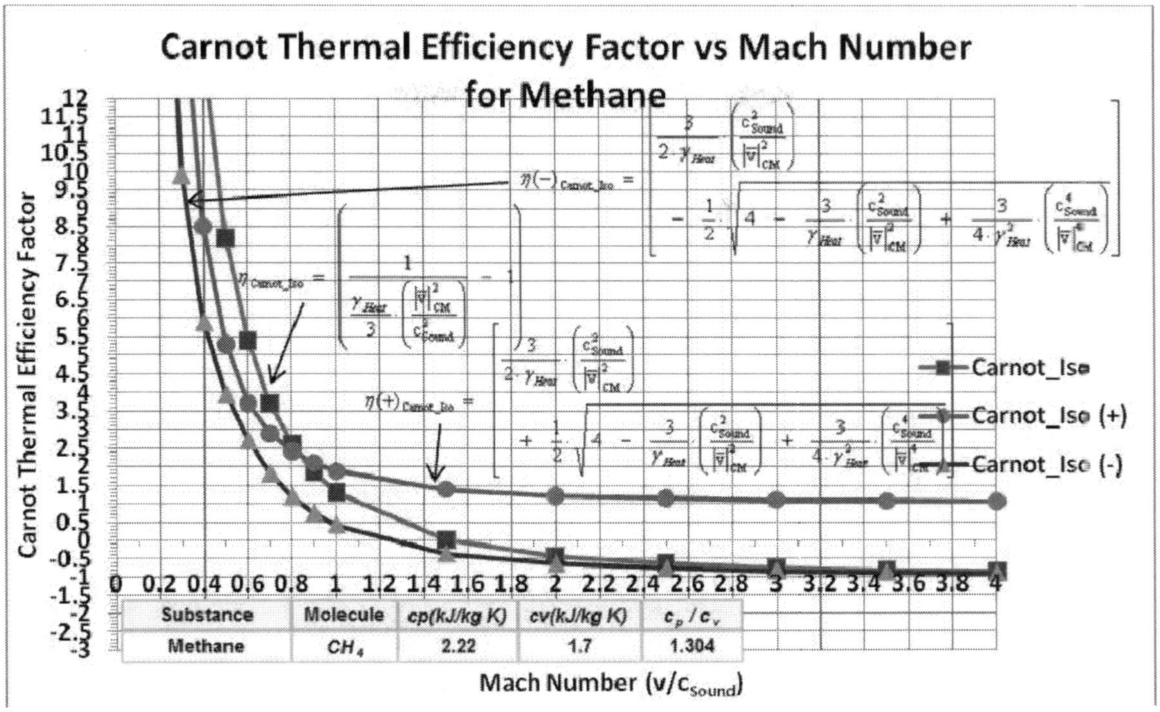

Figure 6.6: Carnot Thermal "Isotropy" Efficiency Factor vs. Mach number for - Methane.

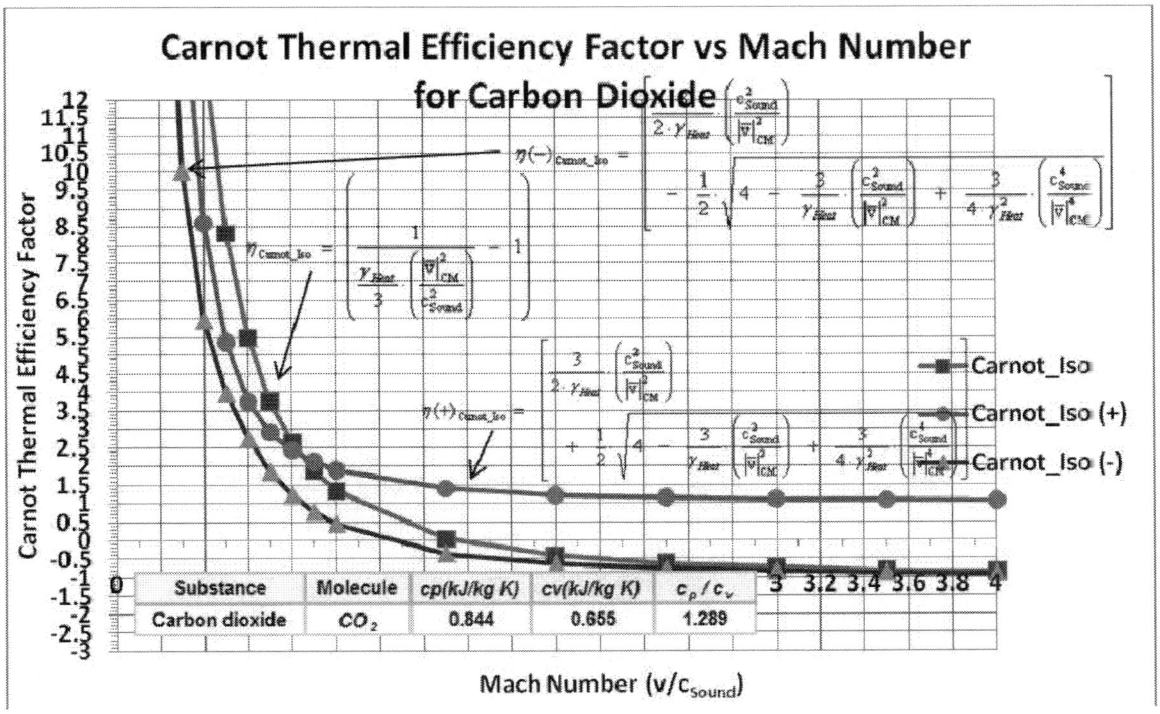

Figure 6.7: Carnot Thermal "Isotropy" Efficiency Factor vs. Mach number for – Carbon Dioxide.

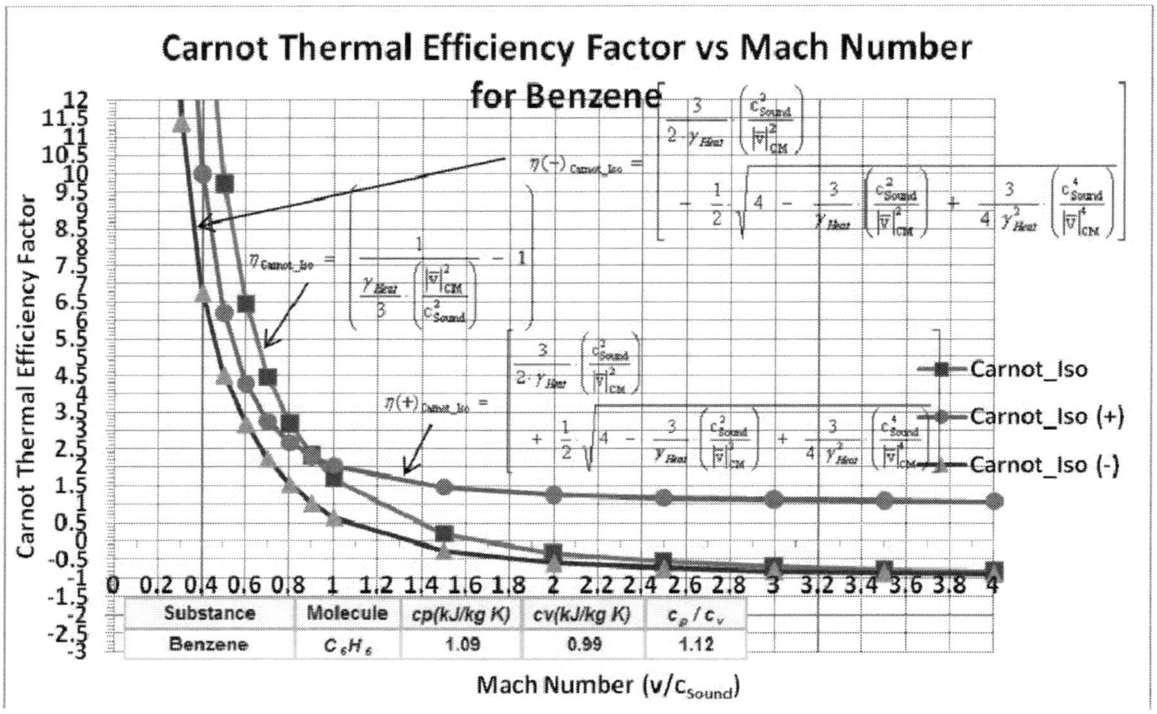

Figure 6.8: Carnot Thermal "Isotropy" Efficiency Factor vs. Mach number for - Benzene.

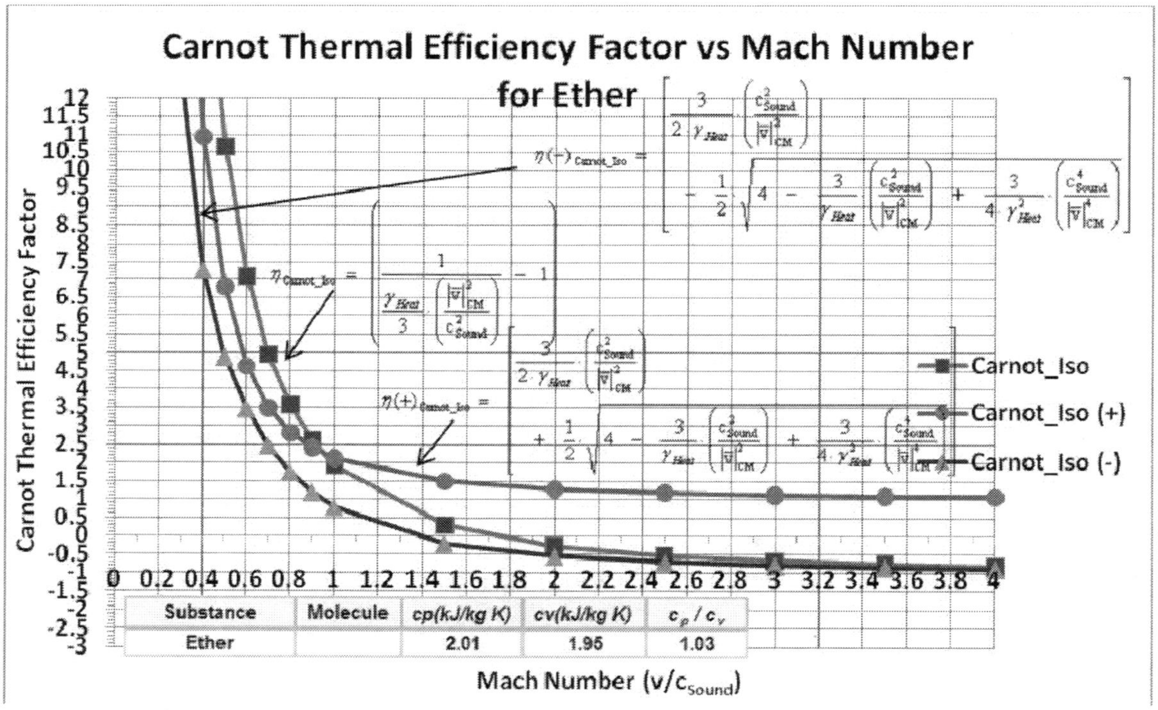

Figure 6.9: Carnot Thermal "Isotropy" Efficiency Factor vs. Mach number for - Ether.

Chapter 7

The Special Theory of Thermodynamics

(Thermodynamic Heat Energy)
(Transfer Mechanisms)

7.1 Thermodynamic Entropic Force & Energy

Any dynamic system mass body under constant external atmospheric environmental medium conditions is observed to change in such a way as to approach a final state called an equilibrium state.

For example, consider two bodies initially at different temperatures, then suddenly are brought together and connected by a thin membrane. And according to the Zeroth Law of Thermodynamics, heat energy "naturally" and spontaneously flows from the hot to the cold body until the temperatures of both bodies become equal, same or identical.

Classically a bound mass body system is at a lower energy level than its unbound constituents. This system only works when its bound mass is "less than" the total mass of its unbound constituents.

For systems with low binding energies, this "lost" mass after binding is very small. But, for systems with high binding energies, the missing mass is a lot easier to measure.

Since all forms of energy have mass, the question of where the missing mass of the binding energy goes is of interest. The answer is that this mass is lost from a system which is not isolated.

This "lost mass-energy" is transformed to heat, light, higher energy states of the nucleus/atom or other forms of energy. These types of energy also have mass, and it is necessary that they be removed from the system before its mass may decrease.

The "mass deficit" from binding energy is therefore removed mass-energy from the system, which corresponds to Einstein's famous Rest Mass-Energy equation ($E_{Rest\_Energy} = m_{Net} \cdot \bar{c}^2_{Light}$).

Another example of "mass deficit" is when a heated system cools to normal temperatures and returns to its ground states in terms of energy levels. The result after cooling is that there is less mass remaining in the system than there was when it was initially heated.

Mass measurements are almost always made at low temperatures with systems in ground states, and this difference between the mass of a system and the sum of the masses of its constituent parts is called a mass deficit.

Thus, if binding mass-energy is transformed into heat, the system must be cooled (the heat removed) before the mass-deficit appears in the cooled system.

In that case, the removed heat represents the aether mass-energy "deficit", which is also heat energy that is removed from the system.

As another example consider two massive objects attracting each other in space through their gravitational field. The attraction force accelerates the objects towards their common center of mass, and the individual masses gain some speed as they move towards the common center of mass; which reduces the potential (gravity) energy and increases the kinetic (movement) energy in the process.

When either of the mass bodies pass through each other without interaction or if they elastically repel during the collision, the gained kinetic energy (related to speed), starts to convert back into potential energy as the particles which collided move apart.

The decelerating particles will return to the initial distance and beyond into infinity or stop and repeat the collision (oscillation takes place).

This shows that any system, which loses no energy, does not combine (bind) into a solid object. Therefore, in order to bind the particles, the kinetic energy gained due to the attraction must be dissipated (by resistive force).

Complex objects in collision ordinarily undergo inelastic collision, transforming some kinetic energy into internal energy (heat content, which is atomic movement), which is further radiated in the form of photons—the light and heat.

Once the energy to escape the gravity is dissipated in the collision, the parts will oscillate at closer distances, thus eventually looking like one solid object. This lost energy, necessary to overcome the potential barrier in order to separate the objects, is the binding energy.

If this binding energy were retained in the system as heat, its mass would not decrease. However, binding energy lost from the system (as heat radiation) would itself have mass, and directly represent of the "mass deficit" of the cold, bound system.

Other analogous considerations apply for chemical and nuclear reactions. Exothermic chemical reactions in closed systems do not change mass, but (in theory) become less massive once the heat of reaction is removed.

In nuclear reactions, the energy that must be radiated or otherwise removed as binding energy, may be in the form of electromagnetic waves, such as gamma radiation, or as heat. Again, however, no mass deficit can in theory appear until this radiation has been emitted and is no longer part of the system.

Typically this mass "deficit" is too small to measure with standard equipment. In nuclear reactions, however, the fraction of mass that may be removed as light or heat, i.e., binding energy, is often a much larger fraction of the system mass.

In chemical reactions the mass "deficit" may be measured directly as a mass difference between rest masses of reactants and products. This is because nuclear forces are comparatively stronger than Coulomb electric attraction forces associated with the interactions between electrons and protons that generate heat in chemistry.

In nuclear reactions the energy given off during either nuclear fusion or nuclear fission is the difference between the binding energies of the fuel and the fusion or fission products.

In practice, this energy may also be calculated from the substantial mass differences between the fuel and products, once evolved heat and radiation have been removed.

When the nucleons are grouped together to form a nucleus, they lose a small amount of mass, i.e., there is "mass deficit." This energy is a measure of the forces that hold the nucleons together, and it represents energy which must be supplied from the environment if the nucleus is to be broken up.

This Mass Deficit or Defect it is known as binding energy, because it simply represents the mass of the energy which has been lost to the environment after binding.

It has been determined by various experiments that the reverse processes of heat energy transfer will never occur if the systems are left solely to themselves. Heat is never observed to flow from the cold body to the hot body spontaneously and naturally. Additional work must be supplied to a system to make heat energy flow from cold to hot; but this does not happen naturally.

Max Planck in 1905 classified all elementary processes into three categories: natural, unnatural, and reversible.

Natural processes do occur, and proceed in a direction toward equilibrium. Unnatural processes move away from equilibrium and never occur. A reversible process is an idealized natural process that passes through a continuous sequence of equilibrium states.

According to the Law of Conservation of Energy, the constituent types of energy for an isolated dynamic system are never destroyed. But in heat transfer between bodies, as in any natural process, some energy is lost.

During the heat transfer process the original energy has been degraded to a less useful form. The energy transferred from a high-temperature body to a lower-temperature body is also in a less useful form.

If another system is used to restore this degraded energy to its original form it is found that the restoring system has degraded the energy even more than the original system had; this process is known as Entropy. Thus, every process occurring in the world is entropic and results in an overall increase in entropy and a corresponding degradation in energy.

The Entropy of a system is a function of the state of a thermodynamic system, whose change in any differential reversible process is equal to the heat absorbed by the system from its surroundings divided by the absolute temperature of the system.

Furthermore, Entropy is the tendency of a system to change from a state of order to a state of disorder, expressed in physics as a measure of the part of the energy in a thermodynamic system that is not available to perform useful work.

According to the principles of evolution, living organisms tend to go from a state of disorder to a state of order in their development and thus appear to reverse entropy which is known as Syntropy.

However, maintaining a living system requires the expenditure of energy, leaving less energy available for work, with the result that the entropy of the system and its surroundings increases.

In physics, a **Thermodynamic Entropic Force** ($\overline{F}_{Entropic\_Force}$) acting in a system is a macroscopic force whose properties are primarily determined not by the character of a particular underlying microscopic force (such as electromagnetism), but by the whole system's statistical tendency to increase its entropy.

In general the **Thermodynamic Entropic Force** ($\overline{F}_{Entropic\_Force}$) is defined by Newton's Second Law of Motion and is defined as the rate of momentum change. The "Entropic Impulse" Momentum Difference is a natural spontaneous process with takes place within a specific amount of time or within a specific frequency; without any external influences besides a system mass body and its atmospheric environment medium.

The **Thermodynamic Entropic Force** ($\overline{F}_{Entropic\_Force}$) is defined as the "Entropic Impulse" which is a Momentum Difference and is the energetic description of the driving force of any natural spontaneous or reaction process between mass, aether, and energy; which is a natural spontaneous process that takes place within a specific amount of time or within a specific frequency.

General - Thermodynamic Entropic Force

7.1

$$\overline{F}_{Entropic\_Force} = \frac{d\overline{p}_{Entropic\_Momentum}}{dt} \rightarrow kg\, m/s^2$$

$$\overline{F}_{Entropic\_Force} = \frac{\left[\overline{p}_{Net\_Iso} - \overline{p}_{Inertial\_Net}\right]}{\Delta t}$$

$$\overline{F}_{Entropic\_Force} = \frac{\left[\overline{p}_{Net\_Iso} - \overline{p}_{Inertial\_Net}\right] \cdot \overline{c}_{Propagation\_Velocity}}{\overline{d}}$$

$$\overline{F}_{Entropic\_Force} = \left(\frac{\Delta\overline{p}^{2}_{Free\_Momentum}}{2 \cdot m_{Net} \cdot \overline{d}}\right) \cdot \left[\frac{2 \cdot m_{Net} \cdot \overline{c}_{Propagation\_Velocity}}{\left[\overline{p}_{Inertial\_Net} + \overline{p}_{Net\_Iso}\right]}\right]$$

$$\overline{F}_{Entropic\_Force} = \overline{F}_{Free\_Work\_Force} \cdot \left[\frac{2 \cdot \overline{c}_{Propagation\_Velocity}}{\left[\left|\overline{v}\right|_{CM} + \sqrt{\left|v^{2}\right|_{Iso}}\right]}\right]$$

$$\overline{F}_{Entropic\_Force} = \frac{\overline{F}_{Thermo\_Pump\_Force}}{\gamma^{4}_{Super\_Compress}} \cdot \left[\frac{2 \cdot \overline{c}_{Propagation\_Velocity}}{\left[\left|\overline{v}\right|_{CM} + \sqrt{\left|v^{2}\right|_{Iso}}\right]}\right]$$

The **Thermodynamic Entropic "Light/Sound" Force** ($\overline{F}_{Entropic\_Force}$) is the energetic description of the driving force of any natural spontaneous or reaction process that takes place at the speed of sound, or the speed of light. What is common for any specific entropic process is the distance ($\overline{d} = \overline{c}_{Light} \cdot \Delta t_{Light} = \overline{c}_{Sound} \cdot \Delta t_{Sound}$) traveled by spontaneous reaction of a fluid as demonstrated in the following equations.

For Light Speed:

$$\overline{d} = \overline{c}_{Light} \cdot \Delta t_{Light} \rightarrow m$$

Thermodynamic Entropic "Light" Force

7.2

$$\overline{F}_{Entropic\_Light\_Force} = \frac{\left[\overline{p}_{Net\_Iso} - \overline{p}_{Inertial\_Net}\right]}{\Delta t_{Light}} \rightarrow kg\, m/s^2$$

$$\overline{F}_{Entropic\_Light\_Force} = \frac{\left[\overline{p}_{Net\_Iso} - \overline{p}_{Inertial\_Net}\right] \cdot \overline{c}_{Light}}{\overline{d}}$$

$$\overline{F}_{Entropic\_Light\_Force} = \left(\frac{\Delta \overline{p}^{2}_{Free\_Momentum}}{2 \cdot m_{Net} \cdot \overline{d}}\right) \cdot \left[\frac{2 \cdot m_{Net} \cdot \overline{c}_{Light}}{\left[\overline{p}_{Inertial\_Net} + \overline{p}_{Net\_Iso}\right]}\right]$$

$$\overline{F}_{Entropic\_Light\_Force} = \overline{F}_{Free\_Work\_Force} \cdot \left[\frac{2 \cdot \overline{c}_{Light}}{\left[\left|\overline{v}\right|_{CM} + \sqrt{\left|v^2\right|_{Iso}}\right]}\right]$$

$$\overline{F}_{Entropic\_Light\_Force} = \frac{\overline{F}_{Thermo\_Pump\_Force}}{\gamma^{4}_{Super\_Compress}} \cdot \left[\frac{2 \cdot \overline{c}_{Light}}{\left[\left|\overline{v}\right|_{CM} + \sqrt{\left|v^2\right|_{Iso}}\right]}\right]$$

For Sound Speed:

$$\overline{d} = \overline{c}_{Sound} \cdot \Delta t_{Sound} \rightarrow m$$

Thermodynamic Entropic "Sound" Force

7.3

$$\overline{F}_{Entropic\_Sound\_Force} = \frac{\left[\overline{p}_{Net\_Iso} - \overline{p}_{Inertial\_Net}\right]}{\Delta t_{Sound}} \rightarrow kg\, m/s^2$$

$$\overline{F}_{Entropic\_Sound\_Force} = \frac{\left[\overline{p}_{Net\_Iso} - \overline{p}_{Inertial\_Net}\right] \cdot \overline{c}_{Sound}}{\overline{d}}$$

$$\overline{F}_{Entropic\_Sound\_Force} = \left(\frac{\Delta\overline{p}^{2}_{Free\_Momentum}}{2 \cdot m_{Net} \cdot \overline{d}}\right) \cdot \left[\frac{2 \cdot m_{Net} \cdot \overline{c}_{Sound}}{\left[\overline{p}_{Inertial\_Net} + \overline{p}_{Net\_Iso}\right]}\right]$$

$$\overline{F}_{Entropic\_Sound\_Force} = \overline{F}_{Free\_Work\_Force} \cdot \left[\frac{2 \cdot \overline{c}_{Sound}}{\left[\left|\overline{v}\right|_{CM} + \sqrt{\left|v^2\right|_{Iso}}\right]}\right]$$

$$\overline{F}_{Entropic\_Sound\_Force} = \frac{\overline{F}_{Thermo\_Pump\_Force}}{\gamma^{4}_{Super\_Compress}} \cdot \left[\frac{2 \cdot \overline{c}_{Sound}}{\left[\left|\overline{v}\right|_{CM} + \sqrt{\left|v^2\right|_{Iso}}\right]}\right]$$

In a generalized sense, a thermodynamic force is a tension whose conjugate extensity is length. The central concept of thermodynamics is that of particle interaction, kinetic energy, and the ability to do work.

According to the Law of Conservation of Energy, an isolated dynamic system mass body and its surroundings is conserved. The energy of the system may be transferred into the mass body by heating, cooling, compression, expansion, addition of matter, or the extraction of matter.

In mechanics, for comparison, energy transfer results from a force which causes displacement, the product of the two being the amount of energy transferred.

In a similar way, thermodynamic systems can be thought of as transferring energy as the result of a "generalized force" causing a "generalized displacement", with the product of the two being the amount of energy transferred. For Light Speed:

Thermodynamic Entropic "Light" Work/Energy

7.4

$$\overline{F}_{Entropic\_Light\_Force} \cdot \overline{d} = \left[\overline{p}_{Net\_Iso} - \overline{p}_{Inertial\_Net}\right] \cdot \overline{c}_{Light}$$

$$\overline{F}_{Entropic\_Light\_Force} \cdot \overline{d} = \Delta\overline{p}^{2}_{Free\_Momentum} \cdot \left[\frac{\overline{c}_{Light}}{\left[\overline{p}_{Inertial\_Net} + \overline{p}_{Net\_Iso}\right]}\right]$$

$$\overline{F}_{Entropic\_Light\_Force} \cdot \overline{d} = \left(\frac{\Delta\overline{p}^{2}_{Free\_Momentum}}{2 \cdot m_{Net}}\right) \cdot \left[\frac{2 \cdot m_{Net} \cdot \overline{c}_{Light}}{\left[\overline{p}_{Inertial\_Net} + \overline{p}_{Net\_Iso}\right]}\right]$$

$$\overline{F}_{Entropic\_Light\_Force} \cdot \overline{d} = W_{Free\_Work\_Energy} \cdot \left[\frac{2 \cdot \overline{c}_{Light}}{\left[\left|\overline{v}\right|_{CM} + \sqrt{\left|v^{2}\right|_{Iso}}\right]}\right]$$

$$\overline{F}_{Entropic\_Light\_Force} \cdot \overline{d} = \frac{W_{Thermo\_Pump\_Energy}}{\gamma^{4}_{Super\_Compress}} \cdot \left[\frac{2 \cdot \overline{c}_{Light}}{\left[\left|\overline{v}\right|_{CM} + \sqrt{\left|v^{2}\right|_{Iso}}\right]}\right] \rightarrow kg \cdot m^{2}/s^{2}$$

For Sound Speed:

Thermodynamic Entropic "Sound" Work/Energy

7.5

$$\overline{F}_{Entropic\_Sound\_Force} \cdot \overline{d} = \left[\overline{p}_{Net\_Iso} - \overline{p}_{Inertial\_Net}\right] \cdot \overline{c}_{Sound} \rightarrow kg \cdot m^2 / s^2$$

$$\overline{F}_{Entropic\_Sound\_Force} \cdot \overline{d} = \Delta\overline{p}^2_{Free\_Momentum} \cdot \left[\frac{\overline{c}_{Sou\,nd}}{\left[\overline{p}_{Inertial\_Net} + \overline{p}_{Net\_Iso}\right]}\right]$$

$$\overline{F}_{Entropic\_Sound\_Force} \cdot \overline{d} = \left(\frac{\Delta\overline{p}^2_{Free\_Momentum}}{2 \cdot m_{Net}}\right) \cdot \left[\frac{2 \cdot m_{Net} \cdot \overline{c}_{Sound}}{\left[\overline{p}_{Inertial\_Net} + \overline{p}_{Net\_Iso}\right]}\right]$$

$$\overline{F}_{Entropic\_Sound\_Force} \cdot \overline{d} = W_{Free\_Work\_Energy} \cdot \left[\frac{2 \cdot \overline{c}_{Sound}}{\left[\left|\overline{v}\right|_{CM} + \sqrt{\left|v^2\right|_{Iso}}\right]}\right]$$

$$\overline{F}_{Entropic\_Sound\_Force} \cdot \overline{d} = \frac{W_{Thermo\_Pump\_Energy}}{\gamma^4_{Super\_Compress}} \cdot \left[\frac{2 \cdot \overline{c}_{Sound}}{\left[\left|\overline{v}\right|_{CM} + \sqrt{\left|v^2\right|_{Iso}}\right]}\right]$$

These thermodynamic force-displacement pairs are known as conjugate variables. The most common conjugate thermodynamic variables are pressure-volume (mechanical parameters), temperature-entropy (thermal parameters), and chemical potential-particle number (material parameters).

The thermodynamic forces that drive most of the "flows" in biological systems are affinities. When affinity is the difference in chemical potential between reactants and products, the corresponding flow is a chemical reaction.

When thermodynamic force is the difference in chemical potential from one location to another, the flow is transport of matter.

A standard example of an entropic force is the elasticity of a freely-jointed polymer molecule. If the polymer molecule is pulled into an extended configuration, the fact that more contracted, randomly coiled configurations are also overwhelmingly more probable (i.e. possess higher entropy) which result in the chain eventually returning (through diffusion) to such configurations.

To the macroscopic observer, the precise origin of the microscopic forces that drive the motion is irrelevant. The observer simply sees the polymer contract into a state of higher entropy, as if driven by an elastic force.

Entropic forces occur in the physics of gases and solutions, where they generate the pressure of an ideal gas, in the osmotic pressure of a dilute solution, and in colloidal suspensions where they are responsible for the crystallization of hard spheres.

For another example, consider a very large collection of water molecules, such that when any two particles come very close together, they merge; and the water trapped in between would be released from the trapped state and join the free bulk water molecules which result in an increase in entropy.

I theorize that an increase in entropy between any two water molecules is the basis of the so called "attraction" between hydrophobic objects in solution.

Now consider when those same two hydrophobic objects attract when they are very close, there may be a mild repulsion when they are about to come close.

This mild repulsion may happen because the water molecules when trapped in between two hydrophobic surfaces coming from different directions causes the anisotropy to suddenly increase and being forced to orient tangential to one surface or the other.

However when the surfaces come even closer together this repulsion is overcome by the great increase in entropy. This means that the attraction force is increased when the water molecules are completely displaced into the environment.

Work can be extracted from bodies colder than its surroundings. When the flow of energy is coming into the body, work is performed by this energy obtained from the surroundings. And, according to the standard definition, Energy is a measure of the ability to do work.

Work can involve the movement of a mass by a force that results from a transformation of energy. If there is an energy transformation, the second principle of energy flow transformations says that this process must involve the dissipation of some energy as heat.

Measuring the amount of heat released is one way of quantifying the energy, or ability to do work and apply a force over a distance.

However, it appears that the ability to do work is relative to the energy transforming mechanism that applies a force. This is to say that some forms of energy perform no work with respects to some mechanisms, but perform work with respects to others.

For example, water does not have a propensity to combust in an internal combustion engine, whereas gasoline does. Relative to the internal combustion engine, water has little ability to do work that provides a motive force.

If “energy” is defined as the ability to do work, then a consequence of this simple example is that water has no energy — according to this definition. Nevertheless, water, raised to a height, does have the ability to do work like driving a turbine, and so does have energy.

This example means to demonstrate that the ability to do work can be considered relative to the mechanism that transforms energy, and through such a conversion applies a force.

Thermodynamic Free Energy is an expendable energy of the Adiabatic Isothermal Thermodynamic Field at absolute temperature establishing equilibrium between the system body and its surroundings.

Thermodynamic Free Energy is work energy in that it can make things happen within finite amounts of time, and is subject to irreversible energy loss; and in the course of doing work is conserved, and may or may not be available for doing useful work.

For an Adiabatic Isentropic Thermodynamic Field each expanding volume of Ideal Gas potentials also represents a certain amount of "kinetic energy" at thermodynamic equilibrium that will be used as the molecules of the system do work on each other when they change from one state to another; changing pressures, densities, and decreasing temperatures distributed across the expanding volumes of the system mass body.

During this potential energy expansion, there will be a certain amount of heat energy loss or dissipation for each potential due to intermolecular friction and collisions as each volume potential establishes thermodynamic equilibrium between itself and the surroundings.

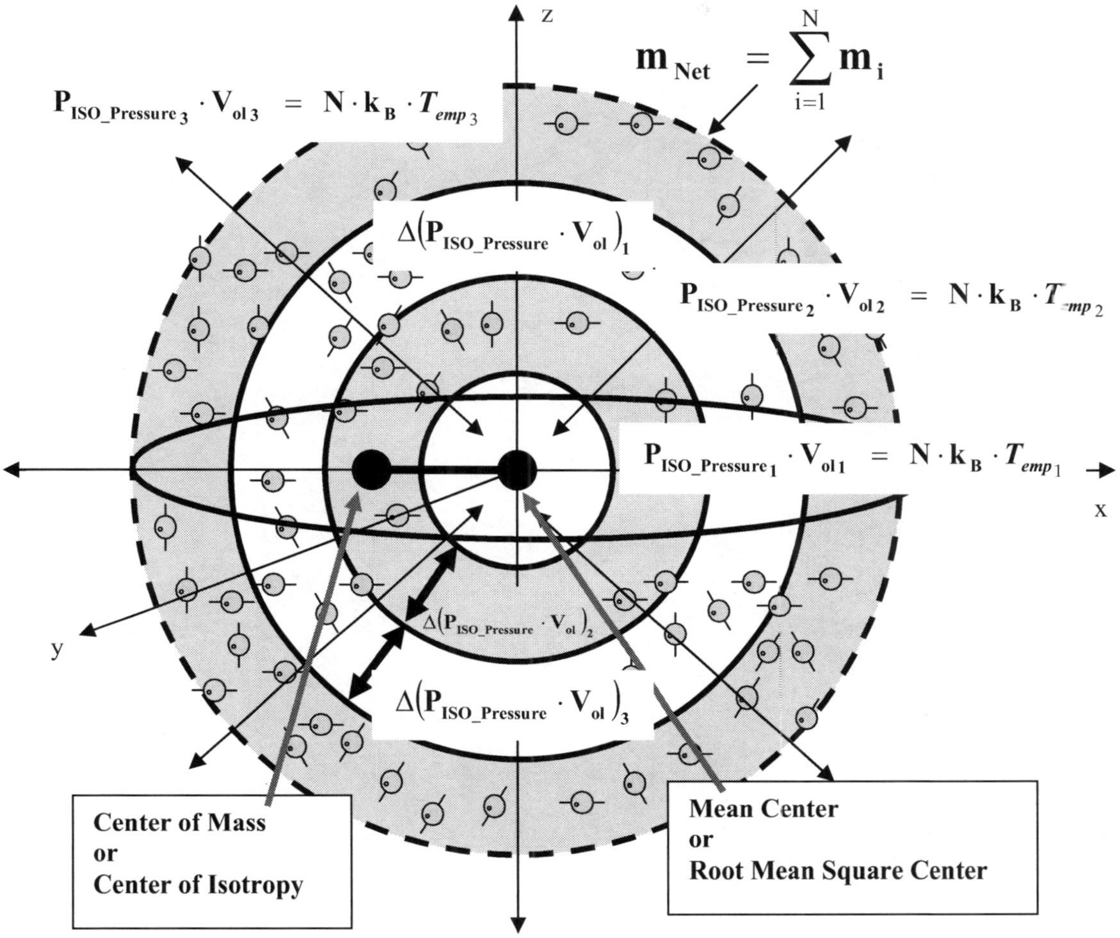

Figure 7.1: An Adiabatic Isentropic Thermodynamic Field is made of finite layer of Isothermal Ideal Gas Potential Energy.

The Thermodynamic Free Energy is repeatable however it may or may not be recoverable if the process is reversed.

Where **Thermodynamic Field Reversible/Irreversible Heat "Free Aether" Work Energy Transfer** ($\Delta(PV)_{\text{Free\_Energy}}$) is the amount of heat absorbed or radiated by the system in an isothermal or isentropic and reversible or irreversible process in which the system goes from one state to another, and (T_{emp}) is the absolute internal temperature at which the process is occurring.

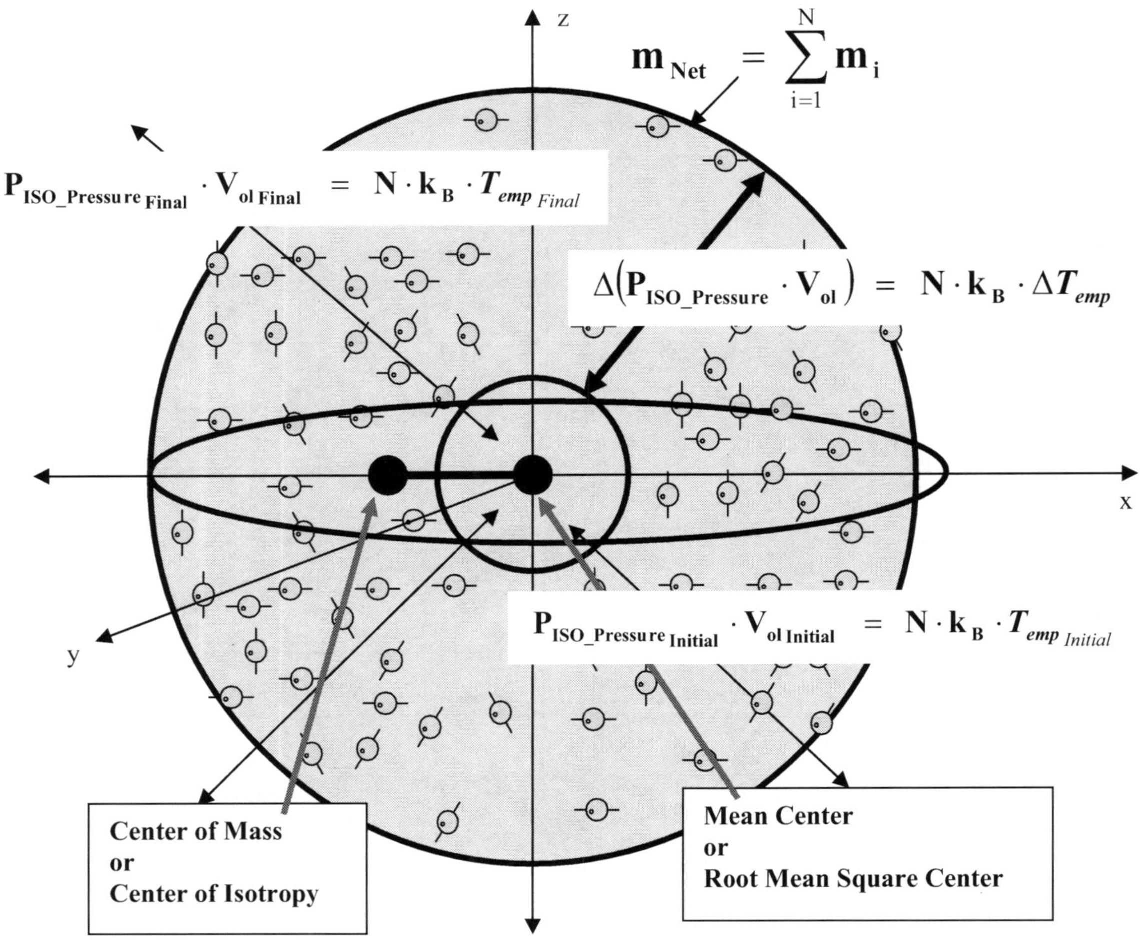

Figure 7.2: Adiabatic Isothermal Thermodynamic Field Ideal Gas Potential Energy exist at constant temperature.

Thermodynamic Field Reversible/Irreversible Heat "Free Aether" Work Energy Transfer

7.6

$$\Delta(PV)_{\text{Free\_Energy}} = \lfloor \delta Q_{\text{Heat\_Pressure\_Free}} - \delta Q_{\text{Heat\_Energy\_Free}} \rfloor \rightarrow kg\, m^2/s^2$$

$$\Delta(PV)_{\text{Free\_Energy}} = \lfloor P_{Iso\_Pressure} - q_{Dynamic\_Pressure} \rfloor \cdot \mathrm{V_{ol}}$$

$$\Delta(PV)_{\text{Free\_Energy}} = \left(1 - \frac{1}{\gamma_{Heat}}\right) \cdot C_{Specific\_Pressure} \cdot \left(\frac{\Delta \overline{\mathrm{p}}^{2}_{\text{Free\_Momentum}}}{3 \cdot \mathrm{N} \cdot \mathrm{k_B}}\right)$$

$$\Delta(PV)_{\text{Free\_Energy}} = \left(1 - \frac{1}{\gamma_{Heat}}\right) \cdot \left(\frac{2 \cdot \mathrm{m_{Net}} \cdot C_{Specific\_Pressure}}{3 \cdot \mathrm{N} \cdot \mathrm{k_B}}\right) \cdot \left[\mathrm{E_{Net\_Iso\_Kinetic}} - \mathrm{E_{Net\_Translational}}\right]$$

$$\Delta(PV)_{\text{Free\_Energy}} = \left(1 - \frac{1}{\gamma_{Heat}}\right) \cdot \left(\frac{C_{Specific\_Pressure}}{3 \cdot \mathrm{N} \cdot \mathrm{k_B}}\right) \cdot \frac{\overline{\mathrm{F}}_{\text{Entropic\_Force}} \cdot \overline{\mathrm{d}}}{\left[\frac{\overline{\mathrm{c}}_{\text{Propagation\_Velocity}}}{\lfloor \overline{\mathrm{p}}_{\text{Inertial\_Net}} + \overline{\mathrm{p}}_{\text{Net\_Iso}} \rfloor}\right]}$$

$$\Delta(PV)_{\text{Free\_Energy}} = \left(1 - \frac{1}{\gamma_{Heat}}\right) \cdot \left(\frac{C_{Specific\_Pressure}}{3 \cdot \mathrm{N} \cdot \mathrm{k_B}}\right) \cdot \frac{\overline{\mathrm{F}}_{\text{Entropic\_Force}} \cdot (\mathrm{m_{Net}} \cdot \overline{\mathrm{d}})}{\left[\frac{\mathrm{m_{Net}} \cdot \overline{\mathrm{c}}_{\text{Propagation\_Velocity}}}{\lfloor \overline{\mathrm{p}}_{\text{Inertial\_Net}} + \overline{\mathrm{p}}_{\text{Net\_Iso}} \rfloor}\right]}$$

As the system mass body and its surroundings radiate Thermodynamic Field "Free Aether" Heat Energy ($\Delta(PV)_{\text{Free\_Energy}}$) then the total change in entropy for the surroundings is given by the following equations.

Adiabatic Isentropic Thermodynamic Field Entropy – Surroundings

7.7

$$\Delta S_{Entropy\ Free\_Surroundings} = \Delta(PV)_{\mathrm{Free\_Energy}} \cdot \left[\frac{1}{T_{emp\_Ext}} - \frac{1}{T_{emp}} \right] \rightarrow kg\, m^2 / s^2 \cdot K$$

$$\Delta S_{Entropy\ Free\_Surroundings} = \left(\frac{1}{\gamma^4_{Super\_Compress}} \right) \cdot \left(\frac{\Delta(PV)_{\mathrm{Free\_Energy}}}{\Delta T_{Free}} \right)$$

$$\Delta S_{Entropy\ Free\_Surroundings} = \frac{1}{\gamma^4_{Super\_Compress}} \cdot \left(\frac{\left[\delta Q_{\mathrm{Heat\_Pressure\_Free}} - \delta Q_{\mathrm{Heat\_Energy\_Free}} \right]}{\left(T_{emp} - T_{emp\_Ext} \right)} \right)$$

$$\Delta S_{Entropy\ Free\_Surroundings} = \left(\frac{1}{\gamma^4_{Super\_Compress}} \right) \cdot \left(\left(1 - \frac{1}{\gamma_{Heat}} \right) \cdot \mathrm{m}_{\mathrm{Net}} \cdot C_{Specific\_Pressure} \right)$$

$$\Delta S_{Entropy\ Free\_Surroundings} = \left(\frac{1}{\gamma^4_{Super\_Compress}} \right) \cdot \left[\frac{\left(\overline{\mathrm{F}}_{\mathrm{Free\_Work\_Force}} \cdot \mathrm{d} \right)}{\left(\frac{3}{2} \cdot \mathrm{N} \cdot \mathrm{k_B} \cdot \Delta T_{Free} \right)} \right] \cdot \left(\left(1 - \frac{1}{\gamma_{Heat}} \right) \cdot \mathrm{m}_{\mathrm{Net}} \cdot C_{Specific\_Pressure} \right)$$

7.2 Thermodynamic Heat Energy Transfer Mechanisms

Based on the above discussion of work and heat energy transfer, a dynamic system mass body and its atmospheric environment perform work via two distinct heat energy transfer mechanisms.

1. Inertial Mass Kinetic Heat Energy Transfer mechanism
    a. Behaves like a ***Carnot Heat Engine***

2. Aether Mass Density and Kinetic Energy Transfer mechanism
    a. Behaves like a ***Carnot Heat Pump***

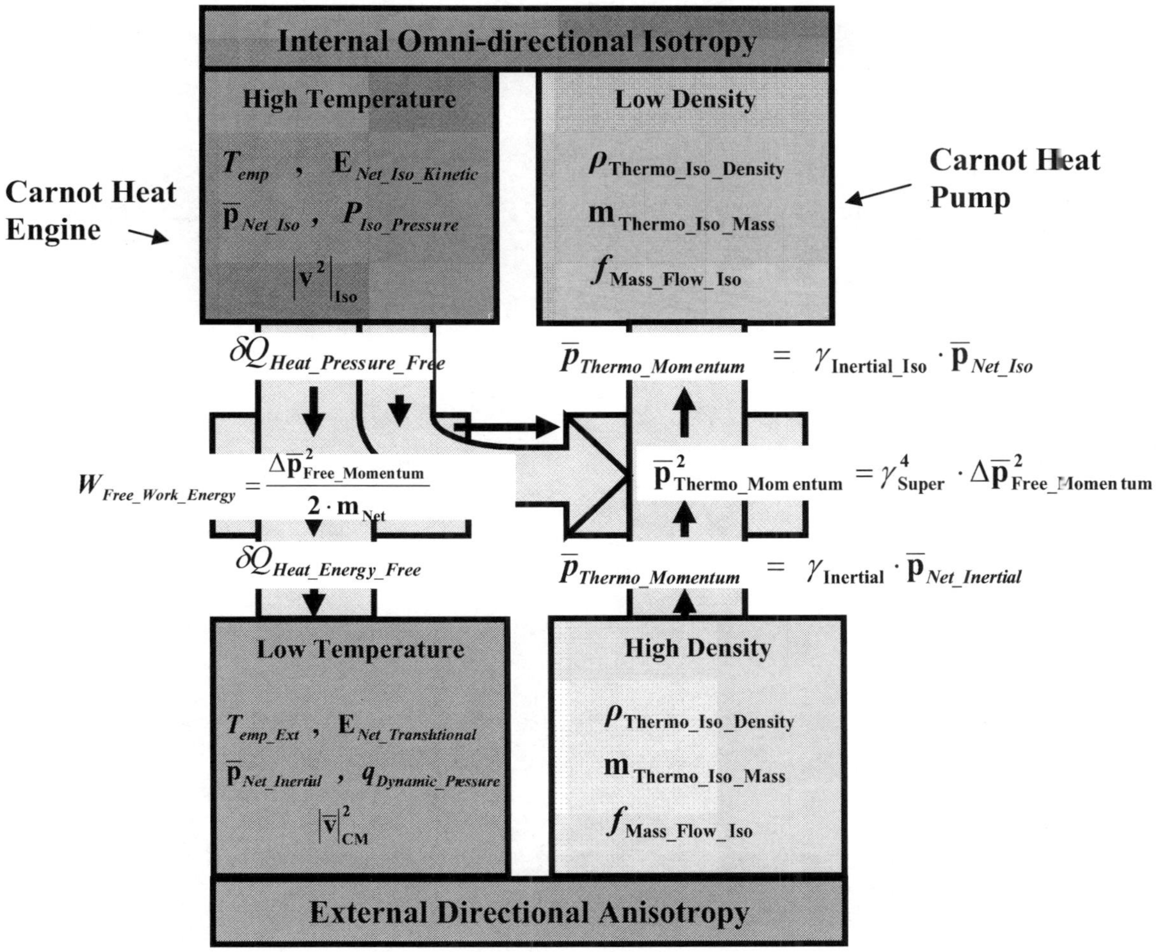

$$\phi_{Respiration} = (\phi_{Loco\_Motion})^{\frac{1}{4}} \leq 1$$

Figure 7.3: Heat Energy Transfer Device – Kemp Thermal Engine.

Carnot Thermal Heat Engine - *Inertial Mass Kinetic Energy Transfer mechanism*

The first heat energy transfer mechanism is called an ***Inertial Mass Kinetic Energy Transfer mechanism***, which behaves like a Carnot Thermal Heat Engine does work transferring heat energy between two different temperature reservoirs, with a goal to establish equilibrium between an inertial net mass body and its atmospheric surroundings medium.

The **Carnot Thermal Heat Engine** - ***Inertial Mass Kinetic Energy Transfer mechanism***, reservoirs are represented by the Anisotropy and Isotropy of the system given by the following two sets of equations:

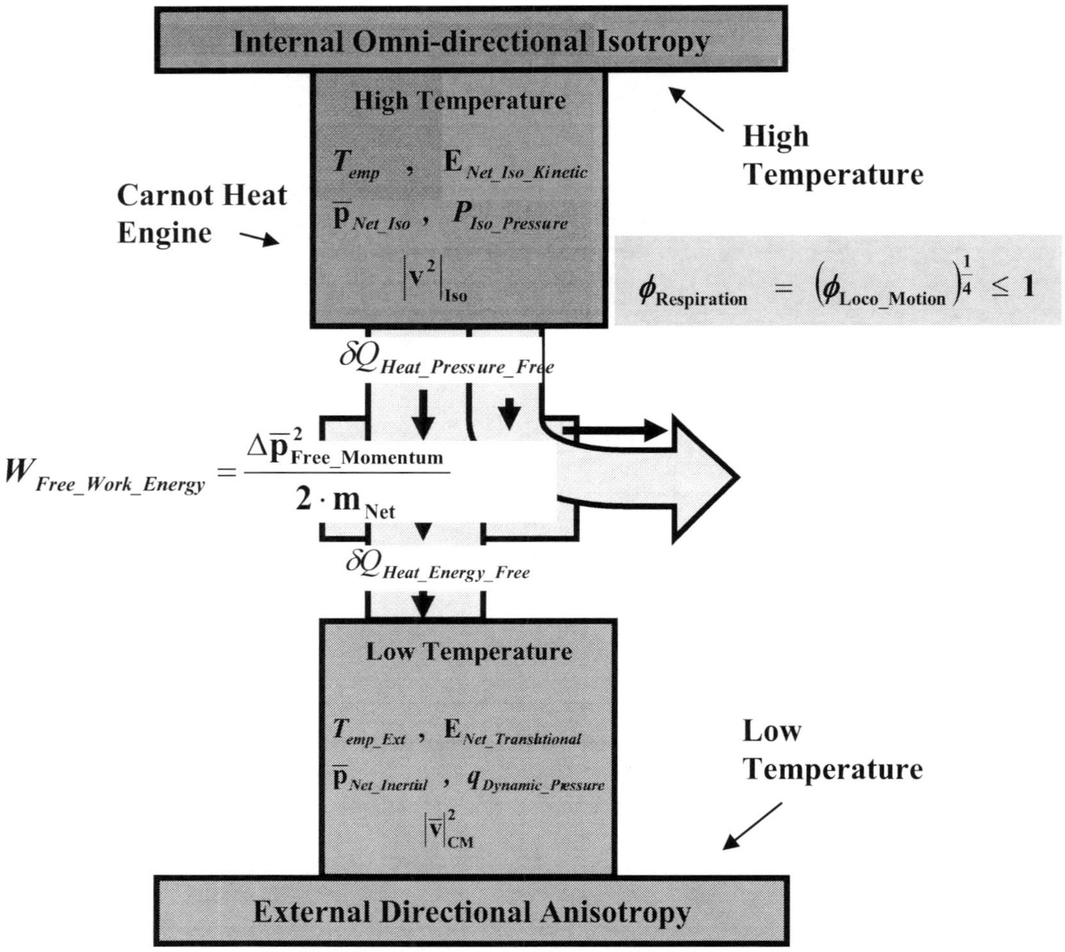

Figure 7.4: Heat Energy Transfer Device – Carnot Thermal Heat Engine.

The Law of Conservation of Energy and Motion, for Anisotropy:

- **External Anisotropic Inertial Linear Net Momentum**

$$\overline{p}_{\text{Inertial\_Net}} = m_{\text{Net}} \cdot |\overline{v}|_{\text{CM}} \rightarrow kg \cdot m/s$$

- **External Anisotropic Translational Kinetic Energy**

$$E_{\text{Net\_Translational}} = \frac{1}{2} \cdot m_{\text{Net}} \cdot |\overline{v}|_{\text{CM}}^2 = \frac{3}{2} \cdot q_{Dynamic\_Pressure} \cdot V_{\text{ol}} = \frac{3}{2} \cdot N \cdot k_B \cdot T_{emp\_Ext} \rightarrow kg\ m^2/s^2$$

- **External Anisotropic Aerodynamic Pressure**

$$q_{\text{Dynamic\_Pressure}} = \frac{2}{3} \cdot \left(\frac{E_{\text{Net\_Translational}}}{V_{\text{ol}}} \right) = \frac{1}{3} \cdot \left(\frac{m_{\text{Net}}}{V_{\text{ol}}} \right) \cdot |\overline{v}|_{\text{CM}}^2 \rightarrow kg/m \cdot s^2$$

- **External Anisotropic Aerodynamic Temperature**

$$T_{emp\_Ext} = \frac{q_{Dynamic\_Pressure} \cdot V_{\text{ol}}}{N \cdot k_B} = \frac{2}{3} \cdot \left(\frac{E_{\text{Net\_Translational}}}{N \cdot k_B} \right) \rightarrow K$$

The Law of Conservation of Energy and Motion, for Isotropy:

- **Internal Isotropic Inertial Linear Net Momentum**

$$\overline{p}_{\text{Iso\_Net}} = m_{\text{Net}} \cdot \sqrt{|v^2|_{\text{Iso}}} \rightarrow kg \cdot m/s$$

- **Internal Isotropic Omni-directional Kinetic Energy**

$$E_{\text{Net\_Iso\_Kinetic}} = \frac{1}{2} \cdot m_{\text{Net}} \cdot |v^2|_{\text{Iso}} = \frac{3}{2} \cdot P_{Iso\_Pressure} \cdot V_{\text{ol}} = \frac{3}{2} \cdot N \cdot k_B \cdot T_{emp} \rightarrow kg \cdot m^2/s^2$$

- **Internal Isotropic Omni-directional Pressure**

$$P_{Iso\_Pressure} = \frac{2}{3} \cdot \left(\frac{E_{\text{Net\_Iso\_Kinetic}}}{V_{\text{ol}}} \right) = \frac{1}{3} \cdot \left(\frac{m_{\text{Net}}}{V_{\text{ol}}} \right) \cdot |v^2|_{\text{Iso}} \rightarrow kg/m \cdot s^2$$

- **Internal Isotropic Absolute Temperature**

$$T_{emp} = \frac{P_{Iso\_Pressure} \cdot V_{\text{ol}}}{N \cdot k_B} = \frac{2}{3} \cdot \left(\frac{E_{\text{Net\_Iso\_Kinetic}}}{N \cdot k_B} \right) \rightarrow K$$

The ***Inertial Mass Kinetic Energy – Carnot Heat Engine- Transfer mechanism*** is used to transfer kinetic heat energy, via kinetic pressure differences between the Isotropy and the Anisotropy of the isolated system and does work in the process.

The Work done by Inertial Mass Kinetic Energy Transfer mechanism of the system mass body is to transfer kinetic energy in the form of heat energy based on: Temperature difference, Kinetic Energy difference, and Pressure differences; between Internal Isotropy and External Anisotropy reservoirs internal and external to an isolated system mass body.

The Aether Kinematic Work "Free" Energy ($W_{Free\_Work\_Energy}$) describes the work done heating and cooling matter, space, and time relative to the Center of Mass and the Center of Isotropy of an isolated system mass body in accordance with the First Law of Motion.

The Aether Kinematic Work "Free" Energy ($W_{Free\_Work\_Energy}$) is also a measure of the Aether "Free" Force doing work creating a dimensional difference between isotropy and anisotropy motions, namely: Momentum Change, Pressure Change, Kinetic Energy Change, Heat Energy Change, and Temperature Change relative to the Center of Mass and the Center of Isotropy of an isolated system mass body.

The Aether Kinematic "Free" Work Energy ($W_{Free\_Work\_Energy}$) is a quantity which represents the work done for free as the Isotropy changes relative to the Anisotropy of an isolated system mass body immersed in any atomic substance material medium, including the Vacuum of Spacetime.

Aether Kinematic Work "Free" Energy – Carnot Heat Engine

7.8

$$W_{Free\_Work\_Energy} = \frac{W_{Thermo\_Pump\_Energy}}{\gamma^4_{Super\_Compress}} \rightarrow kg \cdot m^2/s^2$$

$$W_{Free\_Work\_Energy} = \left(\frac{\Delta \overline{p}^2_{Free\_Momentum}}{2 \cdot m_{Net}} \right) = \left(\frac{\left[\overline{p}^2_{Net\_Iso} - \overline{p}^2_{Inertial\_Net} \right]}{2 \cdot m_{Net}} \right)$$

$$W_{Free\_Work\_Energy} = \frac{1}{\gamma^4_{Super\_Compress}} \cdot \left(\frac{\overline{p}^2_{Thermo\_Momentum}}{2 \cdot m_{Net}} \right)$$

The ability of the ***Inertial Mass Kinetic Energy – Carnot Heat Engine-Transfer mechanism*** of an isolated system mass body is to transfer heat from the external anisotropy to the internal isotropy or vice versa depending on the Inertial Locomotion Foreshortening Factor increases or decreases ($\phi_{\text{Loco\_Motion}} = \left(\frac{|\overline{\text{v}}|^2_{\text{CM}}}{|\text{v}^2|_{\text{Iso}}}\right) = \left(\frac{T_{emp\_Ext}}{T_{emp}}\right)$), and as the square of the Respiration Spacetime Factor ($\phi^2_{\text{Re}\,spiration} = \left(\frac{\gamma_{Inertial\_Iso}}{\gamma_{Inertial}}\right) = \left(\frac{\rho_{\text{Thermo\_Iso\_Mass}}}{\rho_{\text{Thermo\_Aniso\_Mass}}}\right)$) increases or decreases, and likewise as the Mach number increases or decreases ($M_{Mach} = \left(\frac{|\overline{\text{v}}|_{\text{CM}}}{\text{c}_{\text{Sound}}}\right)$); which will be later described.

Carnot Thermal Heat Pump - ***Aether Mass Kinetic Energy Transfer mechanism***

The second heat energy transfer mechanism inherent to an isolated system mass body is called a **Carnot Thermal Heat Pump** - ***Aether Mass Kinetic Energy Transfer mechanism***, which functions like a Thermal Heat Pump doing work transferring heat energy between two different density reservoirs, with a goal to establish equilibrium between an inertial net mass body and its atmospheric surroundings medium.

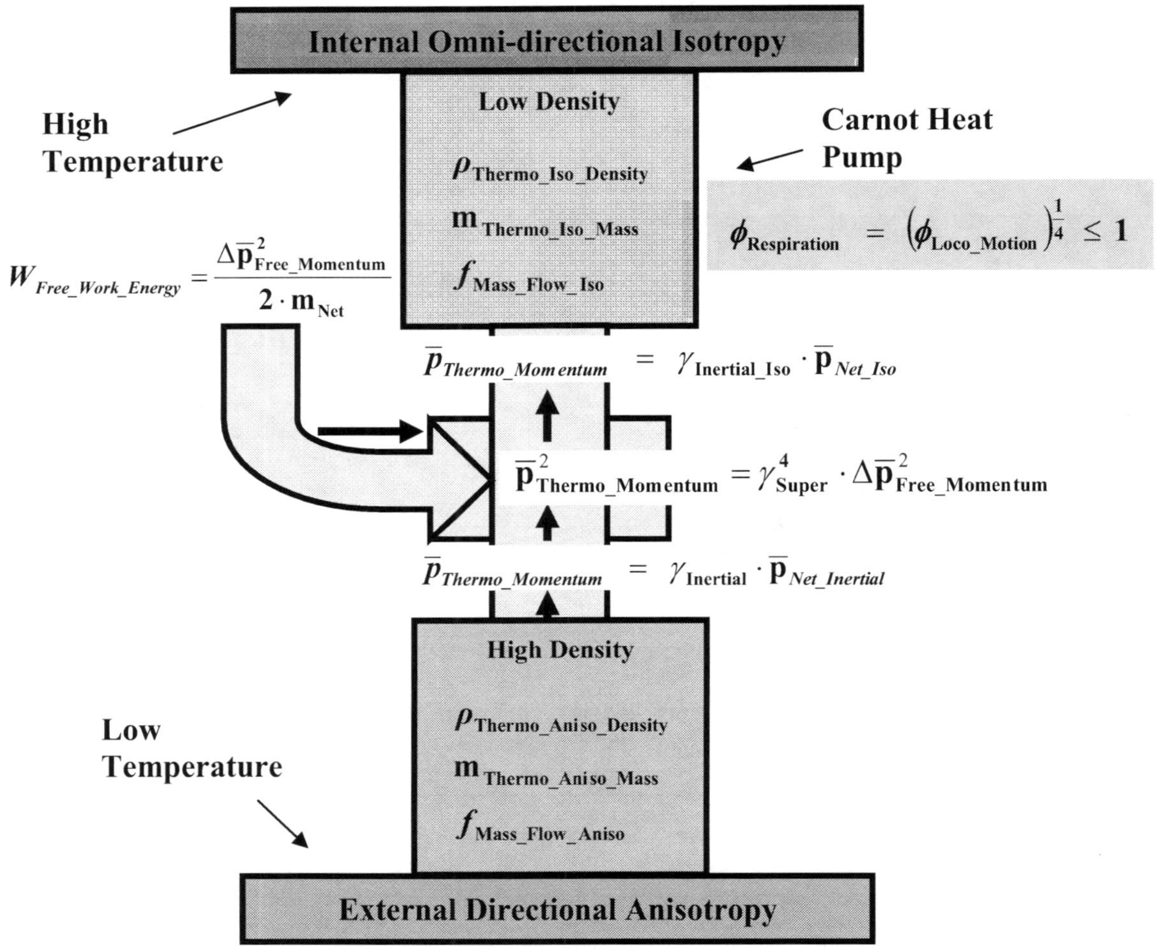

Figure 7.5: Heat Energy Transfer Device – Carnot Thermal Heat Pump.

The ***Aether Mass Kinetic Energy Transfer*** is also the work of a Rarefaction/Condensing mechanism of the system mass body used to reduce or increase the aether mass density.

The Work done by ***Aether Mass Kinetic Energy –Carnot Heat Pump - Transfer mechanism*** is to transfer kinetic energy in the form of heat energy based on Aether mass-energy difference, Aether density difference , and Aether Pressure differences between Internal Isotropy and External Anisotropy reservoirs internal and external to an isolated system mass body.

The Thermodynamic Pump Energy ($W_{Thermo\_Pump\_Energy}$) which behaves like a diffusing diaphragm lets in one type of matter energy into the interior of the body and prevents another type of matter from escaping the body.

The Thermodynamic Pump Energy ($W_{Thermo\_Pump\_Energy}$) describes the work done Rarefying/Condensing matter, space, and time relative to the Center of Mass and the Center of Isotropy of an isolated system mass body in accordance with the First Law of Motion.

The ***Aether Mass Kinetic Energy Transfer mechanism*** uses the Thermodynamic Pump Energy ($W_{Thermo\_Pump\_Energy}$) of a system for maximum useful work possible during a process that brings the dynamic system mass body into mass density equilibrium with a heat reservoir atmosphere environment.

Relative to an inertial frame, the Mean Center, the Center of Mass, and the Center of Isotropy of an isolated system mass body, the **Thermodynamic Pump Work Energy (** $W_{Thermo\_Pump\_Energy}$ **)** is constant, and conserved, and is the measure of the work done by the kinetic energy diaphragm which causes an increase and decrease in volume caused by the action of varying the density by transporting Aether fluid out the one reservoir and transports energy into the other reservoir and vice versa; in any atomic substance material medium including the Vacuum of Spacetime.

Thermodynamic Pump Work Energy- Carnot Heat Pump

7.9

$$W_{Thermo\_Pump\_Energy} = \left(\frac{\overline{p}^{2}_{Thermo\_Momentum}}{2 \cdot m_{Net}} \right) \rightarrow kg \cdot m^2 / s^2$$

$$W_{Thermo\_Pump\_Energy} = \gamma^{4}_{Super\_Compress} \cdot W_{Free\_Work\_Energy}$$

$$W_{Thermo\_Pump\_Energy} = \left(\frac{\overline{p}^{2}_{Thermo\_Momentum}}{2 \cdot m_{Net}} \right) = \gamma^{4}_{Super\_Compress} \cdot \left(\frac{\Delta\overline{p}^{2}_{Free\_Momentum}}{2 \cdot m_{Net}} \right)$$

Stated another way, the ***Aether Mass Kinetic Energy Transfer mechanism*** is used to reduce or increase the Aether Mass Density of either the external anisotropy atmospheric environment or the internal isotropy atmospheric environment of an isolated system mass body.

The ***Aether Mass Kinetic Energy Transfer mechanism*** serves as a barrier between two reservoirs, moving slightly up into one chamber or down into the other allowing aether mass to flow in a particular direction depending differences in density and pressure between a dynamic system mass body and its surroundings.

The Carnot Thermal Heat Pump is a heat energy transfer mechanism used to transfer kinetic heat energy, via aether mass density differences between the Isotropy and the Anisotropy of the isolated system and does work in the process.

The ***Aether Mass Kinetic Energy Transfer mechanism*** is used to transfer kinetic heat energy, via aether mass density differences between the Isotropy and the Anisotropy of the isolated system which does work rarefying and condensing matter, space, and time in the process.

The **Carnot Thermal Heat Pump** - ***Aether Mass Kinetic Energy Transfer mechanism***, reservoirs are represented by the Anisotropy and Isotropy of the system given by the following two sets of equations:

The Law of Conservation of Energy and Motion, for External Anisotropy:

- **Thermodynamic Anisotropy Surplus Density**

7.10

$$\rho_{\text{Thermo\_Aniso\_Density}} = \gamma_{Inertial} \cdot \rho_{\text{Net}} = \frac{\rho_{\text{Net}}}{\sqrt{1 - \frac{\gamma_{Heat}}{3} \cdot \left(\frac{|\overline{v}|^2_{\text{CM}}}{c^2_{\text{Sound}}} \right)}}$$

$$\rho_{\text{Thermo\_Aniso\_Density}} = \frac{1}{|\overline{v}|_{\text{CM}}} \cdot \left(\frac{\overline{p}_{\text{Thermo\_Momentum}}}{V_{\text{ol}}} \right) = \frac{\left(\frac{m_{\text{Net}}}{V_{\text{ol}}} \right)}{\sqrt{1 - \frac{\gamma_{Heat}}{3} \cdot \left(\frac{|\overline{v}|^2_{\text{CM}}}{c^2_{\text{Sound}}} \right)}} \rightarrow {}^{kg}\!/_{m^3}$$

- **Thermodynamic Anisotropy Surplus Aether Dynamic Pressure**

7.11

$$q_{\text{Aether\_Dynamic\_Pressure}} = \frac{1}{3} \cdot \rho_{\text{Thermo\_Aniso\_Density}} \cdot |\overline{v}|^2_{\text{CM}} = \frac{1}{3} \cdot \left(\frac{m_{\text{Thermo\_Aniso\_Mass}}}{V_{\text{ol}}} \right) \cdot |\overline{v}|^2_{\text{CM}} \rightarrow {}^{kg}\!/_{m \cdot s^2}$$

- **Thermodynamic Anisotropy Surplus Inertia Mass and Resistance**

7.12

$$m_{\text{Thermo\_Aniso\_Mass}} = \gamma_{Inertial} \cdot m_{\text{Net}}$$

$$m_{\text{Thermo\_Aniso\_Mass}} = \frac{\overline{p}_{\text{Thermo\_Momentum}}}{|\overline{v}|_{\text{CM}}} = \frac{m_{\text{Net}}}{\sqrt{1 - \frac{\gamma_{Heat}}{3} \cdot \left(\frac{|\overline{v}|^2_{\text{CM}}}{c^2_{\text{Sound}}} \right)}} \rightarrow kg$$

The Law of Conservation of Energy and Motion, for Internal Isotropy:

- **Thermodynamic Isotropy Surplus Density**

7.13

$$\rho_{\text{Thermo\_Iso\_Density}} = \gamma_{Inertial\_Iso} \cdot \rho_{\text{Net}} = \frac{\rho_{\text{Net}}}{\sqrt{\left(\frac{3}{\gamma_{Heat}} \cdot \left(\frac{c^2_{\text{Sound}}}{|\overline{v}|^2_{\text{CM}}}\right) - 1\right)}}$$

$$\rho_{\text{Thermo\_Iso\_Density}} = \frac{\overline{p}_{\text{Thermo\_Momentum}}}{V_{ol} \cdot \sqrt{|v^2|_{\text{Iso}}}} = \frac{\left(\frac{m_{\text{Net}}}{V_{ol}}\right)}{\sqrt{\left(\frac{3}{\gamma_{Heat}} \cdot \left(\frac{c^2_{\text{Sound}}}{|\overline{v}|^2_{\text{CM}}}\right) - 1\right)}} \rightarrow {}^{kg}\!/_{m^3}$$

- **Thermodynamic Isotropy Surplus Aether Static Pressure**

7.14

$$P_{\text{Aether\_Pressure\_Static}} = \frac{1}{3} \cdot \rho_{\text{Thermo\_Iso\_Density}} \cdot |v^2|_{\text{Iso}} = \frac{1}{3} \cdot \left(\frac{m_{\text{Thermo\_Iso\_Mass}}}{V_{ol}}\right) \cdot |v^2|_{\text{Iso}} \rightarrow {}^{kg}\!/_{m \cdot s^2}$$

- **Thermodynamic Isotropy Surplus Inertia Mass and Resistance**

7.15

$$m_{\text{Thermo\_Iso\_Mass}} = \gamma_{Inertial\_Iso} \cdot m_{\text{Net}}$$

$$m_{\text{Thermo\_Iso\_Mass}} = \frac{\overline{p}_{\text{Thermo\_Momentum}}}{\sqrt{|v^2|_{\text{Iso}}}} = \frac{m_{\text{Net}}}{\sqrt{\left(\frac{3}{\gamma_{Heat}} \cdot \left(\frac{c^2_{\text{Sound}}}{|\overline{v}|^2_{\text{CM}}}\right) - 1\right)}} \rightarrow kg$$

Now, let's consider an isolated dynamic system mass body such as an atomic electron unit or a cosmic body such as planet earth or even a galaxy; where each body via its own Thermodynamic Momentum "pump" blows out or sucks in Aether fluid from the external environment.

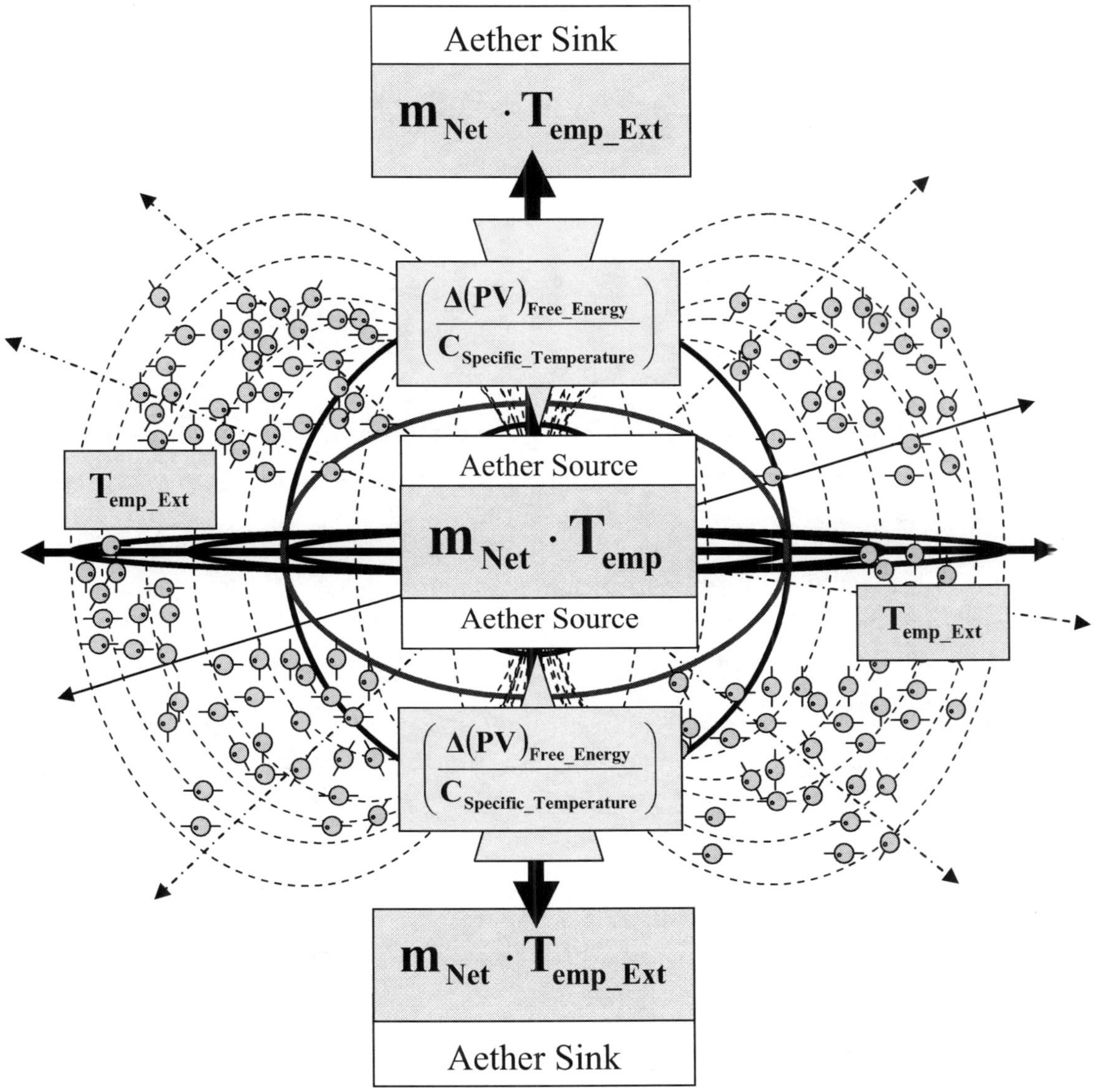

Figure 7.6: A model of the Thermodynamic Pump Energy which is used to create an aether density difference between the system mass body and its environment.

The Thermodynamic Heat Pump transports the Aetherons which behave like a fluid that diffuses into the volume of the pump and imparts surplus or deficit mass, density, pressure, and kinetic energy relative to the isotropy to anisotropy ratio of the system as a whole.

Thus, the Thermodynamic Heat Pump can be thought of as a Carnot heat engine which is operating in reverse.

Positive displacement and momentum transfer pumps have a constant volume flow rate, and pumping speed, so that, as the volume expands the pressure drops; in which the mass, mass density, hydrodynamic pressure, and hydrodynamic force increases.

The Thermodynamic Momentum Pumping speed refers to the mass flow rate and mass density increase or decrease at its pump inlet.

Thermodynamic Momentum Pumping and energy transfer are more effective on some gases than others, so the pumping rate can be different for each of the various gases of material substances and aether being pumped, and the average mass flow rate of the pump will vary depending on the chemical composition of the gases.

The specific heat capacity index (γ_{Heat}) and the speed of sound (c^2_{Sound}) are the theoretical and experimental parameters or physical quantities which allows for different material substances to be evaluated.

The main action of the Thermodynamic Pump energy is to pressurize and transport fluids. Since gases are compressible, the compressor mechanism also reduces the volume of a gas.

When a gas is compressed it increases the pressure of a gas by reducing its volume. Compressors are similar to pumps in that both increase the pressure on a fluid and both can transport the fluid.

In general the Thermodynamic Momentum and Energy function like a vacuum pump. Since fluids can only be pushed and cannot be pulled, it is technically impossible to create a vacuum by what is "typically imaged" as suction.

Suction is the movement of fluids into a vacuum under the effect of a pressure difference. Suction is created whenever there is a pressure or density difference between its internal and external atmospheric conditions.

More specifically "Suction" occurs for any system mass body whenever there is an Isotropy to Anisotropy kinetic energy, pressure, and density difference between its internal and external atmospheric conditions.

Vacuum Suction Pressure is thus a flow from higher pressure to lower pressure or higher density to lower density.

The easiest way to create an artificial vacuum is to expand the volume of a container. For example, consider a human body where the diaphragm muscle expands the chest cavity, which causes the volume of the lungs to increase.

This expansion in the chest cavity representing volume reduces the pressure and creates a partial vacuum, which is soon filled by a specific quantity of air. Once the cavity or volume is filled with air to its maximum capacity it is pushed or rather "Sucked" in by the external atmospheric pressure.

This cycle is repeated in the human body many times and is the source of existence. If this pumping cycle in the chest cavity were to be stalled or ceased. Then life itself for that person would also cease.

Similarly, the Thermodynamic Momentum pump is a mechanism which expands the volume to create a vacuum; or rarefied state. This rarefied expanded vacuum within a finite amount of time fills up with Matter and Aetherons.

Now, consider a membrane volume that has reached its maximum spherical volume state, at the same time the external atmospheric state also reaches maximum external pressure which pushes inward normal to the surface of the membrane volume.

If the internal isotropic pressure is reduced, in an equal finite amount of time the volume condenses by increasing external pressure, which eventually decreases the volume of the system mass body. This cycle is repeated as long as the isolated dynamic system mass body is in existence.

Let's go further, for example when the spherical volume of the membrane is condensed, or compressed, it is condensed into an oblate ellipsoid or prolate ellipsoid shape.

When the system mass body is expanding it is growing from ellipsoidal to spheroid shape. The natural state of a mass body system is spheroid in shape.

The above system can also be used to describe gravitational effects, star formation, galaxy formation, and opens the door for a cyclic universe model.

A gradient density field for this system mass body is visualized to expand into concentric spherical rings of Ideal Gas (Pressure-Volume) Energy, where each ring has contained within an elliptical condensed state.

Because of the pressure differential, between the ellipsoidal and spheroid shape of matter, space and time, specific quantities of fluid are pushed into the pump's internal or external environment.

The pump's role during cooling is to pump aether energy into the external atmosphere, and condense in volume. Within a finite amount of time the spheroid will condense into an ellipsoid, due to external kinetic energy, pressure, and aether mass density increase.

The pump's role during heating is to pump aether energy internally into the interior of the body, and expand in volume. Within a finite amount of time the ellipsoid will expand into a spheroid, due to external kinetic energy, pressure, and aether mass density decrease.

The Thermodynamic Pump Energy is used by the system to ***heat the body and expand*** by:

- Increasing the Thermo Isotropic inertia mass ($\mathrm{m}_{\mathrm{Thermo\_Iso\_Mass}}$), Aether Isotropy density ($\rho_{\mathrm{Thermo\_Iso\_Density}}$), Aether Isotropy mass flow rate ($f_{Mass\_Flow\_Iso}$), and Internal Aether Static Pressure ($\mathrm{P}_{\mathrm{Aether\_Pressure\_Static}}$)

- Decreasing the Thermo Anisotropic inertia mass ($\mathrm{m}_{\mathrm{Thermo\_Aniso\_Mass}}$), Aether Anisotropy density ($\rho_{\mathrm{Thermo\_Aniso\_Density}}$), Aether Anisotropy mass flow rate ($f_{Mass\_Flow\_Aniso}$), and Dynamic Aether Pressure ($\mathrm{q}_{\mathrm{Aether\_Dynamic\_Pressure}}$)

This creates mass-energy flow towards the low pressure body and away from the high pressure atmospheric environment.

The Thermodynamic Pump Energy is used by the system to ***cool the body and condense*** by:

- Decreasing the Thermo Isotropic inertia mass ($\mathrm{m}_{\mathrm{Thermo\_Iso\_Mass}}$), Aether Isotropy density ($\rho_{\mathrm{Thermo\_Iso\_Density}}$), Aether Isotropy mass flow rate ($f_{Mass\_Flow\_Iso}$), and Internal Aether Static Pressure ($\mathrm{P}_{\mathrm{Aether\_Pressure\_Static}}$

- Increasing the Thermo Anisotropic inertia mass ($\mathrm{m}_{\mathrm{Thermo\_Aniso\_Mass}}$), Aether Anisotropy density ($\rho_{\mathrm{Thermo\_Aniso\_Density}}$), Aether Anisotropy mass flow rate ($f_{Mass\_Flow\_Aniso}$), and Dynamic Aether Pressure ($\mathrm{q}_{\mathrm{Aether\_Dynamic\_Pressure}}$)

This creates mass-energy flow away from the high pressure body and towards the rarified low pressure atmospheric environment.

Thus, the energy input and output of a mass body system must always balance according to the First Law of Thermodynamics or the energy conservation principle.

The Thermodynamic Pump Energy functions as a device that moves heat from one location (the 'source') to another location (the 'sink' or 'heat sink') doing mechanical work via aether mass density differences within and without the system as a whole.

This is similar to gravity in that smaller units of mass are gravitationally attracted to larger units of mass. The Thermodynamic Heat Pump transports aether mass-energy inward towards the interior of the body during cooling and outward towards the exterior of the body during heating.

Thus, whenever the external anisotropy atmosphere of the system body is high density or very dense the body condenses and compresses. And wherever the external anisotropy of the system body is low density or rare the body becomes expanded.

A natural example of this is a phase in a sound wave or phonon. Half of a sound wave is made up of the compression of the medium, and the other half is the decompression or rarefaction of the medium.

Another natural example of rarefaction is in the layers of our atmosphere. Because what constitutes our atmosphere has mass, it is definite that most of the atmospheric matter will be nearer to the Earth. Therefore, air at higher layers of the atmosphere has less pressure, or is rarefied in relation to air at lower layers.

Rarefaction can also be easily observed by compressing a spring and releasing. Rarefaction waves expand with time; for most gases the rarefaction wave keeps the same overall profile at all times (it is a 'self-similar expansion'). Rarefaction can also refer to an area of low relative pressure following a shockwave.

Each part of the wave travels at the local speed of sound, in the local medium. This expansion behavior is in contrast to the behavior of pressure increases, which get narrower with time, until they steepen into shock waves.

Consider an aircraft as it moves through the air, the air molecules near the aircraft are disturbed and move around the aircraft. If the aircraft passes at a low speed, typically less than ($|\overline{\mathrm{v}}|_{\mathrm{CM}} \leq$ 250 mph), the density of the air remains constant, and the external Anisotropic Aerodynamic Temperature (T_{emp_Ext}) is considered low temperature.

But for higher speeds ($|\overline{v}|_{CM}$ ≥ 700 mph) the aerodynamic temperature is increased and is considered high temperature, and some of the energy of the aircraft goes into compressing the air and locally changing the density of the air.

This "super compressibility" effect alters the amount of resulting force on the aircraft.

The effect becomes more important as speed increases. Near and beyond the speed of sound, about ($|\overline{v}|_{CM} \cong$ 330 m/s or 760 mph), small disturbances in the flow are transmitted to other locations with isentropic flow or with constant entropy.

When the external atmosphere surroundings behave as a reservoir, and the internal atmosphere of the mass body also behave as a reservoir, the Thermodynamic Pump Energy ($W_{Thermo\_Pump\_Energy}$) is the potential of a system to cause an aether density change, as the system achieves equilibrium between the two reservoirs.

The action of the Thermodynamic Pump energy is very similar to the action of a plunger with the exception that Thermodynamic Pump energy responds to changes in aether mass density and pressure rather than the mechanical force of the shaft.

The increase and decrease in surplus density caused by the action of the Thermodynamic Pump energy alternately forces aether mass fluid out the one reservoir and transports aether mass energy into the other reservoir.

The Thermodynamic Pump Energy is a combination property of a system and its environment because it depends on the state of both the system and its atmospheric environment.

The atmospheric environment can be the Vacuum of Spacetime, or any atomic substance material, such as air, hydrogen, carbon dioxide, etc.

The Thermodynamic Pump Energy transfers aether mass between the isotropy and anisotropy reservoirs of the body via the aether density differences within and without the dynamic system mass body given by the following equation.

Thermodynamic Super Surplus Density

7.16

$$\rho_{\text{Thermo\_Pump\_Density}} = \gamma_{Super\_Compress} \cdot \rho_{\text{Net}} = \gamma_{Super\_Compress} \cdot \left(\frac{m_{\text{Net}}}{V_{\text{ol}}} \right)$$

$$\rho_{\text{Thermo\_Pump\_Density}} = \frac{\rho_{\text{Net}}}{\left(\frac{3}{\gamma_{Heat}} \cdot \left(\frac{c_{\text{Sound}}^2}{\left|\overline{v}\right|_{\text{CM}}^2} \right) - 2 + \frac{\gamma_{Heat}}{3} \cdot \left(\frac{\left|\overline{v}\right|_{\text{CM}}^2}{c_{\text{Sound}}^2} \right) \right)^{\frac{1}{4}}}$$

$$\rho_{\text{Thermo\_Pump\_Density}} = \frac{\left(\frac{m_{\text{Net}}}{V_{\text{ol}}} \right)}{\left(\frac{3}{\gamma_{Heat}} \cdot \left(\frac{c_{\text{Sound}}^2}{\left|\overline{v}\right|_{\text{CM}}^2} \right) - 2 + \frac{\gamma_{Heat}}{3} \cdot \left(\frac{\left|\overline{v}\right|_{\text{CM}}^2}{c_{\text{Sound}}^2} \right) \right)^{\frac{1}{4}}} \rightarrow {}^{kg}\!/_{m^3}$$

Let's consider again what in general a heat pump does.

A heat pump is a machine or device that moves heat from one location (the 'source') to another location (the 'sink' or 'heat sink') using mechanical work. Most heat pump technology moves heat from a low temperature heat source to a higher temperature heat sink.

During the heating cycle of a dynamic system mass body, heat is taken from the external anisotropy environment and "pumped" internally into the interior isotropy of the system mass body.

Heat pumps can be thought of as a heat engine which is operating in reverse. Common examples are food refrigerators and freezers, air conditioners.

During the cooling cycle of a dynamic system mass body, heat is taken from the internal interior isotropy environment and "pumped" outside to the external anisotropy environment of the system mass body.

Cooling and Refrigeration is the process of removing heat from an enclosed space, or from a substance, and moving it to a place where it is high speed kinetic energy and pressure to a place where it is low speed kinetic energy and pressure.

Cooling and Refrigeration is also the process of removing heat from an enclosed space, or from a substance, and moving it to a place where it is high aether mass density and pressure, or to a place where it is low aether mass density and pressure.

The primary purpose of refrigeration is lowering the temperature of the enclosed space or substance and then maintaining that lower temperature.

The term cooling refers generally to any natural or artificial process by which heat is dissipated.

Cold is the absence of heat, hence in order to decrease a temperature, one "removes heat", rather than "adding cold." In order to satisfy the Second Law of Thermodynamics, some form of work must be performed to accomplish this.

The ability of the Thermodynamic Momentum heat pump mechanism of an isolated system mass body is to transfer heat from the external anisotropy to the internal isotropy or vice versa depending on the Inertial Locomotion Foreshortening Factor increases or decreases ($\phi_{\text{Loco\_Motion}} = \left(\frac{|\overline{\text{v}}|^2_{\text{CM}}}{|\text{v}^2|_{\text{Iso}}} \right) = \left(\frac{T_{emp\_Ext}}{T_{emp}} \right)$), and as the square of the Respiration Spacetime Factor ($\phi^2_{\text{Re}\,spiration} = \left(\frac{\gamma_{Inertial\_Iso}}{\gamma_{Inertial}} \right) = \left(\frac{\rho_{\text{Thermo\_Iso\_Mass}}}{\rho_{\text{Thermo\_Aniso\_Mass}}} \right)$) increases or decreases, and likewise as the Mach number increases or decreases ($M_{Mach} = \left(\frac{|\overline{\text{v}}|_{\text{CM}}}{\text{c}_{\text{Sound}}} \right)$).

7.3 The Kemp Matter-Aether Engine

Carnot Heat Pump Relation

Universal Geometric Mean Ratio – Respiration Spacetime Factor

7.17

$$\phi^2_{Respiration} = \left(\frac{\gamma_{Inertial\_Iso}}{\gamma_{Inertial}}\right) = \left(\frac{\rho_{\text{Thermo\_Iso\_Density}}}{\rho_{\text{Thermo\_Aniso\_Density}}}\right)$$

$$\phi^2_{Respiration} = \sqrt{\phi_{\text{Loco\_Motion}}} \rightarrow Unitless$$

Carnot Heat Engine Relation

Universal Geometric Mean Ratio – Inertial Locomotion Foreshortening Factor

7.18

$$\phi_{\text{Loco\_Motion}} = \phi^4_{Respiration} = \left(\frac{\left|\overline{\mathrm{v}}\right|^2_{\mathrm{CM}}}{\left|\mathrm{v}^2\right|_{\mathrm{Iso}}}\right) = \frac{\gamma_{Heat}}{3} \cdot \left(\frac{\left|\overline{\mathrm{v}}\right|^2_{\mathrm{CM}}}{\mathrm{c}^2_{\mathrm{Sound}}}\right)$$

$$\phi_{\text{Loco\_Motion}} = \left(\frac{\mathrm{E}_{\text{Net\_Translational}}}{\mathrm{E}_{\text{Net\_Iso\_Kinetic}}}\right) = \left(\frac{q_{Dynamic\_Pressure}}{P_{Iso\_Pressure}}\right) \rightarrow Unitless$$

External Environment *Heating* & Mass Body Expansion

- **Carnot Thermal Engine**
  - Thermodynamic Pump Respiration Ratio

$$\phi^2_{Respiration} = \left(\frac{\gamma_{Inertial\_Iso}}{\gamma_{Inertial}}\right) = \left(\frac{\rho_{\text{Thermo\_Iso\_Mass}}}{\rho_{\text{Thermo\_Aniso\_Mass}}}\right) \geq 1$$

- **Carnot Thermal Heat Pump**
  - Inertial Matter Inertial Locomotion Foreshortening Factor

$$\phi_{\text{Loco\_Motion}} = \frac{\gamma_{Heat}}{3}\cdot\left(\frac{\left|\overline{\text{v}}\right|^2_{\text{CM}}}{\text{c}^2_{\text{Sound}}}\right) = \left(\frac{q_{Dynamic\_Pressure}}{P_{Iso\_Pressure}}\right) \geq 1$$

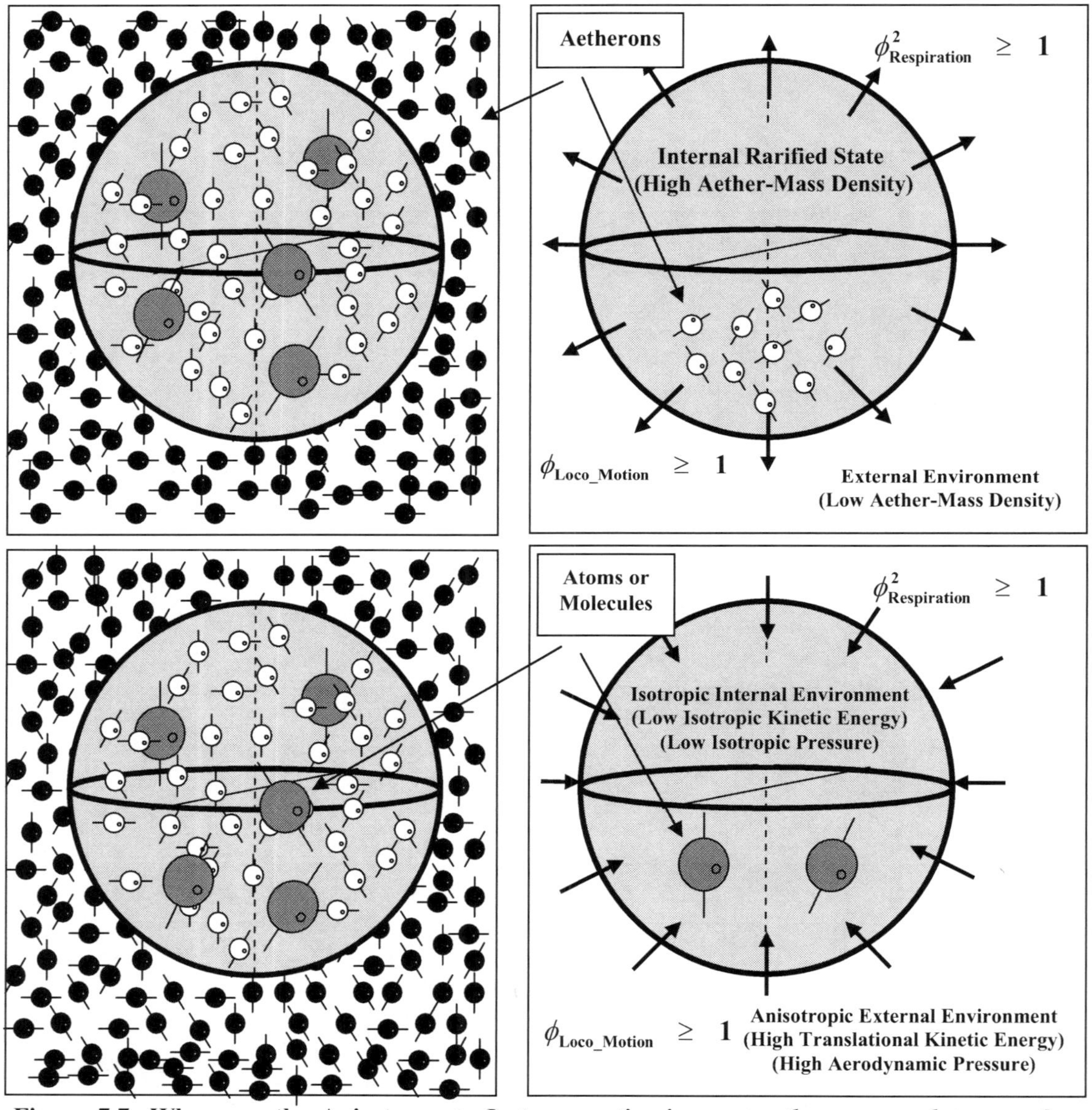

Figure 7.7: Whenever the Anisotropy to Isotropy ratios is *greater than or equal to one*, the system mass body is *Heated and Expanded.*

External Environment *Cooling* and Mass Body Condensing

- **Carnot Thermal Engine**
  - Thermodynamic Pump Respiration Ratio

$$\phi^2_{Respiration} = \left(\frac{\gamma_{Inertial\_Iso}}{\gamma_{Inertial}}\right) = \left(\frac{\rho_{Thermo\_Iso\_Mass}}{\rho_{Thermo\_Aniso\_Mass}}\right) \leq 1$$

- **Carnot Thermal Heat Pump**
  - Inertial Matter Inertial Locomotion Foreshortening Factor

$$\phi_{Loco\_Motion} = \frac{\gamma_{Heat}}{3} \cdot \left(\frac{|\overline{v}|^2_{CM}}{c^2_{Sound}}\right) = \left(\frac{q_{Dynamic\_Pressure}}{P_{Iso\_Pressure}}\right) \leq 1$$

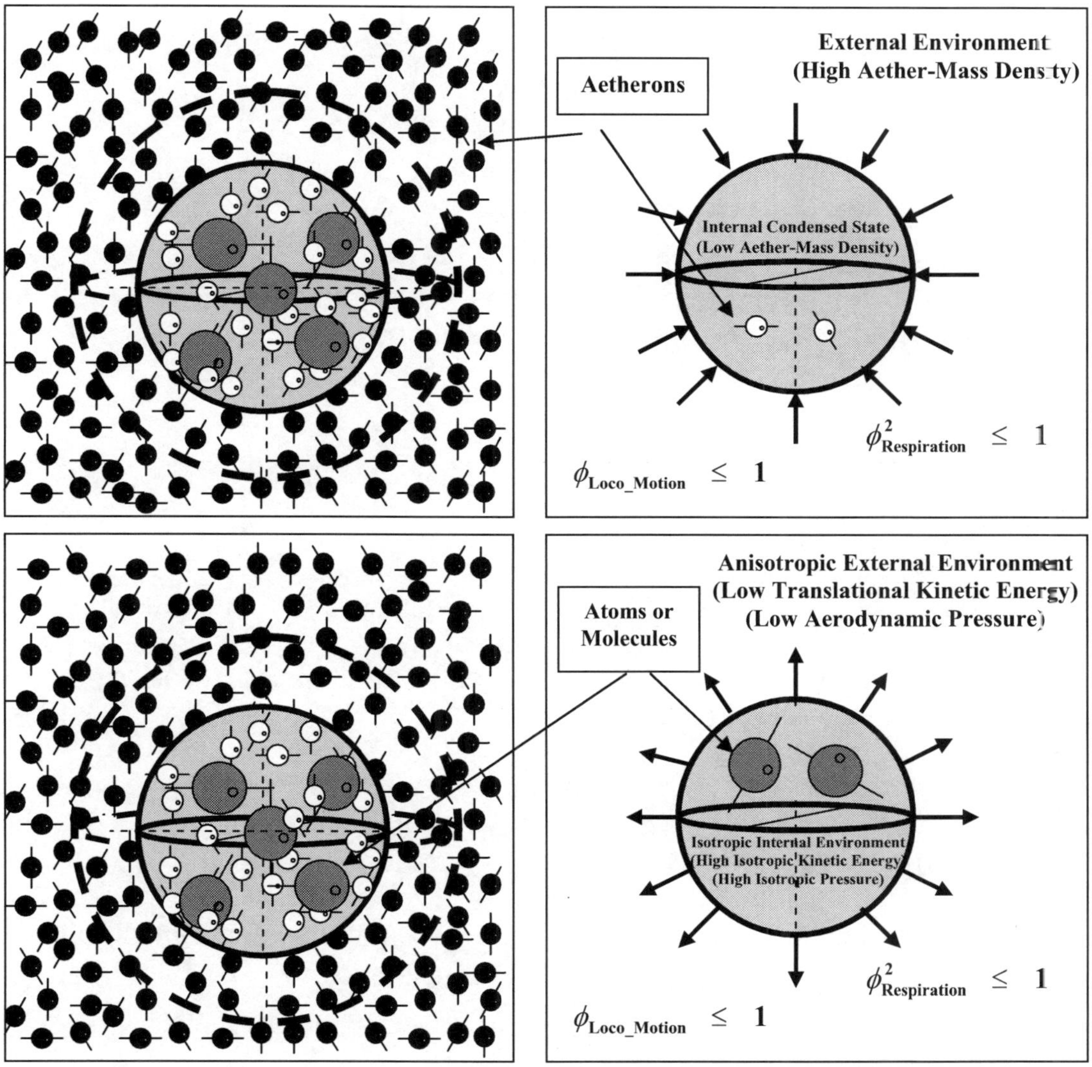

Figure 7.8: Whenever the Anisotropy to Isotropy ratios is *less than or equal to one*, the system mass body is *Cooled and Condensed.*

The Thermodynamic Momentum of an isolated dynamic system mass body in general is a positive displacement vacuum pump which moves the same volume of gas with each cycle, so its pumping speed is constant unless it is overcome by backstreaming.

Backstreaming occurs whenever the ratio of Anisotropy to Isotropy is stalled or the loco motion foreshortening factor is approximately or equal to one $\left(\phi_{\text{Loco\_Motion}} = \left(\frac{|\overline{v}|^2_{\text{CM}}}{|v^2|_{\text{Iso}}}\right) = \frac{\gamma_{Heat}}{3}\cdot\left(\frac{|\overline{v}|^2_{\text{CM}}}{c^2_{\text{Sound}}}\right) \approx 1\right)$.

The Carnot Heat engine external to internal atmospheric balance point is reached when the isotropic kinetic energy and pressure are equal to the anisotropic kinetic energy and pressure.

7.19

$$\phi_{\text{Loco\_Motion}} = \left(\frac{|\overline{v}|^2_{\text{CM}}}{|v^2|_{\text{Iso}}}\right) = \frac{\gamma_{Heat}}{3}\cdot\left(\frac{|\overline{v}|^2_{\text{CM}}}{c^2_{\text{Sound}}}\right) = \left(\frac{q_{Dynamic\_Pressure}}{P_{Iso\_Pressure}}\right) = 1$$

And likewise the Carnot Heat pump external to internal atmospheric balance point is reached when Thermodynamic Momentum heat pump's heat gain is equal to the heat loss of the system mass body.

7.20

$$\phi^2_{Respiration} = \left(\frac{\gamma_{Inertial\_Iso}}{\gamma_{Inertial}}\right) = \left(\frac{\rho_{\text{Thermo\_Iso\_Mass}}}{\rho_{\text{Thermo\_Aniso\_Mass}}}\right) = 1$$

Below this external atmospheric aerodynamic temperature, the Thermodynamic Momentum heat pump can supply only part of the heat required to keep the circulation moving, and keep the system balanced, thus supplementary heat is required.

The heat delivered by the Thermodynamic Momentum heat pump mechanism of an isolated system mass body is theoretically the sum of the heat extracted from the heat source, and the energy needed to drive the cycle.

The cycle described above is reversed to cool the system mass body. The system takes heat out of the internal isotropy interior (T_{emp}) environment and rejects it to the outside anisotropic external environment (T_{emp_Ext}).

Isolated System Mass Body & Environment Cooling

$$\phi_{\mathrm{Locomotion}} = \frac{\left|\overline{\mathbf{v}}\right|^2_{\mathrm{CM}}}{\left|\mathbf{v}^2\right|_{\mathrm{Iso}}} = \left(\frac{q_{Dynamic\_Pressure}}{P_{Iso\_Pressure}}\right) \leq 1$$

Isolated System Mass Body & Environment Condensing

$$\phi^2_{\mathrm{Respiration}} = \left(\frac{\rho_{\mathrm{Thermo\_Iso\_Density}}}{\rho_{\mathrm{Thermo\_Aniso\_Density}}}\right) \leq 1$$

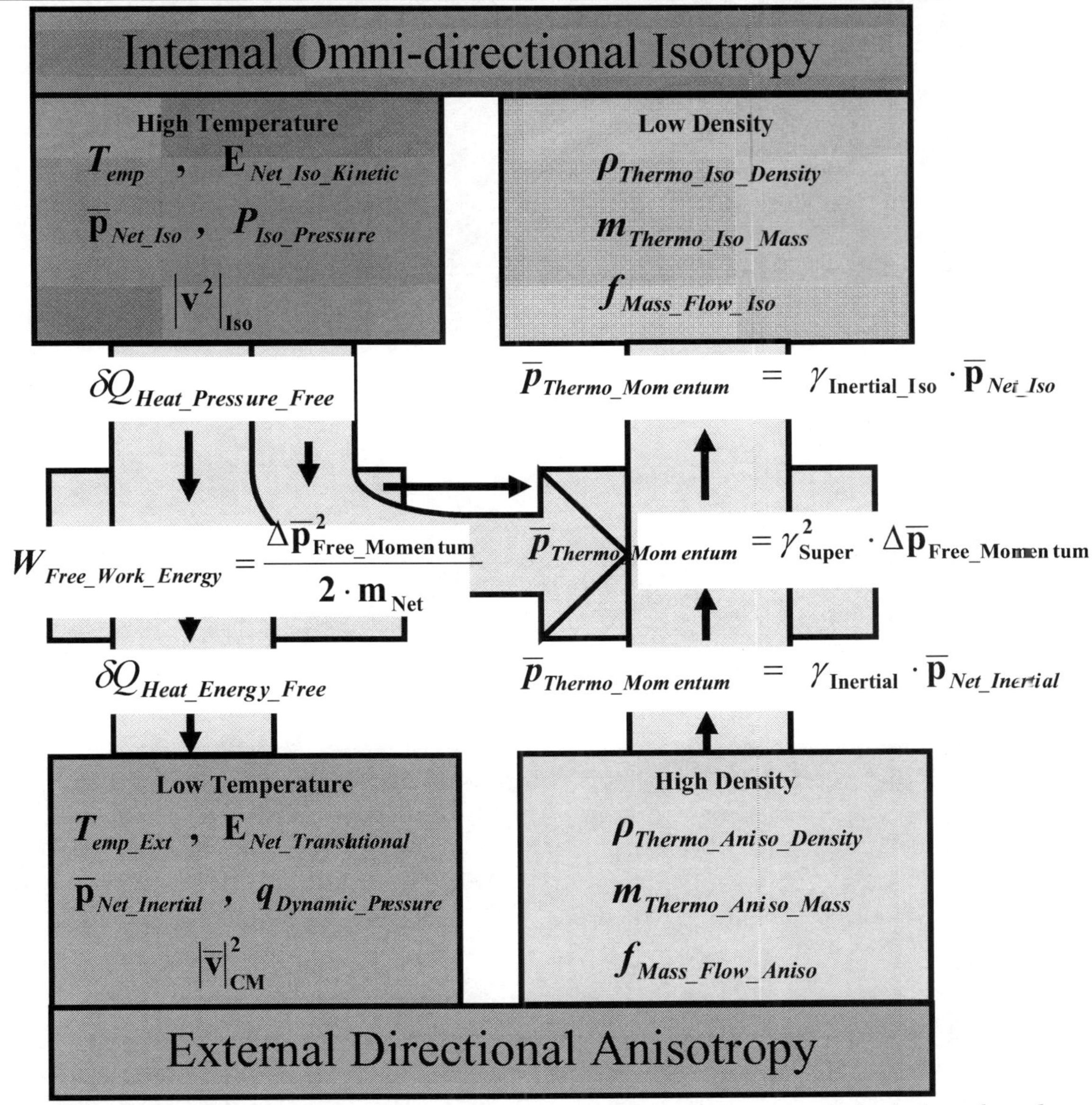

Figure 7.9: The Thermodynamic Pump Energy is used by the system to *cool the body and condense* via a difference in the anisotropy to isotropy of an isolated dynamic system.

Isolated System Mass Body & Environment Heating

$$\phi_{\text{Locomotion}} = \left(\frac{|\overline{\mathbf{v}}|^2_{\text{CM}}}{|\mathbf{v}^2|_{\text{Iso}}} \right) = \left(\frac{q_{Dynamic\_Pressure}}{P_{Iso\_Pressure}} \right) \geq 1$$

Isolated System Mass Body & Environment Expansion

$$\phi^2_{\text{Repiration}} = \left(\frac{\rho_{\text{Thermo\_Iso\_Density}}}{\rho_{\text{Thermo\_Aniso\_Density}}} \right) \geq \mathbf{1}$$

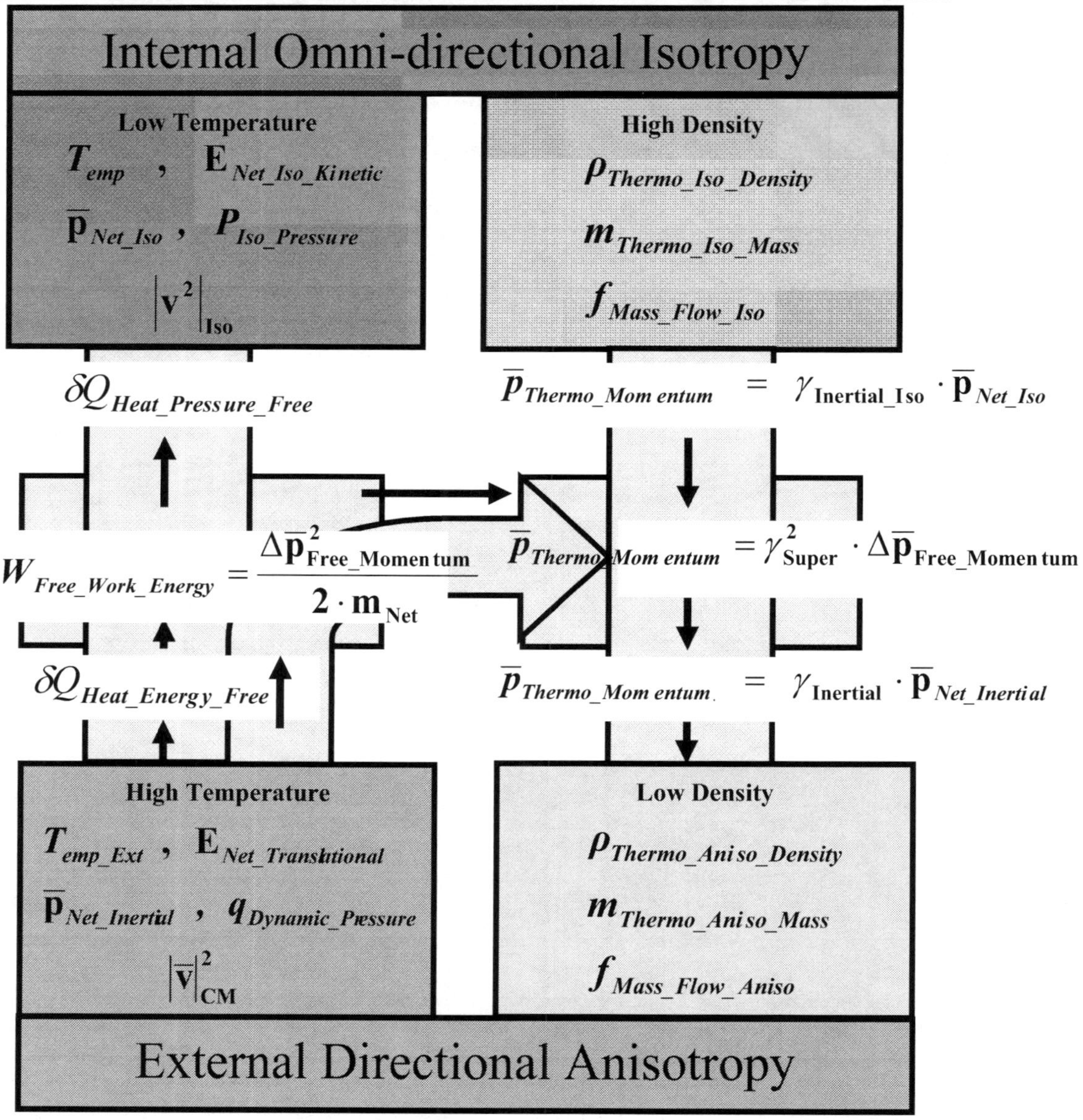

Figure 7.10: The Thermodynamic Pump Energy is used by the system to *heat the body and expand* via a difference in the anisotropy to isotropy of an isolated dynamic system.

The Thermodynamic Momentum heat pump ($\overline{p}_{Thermo\_Momentum}$) divided by the Internal Isotropic Omni-directional Momentum ($\overline{p}_{Net\_Iso}$) or the External Anisotropic Rectilinear Momentum ($\overline{p}_{Inertial\_Net}$) of the system is the measure of the ratio of heat delivered by the heat pump, and the Aether-Kinematic Work "Free" Momentum ($\Delta\overline{p}_{Free\_Momentum}$) required to transfer heat energy between any two or more mass density reservoirs, and between any two or more temperature reservoir bodies in a single isolated dynamic mass body system.

The presence of an isolated dynamic molecular fluid environment continually adds and removes Aether kinetic energy from the system, causing an ensemble of mass and Aether to mix together through many, many collisional interactions, which spread out, and expand or condense within the volume of a system mass body and its surroundings.

The system and its surroundings eventually reach thermal equilibrium between a mass body and its environment medium.

The Thermodynamic Momentum of an isolated dynamic system body acts as vacuum pump that removes and adds Aether kinetic energy through volume, area, and distance of space in order to leave behind a partial vacuum; or rarefied state.

Isolated System Mass Body & Environment Cooling

$$\phi_{\text{Locomotion}} = \frac{\left|\overline{\mathbf{v}}\right|^2_{\text{CM}}}{\left|\mathbf{v}^2\right|_{\text{Iso}}} = \left(\frac{q_{Dynamic\_Pressure}}{P_{Iso\_Pressure}}\right) \leq 1$$

Isolated System Mass Body & Environment Condensing

$$\phi^2_{\text{Respiration}} = \left(\frac{\rho_{\text{Thermo\_Iso\_Density}}}{\rho_{\text{Thermo\_Aniso\_Density}}}\right) \leq 1$$

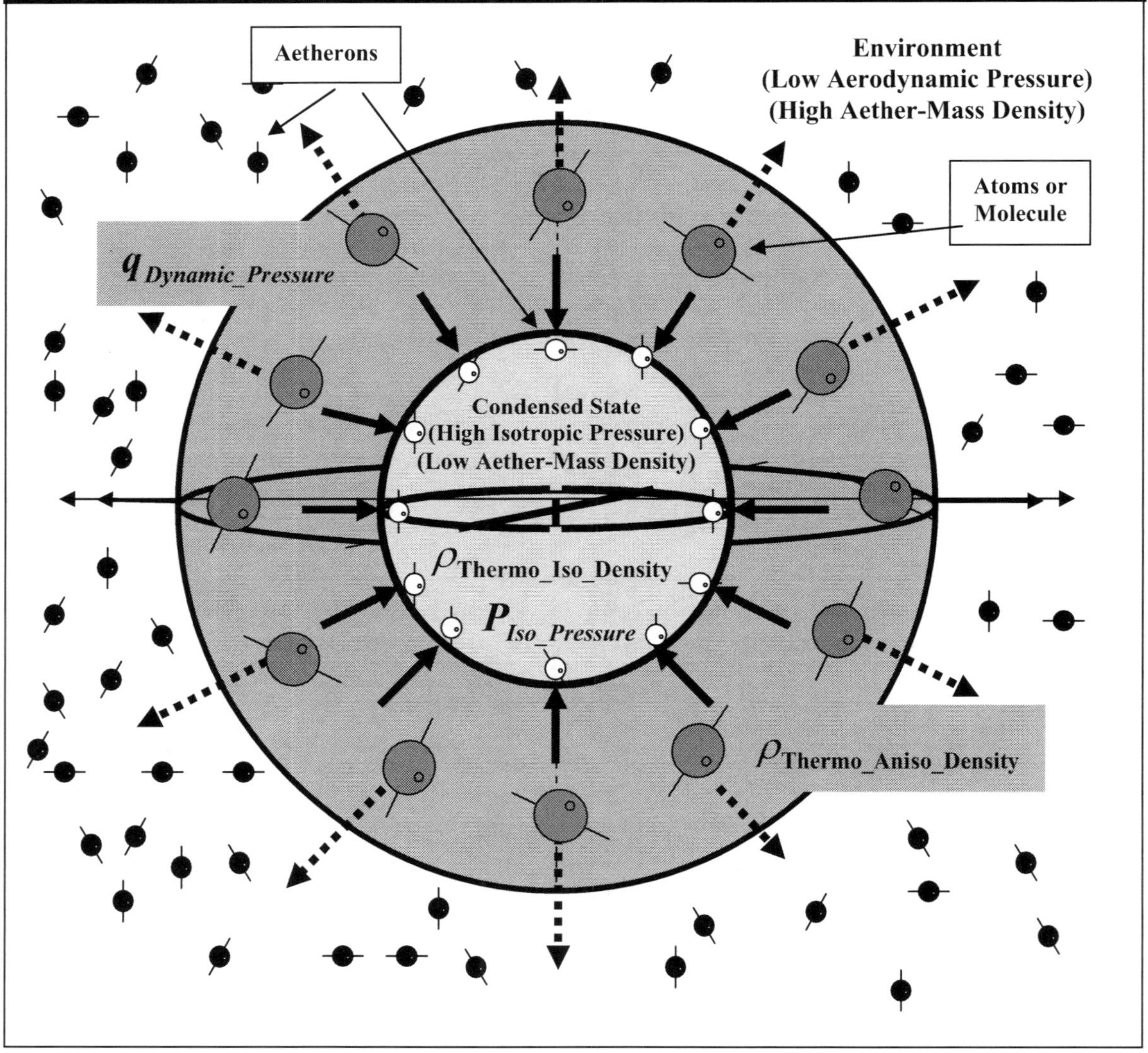

Figure 7.11: The Thermodynamic Pump Energy is used by the system to *heat the body and expand* via a difference in the anisotropy to isotropy of an isolated dynamic system.

Isolated System Mass Body & Environment Heating

$$\phi_{\text{Locomotion}} = \frac{\left|\bar{\mathbf{v}}\right|_{\text{CM}}^{2}}{\left|\mathbf{v}^{2}\right|_{\text{Iso}}} = \left(\frac{q_{Dynamic\_Pressure}}{P_{Iso\_Pressure}}\right) \geq 1$$

Isolated System Mass Body & Environment Expansion

$$\phi_{\text{Respiration}}^{2} = \left(\frac{\rho_{\text{Thermo\_Iso\_Density}}}{\rho_{\text{Thermo\_Aniso\_Density}}}\right) \geq 1$$

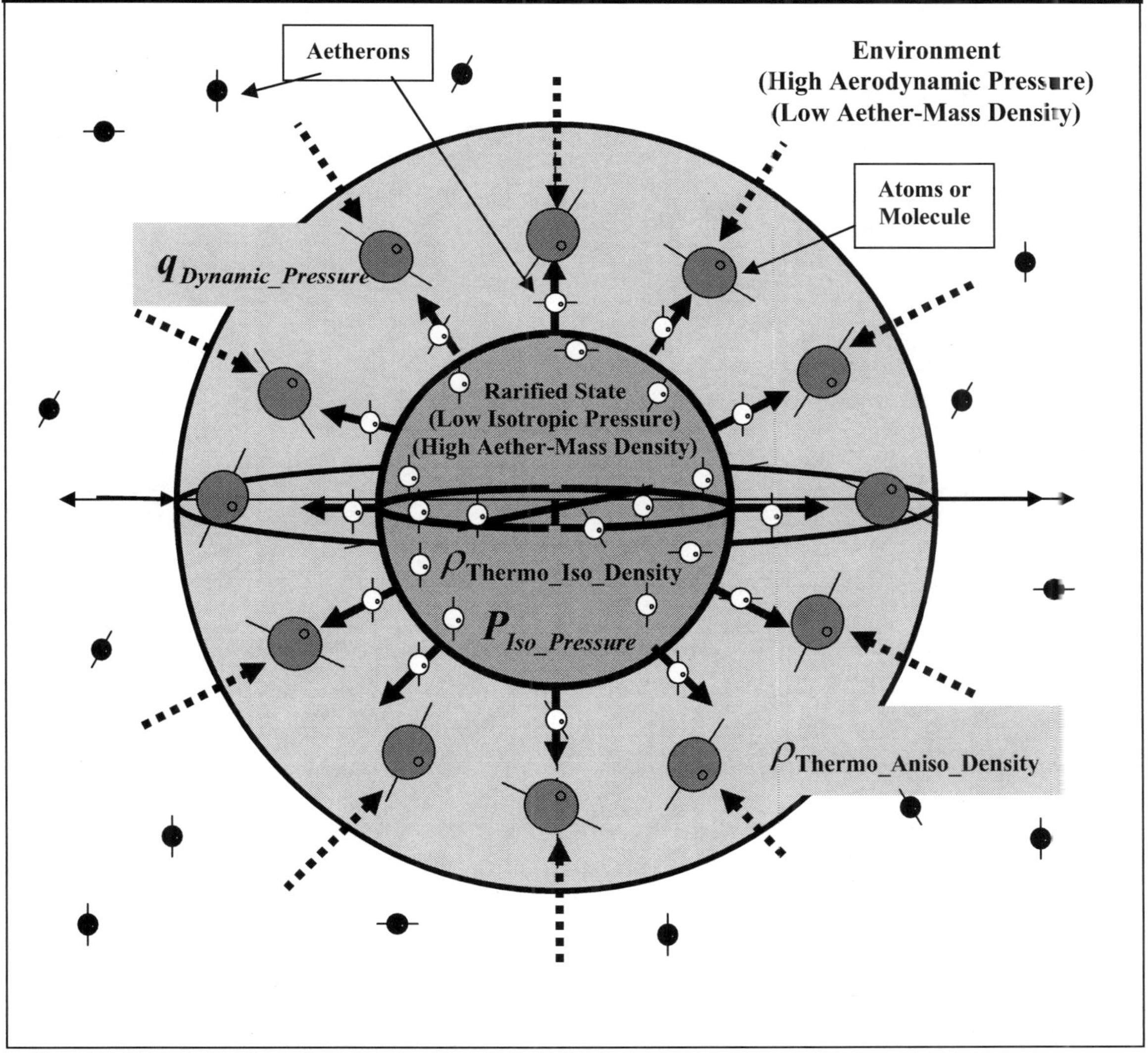

Figure 7.12: The Thermodynamic Pump Energy is used by the system to *heat the body and expand* via a difference in the anisotropy to isotropy of an isolated dynamic system.

7.4 The Super Carnot Efficiency Factor & Coefficient of Entropy

For an isolated system mass body, Carnot machine functioning as a Heat Engine and Heat Pump in which the "effectiveness" or work done is the ratio of the product of the energy delivered to the high-temperature reservoir multiplied by the energy from the low-temperature reservoir divided by the square of the work required to force the machine around its cycle.

Super Carnot Efficiency Factor

7.21

$$\eta_{\text{Canot\_Super}} = \eta_{\text{Carnot}} \cdot \eta_{\text{Carnot\_Iso}} \rightarrow Unitless$$

$$\eta_{\text{Canot\_Super}} = \left(\frac{\Delta S_{Entropy\ Free\_Surroundings}}{\left(\dfrac{\Delta(PV)_{\text{Free\_Energy}}}{\Delta T_{Free}} \right)} \right) = \left[\frac{\left(T_{emp} - T_{emp\_Ext}\right)^2}{T_{emp} \cdot T_{emp\_Ext}} \right]$$

$$\eta_{\text{Canot\_Super}} = \eta_{\text{Carnot}} \cdot \eta_{\text{Carnot\_Iso}} = \left(\frac{1}{\gamma^2_{Inertial}} \right) \cdot \left(\frac{1}{\gamma^2_{Inertial\_Iso}} \right) = \frac{1}{\gamma^4_{Super\_Compress}}$$

The effectiveness of Thermodynamic Momentum Pumping mechanism of an isolated dynamic mass body system is that it behaves like a diffusing diaphragm that lets some energy in and prevents another type of energy from leaving, and is called the **Coefficient of Entropy**.

The measure of the **Coefficient of Entropy** is the pumping performance factor ($COP_{Entropy}$) and is the measure of the work-energy that transfers aether mass, density, and energy, between two mass-energy reservoirs of an isolated dynamic system, and is equal to the **Fourth Power of the Inertial Super Compressibility Factor.**

Definition 7.1: The **Fourth power of the Super Compressibility Factor** ($\gamma^4_{Super\_Compress} = \gamma^2_{Inertial} \cdot \gamma^2_{Inertial\_Iso} = \frac{1}{\eta_{\text{Carnot}} \cdot \eta_{\text{Carnot\_Iso}}}$) is a measure of the **Coefficient of Entropy** between a system mass body and its environment defined as the inverse of the product of the **Anisotropic Carnot Thermodynamic Efficiency factor** (η_{Carnot}) multiplied by the **Isotropic Carnot Efficiency factor** ($\eta_{\text{Carnot\_Iso}}$); and likewise is equal to the product of the **Square Anisotropic Carnot Thermodynamic Efficiency factor** ($\gamma^2_{Inertial}$) multiplied by the **Square Isotropic Carnot Efficiency factor** ($\gamma^2_{Inertial\_Iso}$).

Definition 7.2: The **Fourth power of the Super Compressibility Factor** ($\gamma^4_{Super\_Compress} = \gamma^2_{Inertial} \cdot \gamma^2_{Inertial\_Iso} = \frac{1}{\eta_{\text{Carnot}} \cdot \eta_{\text{Carnot\_Iso}}}$) is a measure of the **Coefficient of Entropy** between a system mass body and its environment defined as the product of the **External Anisotropic Inertial Rectilinear Translational Kinetic Energy** ($\text{E}_{\text{Net\_Translational}}$) multiplied by the **Internal Isotropic Omni-directional Kinetic Energy** ($\text{E}_{\text{Net\_Iso\_Kinetic}}$) divided by the **Square of the Aether-Kinematic Work "Free" Energy** ($\text{W}^2_{\text{Work\_Free}}$).

Fourth power of the Super Compressibility Factor

7.22

$$\gamma^4_{Super\_Compress} = \gamma^2_{Inertial} \cdot \gamma^2_{Inertial\_Iso} = \frac{1}{\eta_{\mathrm{Carnot}} \cdot \eta_{\mathrm{Carnot\_Iso}}} \rightarrow Unitless$$

$$\gamma^4_{Super\_Compress} = \left[\frac{\mathrm{E}_{\mathrm{Net\_Translational}} \cdot \mathrm{E}_{\mathrm{Net\_Iso\_Kinetic}}}{\mathrm{W}^2_{\mathrm{Work\_Free}}}\right] = \left[\frac{\mathrm{E}_{\mathrm{Net\_Translational}} \cdot \mathrm{E}_{\mathrm{Net\_Iso\_Kinetic}}}{\left(\mathrm{E}_{\mathrm{Net\_Iso\_Kinetic}} - \mathrm{E}_{\mathrm{Net\_Translational}}\right)^2}\right]$$

$$\gamma^4_{Super\_Compress} = \left(\frac{\overline{\mathrm{p}}_{\mathrm{Thermo\_Momentum}}}{\Delta\overline{\mathrm{p}}_{\mathrm{Free\_Momentum}}}\right)^2 = \left[\frac{Low\ \mathrm{Temperature} \cdot High\ \mathrm{Temperature}}{\left(High\ \mathrm{Temperature} - Low\ \mathrm{Temperature}\right)^2}\right]$$

Fourth power of the Super Compressibility Factor

7.23

$$\gamma^4_{Super\_Compress} = \left[\frac{T_{emp\_Ext} \cdot T_{emp}}{\left(T_{emp} - T_{emp\_Ext}\right)^2}\right] = \left[\frac{\left|\overline{\mathrm{v}}\right|^2_{\mathrm{CM}} \cdot \left|\mathrm{v}^2\right|_{\mathrm{Iso}}}{\left(\left|\mathrm{v}^2\right|_{\mathrm{Iso}} - \left|\overline{\mathrm{v}}\right|^2_{\mathrm{CM}}\right)^2}\right]$$

$$\gamma^4_{Super\_Compress} = \frac{1}{\Delta S_{Entropy\ Free\_Surroundings}} \cdot \left(\frac{\Delta(PV)_{\mathrm{Free\_Energy}}}{\Delta T_{Free}}\right) \rightarrow Unitless$$

$$\gamma^4_{Super\_Compress} = \frac{1}{\Delta S_{Entropy\ Free\_Surroundings}} \cdot \left(\frac{\left[\delta Q_{\mathrm{Heat\_Pressure\_Free}} - \delta Q_{\mathrm{Heat\_Energy\_Free}}\right]}{\left(T_{emp} - T_{emp\_Ext}\right)}\right)$$

Chapter 8

The Special Theory of Thermodynamics

(Thermodynamic "Pump")
(Momentum & Energy)

8.1 Thermodynamic "Pump" Momentum

A system is in thermodynamic equilibrium if the state variables associated with the system do not change with time. In classical terms, that means that the temperature of the interior system equals that of the exterior.

Thermodynamics describes the exchange of heat (and energy) from the system mass body interior to the exterior.

Thermo-statics describes systems in thermodynamic equilibrium, and considers the transition from one state of equilibrium, $\mathcal{E}_1$, to another state of equilibrium, $\mathcal{E}_2$. These have obvious analogues in basic mechanics. A structure in static equilibrium can move to another equilibrium state under a change of forces.

Thermodynamics describes systems that are evolving towards a state of thermodynamic equilibrium. The movement of energy between the isotropic states to the anisotropic state is dynamic process.

Similarly, a system in thermostatic equilibrium can move to another state of thermostatic equilibrium under a change in thermodynamic forces and conditions. The transformation between states is a thermodynamic process.

This simple thermodynamic result can be easily described from the point of view of the kinetic theory. Consider a spherical container divided into two parts having an interior and an exterior separated by a semi-permeable Q-sphere membrane with Aetherons on each side. However there is a great deal more Aetherons on the outside than on the inside of the Q-sphere.

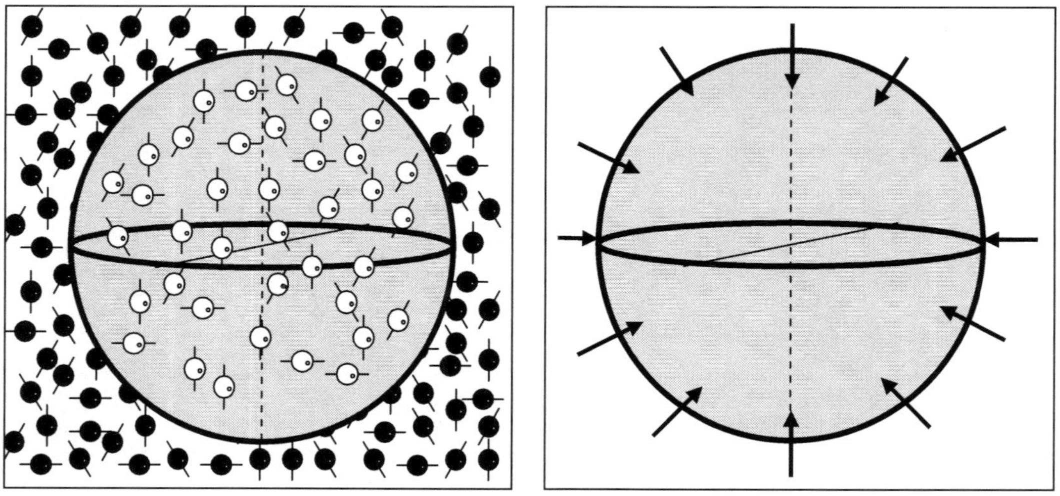

Figure 8.1: The Thermodynamic Pump Energy is used by the system to *heat the body and expand* by creating mass-energy flow towards the low pressure body and away from the high pressure atmospheric environment.

Since the Aetherons can pass freely through the semi-permeable membrane, the pressure on both sides of the membrane will be the same.

Now, let an atomic substance such as air molecules dissolve in the interior of the Q-Sphere membrane, mixing with the Aetherons on the inside.

What follows is that the pressure on the outside of the membrane facing the interior will be increased by the impacts against it of the molecules of the dissolved atomic air substance, which cannot pass through to the outside of the membrane.

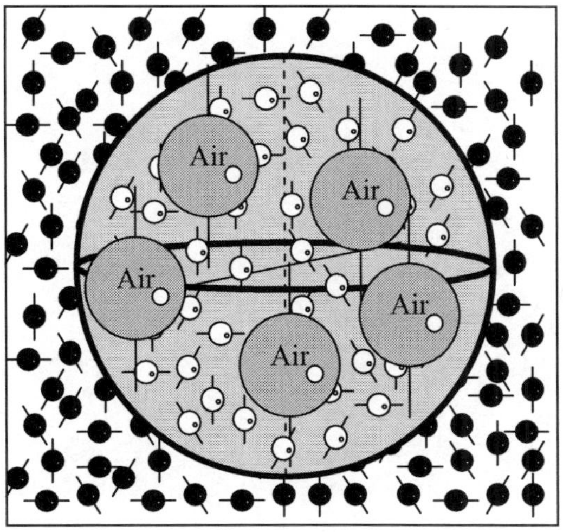

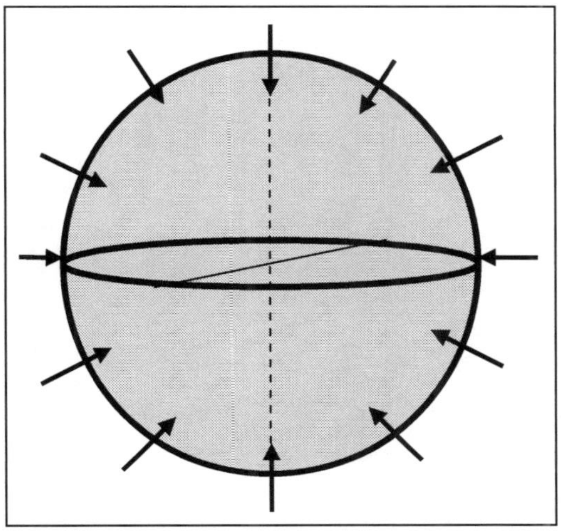

Figure 8.2: The Thermodynamic Pump Energy is used by the system to *heat the body and expand* by creating mass-energy flow towards the low pressure body and away from the high pressure atmospheric environment.

The internal isotropic Omni-directional momentum ($\overline{p}_{Net\_Iso}$) and velocity ($\left|v^2\right|_{Iso}$) of the air molecules which move about within the interior depend on the Internal Isotropic Omni-directional Kinetic Energy ($E_{Net\_Iso\_Kinetic}$), and Absolute Temperature (T_{emp}) of the system.

The larger the number (N) of air molecules dissolved in the Aetherons and the higher the Absolute Temperature (T_{emp}) of the system, the larger will be the number of impacts per unit time and, hence, the greater the Isotropic Omni-directional pressure.

It can be shown from the kinetic theory that the velocities of air molecules are not affected by the Aetherons in the interior, but are equal to the velocities that they would have if they were in a gaseous state.

Therefore, both the number and the intensity of the impacts of the molecules against the Q-Sphere membrane are equal to the number and intensity of the impacts that one expects for a gas.

Fundamental to thermodynamic processes are the concepts of work and heat energy transfer.

8.1

$$W_{\text{Free\_Work\_Energy}} = \frac{\Delta\bar{p}^2_{\text{Free\_Momentum}}}{2 \cdot m_{\text{Net}}} = \frac{W_{\text{Thermo\_Pump\_Energy}}}{\gamma^4_{Super}} \rightarrow kg \cdot m^2 / s^2$$

$$W_{\text{Free\_Work\_Energy}} = \frac{3 \cdot N \cdot k_B}{2 \cdot m_{\text{Net}}} \cdot \left(\frac{\Delta(PV)_{\text{Free\_Energy}}}{\left(1 - \frac{1}{\gamma_{Heat}}\right) \cdot C_{Specific\_Pressure}} \right)$$

Osmosis is the flow of a fluid (gas or liquid) via a semi-permeable membrane that blocks the passage of particles dissolved within it. The flow direction and rate are controlled by pressure on the fluid solution.

Once again consider the Q-Sphere semi-permeable membrane that contains the Aether gas fluid where the Aetherons are free to cross the membrane; however the Atoms or molecules within the Q-Sphere are not allowed to cross the membrane.

It is also assumed that there is no chemical interaction between the Atoms or molecules and the Aether particles within the Aether and collection of atoms system.

The work of fluid expansion is equal to the work of fluid compression Therefore, by the second law of thermodynamics; the work of particle compression within the fluid must be equal to the work of particle expansion within the gas.

When static pressure is built up within a fluid, Pascal law postulates that the pressure will be distributed equally within the fluid volume and will act perpendicular or normal to the surface of the Q-sphere fluid boundary. This static isotropic pressure is independent of Gravitation; Gravity is not included.

The laws that govern this action are a direct consequence of the conservation laws, which include the conservation law of linear momentum and the conservation law of kinetic energy.

Since the Atoms within the Q-Sphere are prevented from leaving the semi-permeable membrane they will exert internal isotropic Omni-directional pressure on it.

A molecule “Air” particle exerts pressure by delivering momentum to the interior of the Q-Sphere membrane. Each air molecule delivers momentum also to its neighbor molecules, so that the time averaged momentum relative to the center of mass of the system delivered to the Q-Sphere membrane and to the neighbors in all directions equals zero; as required by conservation of linear momentum.

The non-zero momentum delivered to the neighboring atoms as they collide with each other, including the Q-sphere membrane, spreads in all directions toward the solution boundaries and generates pressure on them.

Whenever an Isotropic Omni-directional pressure ($P_{Iso\_Pressure}$) is built-up, according to Pascal’s law it will act equally on all of the internal system boundaries. It will then push any boundary that is free to move.

The Isotropic Omni-directional pressure ($P_{Iso\_Pressure}$) acts on a moving boundary from inside the solution, and in order to keep the system under balance a similar extra external pressure must act on the boundary from outside or externally.

If the external Aerodynamic pressure ($q_{Dynamic\_Pressure}$) is reduced below the Isotropic Omni-directional pressure ($P_{Iso\_Pressure}$), then the Aetherons will flow via the Q-Sphere membrane ***into the Atomic solution***, and if the external Aerodynamic pressure ($q_{Dynamic\_Pressure}$) is increased above the Isotropic Omni-directional pressure, then the Aetherons will flow via the Q-Sphere membrane ***out of the Atomic solution***.

The above conclusions are direct consequences of fundamental laws of conservation of momentum and energy. The role of the Q-Sphere membrane is to act as a Thermodynamic Momentum Heat Pump in osmosis that does work letting the Aetherons pass through while at the same time block the Atoms from escaping.

Isolated System Mass Body & Environment Cooling

$$\phi_{\text{Locomotion}} = \frac{|\bar{\mathbf{v}}|^2_{\text{CM}}}{|\mathbf{v}^2|_{\text{Iso}}} = \left(\frac{q_{Dynamic\_Pressure}}{P_{Iso\_Pressure}}\right) \le 1$$

Isolated System Mass Body & Environment Condensing

$$\phi^2_{\text{Respiration}} = \left(\frac{\rho_{\text{Thermo\_Iso\_Mass}}}{\rho_{\text{Thermo\_Aniso\_Mass}}}\right) \le 1$$

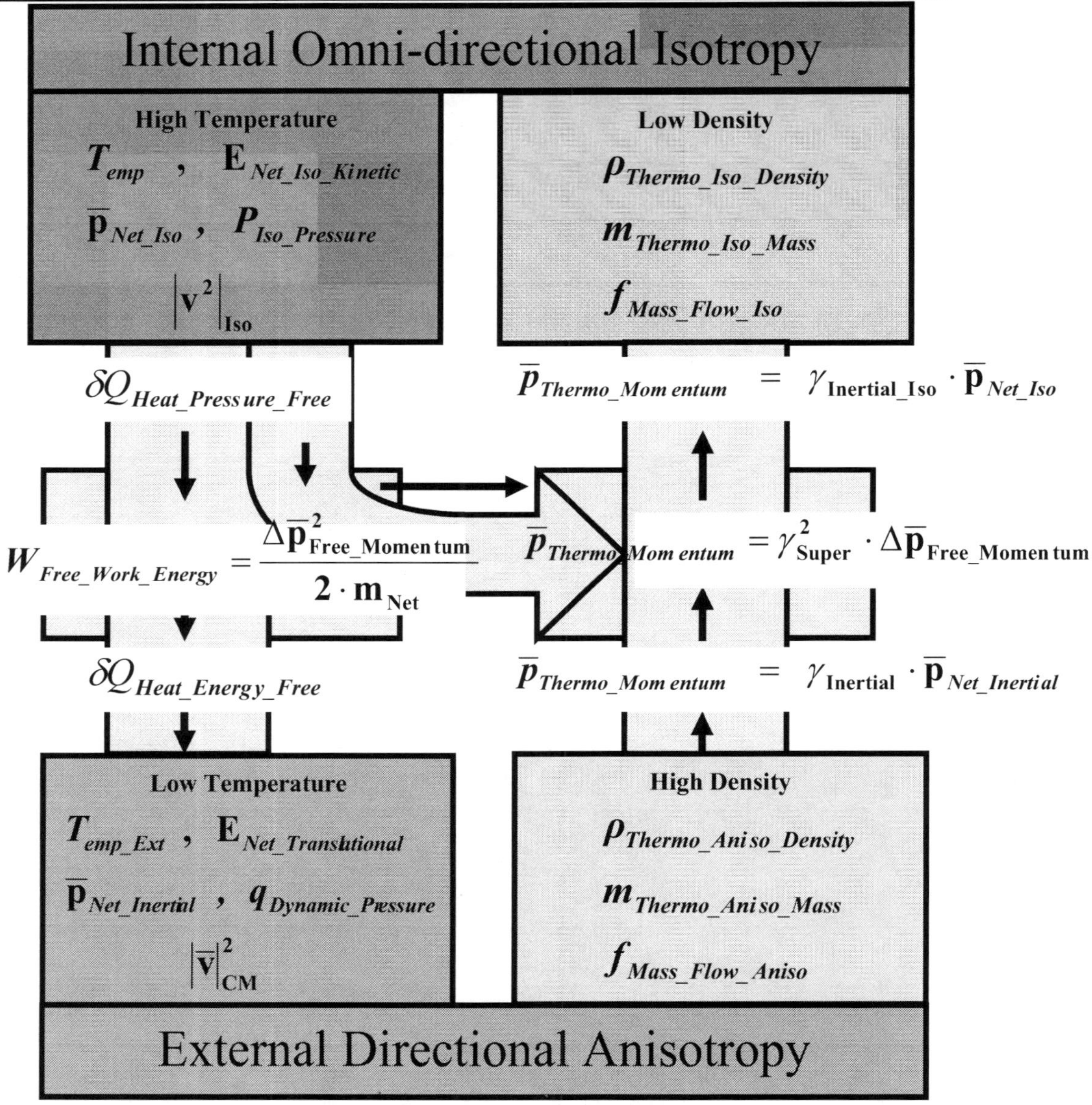

Figure 8.3: The Thermodynamic Pump Energy is used by the system to *cool the body and condense* via a difference in the anisotropy to isotropy of an isolated dynamic system.

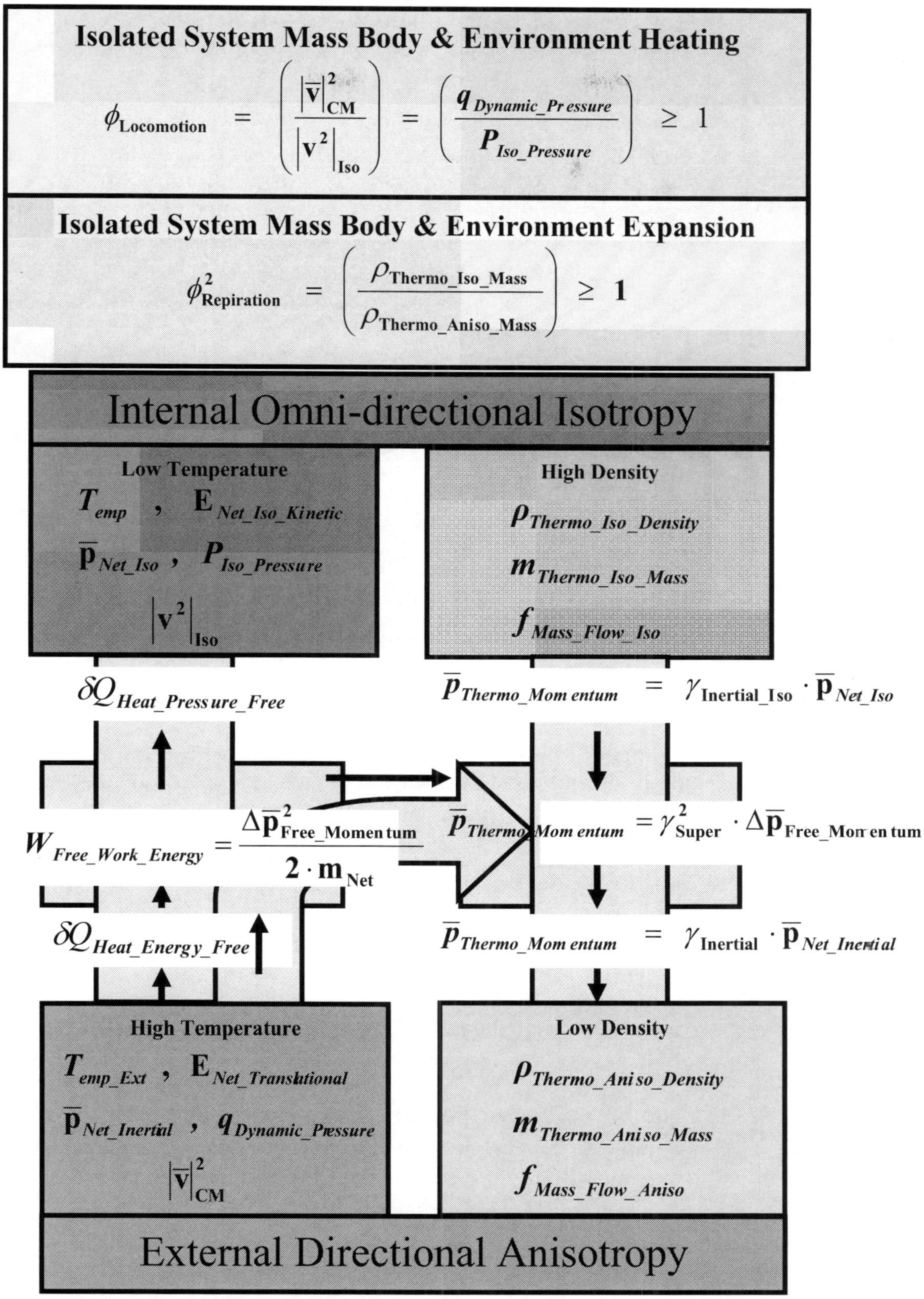

Figure 8.4: The Thermodynamic Pump Energy is used by the system to *heat the body and expand* via a difference in the anisotropy to isotropy of an isolated dynamic system.

Definition 8.1: Relative to an inertial frame, the Mean Center, the Center of Mass, and the Center of Isotropy of an isolated system mass body, the **Thermodynamic "Pump" Momentum** ($\overline{\mathrm{p}}_{\mathrm{Thermo\_Momentum}} = \gamma_{Inertial} \cdot \overline{\mathrm{p}}_{\mathrm{Inertial\_Net}} = \gamma_{Inertial\_Iso} \cdot \overline{\mathrm{p}}_{\mathrm{Net\_Iso}}$) is constant, and conserved, and is the measure of the diaphragm pumping resistance which prevents the atoms from escaping the Q-Sphere membrane which causes an increase and decrease in volume caused by the action of the density differences creating transporting Aether fluid out the inertial mass body reservoir and transports energy into the atmosphere environment, and vice versa; and is defined as the product of the **Internal Isotropic Inertial/Aether Omni-directional Center of Isotropy Net Momentum** ($\overline{\mathrm{p}}_{\mathrm{Net\_Iso}} = \mathrm{m}_{\mathrm{Net}} \cdot \sqrt{\left|\mathrm{v}^2\right|_{\mathrm{Iso}}}$) multiplied by the **Isotropic Inertial Motion Compressibility Factor** ($\gamma_{Inertial\_Iso}$); and likewise is equal to the **External Anisotropic Inertial Net Linear Momentum** ($\overline{\mathrm{p}}_{\mathrm{Inertial\_Net}} = \mathrm{m}_{\mathrm{Net}} \cdot \left|\overline{\mathrm{v}}\right|_{\mathrm{CM}}$) multiplied by the **Inertial Motion Compressibility Factor** ($\gamma_{Inertial}$) in any atomic substance material medium including the Vacuum of Spacetime.

Definition 8.2: Relative to an inertial frame, the Mean Center, the Center of Mass, and the Center of Isotropy of an isolated system mass body, the **Thermodynamic "Pump" Momentum** ($\overline{\mathrm{p}}_{\mathrm{Thermo\_Momentum}} = \gamma^2_{Super\_Compress} \cdot \Delta\overline{\mathrm{p}}_{\mathrm{Free\_Momentum}}$) is constant, and conserved, and is the measure of the work done by the diaphragm which causes an increase and decrease in volume caused by the action of the varying the density by transporting Aether fluid out the one reservoir and transports energy into the other reservoir; and is defined as the product of the **Aether-Kinematic Work "Free" Momentum** ($\Delta\overline{\mathrm{p}}_{\mathrm{Free\_Momentum}}$) multiplied by the **Square Super Compressibility Factor** ($\gamma^2_{Super\_Compress}$) in any atomic substance material medium including the Vacuum of Spacetime.

Thermodynamic "Pump" Momentum

8.2

$$\overline{p}_{\text{Thermo\_Momentum}} = \frac{\overline{p}^{2}_{\text{Conic\_Momentum}}}{\Delta\overline{p}_{\text{Free\_Momentum}}} \rightarrow kg \cdot m/s$$

$$\overline{p}_{\text{Thermo\_Momentum}} = \gamma_{Inertial} \cdot \overline{p}_{\text{Inertial\_Net}} = \gamma_{Inertial\_Iso} \cdot \overline{p}_{\text{Net\_Iso}}$$

$$\overline{p}_{\text{Thermo\_Momentum}} = \frac{\overline{p}_{\text{Inertial\_Net}}}{\left(\sqrt{1 - \frac{E_{\text{Net\_Translational}}}{E_{\text{Net\_Iso\_Kinetic}}}}\right)} = \frac{\overline{p}_{\text{Net\_Iso}}}{\left(\sqrt{\frac{E_{\text{Net\_Iso\_Kinetic}}}{E_{\text{Net\_Translational}}} - 1}\right)}$$

$$\overline{p}_{\text{Thermo\_Momentum}} = \frac{m_{\text{Net}} \cdot |\overline{v}|_{\text{CM}}}{\sqrt{1 - \frac{|\overline{v}|^2_{\text{CM}}}{|v^2|_{\text{Iso}}}}} = \frac{m_{\text{Net}} \cdot \sqrt{|v^2|_{\text{Iso}}}}{\sqrt{\left(\frac{|v^2|_{\text{Iso}}}{|\overline{v}|^2_{\text{CM}}}\right) - 1}}$$

$$\overline{p}_{\text{Thermo\_Momentum}} = \frac{m_{\text{Net}} \cdot |\overline{v}|_{\text{CM}}}{\sqrt{1 - \frac{\gamma_{Heat}}{3} \cdot \left(\frac{|\overline{v}|^2_{\text{CM}}}{c^2_{\text{Sound}}}\right)}} = \frac{m_{\text{Net}} \cdot \sqrt{|v^2|_{\text{Iso}}}}{\sqrt{\frac{3}{\gamma_{Heat}} \cdot \left(\frac{c^2_{\text{Sound}}}{|\overline{v}|^2_{\text{CM}}}\right) - 1}}$$

$$\overline{p}_{\text{Thermo\_Momentum}} = \frac{\left(m_{\text{Net}} \cdot c_{\text{Sound}}\right) \cdot \sqrt{\frac{3}{\gamma_{Heat}}}}{\sqrt{\frac{3}{\gamma_{Heat}} \cdot \left(\frac{c^2_{\text{Sound}}}{|\overline{v}|^2_{\text{CM}}}\right) - 1}}$$

Thermodynamic "Pump" Momentum

8.3

$$\overline{p}_{\text{Thermo\_Momentum}} = \gamma_{Inertial} \cdot \overline{p}_{\text{Inertial\_Net}} = \gamma_{Inertial\_Iso} \cdot \overline{p}_{\text{Net\_Iso}}$$

$$\overline{p}_{\text{Thermo\_Momentum}} = \gamma^2_{Super\_Compress} \cdot \Delta\overline{p}_{\text{Free\_Momentum}}$$

$$\overline{p}_{\text{Thermo\_Momentum}} = \sqrt{\gamma_{Inertial} \cdot \gamma_{Inertial\_Iso} \cdot \overline{p}_{\text{Net\_Iso}} \cdot \overline{p}_{\text{Inertial\_Net}}}$$

$$\overline{p}_{\text{Thermo\_Momentum}} = \gamma_{Super\_Compress} \cdot \overline{p}_{\text{Conic\_Momentum}}$$

$$\overline{p}_{\text{Thermo\_Momentum}} = \frac{\overline{p}^2_{\text{Conic\_Momentum}}}{\Delta\overline{p}_{\text{Free\_Momentum}}} = \frac{\overline{p}_{\text{Net\_Iso}} \cdot \overline{p}_{\text{Inertial\_Net}}}{\sqrt{\left[\overline{p}^2_{\text{Net\_Iso}} - \overline{p}^2_{\text{Inertial\_Net}}\right]}}$$

$$\overline{p}_{\text{Thermo\_Momentum}} = \frac{\overline{p}_{\text{Inertial\_Net}}}{\sqrt{1 - \frac{\left|\overline{v}\right|^2_{\text{CM}}}{\left|v^2\right|_{\text{Iso}}}}} = \frac{\overline{p}_{\text{Net\_Iso}}}{\sqrt{\left(\frac{\left|v^2\right|_{\text{Iso}}}{\left|\overline{v}\right|^2_{\text{CM}}} - 1\right)}}$$

$$\overline{p}_{\text{Thermo\_Momentum}} = \frac{m_{\text{Net}} \cdot \left|\overline{v}\right|_{\text{CM}}}{\sqrt{1 - \frac{\left|\overline{v}\right|^2_{\text{CM}}}{\left|v^2\right|_{\text{Iso}}}}} = \frac{m_{\text{Net}} \cdot \sqrt{\left|v^2\right|_{\text{Iso}}}}{\sqrt{\left(\frac{\left|v^2\right|_{\text{Iso}}}{\left|\overline{v}\right|^2_{\text{CM}}} - 1\right)}} \rightarrow kg \cdot m/s$$

Thermodynamic “Pump” Momentum

8.4

$$\bar{p}_{Thermo\_Momentum} = \frac{\bar{p}^2_{Conic\_Momentum}}{\Delta\bar{p}_{Free\_Momentum}} \rightarrow kg \cdot m/s$$

$$\bar{p}_{Thermo\_Momentum} = \frac{\bar{p}_{Net\_Iso} \cdot \bar{p}_{Inertial\_Net}}{\Delta\bar{p}_{Free\_Momentum}} = \frac{\bar{p}_{Net\_Iso} \cdot \bar{p}_{Inertial\_Net}}{\sqrt{\left[\bar{p}^2_{Net\_Iso} - \bar{p}^2_{Inertial\_Net}\right]}}$$

$$\bar{p}_{Thermo\_Momentum} = \frac{\left\|\bar{p}_{Net\_Iso}\right\| \cdot \left\|\bar{p}_{Inertial\_Net}\right\| \cdot \cos\theta}{\Delta\bar{p}_{Free\_Momentum}} = \frac{\left\|\bar{p}_{Net\_Iso}\right\| \cdot \left\|\bar{p}_{Inertial\_Net}\right\| \cdot \cos\theta}{\sqrt{\left[\bar{p}^2_{Net\_Iso} - \bar{p}^2_{Inertial\_Net}\right]}}$$

$$\bar{p}_{Thermo\_Momentum} = m_{Net} \cdot \sqrt{\frac{\left|v^2\right|_{Iso} \cdot \left|\bar{v}\right|^2_{CM}}{\left[\left|v^2\right|_{Iso} - \left|\bar{v}\right|^2_{CM}\right]}}$$

$$\bar{p}_{Thermo\_Momentum} = \frac{m_{Net} \cdot \left|\bar{v}\right|_{CM}}{\sqrt{1 - \frac{\gamma_{Heat}}{3} \cdot \left(\frac{\left|\bar{v}\right|^2_{CM}}{c^2_{Sound}}\right)}}$$

$$\bar{p}_{Thermo\_Momentum} = \frac{\bar{p}_{Inertial\_Net}}{\sqrt{1 - \frac{\gamma_{Heat}}{3} \cdot \left(\frac{\left|\bar{v}\right|^2_{CM}}{c^2_{Sound}}\right)}}$$

$$\bar{p}_{Thermo\_Momentum} = \frac{\bar{p}_{Inertial\_Net}}{\sqrt{1 - \frac{\gamma_{Heat}}{3} \cdot \left(\frac{\left|\bar{v}\right|^2_{CM}}{c^2_{Sound}}\right)}} = \frac{m_{Net} \cdot \left|\bar{v}\right|_{CM}}{\sqrt{1 - \left(\frac{m_{Net} \cdot \left|\bar{v}\right|^2_{CM}}{3 \cdot \left(N \cdot k_B \cdot T_{emp}\right)}\right)}}$$

Thermodynamic "Pump" Momentum

8.5

$$\overline{p}_{Thermo\_Momentum} = \frac{m_{Net} \cdot \left|\overline{v}\right|_{CM}}{\sqrt{1 - \dfrac{T_{emp\_Ext}}{T_{emp}}}} = \frac{m_{Net} \cdot \sqrt{\left|v^2\right|_{Iso}}}{\left(\sqrt{\dfrac{T_{emp}}{T_{emp\_Ext}}} - 1\right)}$$

$$\overline{p}_{Thermo\_Momentum} = \frac{m_{Net} \cdot \sqrt{\dfrac{3 \cdot N \cdot k_B \cdot T_{emp\_Ext}}{m_{Net}}}}{\sqrt{1 - \dfrac{T_{emp\_Ext}}{T_{emp}}}} = \frac{m_{Net} \cdot \sqrt{\dfrac{3 \cdot N \cdot k_B \cdot T_{emp}}{m_{Net}}}}{\left(\sqrt{\dfrac{T_{emp}}{T_{emp\_Ext}}} - 1\right)}$$

Thermodynamic "Pump" Momentum

8.6

$$\overline{p}_{Thermo\_Momentum} = \frac{\left(m_1\overline{v}_1 + m_2\overline{v}_2 + m_3\overline{v}_3 + \cdots m_N\overline{v}_N\right)}{\sqrt{1 - \dfrac{1}{m_{Net}} \cdot \left[\dfrac{\left(m_1\overline{v}_1 + m_2\overline{v}_2 + m_3\overline{v}_3 + \cdots m_N\overline{v}_N\right)^2}{\left[m_1\overline{v}_1^2 + m_2\overline{v}_2^2 + m_3\overline{v}_3^2 + \cdots m_N\overline{v}_N^2\right]}\right]}}$$

$$\overline{p}_{Thermo\_Momentum} = \frac{\left(m_1\overline{v}_1 + m_2\overline{v}_2 + m_3\overline{v}_3 + \cdots m_N\overline{v}_N\right)}{\sqrt{1 - \dfrac{\gamma_{Heat}}{3 \cdot m_{Net}^2} \cdot \left[\dfrac{\left(m_1\overline{v}_1 + m_2\overline{v}_2 + m_3\overline{v}_3 + \cdots m_N\overline{v}_N\right)^2}{c_{Sound}^2}\right]}}$$

$$\overline{p}_{Thermo\_Momentum} = \frac{\left(m_1\overline{v}_1 + m_2\overline{v}_2 + m_3\overline{v}_3 + \cdots m_N\overline{v}_N\right)}{\sqrt{1 - \dfrac{1}{3 \cdot m_{Net}} \cdot \left[\dfrac{\left(m_1\overline{v}_1 + m_2\overline{v}_2 + m_3\overline{v}_3 + \cdots m_N\overline{v}_N\right)^2}{N \cdot k_B \cdot T_{emp}}\right]}}$$

Two Body – Thermodynamic "Pump" Momentum

8.7

$$\overline{p}_{\text{Thermo\_Momentum}} = \frac{(m_1\overline{v}_1 + m_2\overline{v}_2)}{\sqrt{1 - \frac{1}{(m_1 + m_2)}\cdot\left[\frac{(m_1\overline{v}_1 + m_2\overline{v}_2)^2}{[m_1\overline{v}_1^2 + m_2\overline{v}_2^2]}\right]}} \rightarrow kg \cdot m/s$$

Two Body – Thermodynamic "Pump" Momentum – V(1) = 0

8.8

$$\overline{p}_{\text{Thermo\_Momentum}} = \frac{(m_2\overline{v}_2)}{\sqrt{1 - \frac{m_2}{(m_1 + m_2)}}} \rightarrow kg \cdot m/s$$

Two Body – Thermodynamic "Pump" Momentum – V(2) = 0

8.9

$$\overline{p}_{\text{Thermo\_Momentum}} = \frac{(m_1\overline{v}_1)}{\sqrt{1 - \frac{m_1}{(m_1 + m_2)}}} \rightarrow kg \cdot m/s$$

Two Body – Thermodynamic "Pump" Momentum – (m1 = m2)

3.10

$$\overline{p}_{\text{Thermo\_Momentum}} = \frac{m\cdot(\overline{v}_1 + \overline{v}_2)}{\sqrt{1 - \frac{1}{2}\cdot\left[\frac{(\overline{v}_1 + \overline{v}_2)^2}{[\overline{v}_1^2 + \overline{v}_2^2]}\right]}} \rightarrow kg \cdot m/s$$

The **Thermodynamic "Pump" Momentum,** in three dimensional Cartesian coordinates x, y, and z, are given by the following,

8.11

$$\overline{p}_{Thermo\_Momentum_x} = \frac{\overline{p}_{Inertial\_Net_x}}{\sqrt{1 - \frac{|\overline{v}|^2_{CM_x}}{|v^2|_{Iso_x}}}} = \frac{\overline{p}_{Net\_Iso_x}}{\sqrt{\left(\frac{|v^2|_{Iso_x}}{|\overline{v}|^2_{CM_x}} - 1\right)}}$$

$$\overline{p}_{Thermo\_Momentum_y} = \frac{\overline{p}_{Inertial\_Net_y}}{\sqrt{1 - \frac{|\overline{v}|^2_{CM_y}}{|v^2|_{Iso_y}}}} = \frac{\overline{p}_{Net\_Iso_y}}{\sqrt{\left(\frac{|v^2|_{Iso_y}}{|\overline{v}|^2_{CM_y}} - 1\right)}}$$

$$\overline{p}_{Thermo\_Momentum_z} = \frac{\overline{p}_{Inertial\_Net_z}}{\sqrt{1 - \frac{|\overline{v}|^2_{CM_z}}{|v^2|_{Iso_z}}}} = \frac{\overline{p}_{Net\_Iso_z}}{\sqrt{\left(\frac{|v^2|_{Iso_z}}{|\overline{v}|^2_{CM_z}} - 1\right)}}$$

8.12

Vector		
Magnitude	Vector	Units
Thermodynamic "Pump" Momentum		
$\overline{p}_{Thermo\_Momentum} = \left[\sqrt{\overline{p}^2_{Thermo\_Momentum_x} + \overline{p}^2_{Thermo\_Momentum_y} + \overline{p}^2_{Thermo\_Momentum_z}}\right]$	$\overline{p}_{Thermo\_Momentum} = \begin{bmatrix} \overline{p}_{Thermo\_Momentum_x} \\ \overline{p}_{Thermo\_Momentum_y} \\ \overline{p}_{Thermo\_Momentum_z} \end{bmatrix}$	$\frac{kg \cdot m}{s}$

8.2 Square of the Thermodynamic "Pump" Momentum

The Square of the Thermodynamic Momentum is derived from the above section.

Square of the Thermodynamic Momentum

8.13

$$\overline{p}^2_{\text{Thermo\_Momentum}} = \gamma^2_{Inertial} \cdot \overline{p}^2_{\text{Inertial\_Net}} = \gamma^2_{Inertial\_Iso} \cdot \overline{p}^2_{\text{Net\_Iso}} \rightarrow kg^2 \cdot m^2 / s^2$$

$$\overline{p}^2_{\text{Thermo\_Momentum}} = \frac{\overline{p}^2_{\text{Inertial\_Net}}}{\left(1 - \frac{E_{\text{Net\_Translational}}}{E_{\text{Net\_Iso\_Kinetic}}}\right)} = \frac{\overline{p}^2_{\text{Net\_Iso}}}{\left(\frac{E_{\text{Net\_Iso\_Kinetic}}}{E_{\text{Net\_Translational}}} - 1\right)}$$

$$\overline{p}^2_{\text{Thermo\_Momentum}} = \frac{\overline{p}^2_{\text{Inertial\_Net}}}{\left(1 - \frac{\left|\overline{v}\right|^2_{\text{CM}}}{\left|v^2\right|_{\text{Iso}}}\right)} = \frac{\overline{p}^2_{\text{Net\_Iso}}}{\left(\frac{\left|v^2\right|_{\text{Iso}}}{\left|\overline{v}\right|^2_{\text{CM}}} - 1\right)}$$

$$\overline{p}^2_{\text{Thermo\_Momentum}} = \frac{m^2_{\text{Net}} \cdot \left|\overline{v}\right|^2_{\text{CM}}}{\left(1 - \frac{\left|\overline{v}\right|^2_{\text{CM}}}{\left|v^2\right|_{\text{Iso}}}\right)} = \frac{m^2_{\text{Net}} \cdot \left|v^2\right|_{\text{Iso}}}{\left(\frac{\left|v^2\right|_{\text{Iso}}}{\left|\overline{v}\right|^2_{\text{CM}}} - 1\right)}$$

Square of the Thermodynamic Momentum

8.14

$$\overline{p}^2_{\text{Thermo\_Momentum}} = \gamma^2_{Inertial} \cdot \overline{p}^2_{\text{Inertial\_Net}} \rightarrow kg^2 \cdot m^2 / s^2$$

$$\overline{p}^2_{\text{Thermo\_Momentum}} = \gamma^2_{Inertial\_Iso} \cdot \overline{p}^2_{\text{Net\_Iso}}$$

$$\overline{p}^2_{\text{Thermo\_Momentum}} = \gamma_{Inertial} \cdot \gamma_{Inertial\_Iso} \cdot \overline{p}_{\text{Net\_Iso}} \cdot \overline{p}_{\text{Inertial\_Net}}$$

$$\overline{p}^2_{\text{Thermo\_Momentum}} = \gamma^2_{Super\_Compress} \cdot \overline{p}^2_{\text{Conic\_Momentum}}$$

$$\overline{p}^2_{\text{Thermo\_Momentum}} = \gamma^2_{Inertial} \cdot \gamma^2_{Inertial\_Iso} \cdot \Delta\overline{p}^2_{\text{Free\_Momentum}}$$

$$\overline{p}^2_{\text{Thermo\_Momentum}} = \gamma^2_{Inertial} \cdot \gamma^2_{Inertial\_Iso} \cdot \left[\Delta\overline{p}^2_{\text{Iso\_Momentum}} + \Delta\overline{p}^2_{\text{Aniso\_Momentum}}\right]$$

$$\overline{p}^2_{\text{Thermo\_Momentum}} = \left[\overline{p}^2_{\text{Inertial\_Net}} + \Delta\overline{p}^2_{\text{Aniso\_Momentum}}\right]$$

$$\overline{p}^2_{\text{Thermo\_Momentum}} = \left[\overline{p}^2_{\text{Net\_Iso}} - \Delta\overline{p}^2_{\text{Iso\_Momentum}}\right]$$

Square of the Thermodynamic Momentum

8.15

$$\overline{p}^{2}_{\text{Thermo\_Momentum}} = \frac{m^{2}_{\text{Net}} \cdot \left|\overline{v}\right|^{2}_{\text{CM}}}{\left(1 - \frac{T_{emp\_Ext}}{T_{emp}}\right)} = \frac{m^{2}_{\text{Net}} \cdot \left|v^{2}\right|_{\text{Iso}}}{\left(\frac{T_{emp}}{T_{emp\_Ext}} - 1\right)} \rightarrow kg^{2} \cdot m^{2} / s^{2}$$

$$\overline{p}^{2}_{\text{Thermo\_Momentum}} = \frac{3 \cdot N \cdot m_{\text{Net}} \cdot k_{B} \cdot T_{emp\_Ext}}{\left(1 - \frac{T_{emp\_Ext}}{T_{emp}}\right)} = \frac{3 \cdot N \cdot m_{\text{Net}} \cdot k_{B} \cdot T_{emp}}{\left(\frac{T_{emp}}{T_{emp\_Ext}} - 1\right)}$$

8.3 Thermodynamic Pump Work Energy

Definition 8.3: Relative to an inertial frame, the Mean Center, the Center of Mass, and the Center of Isotropy of an isolated system mass body, the **Thermodynamic Pump Work Energy** ($W_{Thermo\_Pump\_Energy} = \left(\frac{\overline{p}^2_{Thermo\_Momentum}}{2 \cdot m_{Net}}\right)$) is constant, and conserved, and is the measure of the work done by the kinetic energy diaphragm which causes an increase and decrease in volume caused by the action of the varying the density by transporting Aether fluid out the one reservoir and transports energy into the other reservoir; and is defined as the **Square of the Thermodynamic Momentum** ($\overline{p}^2_{Thermo\_Momentum}$) divided by two times the **Inertial Net Mass** (m_{Net}) in any atomic substance material medium including the Vacuum of Spacetime.

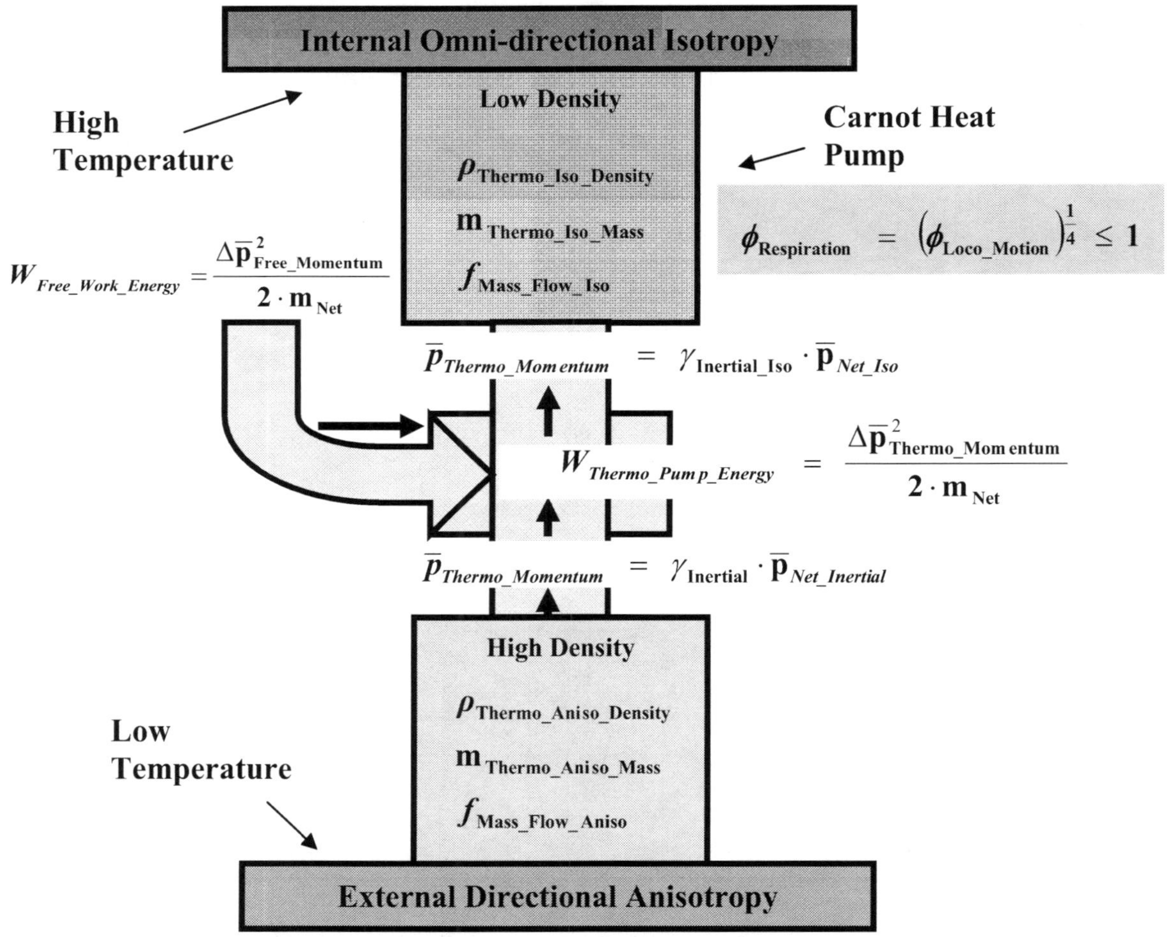

Figure 8.5: The Thermodynamic Pump Energy is used by the system to do work which *cools the body and condenses* by creating aether density difference without and within the body.

Thermodynamic Pump Work Energy

8.16

$$W_{\text{Thermo\_Pump\_Energy}} = \left(\frac{\overline{p}^{2}_{\text{Thermo\_Momentum}}}{2 \cdot m_{\text{Net}}} \right) \rightarrow kg \cdot m^2 / s^2$$

$$W_{\text{Thermo\_Pump\_Energy}} = \gamma^2_{Inertial} \cdot \gamma^2_{Inertial\_Iso} \cdot \left(\frac{\Delta\overline{p}^{2}_{\text{Free\_Momentum}}}{2 \cdot m_{\text{Net}}} \right)$$

$$W_{\text{Thermo\_Pump\_Energy}} = \gamma^2_{Inertial} \cdot \gamma^2_{Inertial\_Iso} \cdot W_{\text{Free\_Work\_Energy}}$$

$$W_{\text{Thermo\_Pump\_Energy}} = \gamma_{Inertial} \cdot \gamma_{Inertial\_Iso} \cdot \left(\frac{\overline{p}_{\text{Net\_Iso}} \cdot \overline{p}_{\text{Inertial\_Net}}}{2 \cdot m_{\text{Net}}} \right)$$

Thermodynamic Pump Work Energy

8.17

$$W_{Thermo\_Pump\_Energy} = \gamma^2_{Inertial} \cdot E_{Net\_Translational} = \gamma^2_{Inertial\_Iso} \cdot E_{Net\_Iso\_Kinetic}$$

$$W_{Thermo\_Pump\_Energy} = \frac{\frac{1}{2} \cdot m_{Net} \cdot \left|\overline{v}\right|^2_{CM}}{\left(1 - \frac{T_{emp\_Ext}}{T_{emp}}\right)} = \frac{\frac{1}{2} \cdot m_{Net} \cdot \left|v^2\right|_{Iso}}{\left(\frac{T_{emp}}{T_{emp\_Ext}} - 1\right)}$$

$$W_{Thermo\_Pump\_Energy} = \frac{\frac{3}{2} \cdot N \cdot k_B \cdot T_{emp\_Ext}}{\left(1 - \frac{T_{emp\_Ext}}{T_{emp}}\right)} = \frac{\frac{3}{2} \cdot N \cdot k_B \cdot T_{emp}}{\left(\frac{T_{emp}}{T_{emp\_Ext}} - 1\right)}$$

$$W_{Thermo\_Pump\_Energy} = \left(\frac{\overline{p}^2_{Thermo\_Momentum}}{2 \cdot m_{Net}}\right)$$

$$W_{Thermo\_Pump\_Energy} = \gamma^4_{Super\_Compress} \cdot W_{Free\_Work\_Energy}$$

$$W_{Thermo\_Pump\_Energy} = \gamma^2_{Inertial} \cdot \gamma^2_{Inertial\_Iso} \cdot W_{Free\_Work\_Energy}$$

$$W_{Thermo\_Pump\_Energy} = \left(\frac{\overline{p}^2_{Thermo\_Momentum}}{2 \cdot m_{Net}}\right) = \gamma^4_{Super\_Compress} \cdot \left(\frac{\Delta\overline{p}^2_{Free\_Momentum}}{2 \cdot m_{Net}}\right)$$

$$W_{Thermo\_Pump\_Energy} = \gamma^4_{Super\_Compress} \cdot \left(\frac{3 \cdot N \cdot k_B \cdot \Delta T_{Temp\_Free}}{2}\right)$$

$$W_{Thermo\_Pump\_Energy} = \gamma^4_{Super\_Compress} \cdot \left(\frac{3 \cdot N \cdot k_B \cdot \left[T_{Temp} - T_{Temp\_Ext}\right]}{2}\right) \rightarrow kg \cdot m^2 / s^2$$

Thermodynamic Pump Work Energy

8.18

$$W_{Thermo\_Pump\_Energy} = \gamma^4_{Super\_Compress} \cdot \left[\frac{\overline{p}^2_{Net\_Iso}}{2 \cdot m_{Net}} - \frac{\left[E'^2_{Relativistic\_Energy} - E^2_{Rest\_Energy}\right]}{2 \cdot E_{Rest\_Energy}} \cdot \left(1 - \frac{|\overline{v}|^2_{CM}}{c^2_{Light}}\right) \right]$$

$$W_{Thermo\_Pump\_Energy} = \gamma^4_{Super\_Compress} \cdot \left[\frac{m_{Net} \cdot |v^2|_{Iso}}{2} - \frac{1}{2 \cdot m_{Net} \cdot c^2_{Light}} \cdot \left[\left[\frac{\left[m^2_{Net} - m^2_{Net} \cdot \left(1 - \frac{|\overline{v}|^2_{CM}}{c^2_{Light}}\right) \right] \cdot G}{\left[R_{\mu\nu} - g_{\mu\nu} \cdot \left(\frac{1}{2} \cdot R \pm \Lambda_{Einstein} \right) \right]} \right] \cdot \left[\pm 8\pi \cdot T_{\mu\nu} \right] \right] \right]$$

Thermodynamic Pump Work Energy

8.19

$$W_{Thermo\_Pump\_Energy} = \left(\frac{\overline{p}^{2}_{Thermo\_Momentum}}{2 \cdot m_{Net}} \right) \rightarrow kg \cdot m^2 / s^2$$

$$W_{Thermo\_Pump\_Energy} = \frac{\left(m_1\overline{v}_1 + m_2\overline{v}_2 + m_3\overline{v}_3 + \cdots m_N\overline{v}_N \right)^2}{2 \cdot m_{Net} \cdot \left[1 - \frac{1}{m_{Net}} \cdot \left[\frac{\left(m_1\overline{v}_1 + m_2\overline{v}_2 + m_3\overline{v}_3 + \cdots m_N\overline{v}_N \right)^2}{\left[m_1\overline{v}_1^2 + m_2\overline{v}_2^2 + m_3\overline{v}_3^2 + \cdots m_N\overline{v}_N^2 \right]} \right] \right]}$$

$$W_{Thermo\_Pump\_Energy} = \frac{\left(m_1\overline{v}_1 + m_2\overline{v}_2 + m_3\overline{v}_3 + \cdots m_N\overline{v}_N \right)^2}{2 \cdot m_{Net} \cdot \left[1 - \frac{\gamma_{Heat}}{3 \cdot m_{Net}^2} \cdot \left[\frac{\left(m_1\overline{v}_1 + m_2\overline{v}_2 + m_3\overline{v}_3 + \cdots m_N\overline{v}_N \right)^2}{c_{Sound}^2} \right] \right]}$$

$$W_{Thermo\_Pump\_Energy} = \frac{\left(m_1\overline{v}_1 + m_2\overline{v}_2 + m_3\overline{v}_3 + \cdots m_N\overline{v}_N \right)^2}{2 \cdot m_{Net} \cdot \left[1 - \frac{1}{3 \cdot m_{Net}} \cdot \left[\frac{\left(m_1\overline{v}_1 + m_2\overline{v}_2 + m_3\overline{v}_3 + \cdots m_N\overline{v}_N \right)^2}{N \cdot k_B \cdot T_{emp}} \right] \right]}$$

The **Thermodynamic Pump Work Energy,** in three dimensional Cartesian coordinates x, y, and z, are given by the following,

8.20

$$W_{Thermo\_Pump\_Energy\,x} = \frac{\overline{p}^2_{Inertial\_Net\,x}}{2 \cdot m_{Net} \cdot \left(1 - \frac{|\overline{v}|^2_{CM\,x}}{|v^2|_{Iso\,x}}\right)} = \frac{\overline{p}^2_{Net\_Iso\,x}}{2 \cdot m_{Net} \cdot \left(\frac{|v^2|_{Iso\,x}}{|\overline{v}|^2_{CM\,x}} - 1\right)}$$

$$W_{Thermo\_Pump\_Energy\,y} = \frac{\overline{p}^2_{Inertial\_Net\,y}}{2 \cdot m_{Net} \cdot \left(1 - \frac{|\overline{v}|^2_{CM\,y}}{|v^2|_{Iso\,y}}\right)} = \frac{\overline{p}^2_{Net\_Iso\,y}}{2 \cdot m_{Net} \cdot \left(\frac{|v^2|_{Iso\,y}}{|\overline{v}|^2_{CM\,y}} - 1\right)}$$

$$W_{Thermo\_Pump\_Energy\,z} = \frac{\overline{p}^2_{Inertial\_Net\,z}}{2 \cdot m_{Net} \cdot \left(1 - \frac{|\overline{v}|^2_{CM\,z}}{|v^2|_{Iso\,z}}\right)} = \frac{\overline{p}^2_{Net\_Iso\,z}}{2 \cdot m_{Net} \cdot \left(\frac{|v^2|_{Iso\,z}}{|\overline{v}|^2_{CM\,z}} - 1\right)}$$

8.21

Scalar		
Magnitude	Scalar	Units
Thermodynamic Pump Work Energy		
$W_{Thermo\_Pump\_Energy} = \begin{bmatrix} W_{Thermo\_Pump\_Energy\,x} \\ + W_{Thermo\_Pump\_Energy\,y} \\ + W_{Thermo\_Pump\_Energy\,z} \end{bmatrix}$	$W_{Thermo\_Pump\_Energy} = \begin{bmatrix} W_{Thermo\_Pump\_Energy\,x} \\ W_{Thermo\_Pump\_Energy\,y} \\ W_{Thermo\_Pump\_Energy\,z} \end{bmatrix}$	$\frac{kg \cdot m^2}{s^2}$

8.4 Aether-Kinematic Work “Free” Momentum

The Aether Kinematic Work “ Free” Momentum ($\Delta\overline{p}_{Free\_Momentum}$) is a quantity of matter in rectilinear motion which represent the Impulse Momentum Transfer between the Isotropy and Anisotropy of an isolated system mass body immersed in any atomic substance material medium including the Vacuum of Spacetime.

The Aether Kinematic Work “Free” Momentum ($\Delta\overline{p}_{Free\_Momentum}$) describes rectilinear condensing and expansion of matter, space, and time relative to the Center of Mass and the Center of Isotropy of an isolated system mass body in accordance with the First Law of Motion.

The Aether Kinematic Work “Free” Momentum ($\Delta\overline{p}_{Free\_Momentum}$) is a measure of the dimensional difference between isotropy and anisotropy rectilinear motions, namely: Momentum Change, Pressure Change, Kinetic Energy Change, Heat Energy Change, and Temperature Change relative to the Center of Mass and the Center of Isotropy of an isolated system mass body.

The Aether Kinematic Work “Free” Momentum ($\Delta\overline{p}_{Free\_Momentum}$) is also a Momentum Tensor that is covariant to the Anisotropy and Isotropy frames of reference.

Definition 8.4: Relative to an inertial frame, the Mean Center, the Center of Mass, and the Center of Isotropy of an isolated system mass body, the **Aether-Kinematic Work “Free” Momentum** ($\Delta\overline{p}_{Free\_Momentum} = \dfrac{\overline{p}_{Net\_Iso}}{\gamma_{Inertial}} = \dfrac{\overline{p}_{Inertial\_Net}}{\gamma_{Inertial\_Iso}}$) is constant, and conserved, and is the measure of an impulse which is a condensing of space and time between Isotropic Motion and Anisotropic Motion; and is defined as the square root of the difference between the **Internal Isotropic Inertial/Aether Omni-directional Center of Isotropy Net Momentum** ($\overline{p}_{Net\_Iso} = m_{Net} \cdot \sqrt{\left|v^2\right|_{Iso}}$) divided by the **Isotropic Inertial Motion Compressibility Factor** ($\gamma_{Inertial\_Iso}$); and likewise is equal to the **External Anisotropic Inertial Net Linear Momentum** ($\overline{p}_{Inertial\_Net} = m_{Net} \cdot \left|\overline{v}\right|_{CM}$) divided by the **Inertial Motion Compressibility Factor** ($\gamma_{Inertial}$) in any atomic substance material medium including the Vacuum of Spacetime.

Aether-Kinematic Work "Free" Momentum

8.22

$$\Delta\overline{p}_{\text{Free\_Momentum}} = \frac{\overline{p}_{\text{Net\_Iso}}}{\gamma_{Inertial}} = \frac{\overline{p}_{\text{Inertial\_Net}}}{\gamma_{Inertial\_Iso}} \rightarrow kg \cdot m/s$$

$$\Delta\overline{p}_{\text{Free\_Momentum}} = \frac{\overline{p}^{2}_{\text{Conic\_Momentum}}}{\overline{p}_{\text{Thermo\_Momentum}}}$$

$$\Delta\overline{p}_{\text{Free\_Momentum}} = \frac{\overline{p}_{\text{Thermo\_Momentum}}}{\gamma_{Inertial} \cdot \gamma_{Inertial\_Iso}} = \frac{\overline{p}_{\text{Net\_Iso}} \cdot \overline{p}_{\text{Inertial\_Net}}}{\overline{p}_{\text{Thermo\_Momentum}}}$$

$$\Delta\overline{p}_{\text{Free\_Momentum}} = \frac{\left\|\overline{p}_{\text{Net\_Iso}}\right\| \cdot \left\|\overline{p}_{\text{Inertial\_Net}}\right\| \cdot \cos\theta}{\overline{p}_{\text{Thermo\_Momentum}}}$$

$$\Delta\overline{p}_{\text{Free\_Momentum}} = \sqrt{\frac{\overline{p}_{\text{Net\_Iso}} \cdot \overline{p}_{\text{Inertial\_Net}}}{\gamma_{Inertial} \cdot \gamma_{Inertial\_Iso}}}$$

$$\Delta\overline{p}_{\text{Free\_Momentum}} = \frac{\overline{p}_{\text{Net\_Iso}}}{\gamma_{Inertial}} = \overline{p}_{\text{Net\_Iso}} \cdot \sqrt{1 - \frac{E_{\text{Net\_Translational}}}{E_{\text{Net\_Iso\_Kinetic}}}}$$

$$\Delta\overline{p}_{\text{Free\_Momentum}} = \frac{\overline{p}_{\text{Inertial\_Net}}}{\gamma_{Inertial\_Iso}} = \overline{p}_{\text{Inertial\_Net}} \cdot \sqrt{\frac{E_{\text{Net\_Iso\_Kinetic}}}{E_{\text{Net\_Translational}}} - 1}$$

Aether-Kinematic Work "Free" Momentum

8.23

$$\Delta\overline{\mathrm{p}}_{\mathrm{Free\_Momentum}} = \frac{\left\|\overline{\mathrm{p}}_{\mathrm{Net\_Iso}}\right\| \cdot \left\|\overline{\mathrm{p}}_{\mathrm{Inertial\_Net}}\right\| \cdot \cos\theta}{\overline{\mathrm{p}}_{\mathrm{Thermo\_Momentum}}} \rightarrow kg \cdot m/s$$

$$\Delta\overline{\mathrm{p}}_{\mathrm{Free\_Momentum}} = \frac{\left\|\overline{\mathrm{p}}_{\mathrm{Net\_Iso}}\right\| \cdot \cos\theta \cdot \sqrt{1 - \dfrac{\gamma_{Heat}}{3} \cdot \left(\dfrac{\left|\overline{\mathrm{v}}\right|^2_{\mathrm{CM}}}{\mathrm{c}^2_{\mathrm{Sound}}}\right)}}{\left[\dfrac{\overline{\mathrm{p}}_{\mathrm{Inertial\_Net}}}{\left\|\overline{\mathrm{p}}_{\mathrm{Inertial\_Net}}\right\|}\right]}$$

$$\Delta\overline{\mathrm{p}}_{\mathrm{Free\_Momentum}} = \frac{\left\|\overline{\mathrm{p}}_{\mathrm{Inertial\_Net}}\right\| \cdot \cos\theta \cdot \sqrt{\left(\dfrac{3}{\gamma_{Heat}} \cdot \left(\dfrac{\mathrm{c}^2_{\mathrm{Sound}}}{\left|\overline{\mathrm{v}}\right|^2_{\mathrm{CM}}}\right) - 1\right)}}{\left[\dfrac{\overline{\mathrm{p}}_{\mathrm{Net\_Iso}}}{\left\|\overline{\mathrm{p}}_{\mathrm{Net\_Iso}}\right\|}\right]}$$

Aether-Kinematic Work "Free" Momentum

8.24

$$\Delta\overline{p}_{Free\_Momentum} = \sqrt{\left[\overline{p}^2_{Net\_Iso} - \overline{p}^2_{Inertial\_Net}\right]} \rightarrow kg \cdot m/s$$

$$\Delta\overline{p}_{Free\_Momentum} = m_{Net} \cdot \sqrt{\left[\left.v^2\right|_{Iso} - \left|\overline{v}\right|^2_{CM}\right]}$$

$$\Delta\overline{p}_{Free\_Momentum} = \frac{\overline{p}_{Inertial\_Net}}{\gamma_{Inertial\_Iso}} = \overline{p}_{Inertial\_Net} \cdot \sqrt{\frac{\left|v^2\right|_{Iso}}{\left|\overline{v}\right|^2_{CM}} - 1}$$

$$\Delta\overline{p}_{Free\_Momentum} = \frac{\overline{p}_{Net\_Iso}}{\gamma_{Inertial}} = \overline{p}_{Net\_Iso} \cdot \sqrt{1 - \frac{\left|\overline{v}\right|^2_{CM}}{\left|v^2\right|_{Iso}}}$$

$$\Delta\overline{p}_{Free\_Momentum} = \frac{\overline{p}_{Net\_Iso}}{\gamma_{Inertial}} = \overline{p}_{Net\_Iso} \cdot \sqrt{1 - \frac{\gamma_{Heat}}{3} \cdot \left(\frac{\left|\overline{v}\right|^2_{CM}}{c^2_{Sound}}\right)}$$

$$\Delta\overline{p}_{Free\_Momentum} = \left(\sqrt{3 \cdot m_{Net} \cdot \left(N \cdot k_B \cdot T_{emp}\right)}\right) \cdot \sqrt{1 - \left(\frac{m_{Net} \cdot \left|\overline{v}\right|^2_{CM}}{3 \cdot \left(N \cdot k_B \cdot T_{emp}\right)}\right)}$$

$$\Delta\overline{p}_{Free\_Momentum} = m_{Net} \cdot \sqrt{\left|v^2\right|_{Iso}} \cdot \sqrt{1 - \frac{\left|\overline{v}\right|^2_{CM}}{\left|v^2\right|_{Iso}}} = \left(m_{Net} \cdot \left|\overline{v}\right|_{CM}\right) \cdot \sqrt{\left[\frac{\left|v^2\right|_{Iso}}{\left|\overline{v}\right|^2_{CM}} - 1\right]}$$

Aether-Kinematic Work “Free” Momentum

8.25

$$\Delta\overline{p}_{Free\_Momentum} = \sqrt{\left[\overline{p}^{2}_{Net\_Iso} - \overline{p}^{2}_{Inertial\_Net}\right]} \rightarrow kg \cdot m/s$$

$$\Delta\overline{p}_{Free\_Momentum} = \sqrt{\left(\overline{p}_{Net\_Iso} + \overline{p}_{Inertial\_Net}\right) \cdot \left(\overline{p}_{Net\_Iso} - \overline{p}_{Inertial\_Net}\right)}$$

$$\Delta\overline{p}_{Free\_Momentum} = \sqrt{\left(\sqrt{\left|v^{2}\right|_{Iso}} + \left|\overline{v}\right|^{2}_{CM}\right) \cdot \left(\sqrt{\left|v^{2}\right|_{Iso}} - \left|\overline{v}\right|^{2}_{CM}\right)}$$

$$\Delta\overline{p}_{Free\_Momentum} = \sqrt{2 \cdot m_{Net} \cdot \left[E_{Net\_Iso\_Kinetic} - E_{Net\_Translational}\right]}$$

$$\Delta\overline{p}_{Free\_Momentum} = \sqrt{3 \cdot N \cdot k_{B} \cdot m_{Net} \cdot \left[T_{Temp} - T_{Temp\_Ext}\right]}$$

$$\Delta\overline{p}_{Free\_Momentum} = \left[\sqrt{\overline{p}^{2}_{Net\_Iso} - \frac{1}{c^{2}_{Light}} \cdot \left[\begin{matrix} E'^{2}_{Relativistic\_Energy} \\ - \quad E^{2}_{Rest\_Energy} \end{matrix}\right] \cdot \left(1 - \frac{\left|\overline{v}\right|^{2}_{CM}}{c^{2}_{Light}}\right)}\right]$$

Aether-Kinematic Work "Free" Momentum

8.26

$$\Delta\overline{p}_{Free\_Momentum} = \left[\sqrt{\frac{m_{Net}^{2}\cdot\left|v^{2}\right|_{Iso}}{\left(1-\frac{\left|\overline{v}\right|_{CM}^{2}}{c_{Light}^{2}}\right)} - \left[\frac{m_{Net}^{2}}{\left(1-\frac{\left|\overline{v}\right|_{CM}^{2}}{c_{Light}^{2}}\right)} - m_{Net}^{2}\right]\cdot c_{Light}^{2}}\right]\cdot\sqrt{1-\frac{\left|\overline{v}\right|_{CM}^{2}}{c_{Light}^{2}}}$$

$$\Delta\overline{p}_{Free\_Momentum} = \overline{p}_{Net\_Iso}\cdot\sqrt{1-\frac{\left|\overline{v}\right|_{CM}^{2}}{\left|v^{2}\right|_{Iso}}} \rightarrow kg\cdot m/s$$

$$\Delta\overline{p}_{Free\_Momentum} = \sqrt{m_{Net}\cdot\left[m_{1}\overline{v}_{1}^{2} + m_{2}\overline{v}_{2}^{2} + m_{3}\overline{v}_{3}^{2} + \cdots m_{N}\overline{v}_{N}^{2}\right]}\cdot\sqrt{1-\frac{1}{m_{Net}}\cdot\left[\frac{\left(m_{1}\overline{v}_{1} + m_{2}\overline{v}_{2} + m_{3}\overline{v}_{3} + \cdots m_{N}\overline{v}_{N}\right)^{2}}{\left[m_{1}\overline{v}_{1}^{2} + m_{2}\overline{v}_{2}^{2} + m_{3}\overline{v}_{3}^{2} + \cdots m_{N}\overline{v}_{N}^{2}\right]}\right]}$$

Aether-Kinematic Work "Free" Momentum

8.27

$$\Delta\overline{p}_{\text{Free\_Momentum}} = \sqrt{m_{\text{Net}} \cdot \begin{bmatrix} m_1\overline{v}_1^2 + m_2\overline{v}_2^2 \\ + m_3\overline{v}_3^2 \\ + \cdots m_N\overline{v}_N^2 \end{bmatrix}} \cdot \sqrt{1 - \frac{\gamma_{Heat}}{3 \cdot m_{\text{Net}}^2} \cdot \left[\frac{\begin{pmatrix} m_1\overline{v}_1 + \\ m_2\overline{v}_2 \\ + m_3\overline{v}_3 \\ + \cdots m_N\overline{v}_N \end{pmatrix}^2}{c_{\text{Sound}}^2} \right]}$$

$$\Delta\overline{p}_{\text{Free\_Momentum}} = \sqrt{m_{\text{Net}} \cdot \begin{bmatrix} m_1\overline{v}_1^2 + m_2\overline{v}_2^2 \\ + m_3\overline{v}_3^2 \\ + \cdots m_N\overline{v}_N^2 \end{bmatrix}} \cdot \sqrt{1 - \frac{1}{3 \cdot m_{\text{Net}}} \cdot \left[\frac{\begin{pmatrix} m_1\overline{v}_1 \\ + m_2\overline{v}_2 \\ + m_3\overline{v}_3 \\ + \cdots m_N\overline{v}_N \end{pmatrix}^2}{N \cdot k_B \cdot T_{emp}} \right]}$$

Two Body – Aether-Kinematic Work “Free” Momentum

8.28

$$\Delta\overline{p}_{\text{Free\_Momentum}} = \sqrt{(m_1 + m_2)\cdot\begin{bmatrix} m_1\overline{v}_1^2 \\ + \quad m_2\overline{v}_2^2 \end{bmatrix}}\cdot\sqrt{1 - \frac{1}{(m_1 + m_2)}\cdot\left[\frac{\begin{pmatrix} m_1\overline{v}_1 \\ + \quad m_2\overline{v}_2 \end{pmatrix}^2}{\begin{bmatrix} m_1\overline{v}_1^2 \\ + \quad m_2\overline{v}_2^2 \end{bmatrix}}\right]}$$

Two Body – Aether-Kinematic Work “Free” Momentum – V(1) = 0

8.29

$$\Delta\overline{p}_{\text{Free\_Momentum}} = \sqrt{(m_1 + m_2)\cdot m_2\overline{v}_2^2}\cdot\sqrt{1 - \frac{m_2}{(m_1 + m_2)}}$$

Two Body – Aether-Kinematic Work “Free” Momentum – V(2) = 0

8.30

$$\Delta\overline{p}_{\text{Free\_Momentum}} = \sqrt{(m_1 + m_2)\cdot m_1\overline{v}_1^2}\cdot\sqrt{1 - \frac{m_1}{(m_1 + m_2)}}$$

Two Body – Aether-Kinematic Work “Free” Momentum – (m1 = m2)

8.31

$$\Delta\overline{p}_{\text{Free\_Momentum}} = m\cdot\sqrt{2\cdot\left[\overline{v}_1^2 + \overline{v}_2^2\right]}\cdot\sqrt{1 - \frac{1}{2}\cdot\left[\frac{(\overline{v}_1 + \overline{v}_2)^2}{\left[\overline{v}_1^2 + \overline{v}_2^2\right]}\right]}$$

Conservation of General Relativistic Vacuum of Spacetime Equation

$$c^4_{Light} = \pm 8\pi \cdot G \cdot \left(\frac{T_{\mu\nu}}{G_{\mu\nu}}\right) = \frac{\pm 2\pi \cdot T_{\mu\nu}}{\frac{1}{4 \cdot G} \cdot \left[R_{\mu\nu} - g_{\mu\nu} \cdot \left(\frac{1}{2} \cdot R \pm \Lambda_{Einstein}\right)\right]} \rightarrow m^4/_{s^4}$$

Aether-Kinematic Work "Free" Momentum

8.32

$$\Delta \overline{p}_{Free\_Momentum} = \sqrt{\left[\overline{p}^2_{Net\_Iso} - \frac{\left[E'^2_{Relativistic\_Energy} - E^2_{Rest\_Energy}\right]}{c^2_{Light}} \cdot \left(1 - \frac{|\overline{v}|^2_{CM}}{c^2_{Light}}\right)\right]}$$

$$\Delta \overline{p}_{Free\_Momentum} = \sqrt{\left[m^2_{Net} \cdot \left|v^2\right|_{Iso} - \frac{\left[m'^2_{Net\_Rel} - m^2_{Net}\right] \cdot c^4_{Light}}{c^2_{Light}} \cdot \left(1 - \frac{|\overline{v}|^2_{CM}}{c^2_{Light}}\right)\right]}$$

$$\Delta \overline{p}_{Free\_Momentum} = \sqrt{\left[\begin{array}{l} m^2_{Net} \cdot \left|v^2\right|_{Iso} \\ - \left[\pm 8\pi \cdot \left(\frac{G}{c^2_{Light}}\right) \cdot \left[m'^2_{Net\_Rel} - m^2_{Net}\right] \cdot \left(\frac{T_{\mu\nu}}{G_{\mu\nu}}\right)\right] \cdot \left(1 - \frac{|\overline{v}|^2_{CM}}{c^2_{Light}}\right) \end{array}\right]}$$

$$\Delta \overline{p}_{Free\_Momentum} = \sqrt{\left[\begin{array}{l} m^2_{Net} \cdot \left|v^2\right|_{Iso} \\ - \frac{1}{c^2_{Light}} \cdot \left[\left[\frac{\left[m^2_{Net} - m^2_{Net} \cdot \left(1 - \frac{|\overline{v}|^2_{CM}}{c^2_{Light}}\right)\right] \cdot G}{\left[R_{\mu\nu} - g_{\mu\nu} \cdot \left(\frac{1}{2} \cdot R \pm \Lambda_{Einstein}\right)\right]}\right] \cdot \left[\pm 8\pi \cdot T_{\mu\nu}\right]\right] \end{array}\right]}$$

The **Aether-Kinematic Work "Free" Momentum,** in three dimensional Cartesian coordinates x, y, and z, are given by the following,

8.33

$$\Delta\overline{p}_{Free\_Momentum_x} = \sqrt{\left[\overline{p}^2_{Net\_Iso_x} - \overline{p}^2_{Inertial\_Net_x}\right]}$$

$$\Delta\overline{p}_{Free\_Momentum_y} = \sqrt{\left[\overline{p}^2_{Net\_Iso_y} - \overline{p}^2_{Inertial\_Net_y}\right]}$$

$$\Delta\overline{p}_{Free\_Momentum_z} = \sqrt{\left[\overline{p}^2_{Net\_Iso_z} - \overline{p}^2_{Inertia\_Net_z}\right]}$$

8.34

Tensor		
Magnitude	Vector	Units
Aether-Kinematic Work "Free" Momentum		
$\Delta\overline{p}_{Free\_Momentum} = \left[\sqrt{\Delta\overline{p}^2_{Free\_Momentum_x} + \Delta\overline{p}^2_{Free\_Momentum_y} + \Delta\overline{p}^2_{Free\_Momentum_z}}\right]$	$\Delta\overline{p}_{Free\_Momentum} = \begin{bmatrix} \Delta\overline{p}_{Free\_Momentum_x} \\ \Delta\overline{p}_{Free\_Momentum_y} \\ \Delta\overline{p}_{Free\_Momentum_z} \end{bmatrix}$	$\frac{kg \cdot m}{s}$

8.5 Square of the Aether-Kinematic Work "Free" Momentum

Square of the Aether-Kinematic Work "Free" Momentum

8.35

$$\Delta\overline{\mathrm{p}}^{2}_{\mathrm{Free\_Momentum}} = \frac{\overline{\mathrm{p}}^{2}_{\mathrm{Net\_Iso}}}{\gamma^{2}_{Inertial}} = \frac{\overline{\mathrm{p}}^{2}_{\mathrm{Inertial\_Net}}}{\gamma^{2}_{Inertial\_Iso}} \rightarrow kg^{2}\cdot m^{2}/s^{2}$$

$$\Delta\overline{\mathrm{p}}^{2}_{\mathrm{Free\_Momentum}} = \frac{\overline{\mathrm{p}}^{2}_{\mathrm{Thermo\_Momentum}}}{\gamma^{2}_{Inertial}\cdot\gamma^{2}_{Inertial\_Iso}} = \frac{\overline{\mathrm{p}}^{2}_{\mathrm{Thermo\_Momentum}}}{\gamma^{4}_{Super\_Compress}}$$

$$\Delta\overline{\mathrm{p}}^{2}_{\mathrm{Free\_Momentum}} = \frac{\overline{\mathrm{p}}^{2}_{\mathrm{Conic\_Momentum}}}{\gamma^{2}_{Super\_Compress}} = \frac{\overline{\mathrm{p}}^{2}_{\mathrm{Conic\_Momentum}}}{\gamma_{Inertial}\cdot\gamma_{Inertial\_Iso}}$$

$$\Delta\overline{\mathrm{p}}^{2}_{\mathrm{Free\_Momentum}} = \overline{\mathrm{p}}^{2}_{\mathrm{Net\_Iso}}\cdot\left(1 - \frac{\mathrm{E}_{\mathrm{Net\_Translational}}}{\mathrm{E}_{\mathrm{Net\_Iso\_Kinetic}}}\right) = \overline{\mathrm{p}}^{2}_{\mathrm{Inertial\_Net}}\cdot\left(\frac{\mathrm{E}_{\mathrm{Net\_Iso\_Kinetic}}}{\mathrm{E}_{\mathrm{Net\_Translational}}} - 1\right)$$

$$\Delta\overline{\mathrm{p}}^{2}_{\mathrm{Free\_Momentum}} = \overline{\mathrm{p}}^{2}_{\mathrm{Net\_Iso}}\cdot\left(1 - \frac{\left|\overline{\mathrm{v}}\right|^{2}_{\mathrm{CM}}}{\left|\mathrm{v}^{2}\right|_{\mathrm{Iso}}}\right) = \overline{\mathrm{p}}^{2}_{\mathrm{Inertial\_Net}}\cdot\left(\frac{\left|\mathrm{v}^{2}\right|_{\mathrm{Iso}}}{\left|\overline{\mathrm{v}}\right|^{2}_{\mathrm{CM}}} - 1\right)$$

$$\Delta\overline{\mathrm{p}}^{2}_{\mathrm{Free\_Momentum}} = \mathrm{m}^{2}_{\mathrm{Net}}\cdot\left|\mathrm{v}^{2}\right|_{\mathrm{Iso}}\cdot\left(1 - \frac{\left|\overline{\mathrm{v}}\right|^{2}_{\mathrm{CM}}}{\left|\mathrm{v}^{2}\right|_{\mathrm{Iso}}}\right) = \mathrm{m}^{2}_{\mathrm{Net}}\cdot\left|\overline{\mathrm{v}}\right|^{2}_{\mathrm{CM}}\cdot\left(\frac{\left|\mathrm{v}^{2}\right|_{\mathrm{Iso}}}{\left|\overline{\mathrm{v}}\right|^{2}_{\mathrm{CM}}} - 1\right)$$

Square of the Aether-Kinematic Work "Free" Momentum

8.36

$$\Delta\overline{p}^2_{Free\_Momentum} = \left[\Delta\overline{p}^2_{Iso\_Momentum} + \Delta\overline{p}^2_{Aniso\_Momentum}\right] \rightarrow kg^2 \cdot m^2 / s^2$$

$$\Delta\overline{p}^2_{Free\_Momentum} = \left[\overline{p}^2_{Net\_Iso} + \overline{p}^2_{Inertial\_Net} - 2 \cdot \overline{p}^2_{Conic\_Momentum}\right]$$

$$\Delta\overline{p}^2_{Free\_Momentum} = \left[\overline{p}^2_{Net\_Iso} + \overline{p}^2_{Inertial\_Net} - 2 \cdot \overline{p}_{Net\_Iso} \cdot \overline{p}_{Inertial\_Net}\right]$$

$$\Delta\overline{p}^2_{Free\_Momentum} = \left[\overline{p}^2_{Net\_Iso} + \overline{p}^2_{Inertial\_Net} - 2 \cdot \left\|\overline{p}_{Net\_Iso}\right\| \cdot \left\|\overline{p}_{Inertial\_Net}\right\| \cdot \cos\theta\right]$$

$$\Delta\overline{p}^2_{Free\_Momentum} = \begin{bmatrix} m^2_{Net} \cdot \left|v^2\right|_{Iso} + m^2_{Net} \cdot \left|\overline{v}\right|^2_{CM} \\ - 2 \cdot m^2_{Net} \cdot \left|\overline{v}\right|_{CM} \cdot \sqrt{\left|v^2\right|_{Iso}} \end{bmatrix}$$

$$\Delta\overline{p}^2_{Free\_Momentum} = \begin{bmatrix} m^2_{Net} \cdot \left|v^2\right|_{Iso} + m^2_{Net} \cdot \left|\overline{v}\right|^2_{CM} \\ - 2 \cdot m^2_{Net} \cdot \left(\left\|\left|\overline{v}\right|_{CM}\right\| \cdot \left\|\sqrt{\left|v^2\right|_{Iso}}\right\|\right) \cdot \cos\theta \end{bmatrix}$$

Square of the Aether-Kinematic Work "Free" Momentum

8.37

$$\Delta\overline{p}^2_{Free\_Momentum} = \left[\begin{array}{l} \overline{p}^2_{Net\_Iso} \\ - \left[\pm 8\pi \cdot \left(\dfrac{G}{c^2_{Light}} \right) \cdot \left[m'^2_{Net\_Rel} - m^2_{Net} \right] \cdot \left(\dfrac{T_{\mu\nu}}{G_{\mu\nu}} \right) \right] \cdot \left(1 - \dfrac{|\overline{v}|^2_{CM}}{c^2_{Light}} \right) \end{array} \right]$$

$$\Delta\overline{p}^2_{Free\_Momentum} = \left[\begin{array}{l} m^2_{Net} \cdot |v^2|_{Iso} \\ - \dfrac{1}{c^2_{Light}} \cdot \left[\left[\dfrac{\left[m^2_{Net} - m^2_{Net} \cdot \left(1 - \dfrac{|\overline{v}|^2_{CM}}{c^2_{Light}} \right) \right] \cdot G}{\left[R_{\mu\nu} - g_{\mu\nu} \cdot \left(\dfrac{1}{2} \cdot R \pm \Lambda_{Einstein} \right) \right]} \right] \cdot \left[\pm 8\pi \cdot T_{\mu\nu} \right] \right] \end{array} \right]$$

The **Square of the Aether-Kinematic Work "Free" Momentum,** in three dimensional Cartesian coordinates x, y, and z, are given by the following,

8.38

$$\Delta\overline{p}^2_{Free\_Momentum_x} = \left[\overline{p}^2_{Net\_Iso_x} - \overline{p}^2_{Inertial\_Net_x}\right]$$

$$\Delta\overline{p}^2_{Free\_Momentum_y} = \left[\overline{p}^2_{Net\_Iso_y} - \overline{p}^2_{Inertial\_Net_y}\right]$$

$$\Delta\overline{p}^2_{Free\_Momentum_z} = \left[\overline{p}^2_{Net\_Iso_z} - \overline{p}^2_{Inertial\_Net_z}\right]$$

8.39

Scalar/Tensor		
Magnitude	Scalar/Tensor	Units
Square of the Aether-Kinematic Work "Free" Momentum		
$\Delta\overline{p}^2_{Free\_Momentum} = \begin{bmatrix} \Delta\overline{p}^2_{Free\_Momentum_x} \\ +\Delta\overline{p}^2_{Free\_Momentum_y} \\ +\Delta\overline{p}^2_{Free\_Momentum_z} \end{bmatrix}$	$\Delta\overline{p}^2_{Free\_Momentum} = \begin{bmatrix} \Delta\overline{p}^2_{Free\_Momentum_x} \\ \Delta\overline{p}^2_{Free\_Momentum_y} \\ \Delta\overline{p}^2_{Free\_Momentum_z} \end{bmatrix}$	$\frac{kg^2 \cdot m^2}{s^2}$

8.6 Aether-Kinematic Work "Free" Energy

The Aether Kinematic Work "Free" Energy ($W_{Free\_Work\_Energy}$) describes the work done condensing and expanding matter, space, and time relative to the Center of Mass and the Center of Isotropy of an isolated system mass body in accordance with the First Law of Motion.

The Aether Kinematic Work "Free" Energy ($W_{Free\_Work\_Energy}$) is a measure of the Aether "Free" Force doing work creating a dimensional difference between isotropy and anisotropy motions, namely: Momentum Change, Pressure Change, Kinetic Energy Change, Heat Energy Change, and Temperature Change relative to the Center of Mass and the Center of Isotropy of an isolated system mass body.

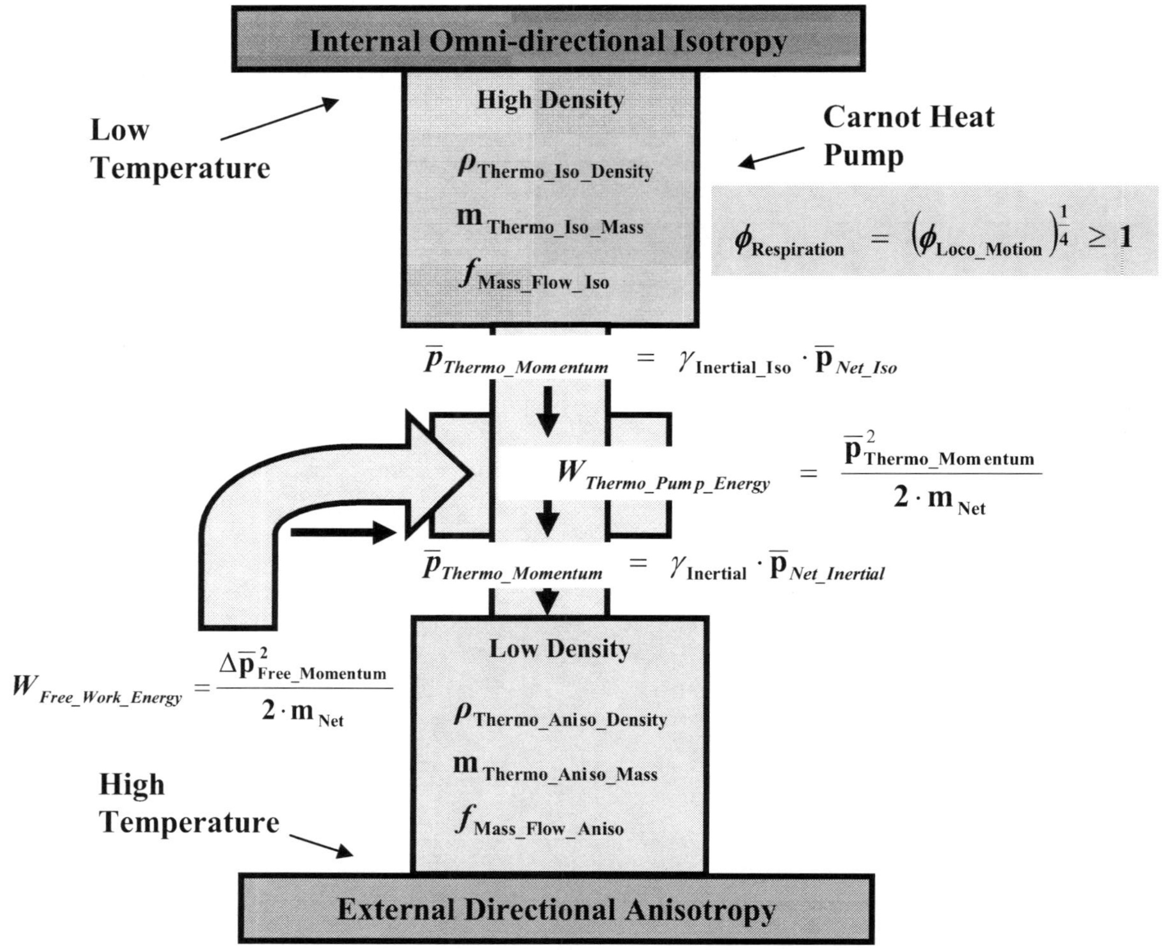

Figure 8.6: The Thermodynamic Pump Energy is used by the system to do work which *heats the body and expands* by creating aether density difference without and within the body.

Definition 8.5: Relative to an inertial frame, the Mean Center, the Center of Mass, and the Center of Isotropy of an isolated system mass body, the **Aether-Kinematic Work "Free" Energy** ($W_{Free\_Work\_Energy} = \frac{\Delta\overline{p}^2_{Free\_Momentum}}{2 \cdot m_{Net}}$) is constant, and conserved, and is the measure of the change in the Isotropy and Anisotropy Kinetic Energy which is equal to the work done condensing and expanding space and time between Isotropic Motion and Anisotropic Motion; and is defined as the difference between the **Internal Isotropic Inertial/Aether Omni-directional Kinetic Energy** ($E_{Net\_Iso\_Kinetic} = \frac{\overline{p}^2_{Net\_Iso}}{2 \cdot m_{Net}} = \frac{m_{Net} \cdot \left|v^2\right|_{Iso}}{2}$) subtracted from the **External Anisotropic Inertial Net Translational Kinetic Energy** ($E_{Net\_Translational} = \frac{\overline{p}^2_{Inertial\_Net}}{2 \cdot m_{Net}} = \frac{m_{Net} \cdot \left|\overline{v}\right|^2_{CM}}{2}$) in any atomic substance material medium including the Vacuum of Spacetime.

Aether-Kinematic Work "Free" Energy

8.40

$$W_{Free\_Work\_Energy} = \left(\frac{\Delta \overline{p}^2_{Free\_Momentum}}{2 \cdot m_{Net}} \right) \rightarrow kg \cdot m^2 / s^2$$

$$W_{Free\_Work\_Energy} = \frac{1}{2 \cdot m_{Net}} \cdot \left[\overline{p}^2_{Net\_Iso} - \overline{p}^2_{Inertial\_Net} \right]$$

$$W_{Free\_Work\_Energy} = \frac{W_{Thermo\_Pump\_Energy}}{\gamma^4_{Super\_Compress}} = \frac{W_{Thermo\_Pump\_Energy}}{\gamma^2_{Inertial} \cdot \gamma^2_{Inertial\_Iso}}$$

$$W_{Free\_Work\_Energy} = \left(\frac{\overline{p}^2_{Net\_Iso}}{2 \cdot m_{Net}} \right) \cdot \left(1 - \frac{\left| \overline{v} \right|^2_{CM}}{\left| v^2 \right|_{Iso}} \right)$$

$$W_{Free\_Work\_Energy} = \left(\frac{m_{Net} \cdot \left| v^2 \right|_{Iso}}{2} \right) \cdot \left(1 - \frac{\left| \overline{v} \right|^2_{CM}}{\left| v^2 \right|_{Iso}} \right)$$

$$W_{Free\_Work\_Energy} = \left(\frac{\overline{p}^2_{Inertial\_Net}}{2 \cdot m_{Net}} \right) \cdot \left(\frac{\left| v^2 \right|_{Iso}}{\left| \overline{v} \right|^2_{CM}} - 1 \right)$$

$$W_{Free\_Work\_Energy} = \left(\frac{m_{Net} \cdot \left| \overline{v} \right|^2_{CM}}{2} \right) \cdot \left(\frac{\left| v^2 \right|_{Iso}}{\left| \overline{v} \right|^2_{CM}} - 1 \right)$$

The **Aether-Kinematic Work “Free” Energy,** in three dimensional Cartesian coordinates x, y, and z, are given by the following,

8.41

$$W_{Free\_Work\_Energy_y} = \frac{\Delta \overline{p}^2_{Free\_Momentum_x}}{2 \cdot m_{Net}} = \frac{1}{2 \cdot m_{Net}} \cdot \left[\overline{p}^2_{Net\_Iso_x} - \overline{p}^2_{Inertial\_Net_x} \right]$$

$$W_{Free\_Work\_Energy_y} = \frac{\Delta \overline{p}^2_{Free\_Momentum_y}}{2 \cdot m_{Net}} = \frac{1}{2 \cdot m_{Net}} \cdot \left[\overline{p}^2_{Net\_Iso_y} - \overline{p}^2_{Inertial\_Net_y} \right]$$

$$W_{Free\_Work\_Energy_z} = \frac{\Delta \overline{p}^2_{Free\_Momentum_z}}{2 \cdot m_{Net}} = \frac{1}{2 \cdot m_{Net}} \cdot \left[\overline{p}^2_{Net\_Iso_z} - \overline{p}^2_{Inertial\_Net_z} \right]$$

8.42

Scalar/Tensor		
Magnitude	Scalar/Tensor	Units
Aether-Kinematic Work “Free” Energy		
$W_{Free\_Work\_Energy} = \begin{bmatrix} W_{Free\_Work\_Energy_x} \\ + W_{Free\_Work\_Energy_y} \\ + W_{Free\_Work\_Energy_z} \end{bmatrix}$	$W_{Free\_Work\_Energy} = \begin{bmatrix} W_{Free\_Work\_Energy_x} \\ W_{Free\_Work\_Energy_y} \\ W_{Free\_Work\_Energy_z} \end{bmatrix}$	$\frac{kg \cdot m^2}{s^2}$

Chapter 9

The Special Theory of Thermodynamics

(Conic Inertial Momentum)
&
(Total Energy Conservation Relation)

9.1 Universal Geometric Mean Law – Thermodynamic Heat Engine & Heat Pump Relation

Universal Geometric Mean Theorem

$$(Mean)^2 = (Max)\cdot(Min)$$

$$(Mean) = \sqrt{(Max)\cdot(Min)}$$

Universal Geometric Mean Theorem – Square Inertial Conic Momentum

9.1

$$\overline{p}^2_{\text{Conic\_Momentum}} = \frac{\overline{p}^2_{\text{Thermo\_Momentum}}}{\gamma^2_{Super\_Compress}} = \overline{p}_{\text{Net\_Iso}}\cdot\overline{p}_{\text{Inertial\_Net}} \rightarrow kg^2\cdot m^2/s^2$$

$$\overline{p}^2_{\text{Conic\_Momentum}} = \frac{\overline{p}^2_{\text{Thermo\_Momentum}}}{\gamma_{Inertial}\cdot\gamma_{Inertial\_Iso}} = \overline{p}_{\text{Net\_Iso}}\cdot\overline{p}_{\text{Inertial\_Net}}$$

$$\overline{p}^2_{\text{Conic\_Momentum}} = \overline{p}_{\text{Thermo\_Momentum}}\cdot\Delta\overline{p}_{\text{Free\_Momentum}}$$

Universal Geometric Mean Theorem – Inertial Conic Momentum

9.2

$$\overline{p}_{\text{Conic\_Momentum}} = \frac{\overline{p}_{\text{Thermo\_Momentum}}}{\gamma_{Super\_Compres}} = \sqrt{\overline{p}_{\text{Net\_Iso}}\cdot\overline{p}_{\text{Inertial\_Net}}} \rightarrow kg\cdot m/s$$

$$\overline{p}_{\text{Conic\_Momentum}} = \sqrt{\overline{p}_{\text{Thermo\_Momentum}}\cdot\Delta\overline{p}_{\text{Free\_Momentum}}}$$

9.2 Conic Inertial Momentum

Definition 9.1: The **Square Conic Inertial Momentum System** ($\overline{p}_{\text{Conic\_Momentum}}$) is constant and conserved for any uniform state of motion and is defined by the Dot Product Law as described by the following.

Square Conic Inertial Momentum

9.3

$$\overline{p}^2_{\text{Conic\_Momentum}} = \overline{p}_{\text{Net\_Iso}} \cdot \overline{p}_{\text{Inertial\_Net}} = \left\|\overline{p}_{\text{Net\_Iso}}\right\| \cdot \left\|\overline{p}_{\text{Inertial\_Net}}\right\| \cdot \cos\theta \rightarrow kg^2 \cdot m^2 / s^2$$

$$\overline{p}^2_{\text{Conic\_Momentum}} = \frac{1}{2} \cdot \left[\overline{p}^2_{\text{Net\_Iso}} + \overline{p}^2_{\text{Inertial\_Net}} - \Delta\overline{p}^2_{\text{Free\_Momentum}}\right]$$

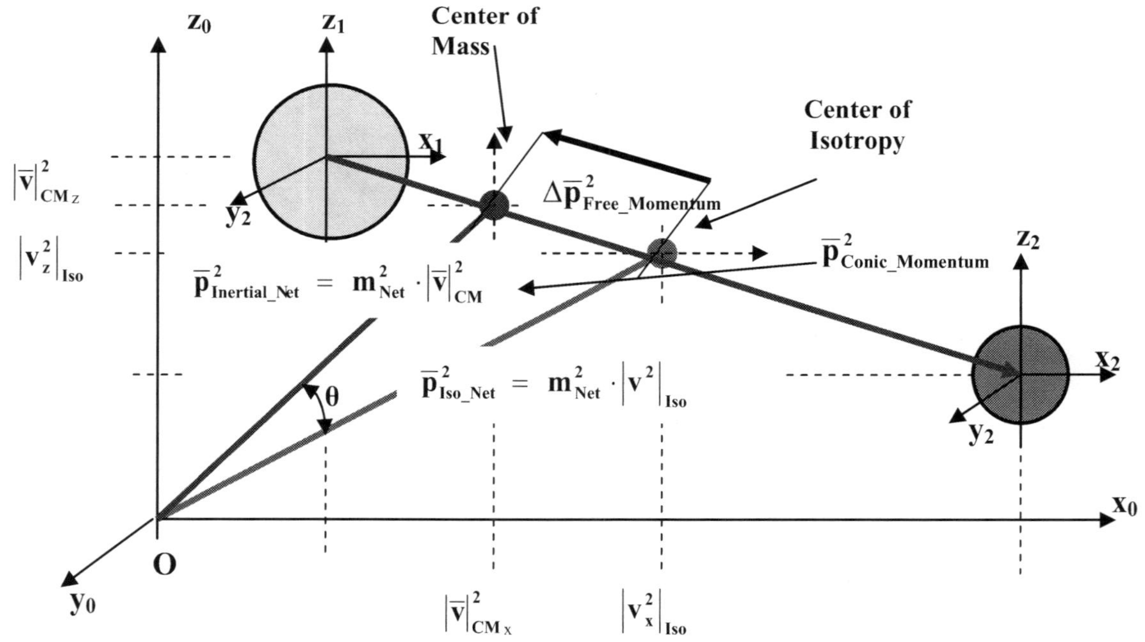

Figure 9.1: The Conic Momentum is defined by the Dot Product Law

Square Conic Inertial Momentum

9.4

$$\overline{p}^2_{\text{Conic\_Momentum}} = \overline{p}_{\text{Net\_Iso}} \cdot \overline{p}_{\text{Inertial\_Net}} \rightarrow kg^2 \cdot m^2 / s^2$$

$$\overline{p}^2_{\text{Conic\_Momentum}} = \overline{p}_{\text{Thermo\_Momentum}} \cdot \Delta\overline{p}_{\text{Free\_Momentum}}$$

$$\overline{p}^2_{\text{Conic\_Momentum}} = \frac{\overline{p}^2_{\text{Thermo\_Momentum}}}{\gamma^2_{Super\_Compress}} = \overline{p}_{\text{Net\_Iso}} \cdot \overline{p}_{\text{Inertial\_Net}}$$

$$\overline{p}^2_{\text{Conic\_Momentum}} = \frac{\overline{p}^2_{\text{Thermo\_Momentum}}}{\gamma_{Inertial} \cdot \gamma_{Inertial\_Iso}} = \overline{p}_{\text{Thermo\_Momentum}} \cdot \Delta\overline{p}_{\text{Free\_Momentum}}$$

$$\overline{p}^2_{\text{Conic\_Momentum}} = \gamma^2_{Super\_Compress} \cdot \Delta\overline{p}^2_{\text{Free\_Momentum}}$$

$$\overline{p}^2_{\text{Conic\_Momentum}} = \overline{p}_{\text{Net\_Iso}} \cdot \overline{p}_{\text{Inertial\_Net}} = \left\|\overline{p}_{\text{Net\_Iso}}\right\| \cdot \left\|\overline{p}_{\text{Inertial\_Net}}\right\| \cdot \cos\theta$$

$$\overline{p}^2_{\text{Conic\_Momentum}} = \frac{1}{2} \cdot \left[\overline{p}^2_{\text{Net\_Iso}} + \overline{p}^2_{\text{Inertial\_Net}} - \Delta\overline{p}^2_{\text{Free\_Momentum}}\right]$$

Square Conic Inertial Momentum

9.5

$$\overline{p}^{2}_{Conic\_Momentum} = m^{2}_{Net} \cdot \left|\overline{v}\right|_{CM} \cdot \sqrt{\left|v^{2}\right|_{Iso}} \rightarrow kg^{2} \cdot m^{2}/s^{2}$$

$$\overline{p}^{2}_{Conic\_Momentum} = \frac{\overline{p}_{Inertial\_Net}}{\sqrt{1 - \frac{\left|\overline{v}\right|^{2}_{CM}}{\left|v^{2}\right|_{Iso}}}} \cdot \sqrt{\left[\overline{p}^{2}_{Net\_Iso} - \overline{p}^{2}_{Inertial\_Net}\right]}$$

$$\overline{p}^{2}_{Conic\_Momentum} = m^{2}_{Net} \cdot \left|\overline{v}\right|_{CM} \cdot \sqrt{\left|v^{2}\right|_{Iso}} = m^{2}_{Net} \cdot \left\|\left|\overline{v}\right|_{CM}\right\| \cdot \left\|\sqrt{\left|v^{2}\right|_{Iso}}\right\| \cdot \cos\theta$$

The **Square Conic Inertial Momentum,** in three dimensional Cartesian coordinates x, y, and z, are given by the following,

9.6

$$\overline{p}^{2}_{Conic\_Momentum_{x}} = \overline{p}_{Thermo\_Momentum_{x}} \cdot \Delta\overline{p}_{Free\_Momentum_{x}} = \overline{p}_{Net\_Iso_{x}} \cdot \overline{p}_{Inertial\_Net_{x}}$$

$$\overline{p}^{2}_{Conic\_Momentum_{y}} = \overline{p}_{Thermo\_Momentum_{y}} \cdot \Delta\overline{p}_{Free\_Momentum_{y}} = \overline{p}_{Net\_Iso_{y}} \cdot \overline{p}_{Inertial\_Net_{y}}$$

$$\overline{p}^{2}_{Conic\_Momentum_{x}} = \overline{p}_{Thermo\_Momentum_{x}} \cdot \Delta\overline{p}_{Free\_Momentum_{x}} = \overline{p}_{Net\_Iso_{x}} \cdot \overline{p}_{Inertial\_Net_{x}}$$

9.7

Scalar/Tensor		
Magnitude	Scalar/Tensor	Units
Square Conic Inertial Momentum		
$\overline{p}^{2}_{Conic\_Momentum} = \begin{bmatrix} \overline{p}^{2}_{Conic\_Momentum_{x}} \\ + \overline{p}^{2}_{Conic\_Momentum_{y}} \\ + \overline{p}^{2}_{Conic\_Momentum_{z}} \end{bmatrix}$	$\overline{p}^{2}_{Conic\_Momentum} = \begin{bmatrix} \overline{p}^{2}_{Conic\_Momentum_{x}} \\ \overline{p}^{2}_{Conic\_Momentum_{y}} \\ \overline{p}^{2}_{Conic\_Momentum_{z}} \end{bmatrix}$	$\frac{kg^{2} \cdot m^{2}}{s^{2}}$

9.3 Isotropic – Aether-Kinematic Work "Free" Momentum Difference

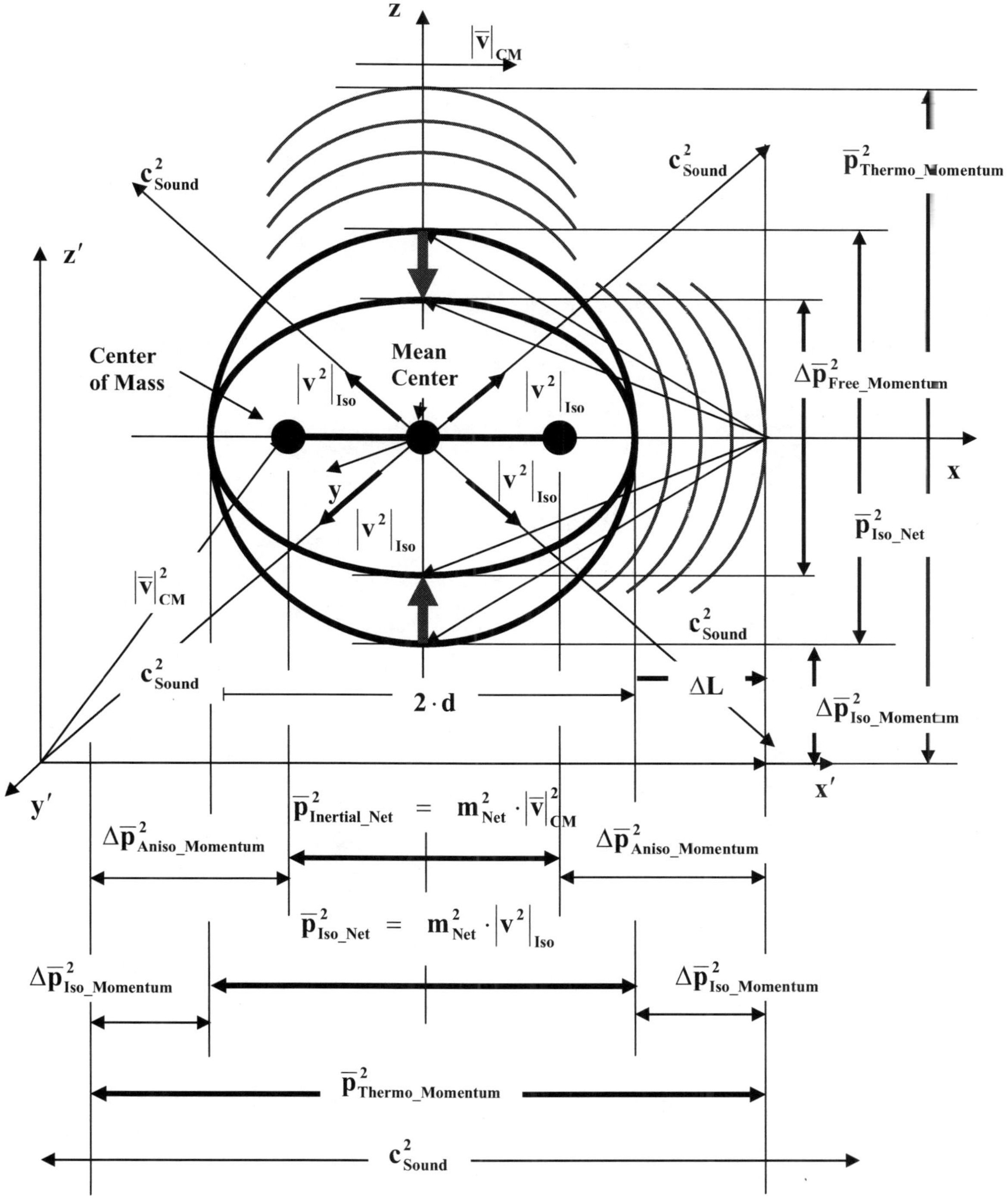

Figure 9.2: Isolated Dynamic System Mass Body.

Isotropic – Aether-Kinematic Work "Free" Momentum Difference

9.8

$$\Delta\overline{p}^2_{Iso\_Momentum} = \left[\overline{p}^2_{Net\_Iso} - \overline{p}^2_{Thermo\_Momentum}\right] \rightarrow kg^2 \cdot m^2 / s^2$$

$$\Delta\overline{p}^2_{Iso\_Momentum} = \overline{p}^2_{Net\_Iso} \cdot \left[1 - \gamma^2_{Inertial\_Iso}\right]$$

$$\Delta\overline{p}^2_{Iso\_Momentum} = \left[\overline{p}^2_{Net\_Iso} - \gamma_{Inertial} \cdot \gamma_{Inertial\_Iso} \cdot \overline{p}_{Net\_Iso} \cdot \overline{p}_{Inertial\_Net}\right]$$

$$\Delta\overline{p}^2_{Iso\_Momentum} = \left[\overline{p}^2_{Net\_Iso} - \gamma^2_{Super\_Compress} \cdot \overline{p}^2_{Conic\_Momentum}\right]$$

$$\Delta\overline{p}^2_{Iso\_Momentum} = \left[\overline{p}^2_{Net\_Iso} - \gamma^2_{Inertial} \cdot \gamma^2_{Inertial\_Iso} \cdot \Delta\overline{p}^2_{Free\_Momentum}\right]$$

$$\Delta\overline{p}^2_{Iso\_Momentum} = \left[\begin{array}{l} \Delta\overline{p}^2_{Free\_Momentum} - \overline{p}^2_{Inertial\_Net} \\ \quad + 2 \cdot \overline{p}^2_{Conic\_Momentum} - \overline{p}^2_{Thermo\_Momentum} \end{array}\right]$$

$$\Delta\overline{p}^2_{Iso\_Momentum} = \left[\begin{array}{l} 2 \cdot \overline{p}^2_{Conic\_Momentum} - \overline{p}^2_{Inertial\_Net} \\ \quad + \Delta\overline{p}^2_{Free\_Momentum} \cdot \left(1 - \gamma^2_{Inertial} \cdot \gamma^2_{Inertial\_Iso}\right) \end{array}\right]$$

Isotropic – Aether-Kinematic Work "Free" Momentum Difference

9.9

$$\Delta\overline{p}^2_{\text{Iso\_Momentum}} = \begin{bmatrix} \Delta\overline{p}^2_{\text{Free\_Momentum}} - \overline{p}^2_{\text{Inertial\_Net}} \\ + 2\cdot\overline{p}^2_{\text{Conic\_Momentum}} - \gamma^2_{\textit{Inertial}}\cdot\overline{p}^2_{\text{Inertial\_Net}} \end{bmatrix}$$

$$\Delta\overline{p}^2_{\text{Iso\_Momentum}} = \begin{bmatrix} \Delta\overline{p}^2_{\text{Free\_Momentum}} + 2\cdot\overline{p}^2_{\text{Conic\_Momentum}} \\ - \overline{p}^2_{\text{Inertial\_Net}}\cdot\left(1 + \gamma^2_{\textit{Inertial}}\right) \end{bmatrix}$$

$$\Delta\overline{p}^2_{\text{Iso\_Momentum}} = \begin{bmatrix} \Delta\overline{p}^2_{\text{Free\_Momentum}} - \overline{p}^2_{\text{Inertial\_Net}} \\ + \overline{p}^2_{\text{Conic\_Momentum}}\cdot\left(2 - \gamma^2_{\textit{Super\_Compress}}\right) \end{bmatrix}$$

$$\Delta\overline{p}^2_{\text{Iso\_Momentum}} = \left[\Delta\overline{p}^2_{\text{Free\_Momentum}} - \Delta\overline{p}^2_{\text{Aniso\_Momentum}}\right]$$

$$\Delta\overline{p}^2_{\text{Iso\_Momentum}} = \left[\Delta\overline{p}^2_{\text{Free\_Momentum}} - \overline{p}^2_{\text{Inertial\_Net}}\cdot\left(\gamma^2_{\textit{Inertial}} - 1\right)\right]$$

$$\Delta\overline{p}^2_{\text{Iso\_Momentum}} = \left[\overline{p}^2_{\text{Net\_Iso}} - \gamma^2_{\textit{Super\_Compress}}\cdot\overline{p}^2_{\text{Conic\_Momentum}}\right]$$

$$\Delta\overline{p}^2_{\text{Iso\_Momentum}} = \overline{p}^2_{\text{Net\_Iso}}\cdot\left[1 - \frac{1}{\left(\dfrac{\left|v^2\right|_{\text{Iso}}}{\left|\overline{v}\right|^2_{\text{CM}}} - 1\right)}\right] \rightarrow kg^2\cdot m^2/s^2$$

The **Isotropic — Aether-Kinematic Work "Free" Momentum Difference,** in three dimensional Cartesian coordinates x, y, and z, are given by the following,

9.10

$$\Delta\overline{p}^2_{Iso\_Momentum_x} = \overline{p}^2_{Net\_Iso_x} \cdot \left[1 - \gamma^2_{Inertial\_Iso_x}\right]$$

$$\Delta\overline{p}^2_{Iso\_Momentum_y} = \overline{p}^2_{Net\_Iso_y} \cdot \left[1 - \gamma^2_{Inertial\_Iso_y}\right]$$

$$\Delta\overline{p}^2_{Iso\_Momentum_z} = \overline{p}^2_{Net\_Iso_z} \cdot \left[1 - \gamma^2_{Inertial\_Iso_z}\right]$$

9.11

Scalar/Tensor		
Magnitude	Scalar/Tensor	Units
Isotropic — Aether-Kinematic Work "Free" Momentum Difference		
$\Delta\overline{p}^2_{Iso\_Momentum} = \begin{bmatrix} \Delta\overline{p}^2_{Iso\_Momentum_x} \\ + \Delta\overline{p}^2_{Iso\_Momentum_y} \\ + \Delta\overline{p}^2_{Iso\_Momentum_z} \end{bmatrix}$	$\Delta\overline{p}^2_{Iso\_Momentum} = \begin{bmatrix} \Delta\overline{p}^2_{Iso\_Momentum_x} \\ \Delta\overline{p}^2_{Iso\_Momentum_y} \\ \Delta\overline{p}^2_{Iso\_Momentum_z} \end{bmatrix}$	$\frac{kg^2 \cdot m^2}{s^2}$

9.4 Anisotropic – Aether-Kinematic Work "Free" Momentum Difference

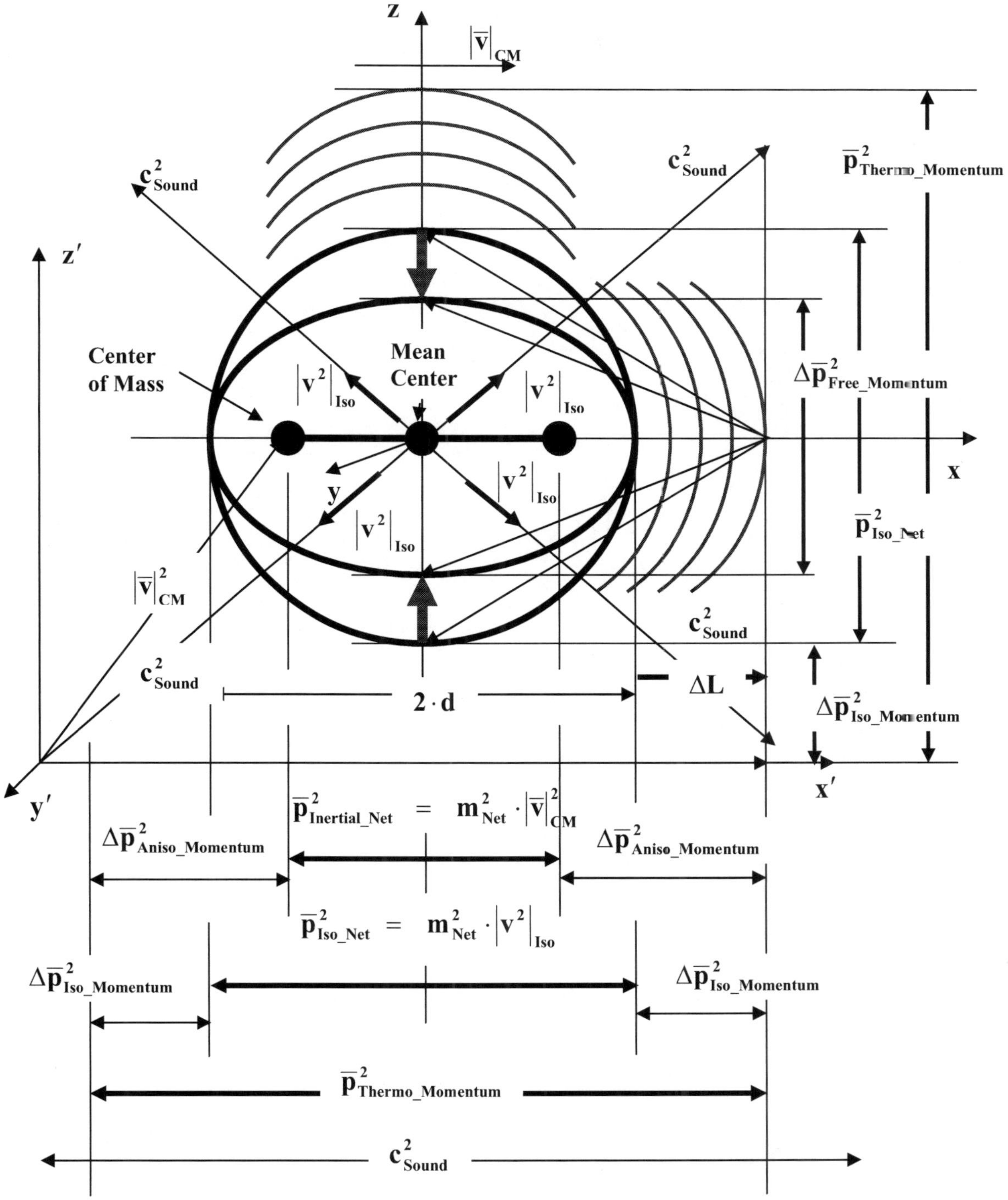

Figure 9.3: Isolated Dynamic System Mass Body.

Anisotropic – Aether-Kinematic Work "Free" Momentum Difference

9.12

$$\Delta\overline{p}^2_{\text{Aniso\_Momentum}} = \left[\overline{p}^2_{\text{Thermo\_Momentum}} - \overline{p}^2_{\text{Inertial\_Net}}\right] \rightarrow kg^2 \cdot m^2 / s^2$$

$$\Delta\overline{p}^2_{\text{Aniso\_Momentum}} = \overline{p}^2_{\text{Inertial\_Net}} \cdot \left(\gamma^2_{Inertial} - 1\right)$$

$$\Delta\overline{p}^2_{\text{Aniso\_Momentum}} = \left[\Delta\overline{p}^2_{\text{Free\_Momentum}} - \Delta\overline{p}^2_{\text{Iso\_Momentum}}\right]$$

$$\Delta\overline{p}^2_{\text{Aniso\_Momentum}} = \left[\overline{p}^2_{\text{Inertial\_Net}} \cdot \left(1 + \gamma^2_{Inertial}\right) - 2 \cdot \overline{p}^2_{\text{Conic\_Momentum}}\right]$$

$$\Delta\overline{p}^2_{\text{Aniso\_Momentum}} = \left[\overline{p}^2_{\text{Inertial\_Net}} + \overline{p}^2_{\text{Conic\_Momentum}} \cdot \left(2 - \gamma^2_{Super\_Compress}\right)\right]$$

$$\Delta\overline{p}^2_{\text{Aniso\_Momentum}} = \left[\Delta\overline{p}^2_{\text{Free\_Momentum}} - \overline{p}^2_{\text{Net\_Iso}} + \gamma^2_{Super\_Compress} \cdot \overline{p}^2_{\text{Conic\_Momentum}}\right]$$

$$\Delta\overline{p}^2_{\text{Aniso\_Momentum}} = \overline{p}^2_{\text{Inertial\_Net}} \cdot \left(\frac{1}{\left(1 - \dfrac{\left|\overline{v}\right|^2_{\text{CM}}}{\left|v^2\right|_{\text{Iso}}}\right)} - 1 \right)$$

Anisotropic – Aether-Kinematic Work "Free" Momentum Difference

9.13

$$\Delta\overline{p}^2_{Aniso\_Momentum} = \left[\Delta\overline{p}^2_{Free\_Momentum} - \Delta\overline{p}^2_{Iso\_Momentum}\right] \rightarrow kg^2 \cdot m^2 / s^2$$

$$\Delta\overline{p}^2_{Aniso\_Momentum} = \left[\begin{array}{l}\Delta\overline{p}^2_{Free\_Momentum} \\ \quad - \overline{p}^2_{Net\_Iso} \cdot \left[1 - \gamma^2_{Inertial\_Iso}\right]\end{array}\right]$$

The **Anisotropic – Aether-Kinematic Work "Free" Momentum Difference,** in three dimensional Cartesian coordinates x, y, and z, are given by the following,

9.14

$$\Delta\overline{p}^2_{Aniso\_Momentum_x} = \overline{p}^2_{Inertial\_Net_x} \cdot \left(\gamma^2_{Inertial_x} - 1\right)$$

$$\Delta\overline{p}^2_{Aniso\_Momentum_y} = \overline{p}^2_{Inertial\_Net_y} \cdot \left(\gamma^2_{Inertial_y} - 1\right)$$

$$\Delta\overline{p}^2_{Aniso\_Momentum_z} = \overline{p}^2_{Inertial\_Net_z} \cdot \left(\gamma^2_{Inertial_z} - 1\right)$$

9.15

Scalar/Tensor		
Magnitude	Scalar/Tensor	Units
Anisotropic – Aether-Kinematic Work "Free" Momentum Difference		
$\Delta\overline{p}^2_{Aniso\_Momentum} = \left[\begin{array}{l}\Delta\overline{p}^2_{Aniso\_Momentum_x} \\ + \Delta\overline{p}^2_{Aniso\_Momentum_y} \\ + \Delta\overline{p}^2_{Aniso\_Momentum_z}\end{array}\right]$	$\Delta\overline{p}^2_{Aniso\_Momentum} = \left[\begin{array}{l}\Delta\overline{p}^2_{Aniso\_Momentum_x} \\ \Delta\overline{p}^2_{Aniso\_Momentum_y} \\ \Delta\overline{p}^2_{Aniso\_Momentum_z}\end{array}\right]$	$\frac{kg^2 \cdot m^2}{s^2}$

9.5 Total Kinetic Energy Conservation Relation

On the molecular level, internal temperature is the result of the motion of particles which make up a substance relative to its environment or external temperature. Many properties of matter are changed with any increases or decreases in temperature.

When a hot object is put in contact with a cold object, heat flows from the high temperature to the low until the two objects reach the same intermediate temperature. The total entropy increases in this process, and the process goes to disorder.

At the beginning of the process, we can distinguish two classes of molecules those with a high average kinetic energy, and those with a low average kinetic energy. After the temperature equilibrium process, all the molecules are in one class with the same average kinetic energy. There is no orderly arrangement of molecules in two classes. Order has gone to disorder; entropy has increased.

Let's consider any rock that we pick up off the ground, the Total Kinetic Energy of the rock at any given moment is an isolated system and can be written as a sum of two terms, the external anisotropic translational kinetic energy of the center of mass of the rock, and the internal isotropic Omni-directional kinetic energy of all the molecules relative to the center of mass and center of isotropy.

The Total Kinetic energy of an isolated system mass body in motion can be expressed as the sum of the External Translational Kinetic Energy and the Internal Isotropic Omni-directional Kinetic Energy.

The first term is the external to internal mechanical translational kinetic energy of the rock as a whole, and it can be considered as the ordered energy (Syntropy) since all of it can be used to do work, i.e. turning a generator, powering a home, or moving something.

The second term is the sum of the internal kinetic energies of all the moving molecules within the rock, it is the energy they would have even if the rock were at rest. The Internal Isotropic kinetic energy represents the random disordered motion (Entropy) of the molecules.

The Isotropic Kinetic Energy can be used to do work; however only a small portion of the energy is available to do work, and only if there is a lower temperature region. Thus, the isolated system body as a whole behaves like a heat engine.

Definition 9.2: The **Total Kinetic Energy (Conservation) of an Isolated System** ($E_{Total\_Kinetic}$) is constant and conserved for any uniform state of motion and is a net thermodynamic measurement of the total energy of motion of the system, and is equal to the net sum of the **Anisotropic External Translational Kinetic Energy** ($E_{Net\_Translational}$), sum the **Isotropic Internal Omni-directional Kinetic Energy** ($E_{Net\_Iso\_Kinetic}$) of a system mass body.

Total Kinetic Energy (Conservation) of an Isolated System

9.16

$$E_{Total\_Kinetic} = [\text{Syntropy} + \text{Entropy}] \rightarrow kg \cdot m^2 / s^2$$

$$E_{Total\_Kinetic} = \left[E_{Net\_Translational} + E_{Net\_Iso\_Kinetic}\right]$$

$$E_{Total\_Kinetic} = \left[\frac{1}{2} \cdot m_{Net} \cdot \left|\overline{v}\right|_{CM}^{2} + \frac{1}{2} \cdot m_{Net} \cdot \left|v^2\right|_{Iso}\right]$$

$$E_{Total\_Kinetic} = \left[\frac{3}{2} \cdot N \cdot k_B \cdot T_{emp\_Ext} + \frac{3}{2} \cdot N \cdot k_B \cdot T_{emp}\right]$$

Total Kinetic Energy of an Isolated System – "Before State"

9.17

$$E_{Total\_Kinetic} = \left[\frac{1}{2 \cdot m_{Net}} \cdot \left(\sum_{i=1}^{N} (m_i \overline{v}_i) \right)^2 + \sum_{i=1}^{N} \left(\frac{m_i \cdot \overline{v}_i^2}{2} \right) \right] \rightarrow kg \cdot m^2 / s^2$$

$$E_{Total\_Kinetic} = \left[\frac{(m_1 \overline{v}_1 + m_2 \overline{v}_2 + m_3 \overline{v}_3 + \cdots m_N \overline{v}_N)^2}{2 \cdot m_{Net}} + \left[\frac{m_1 \overline{v}_1^2}{2} + \frac{m_2 \overline{v}_2^2}{2} + \frac{m_3 \overline{v}_3^2}{2} + \cdots \frac{m_N \overline{v}_N^2}{2} \right] \right]$$

Total Kinetic Energy of an Isolated System – "After State"

9.18

$$E_{Total\_Kinetic} = \left[\frac{1}{2 \cdot m_{Net}} \cdot \left(\sum_{i=1}^{N} (m_i \overline{v}_i') \right)^2 + \sum_{i=1}^{N} \left(\frac{m_i \cdot \overline{v}_{Ai}'^2}{2} \right) \right] \rightarrow kg \cdot m^2 / s^2$$

$$E_{Total\_Kinetic} = \left[\frac{(m_1 \cdot \overline{v}_1' + m_2 \cdot \overline{v}_2' + m_3 \cdot \overline{v}_3' + \cdots m_N \cdot \overline{v}_N')^2}{2 \cdot m_{Net}} + \left[\frac{m_1 \cdot \overline{v}_{A1}'^2}{2} + \frac{m_2 \cdot \overline{v}_{A2}'^2}{2} + \frac{m_3 \cdot \overline{v}_{A3}'^2}{2} + \cdots \frac{m_N \cdot \overline{v}_{AN}'^2}{2} \right] \right]$$

Total Kinetic Energy of an Isolated System

9.19

$$E_{Total\_Kinetic} = \left\lfloor E_{Net\_Iso\_Kinetic} + E_{Net\_Translational} \right\rfloor \rightarrow kg \cdot m^2 / s^2$$

$$E_{Total\_Kinetic} = \frac{1}{2} \cdot m_{Net} \cdot \left[\left| v^2 \right|_{Iso} + \left| \overline{v} \right|^2_{CM} \right]$$

$$E_{Total\_Kinetic} = \frac{3}{2} \cdot N \cdot k_B \cdot \left[T_{emp} + T_{emp\_Ext} \right]$$

$$E_{Total\_Kinetic} = \frac{1}{2} \cdot \left(m_{Net} + m_{Inertia\_Mass} \right) \cdot \left| v^2 \right|_{Iso}$$

$$E_{Total\_Kinetic} = \frac{1}{2} \cdot \left(\left(\frac{m^2_{Net}}{m_{Inertia\_Mass}} \right) + m_{Net} \right) \cdot \left| \overline{v} \right|^2_{CM}$$

$$E_{Total\_Kinetic} = \left(\frac{1}{m_{Inertia\_Mass}} + \frac{1}{m_{Net}} \right) \cdot \frac{\overline{p}^2_{Inertial\_Net}}{2}$$

$$E_{Total\_Kinetic} = \left(\frac{1}{m_{Net}} + \frac{m_{Inertia\_Mass}}{m^2_{Net}} \right) \cdot \frac{\overline{p}^2_{Iso\_Net}}{2}$$

$$E_{Total\_Kinetic} = \frac{\overline{p}^2_{Inertial\_Net}}{2 \cdot \left(\frac{m_{Internal\_Mass} \cdot m_{Net}}{m_{Net} + m_{Internal\_Mass}} \right)} = \left(\frac{\overline{p}^2_{Inertial\_Net}}{2 \cdot m_{Net}} \right) \cdot \left[\frac{\left(1 + \left(\frac{\left| \overline{v} \right|^2_{CM}}{\left| v^2 \right|_{Iso}} \right) \right)}{\left(\frac{\left| \overline{v} \right|^2_{CM}}{\left| v^2 \right|_{Iso}} \right)} \right]$$

Total Kinetic Energy Conservation

9.20

$$E_{Total\_Kinetic} = \left[\frac{\overline{p}^2_{Conic\_Momentum}}{m_{Net}} + W_{Free\_Work\_Energy} \right] = \left[E_{Net\_Iso\_Kinetic} + E_{Net\_Translational} \right] \rightarrow kg \cdot m^2 / s^2$$

$$E_{Total\_Kinetic} = \left[\frac{\overline{p}^2_{Conic\_Momentum}}{m_{Net}} + \frac{\Delta\overline{p}^2_{Free\_Momentum}}{2 \cdot m_{Net}} \right] = \left[\frac{\overline{p}^2_{Net\_Iso}}{2 \cdot m_{Net}} + \frac{\overline{p}^2_{Inertial\_Net}}{2 \cdot m_{Net}} \right]$$

$$E_{Total\_Kinetic} = \left[\frac{\overline{p}^2_{Conic\_Momentum}}{m_{Net}} + \frac{\Delta\overline{p}^2_{Free\_Momentum}}{2 \cdot m_{Net}} \right] = \left[\frac{\overline{p}_{Thermo\_Momentum} \cdot \Delta\overline{p}_{Free\_Momentum}}{m_{Net}} + \frac{\Delta\overline{p}^2_{Free\_Momentum}}{2 \cdot m_{Net}} \right]$$

$$E_{Total\_Kinetic} = \left[\frac{\overline{p}^2_{Conic\_Momentum}}{m_{Net}} + \frac{\Delta\overline{p}^2_{Free\_Momentum}}{2 \cdot m_{Net}} \right] = \left[\frac{\overline{p}_{Net\_Iso} \cdot \overline{p}_{Inertial\_Net}}{m_{Net}} + \left[\frac{\overline{p}^2_{Net\_Iso}}{2 \cdot m_{Net}} - \frac{\overline{p}^2_{Inertial\_Net}}{2 \cdot m_{Net}} \right] \right]$$

Total Kinetic Energy Conservation – "Before State"

9.21

$$E_{Total\_Kinetic} = \frac{1}{2}\cdot m_{Net}\cdot\left[\left|v^2\right|_{Iso} + \left|\overline{v}\right|_{CM}^2\right] \rightarrow kg\cdot m^2/s^2$$

$$E_{Total\_Kinetic} = \frac{m_{Net}}{4\cdot N}\cdot\left[\left[\overline{v}'^2_{A\_Iso\_Total} + \overline{v}^2_{Iso\_Total}\right] + 6\cdot N\cdot\left|\overline{v}\right|_{CM}^2 - 4\cdot\left|\overline{v}\right|_{CM}\cdot\overline{v}_{Inertial\_Net}\right]$$

$$E_{Total\_Kinetic} = \frac{m_{Net}}{4\cdot N}\cdot\left[\left[2\cdot N\cdot\left|v^2\right|_{Iso} - \left(\overline{v}'^2_{Iso\_Total} - \overline{v}^2_{Iso\_Total}\right)\right] + 6\cdot N\cdot\left|\overline{v}\right|_{CM}^2 - 4\cdot\left|\overline{v}\right|_{CM}\cdot\overline{v}_{Inertial\_Net}\right]$$

Total Kinetic Energy Conservation – "After State" Relation (1)

9.22

$$E_{Total\_Kinetic} = \frac{1}{2}\cdot m_{Net}\cdot\left[\left|v^2\right|_{Iso} + \left|\overline{v}\right|_{CM}^2\right] \rightarrow kg\cdot m^2/s^2$$

$$E_{Total\_Kinetic} = \frac{m_{Net}}{4\cdot N}\cdot\left[\left[\overline{v}^2_{A\_Iso\_Total} + \overline{v}'^2_{Iso\_Total}\right] + 6\cdot N\cdot\left|\overline{v}\right|_{CM}^2 - 4\cdot\left|\overline{v}\right|_{CM}\cdot\overline{v}'_{Inertial\_Net}\right]$$

$$E_{Total\_Kinetic} = \frac{m_{Net}}{4\cdot N}\cdot\left[\left[2\cdot N\cdot\left|v^2\right|_{Iso} + \left(\overline{v}'^2_{Iso\_Total} - \overline{v}^2_{Iso\_Total}\right)\right] + 6\cdot N\cdot\left|\overline{v}\right|_{CM}^2 - 4\cdot\left|\overline{v}\right|_{CM}\cdot\overline{v}'_{Inertial\_Net}\right]$$

Total Kinetic Energy Conservation

9.23

$$E_{Total\_Kinetic} = \frac{1}{2} \cdot m_{Net} \cdot \left[\left| v^2 \right|_{Iso} + \left| \overline{v} \right|_{CM}^2 \right] \rightarrow kg \cdot m^2 / s^2$$

$$E_{Total\_Kinetic} = \frac{1}{2} \cdot m_{Net} \cdot \left[\frac{\left[m_1 \overline{v}_1^2 + m_2 \overline{v}_2^2 + m_3 \overline{v}_3^2 + \cdots m_N \overline{v}_N^2 \right]}{m_{Net}} + \frac{\left(m_1 \overline{v}_1 + m_2 \overline{v}_2 + m_3 \overline{v}_3 + \cdots m_N \overline{v}_N \right)^2}{m_{Net}} \right]$$

The **Total Kinetic Energy Conservation,** in three dimensional Cartesian coordinates x, y, and z, are given by the following,

9.24

$$E_{Total\_Kinetic_x} = \frac{1}{2} \cdot m_{Net} \cdot \left[\left| v^2 \right|_{Iso\,x} + \left| \overline{v} \right|_{CM\,x}^2 \right]$$

$$E_{Total\_Kinetic_y} = \frac{1}{2} \cdot m_{Net} \cdot \left[\left| v^2 \right|_{Iso\,y} + \left| \overline{v} \right|_{CM\,y}^2 \right]$$

$$E_{Total\_Kinetic_z} = \frac{1}{2} \cdot m_{Net} \cdot \left[\left| v^2 \right|_{Iso\,z} + \left| \overline{v} \right|_{CM\,z}^2 \right]$$

9.25

Scalar/Tensor		
Magnitude	Scalar/Tensor	Units
Total Kinetic Energy Conservation		
$E_{Total\_Kinetic} = \begin{bmatrix} E_{Total\_Kinetic_x} \\ + E_{Total\_Kinetic_y} \\ + E_{Total\_Kinetic_z} \end{bmatrix}$	$E_{Total\_Kinetic} = \begin{bmatrix} E_{Total\_Kinetic_x} \\ E_{Total\_Kinetic_y} \\ E_{Total\_Kinetic_z} \end{bmatrix}$	$\frac{kg \cdot m^2}{s^2}$

9.6 Total Squared Momentum Conservation Relation

Definition 9.3: The **Total Squared Momentum (Conservation) of an Isolated System** ($\overline{p}^2_{Total\_Momentum}$) is constant and conserved for any uniform state of motion and is a net thermodynamic measurement of the total energy of motion of the system, and is equal to the net sum of the **Anisotropic External Squared Inertial Translation Momentum** ($\overline{p}^2_{Inertial\_Net}$), sum the **Isotropic Internal Omni-directional Squared Momentum** ($\overline{p}^2_{Net\_Iso}$) of a system mass body.

Squared Total Momentum Conservation

9.26

$$\frac{\overline{p}^2_{Total\_Momentum}}{2} = \frac{1}{2}\cdot\left[\overline{p}^2_{Net\_Iso} + \overline{p}^2_{Inertial\_Net}\right] \rightarrow kg^2\cdot m^2/s^2$$

$$\frac{\overline{p}^2_{Total\_Momentum}}{2} = \left[\overline{p}^2_{Conic\_Momentum} + \frac{1}{2}\cdot\Delta\overline{p}^2_{Free\_Momentum}\right]$$

$$\frac{\overline{p}^2_{Total\_Momentum}}{2} = \left[\overline{p}_{Thermo\_Momentum}\cdot\Delta\overline{p}_{Free\_Momentum} + \frac{1}{2}\cdot\Delta\overline{p}^2_{Free\_Momentum}\right]$$

$$\frac{\overline{p}^2_{Total\_Momentum}}{2} = \left[\overline{p}_{Net\_Iso}\cdot\overline{p}_{Inertial\_Net} + \frac{1}{2}\cdot\Delta\overline{p}^2_{Free\_Momentum}\right]$$

Starting with the above Total Momentum Conservation equation

9.27

$$0 = \left[\frac{1}{2} \cdot \Delta\overline{p}^2_{Free\_Momentum} + \overline{p}_{Thermo\_Momentum} \cdot \Delta\overline{p}_{Free\_Momentum} - \frac{\overline{p}^2_{Total\_Momentum}}{2} \right]$$

Obtaining the Quadratic Equation Coefficients

$$a = \left(\frac{1}{2} \right)$$

$$b = +\overline{p}_{Thermo\_Momentum}$$

$$c = -\left(\frac{\overline{p}^2_{Total\_Momentum}}{2} \right)$$

$$0 = \left[a \cdot \Delta\overline{p}^2_{Free\_Momentum} + b \cdot \Delta\overline{p}_{Free\_Momentum} + c \right]$$

$$\Delta\overline{p}_{Free\_Momentum} = \frac{-b \pm \sqrt{b^2 - 4 \cdot a \cdot c}}{2 \cdot a}$$

Aether-Kinematic Work "Free" Momentum

9.28

$$\Delta\overline{p}_{Free\_Momentum} = \left[-\overline{p}_{Thermo\_Momentum} \pm \sqrt{\overline{p}^2_{Thermo\_Momentum} + \overline{p}^2_{Total\_Momentum}} \right] \rightarrow kg \cdot m/s$$

Net Work Momentum Conservation

9.29

$$\left[\Delta\overline{p}_{Free\_Momentum} + \overline{p}_{Thermo\_Momentum} \right] = \pm \sqrt{\overline{p}^2_{Thermo\_Momentum} + \overline{p}^2_{Total\_Momentum}} \rightarrow kg \cdot m/s$$

Net Work Momentum Conservation

9.30

$$\begin{bmatrix} \Delta\overline{p}^2_{Free\_Momentum} \\ + \quad \overline{p}^2_{Thermo\_Momentum} \\ + \quad 2\cdot\overline{p}_{Thermo\_Momentum}\cdot\Delta\overline{p}_{Free\_Momentum} \end{bmatrix} = \begin{bmatrix} \overline{p}^2_{Thermo\_Momentum} \\ + \quad \overline{p}^2_{Total\_Momentum} \end{bmatrix} \rightarrow kg^2\cdot m^2/s^2$$

$$\begin{bmatrix} \Delta\overline{p}^2_{Free\_Momentum} \\ + \quad \overline{p}^2_{Thermo\_Momentum} \\ + \quad 2\cdot\overline{p}^2_{Conic\_Momentum} \end{bmatrix} = \begin{bmatrix} \overline{p}^2_{Thermo\_Momentum} \\ + \quad \overline{p}^2_{Total\_Momentum} \end{bmatrix}$$

$$\begin{bmatrix} \Delta\overline{p}^2_{Free\_Momentum} \\ + \quad \overline{p}^2_{Thermo\_Momentum} \\ + \quad 2\cdot\overline{p}_{Net\_Iso}\cdot\overline{p}_{Inertial\_Net} \end{bmatrix} = \begin{bmatrix} \overline{p}^2_{Thermo\_Momentum} \\ + \quad \overline{p}^2_{Total\_Momentum} \end{bmatrix}$$

Chapter 10

The Special Theory of Thermodynamics

(Respiration Spacetime Pump Factor)

10.1 Universal Geometric Mean Ratio – Respiration Curvature of Spacetime Pump Factor

Respiration Spacetime Pump Factor

Definition 10.1: The **Respiration Curvature of Spacetime Pump Factor** ($\phi_{Respiration} = (\phi_{Loco\_Motion})^{\frac{1}{4}} = \sqrt{\frac{\gamma_{Inertial\_Iso}}{\gamma_{Inertial}}}$) is the ***non linear*** Carnot Heat Pump mechanism which is an aether mass density ratio defined as fourth root of the **Inertial Locomotion Foreshortening Factor** ($(\phi_{Loco\_Motion})^{\frac{1}{4}}$); and likewise is equal to the square root of the ratio of the **Internal (Isotropy) Inertial Motion Compressibility Factor** ($\gamma_{Inertial\_Iso}$) divided by the **External (Anisotropy) Inertial Motion Compressibility Factor** ($\gamma_{Inertial}$) of an isolated dynamic mass system, and is the measure of the non linear aether flow between matter and an atmospheric environment medium.

Respiration Curvature of Spacetime Pump Factor

10.1

$$\phi_{Respiration} = \left(\frac{E_{Net\_Translational}}{E_{Net\_Iso\_Kinetic}}\right)^{\frac{1}{4}} = \left(\frac{\text{External Anisotropic Tranlational Kinetic Energy}}{\text{Internal Isotropic Omni - directional Kinetic Energy}}\right)^{\frac{1}{4}}$$

$$\phi_{Respiration} = \left(\frac{\overline{p}^2_{Inertial\_Net}}{\overline{p}^2_{Net\_Iso}}\right)^{\frac{1}{4}} = \left(\frac{|\overline{v}|^2_{CM}}{|v^2|_{Iso}}\right)^{\frac{1}{4}} = \left(\frac{\begin{array}{c}\text{External Anisotropic}\\ \text{Center of Mass Squared Velocity Inertia}\end{array}}{\begin{array}{c}\text{Internal Isotropic Omni - directional}\\ \text{Center of Isotropy Squared Velocity Inertia}\end{array}}\right)^{\frac{1}{4}}$$

$$\phi_{Respiration} = \left(\frac{T_{emp\_Ext}}{T_{emp}}\right)^{\frac{1}{4}} = \left(\frac{\begin{array}{c}\text{External Anisotropic Tranlational}\\ \text{Aerodynamic Temperature}\end{array}}{\begin{array}{c}\text{Internal Isotropic Omni - directional}\\ \text{Absolute Temperature}\end{array}}\right)^{\frac{1}{4}}$$

Aphorism 10.1: The **Respiration Curvature of Spacetime Pump Factor** ($\phi_{Respiration}$) is a Carnot Heat Pump mechanism that is a measure of the non linear aether mass density Anisotropy to Isotropy ratio of an isolated dynamic mass system, and is directly proportional to the product of the fourth root of the one third the atomic substance Specific Heat Capacity Index ($\left(\frac{\gamma_{Heat}}{3}\right)^{\frac{1}{4}}$) multiplied by the square root of the Mach number ($\left(\sqrt{\frac{|\overline{v}|_{CM}}{c_{Sound}}}\right)$).

$$\phi_{Respiration} = \left(\phi_{Loco\_Motion}\right)^{\frac{1}{4}} = \sqrt{\frac{|\overline{v}|_{CM}}{\sqrt{|v^2|_{Iso}}}} = \left(\frac{\gamma_{Heat}}{3}\right)^{\frac{1}{4}} \cdot \sqrt{\left(\frac{|\overline{v}|_{CM}}{c_{Sound}}\right)} \quad 10.2$$

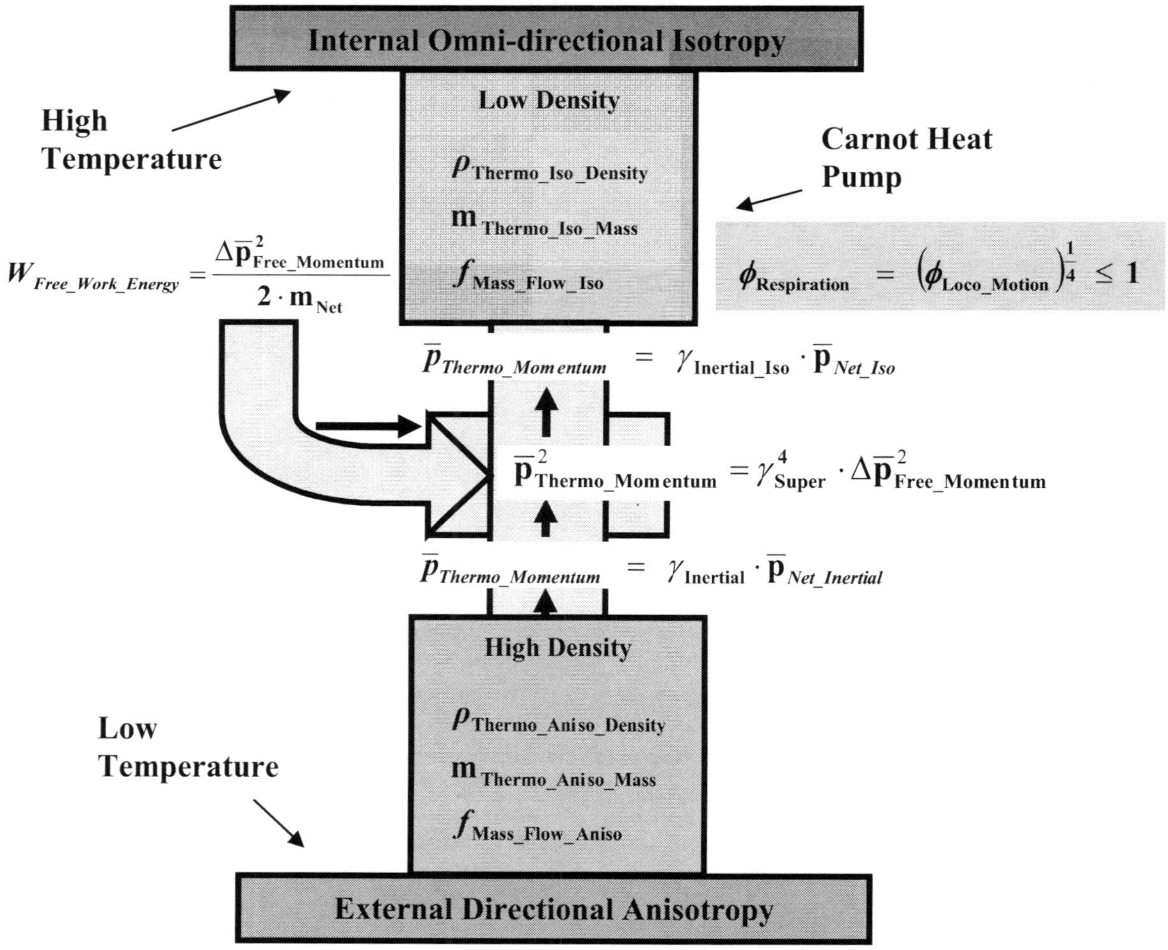

Figure 10.1: Heat Energy Transfer Device – Carnot Thermal Heat Pump – During Cooling.

Respiration Curvature of Spacetime Ratio – Carnot Thermal Heat Pump

10.3

$$\phi_{Respiration} = \left(\phi_{Loco\_Motion}\right)^{\frac{1}{4}} = \left(\frac{\left|\overline{v}\right|^2_{CM}}{\left|v^2\right|_{Iso}}\right)^{\frac{1}{4}} = \left(\frac{\gamma_{Heat}}{3}\cdot\left(\frac{\left|\overline{v}\right|^2_{CM}}{c^2_{Sound}}\right)\right)^{\frac{1}{4}}$$

$$\phi_{Respiration} = \left(\phi_{Loco\_Motion}\right)^{\frac{1}{4}} = \sqrt{\frac{\gamma_{Inertial\_Iso}}{\gamma_{Inertial}}} = \sqrt{\frac{\overline{p}_{Inertial\_Net}}{\overline{p}_{Net\_Iso}}}$$

$$\phi_{Respiration} = \left(\phi_{Loco\_Motion}\right)^{\frac{1}{4}} = \sqrt{\frac{\gamma_{Inertial\_Iso}}{\gamma_{Inertial}}} = \sqrt{\frac{\rho_{Thermo\_Iso\_Mass}}{\rho_{Thermo\_Aniso\_Mass}}}$$

$$\phi_{Respiration} = \left(\phi_{Loco\_Motion}\right)^{\frac{1}{4}} = \sqrt{\frac{\gamma_{Inertial\_Iso}}{\gamma_{Inertial}}} = \sqrt{\frac{m_{Thermo\_Iso\_Mass}}{m_{Thermo\_Aniso\_Mass}}}$$

Universal Geometric Mean Ratio

10.4

$$\frac{(Min)}{(Mean)} = \frac{(Mean)}{(Max)} = \sqrt{\frac{(Min)}{(Max)}} = \left(\frac{(Mean) - (Min)}{(Max) - (Mean)}\right)$$

Universal Geometric Mean Ratio – Respiration Curvature of Spacetime Ratio

10.5

$$\phi_{\text{Re}\,spiration} = \sqrt{\frac{\gamma_{Inertial\_Iso}}{\gamma_{Inertial}}} = \frac{\overline{\mathrm{p}}_{\mathrm{Inertial\_Net}}}{\overline{\mathrm{p}}_{\mathrm{Conic\_Momentum}}} = \frac{\overline{\mathrm{p}}_{\mathrm{Conic\_Momentum}}}{\overline{\mathrm{p}}_{\mathrm{Net\_Iso}}} \rightarrow Unitless$$

$$\phi_{\text{Re}\,spiration} = \sqrt{\frac{\gamma_{Inertial\_Iso}}{\gamma_{Inertial}}} = \frac{\overline{\mathrm{p}}_{\mathrm{Inertial\_Net}}}{\dfrac{\overline{\mathrm{p}}_{\mathrm{Thermo\_Momentum}}}{\sqrt{\gamma_{Inertial} \cdot \gamma_{Inertial\_Iso}}}} = \frac{\dfrac{\overline{\mathrm{p}}_{\mathrm{Thermo\_Momentum}}}{\sqrt{\gamma_{Inertial} \cdot \gamma_{Inertial\_Iso}}}}{\overline{\mathrm{p}}_{\mathrm{Net\_Iso}}}$$

$$\phi_{\text{Re}\,spiration} = \left(\phi_{\mathrm{Loco\_Motion}}\right)^{\frac{1}{4}} = \sqrt{\frac{\gamma_{Inertial\_Iso}}{\gamma_{Inertial}}} = \sqrt{\frac{\overline{\mathrm{p}}_{\mathrm{Inertial\_Net}}}{\overline{\mathrm{p}}_{\mathrm{Net\_Iso}}}}$$

$$\phi_{\text{Re}\,spiration} = \left(\phi_{\mathrm{Loco\_Motion}}\right)^{\frac{1}{4}} = \sqrt{\frac{\gamma_{Inertial\_Iso}}{\gamma_{Inertial}}} = \left(\frac{\overline{\mathrm{p}}_{\mathrm{Conic\_Momentum}} - \overline{\mathrm{p}}_{\mathrm{Inertial\_Net}}}{\overline{\mathrm{p}}_{\mathrm{Net\_Iso}} - \overline{\mathrm{p}}_{\mathrm{Conic\_Momentum}}}\right)$$

$$\phi_{\text{Re}\,spiration} = \sqrt{\frac{\gamma_{Inertial\_Iso}}{\gamma_{Inertial}}} = \left(\frac{\dfrac{\overline{\mathrm{p}}_{\mathrm{Thermo\_Momentum}}}{\sqrt{\gamma_{Inertial} \cdot \gamma_{Inertial\_Iso}}} - \overline{\mathrm{p}}_{\mathrm{Inertial\_Net}}}{\overline{\mathrm{p}}_{\mathrm{Net\_Iso}} - \dfrac{\overline{\mathrm{p}}_{\mathrm{Thermo\_Momentum}}}{\sqrt{\gamma_{Inertial} \cdot \gamma_{Inertial\_Iso}}}}\right)$$

Universal Geometric Mean Ratio

10.6

$$\frac{(Min)}{(Mean)} = \frac{(Mean)}{(Max)} = \sqrt{\frac{(Min)}{(Max)}} = \left(\frac{(Mean) - (Min)}{(Max) - (Mean)}\right)$$

Universal Geometric Mean Ratio – Respiration Curvature of Spacetime Pump Factor

10.7

$$\phi_{Respiration} = \sqrt{\frac{\gamma_{Inertial\_Iso}}{\gamma_{Inertial}}} = \frac{\Delta\overline{p}_{Free\_Momentum}}{\overline{p}_{Conic\_Momentum}} = \frac{\overline{p}_{Conic\_Momentum}}{\overline{p}_{Thermo\_Momentum}} \rightarrow Unitless$$

$$\phi_{Respiration} = \sqrt{\frac{\Delta\overline{p}_{Free\_Momentum}}{\overline{p}_{Thermo\_Momentum}}} = \left(\frac{\overline{p}_{Conic\_Momentum} - \Delta\overline{p}_{Free\_Momentum}}{\overline{p}_{Thermo\_Momentum} - \overline{p}_{Conic\_Momentum}}\right)$$

$$\phi_{Respiration} = \left(\phi_{Loco\_Motion}\right)^{\frac{1}{4}} = \sqrt{\frac{\gamma_{Inertial\_Iso}}{\gamma_{Inertial}}} = \sqrt{\frac{\overline{p}_{Inertial\_Net}}{\overline{p}_{Net\_Iso}}}$$

$$\phi_{Respiration} = \sqrt{\frac{\gamma_{Inertial\_Iso}}{\gamma_{Inertial}}} = \sqrt{\frac{\rho_{Thermo\_Iso\_Mass}}{\rho_{Thermo\_Aniso\_Mass}}}$$

$$\phi_{Respiration} = \left(\phi_{Loco\_Motion}\right)^{\frac{1}{4}} = \sqrt{\frac{m_{Thermo\_Iso\_Mass}}{m_{Thermo\_Aniso\_Mass}}}$$

Universal Geometric Mean Ratio – Respiration Curvature of Spacetime Pump Factor

10.8

$$\phi_{\mathrm{Re}\,spiration} = \sqrt{\frac{\gamma_{Inertial\_Iso}}{\gamma_{Inertial}}} = \left(\frac{\overline{\mathrm{p}}_{\mathrm{Inertial\_Net}}}{\overline{\mathrm{p}}_{\mathrm{Thermo\_Momentum}}}\right)\cdot\sqrt{\gamma_{Inertial}\cdot\gamma_{Inertial\_Iso}} \quad \rightarrow Unitless$$

$$\phi_{\mathrm{Re}\,spiration} = \sqrt{\frac{\gamma_{Inertial\_Iso}}{\gamma_{Inertial}}} = \frac{1}{\sqrt{\gamma_{Inertial}\cdot\gamma_{Inertial\_Iso}}}\cdot\left(\frac{\overline{\mathrm{p}}_{\mathrm{Thermo\_Momentum}}}{\overline{\mathrm{p}}_{\mathrm{Net\_Iso}}}\right)$$

$$\phi_{\mathrm{Re}\,spiration} = \sqrt{\frac{\gamma_{Inertial\_Iso}}{\gamma_{Inertial}}} = \gamma_{Super\_Compres}\cdot\left(\frac{\overline{\mathrm{p}}_{\mathrm{Inertial\_Net}}}{\overline{\mathrm{p}}_{\mathrm{Thermo\_Momentum}}}\right)$$

$$\phi_{\mathrm{Re}\,spiration} = \sqrt{\frac{\gamma_{Inertial\_Iso}}{\gamma_{Inertial}}} = \left(\frac{\overline{\mathrm{p}}_{\mathrm{Inertial\_Net}}}{\overline{\mathrm{p}}_{\mathrm{Thermo\_Momentum}}}\right)\cdot\sqrt{\gamma_{Inertial}\cdot\gamma_{Inertial\_Iso}}$$

$$\phi_{\mathrm{Re}\,spiration} = \sqrt{\frac{\gamma_{Inertial\_Iso}}{\gamma_{Inertial}}} = \frac{1}{\gamma_{Super\_Compres}}\cdot\left(\frac{\overline{\mathrm{p}}_{\mathrm{Thermo\_Momentum}}}{\overline{\mathrm{p}}_{\mathrm{Net\_Iso}}}\right)$$

$$\phi_{\mathrm{Re}\,spiration} = \sqrt{\frac{\gamma_{Inertial\_Iso}}{\gamma_{Inertial}}} = \frac{1}{\gamma_{Super\_Compres}}\cdot\left(\frac{\overline{\mathrm{p}}_{\mathrm{Inertial\_Net}}}{\Delta\overline{\mathrm{p}}_{\mathrm{Free\_Momentum}}}\right)$$

$$\phi_{\mathrm{Re}\,spiration} = \sqrt{\frac{\gamma_{Inertial\_Iso}}{\gamma_{Inertial}}} = \gamma_{Super\_Compres}\cdot\left(\frac{\Delta\overline{\mathrm{p}}_{\mathrm{Free\_Momentum}}}{\overline{\mathrm{p}}_{\mathrm{Net\_Iso}}}\right)$$

Universal Geometric Mean Ratio – Respiration Curvature of Spacetime Pump Factor

10.9

$$\phi_{\text{Respiration}} = \sqrt{\frac{\gamma_{Inertial\_Iso}}{\gamma_{Inertial}}} = \sqrt{\frac{\rho_{\text{Thermo\_Iso\_Mass}}}{\rho_{\text{Thermo\_Aniso\_Mass}}}} \rightarrow Unitless$$

$$\phi_{\text{Respiration}} = \left(\phi_{\text{Loco\_Motion}}\right)^{\frac{1}{4}} = \sqrt{\frac{m_{\text{Thermo\_Iso\_Mass}}}{m_{\text{Thermo\_Aniso\_Mass}}}}$$

$$\phi_{\text{Respiration}} = \sqrt{\frac{\gamma_{Inertial\_Iso}}{\gamma_{Inertial}}} = \frac{\gamma_{Super\_Compress}}{\gamma_{Inertial}} = \frac{\sqrt{1 - \dfrac{\left|\overline{v}\right|^2_{\text{CM}}}{\left|v^2\right|_{\text{Iso}}}}}{\left(\dfrac{\left|v^2\right|_{\text{Iso}}}{\left|\overline{v}\right|^2_{\text{CM}}} - 2 + \dfrac{\left|\overline{v}\right|^2_{\text{CM}}}{\left|v^2\right|_{\text{Iso}}}\right)^{\frac{1}{4}}}$$

$$\phi_{\text{Respiration}} = \sqrt{\frac{\gamma_{Inertial\_Iso}}{\gamma_{Inertial}}} = \frac{\gamma_{Inertial\_Iso}}{\gamma_{Super\_Compress}} = \frac{\left(\dfrac{\left|v^2\right|_{\text{Iso}}}{\left|\overline{v}\right|^2_{\text{CM}}} - 2 + \dfrac{\left|\overline{v}\right|^2_{\text{CM}}}{\left|v^2\right|_{\text{Iso}}}\right)^{\frac{1}{4}}}{\sqrt{\left(\dfrac{\left|v^2\right|_{\text{Iso}}}{\left|\overline{v}\right|^2_{\text{CM}}} - 1\right)}}$$

$$\phi_{\text{Respiration}} = \left(\phi_{\text{Loco\_Motion}}\right)^{\frac{1}{4}} = \sqrt{\frac{\gamma_{Inertial\_Iso}}{\gamma_{Inertial}}} = \sqrt{\left(\frac{\overline{p}_{\text{Inertial\_Net}}}{\overline{p}_{\text{Net\_Iso}}}\right)}$$

$$\phi_{\text{Respiration}} = \left(\phi_{\text{Loco\_Motion}}\right)^{\frac{1}{4}} = \sqrt{\frac{\left|\overline{v}\right|_{\text{CM}}}{\sqrt{\left|v^2\right|_{\text{Iso}}}}} = \sqrt{\left(\frac{\left|\overline{v}\right|_{\text{CM}}}{c_{\text{Sound}}}\right) \cdot \sqrt{\frac{\gamma_{Heat}}{3}}}$$

Substance	Molecule	Specific Heat @ Constant Pressure cp(kJ/kg K)	Specific Heat @ Constant Volume cv(kJ/kg K)	$(c_p / c_v) = \gamma_{Heat}$
Helium	He	5.19	3.12	1.667
Hydrogen	H_2	14.32	10.16	1.405
Air		1.01	0.718	1.4
Oxygen	O_2	0.919	0.659	1.395
Methane	CH_4	2.22	1.7	1.304
Carbon dioxide	CO_2	0.844	0.655	1.289
Benzene	C_6H_6	1.09	0.99	1.12
Ether		2.01	1.95	1.03

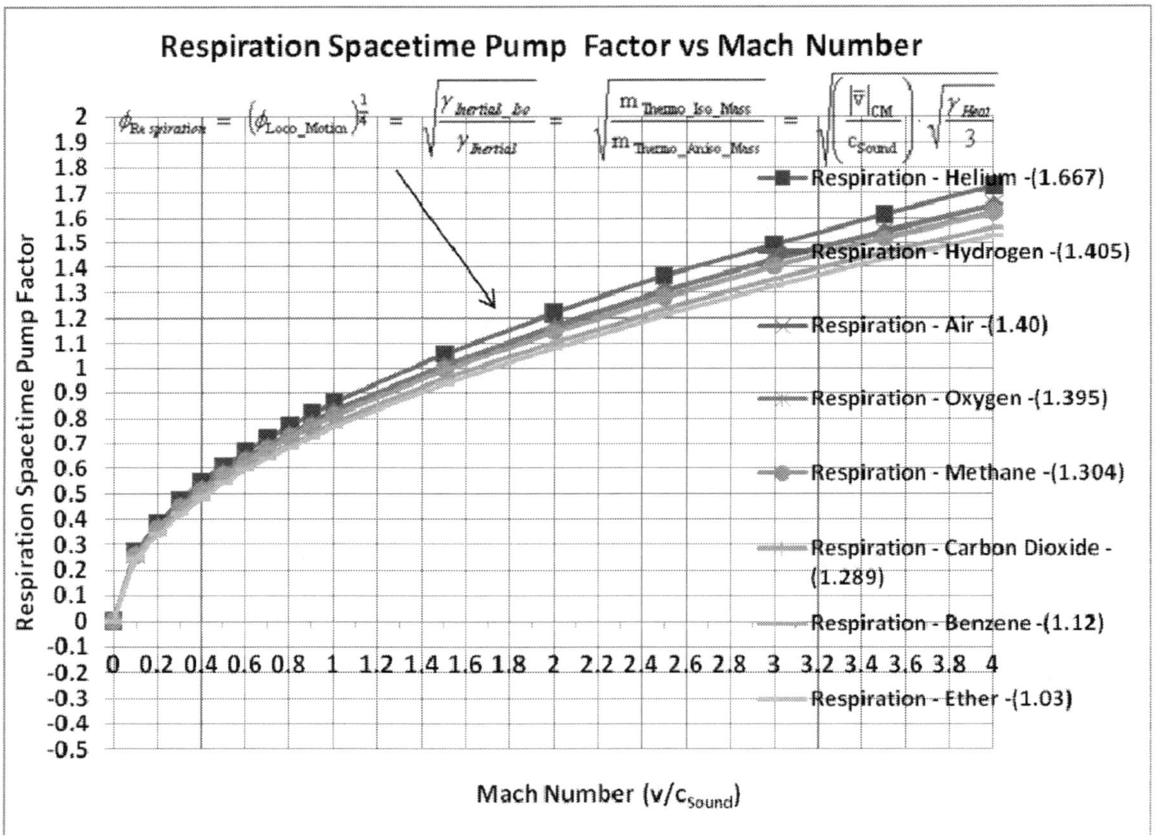

Figure 10.2: Non-linear - Respiration Curvature of Spacetime Pump Factor vs. Mach number for various atomic substances.

10.2 Universal Geometric Mean Ratio — Square Respiration Spacetime Pump Factor

Square Respiration Spacetime Pump Factor

Definition 10.2: The **Square Respiration Spacetime Pump Factor** ($\phi^2_{\text{Re}\,spiration} = \sqrt{\phi_{\text{Loco\_Motion}}} = \left(\dfrac{\gamma_{Inertial\_Iso}}{\gamma_{Inertial}}\right)$) is the ***linear*** Carnot Heat Pump mechanism which is an aether mass density ratio defined as square root of the **Inertial Locomotion Foreshortening Factor** ($\sqrt{\phi_{\text{Loco\_Motion}}}$); and likewise is equal to the ratio of the **Internal (Isotropy) Inertial Motion Compressibility Factor** ($\gamma_{Inertial\_Iso}$) divided by the **External (Anisotropy) Inertial Motion Compressibility Factor** ($\gamma_{Inertial}$) of an isolated dynamic mass system, and is the measure of the non linear aether mass density flow between matter and an atmospheric environment medium.

Square Respiration Spacetime Pump Factor

10.10

$$\phi^2_{\text{Re}\,spiration} = \sqrt{\left(\frac{\text{E}_{\text{Net\_Translational}}}{\text{E}_{\text{Net\_Iso\_Kinetic}}}\right)} = \sqrt{\left(\frac{\text{External Anisotropic Tranlational Kinetic Energy}}{\text{Internal Isotropic Omni - directional Kinetic Energy}}\right)}$$

$$\phi^2_{\text{Re}\,spiration} = \left(\frac{\overline{\text{p}}_{\text{Inertial\_Net}}}{\overline{\text{p}}_{\text{Net\_Iso}}}\right) = \left(\frac{\left|\overline{\text{v}}\right|_{\text{CM}}}{\sqrt{\left|\text{v}^2\right|_{\text{Iso}}}}\right) = \sqrt{\left(\frac{\begin{array}{c}\text{External Anisotropic}\\ \text{Center of Mass Squared Velocity Inertia}\end{array}}{\begin{array}{c}\text{Internal Isotropic Omni - directional}\\ \text{Center of Isotropy Squared Velocity Inertia}\end{array}}\right)}$$

$$\phi^2_{\text{Re}\,spiration} = \sqrt{\left(\frac{T_{emp\_Ext}}{T_{emp}}\right)} = \sqrt{\left(\frac{\begin{array}{c}\text{External Anisotropic Tranlational}\\ \text{Aerodynamic Temperature}\end{array}}{\begin{array}{c}\text{Internal Isotropic Omni - directional}\\ \text{Absolute Temperature}\end{array}}\right)}$$

Aphorism 10.2: The **Respiration Curvature of Spacetime Pump Factor** ($\phi_{Respiration}$) is a Carnot Heat Pump mechanism that is the measure of the non linear aether mass density Anisotropy to Isotropy ratio of an isolated dynamic mass system, and is directly proportional to the product of the square root of one third the atomic substance Specific Heat Capacity Index ($\sqrt{\frac{\gamma_{Heat}}{3}}$) multiplied by the Mach number ($\left(\frac{|\overline{\mathrm{v}}|_{\mathrm{CM}}}{\mathrm{c}_{\mathrm{Sound}}}\right)$).

10.11

$$\phi^2_{Respiration} = \sqrt{\phi_{\text{Loco\_Motion}}} = \left(\frac{|\overline{\mathrm{v}}|_{\mathrm{CM}}}{\sqrt{|\mathrm{v}^2|_{\mathrm{Iso}}}}\right) = \left(\frac{|\overline{\mathrm{v}}|_{\mathrm{CM}}}{\mathrm{c}_{\mathrm{Sound}}}\right)\cdot\sqrt{\frac{\gamma_{Heat}}{3}}$$

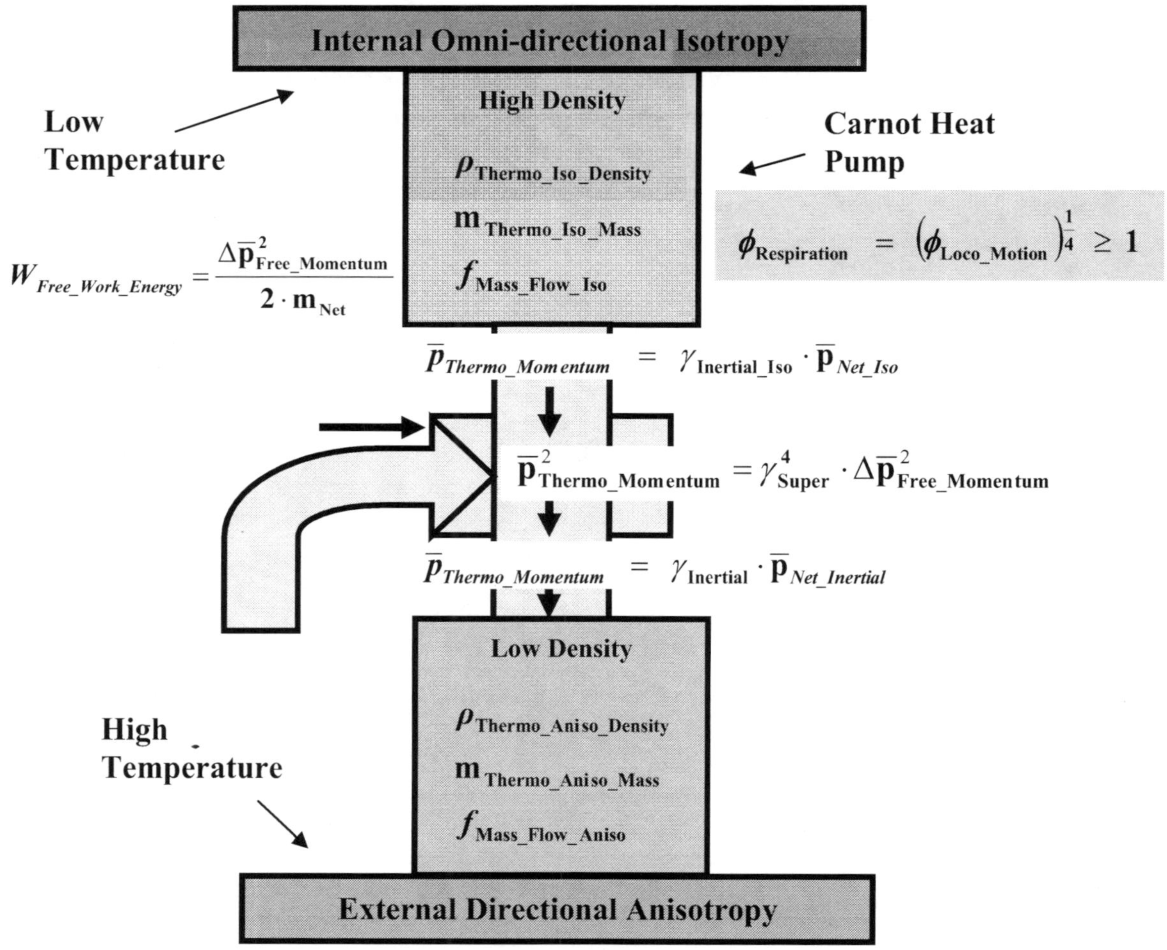

Figure 10.3: Heat Energy Transfer Device – Carnot Thermal Heat Pump – During Heating.

Square Respiration Spacetime Pump Factor – Carnot Thermal Heat Pump

10.12

$$\phi^2_{Respiration} = \sqrt{\phi_{Loco\_Motion}} = \left(\frac{|\overline{v}|_{CM}}{\sqrt{|v^2|_{Iso}}} \right) = \left(\frac{|\overline{v}|_{CM}}{c_{Sound}} \right) \cdot \sqrt{\frac{\gamma_{Heat}}{3}}$$

$$\phi^2_{Respiration} = \sqrt{\phi_{Loco\_Motion}} = \left(\frac{\gamma_{Inertial\_Iso}}{\gamma_{Inertial}} \right) = \left(\frac{\overline{p}_{Inertial\_Net}}{\overline{p}_{Net\_Iso}} \right)$$

$$\phi^2_{Respiration} = \sqrt{\phi_{Loco\_Motion}} = \left(\frac{\gamma_{Inertial\_Iso}}{\gamma_{Inertial}} \right) = \left(\frac{\rho_{Thermo\_Iso\_Mass}}{\rho_{Thermo\_Aniso\_Mass}} \right)$$

$$\phi^2_{Respiration} = \sqrt{\phi_{Loco\_Motion}} = \left(\frac{\gamma_{Inertial\_Iso}}{\gamma_{Inertial}} \right) = \left(\frac{m_{Thermo\_Iso\_Mass}}{m_{Thermo\_Aniso\_Mass}} \right)$$

Square Universal Geometric Mean Ratio

10.13

$$\left(\frac{(Min)}{(Mean)}\right)^2 = \left(\frac{(Mean)}{(Max)}\right)^2 = \left(\frac{(Min)}{(Max)}\right) = \left(\frac{(Mean) - (Min)}{(Max) - (Mean)}\right)^2$$

Universal Geometric Mean Ratio – Square Respiration Spacetime Pump Factor

10.14

$$\phi^2_{\mathrm{Re}\,spiration} = \left(\frac{\gamma_{Inertial\_Iso}}{\gamma_{Inertial}}\right) = \left(\frac{\overline{\mathrm{p}}_{\mathrm{Inertial\_Net}}}{\overline{\mathrm{p}}_{\mathrm{Conic\_Momentum}}}\right)^2 = \left(\frac{\overline{\mathrm{p}}_{\mathrm{Conic\_Momentum}}}{\overline{\mathrm{p}}_{\mathrm{Net\_Iso}}}\right)^2 \rightarrow Unitless$$

$$\phi^2_{\mathrm{Re}\,spiration} = \left(\frac{\gamma_{Inertial\_Iso}}{\gamma_{Inertial}}\right) = \frac{\overline{\mathrm{p}}^2_{\mathrm{Inertial\_Net}}}{\dfrac{\overline{\mathrm{p}}^2_{\mathrm{Thermo\_Momentum}}}{\left(\gamma_{Inertial}\cdot\gamma_{Inertial\_Iso}\right)}} = \frac{\dfrac{\overline{\mathrm{p}}^2_{\mathrm{Thermo\_Momentum}}}{\left(\gamma_{Inertial}\cdot\gamma_{Inertial\_Iso}\right)}}{\overline{\mathrm{p}}^2_{\mathrm{Net\_Iso}}}$$

$$\phi^2_{\mathrm{Re}\,spiration} = \sqrt{\phi_{\mathrm{Loco\_Motion}}} = \left(\frac{\gamma_{Inertial\_Iso}}{\gamma_{Inertial}}\right) = \left(\frac{\overline{\mathrm{p}}_{\mathrm{Inertial\_Net}}}{\overline{\mathrm{p}}_{\mathrm{Net\_Iso}}}\right)$$

$$\phi^2_{\mathrm{Re}\,spiration} = \sqrt{\phi_{\mathrm{Loco\_Motion}}} = \left(\frac{\gamma_{Inertial\_Iso}}{\gamma_{Inertial}}\right) = \left(\frac{\overline{\mathrm{p}}_{\mathrm{Conic\_Momentum}} - \overline{\mathrm{p}}_{\mathrm{Inertial\_Net}}}{\overline{\mathrm{p}}_{\mathrm{Net\_Iso}} - \overline{\mathrm{p}}_{\mathrm{Conic\_Momentum}}}\right)^2$$

$$\phi^2_{\mathrm{Re}\,spiration} = \left(\frac{\gamma_{Inertial\_Iso}}{\gamma_{Inertial}}\right) = \left(\frac{\dfrac{\overline{\mathrm{p}}_{\mathrm{Thermo\_Momentum}}}{\sqrt{\gamma_{Inertial}\cdot\gamma_{Inertial\_Iso}}} - \overline{\mathrm{p}}_{\mathrm{Inertial\_Net}}}{\overline{\mathrm{p}}_{\mathrm{Net\_Iso}} - \dfrac{\overline{\mathrm{p}}_{\mathrm{Thermo\_Momentum}}}{\sqrt{\gamma_{Inertial}\cdot\gamma_{Inertial\_Iso}}}}\right)^2$$

Square Universal Geometric Mean Ratio

10.15

$$\left(\frac{(Min)}{(Mean)}\right)^2 = \left(\frac{(Mean)}{(Max)}\right)^2 = \left(\frac{(Min)}{(Max)}\right) = \left(\frac{(Mean) - (Min)}{(Max) - (Mean)}\right)^2$$

Universal Geometric Mean Ratio – Square Respiration Spacetime Pump Factor

10.16

$$\phi^2_{Respiration} = \left(\frac{\gamma_{Inertial\_Iso}}{\gamma_{Inertial}}\right) = \left(\frac{\Delta\overline{p}_{Free\_Momentum}}{\overline{p}_{Conic\_Momentum}}\right)^2 = \left(\frac{\overline{p}_{Conic\_Momentum}}{\overline{p}_{Thermo\_Momentum}}\right)^2$$

$$\phi^2_{Respiration} = \left(\frac{\Delta\overline{p}_{Free\_Momentum}}{\overline{p}_{Thermo\_Momentum}}\right) = \left(\frac{\overline{p}_{Conic\_Momentum} - \Delta\overline{p}_{Free\_Momentum}}{\overline{p}_{Thermo\_Momentum} - \overline{p}_{Conic\_Momentum}}\right)^2$$

$$\phi^2_{Respiration} = \sqrt{\phi_{Loco\_Motion}} = \left(\frac{\gamma_{Inertial\_Iso}}{\gamma_{Inertial}}\right) = \left(\frac{\overline{p}_{Inertial\_Net}}{\overline{p}_{Net\_Iso}}\right)$$

$$\phi^2_{Respiration} = \left(\frac{\gamma_{Inertial\_Iso}}{\gamma_{Inertial}}\right) = \frac{\rho_{Thermo\_Iso\_Mass}}{\rho_{Thermo\_Aniso\_Mass}}$$

$$\phi^2_{Respiration} = \sqrt{\phi_{Loco\_Motion}} = \left(\frac{m_{Thermo\_Iso\_Mass}}{m_{Thermo\_Aniso\_Mass}}\right) \rightarrow Unitless$$

Universal Geometric Mean Ratio – Square Respiration Spacetime Pump Factor

10.17

$$\phi^2_{Respiration} = \left(\frac{\gamma_{Inertial\_Iso}}{\gamma_{Inertial}}\right) = \left(\frac{\overline{p}_{Inertial\_Net}}{\overline{p}_{Thermo\_Momentum}}\right) \cdot \gamma_{Inertial} \cdot \gamma_{Inertial\_Iso} \rightarrow Unitless$$

$$\phi^2_{Respiration} = \left(\frac{\gamma_{Inertial\_Iso}}{\gamma_{Inertial}}\right) = \frac{1}{\gamma_{Inertial} \cdot \gamma_{Inertial\_Iso}} \cdot \left(\frac{\overline{p}_{Thermo\_Momentum}}{\overline{p}_{Net\_Iso}}\right)^2$$

$$\phi^2_{Respiration} = \left(\frac{\gamma_{Inertial\_Iso}}{\gamma_{Inertial}}\right)^2 = \gamma^2_{Super\_Compres} \cdot \left(\frac{\overline{p}_{Inertial\_Net}}{\overline{p}_{Thermo\_Momentum}}\right)^2$$

$$\phi^2_{Respiration} = \left(\frac{\gamma_{Inertial\_Iso}}{\gamma_{Inertial}}\right) = \left(\frac{\overline{p}_{Inertial\_Net}}{\overline{p}_{Thermo\_Momentum}}\right)^2 \cdot \gamma_{Inertial} \cdot \gamma_{Inertial\_Iso}$$

$$\phi^2_{Respiration} = \left(\frac{\gamma_{Inertial\_Iso}}{\gamma_{Inertial}}\right) = \frac{1}{\gamma^2_{Super\_Compres}} \cdot \left(\frac{\overline{p}_{Thermo\_Momentum}}{\overline{p}_{Net\_Iso}}\right)^2$$

$$\phi^2_{Respiration} = \left(\frac{\gamma_{Inertial\_Iso}}{\gamma_{Inertial}}\right) = \frac{1}{\gamma^2_{Super\_Compres}} \cdot \left(\frac{\overline{p}_{Inertial\_Net}}{\Delta\overline{p}_{Free\_Momentum}}\right)^2$$

$$\phi^2_{Respiration} = \left(\frac{\gamma_{Inertial\_Iso}}{\gamma_{Inertial}}\right) = \gamma^2_{Super\_Compres} \cdot \left(\frac{\Delta\overline{p}_{Free\_Momentum}}{\overline{p}_{Net\_Iso}}\right)^2$$

Universal Geometric Mean Ratio – Square Respiration Spacetime Pump Factor

10.18

$$\phi^2_{Respiration} = \left(\frac{\gamma_{Inertial\_Iso}}{\gamma_{Inertial}} \right) = \left(\frac{\rho_{Thermo\_Iso\_Mass}}{\rho_{Thermo\_Aniso\_Mass}} \right) \rightarrow Unitless$$

$$\phi^2_{Respiration} = \sqrt{\phi_{Loco\_Motion}} = \left(\frac{m_{Thermo\_Iso\_Mass}}{m_{Thermo\_Aniso\_Mass}} \right)$$

$$\phi^2_{Respiration} = \left(\frac{\gamma_{Inertial\_Iso}}{\gamma_{Inertial}} \right) = \left(\frac{\gamma_{Super\_Compress}}{\gamma_{Inertial}} \right)^2 = \frac{\left(1 - \frac{\left|\overline{v}\right|^2_{CM}}{\left|v^2\right|_{Iso}} \right)}{\sqrt{\left(\frac{\left|v^2\right|_{Iso}}{\left|\overline{v}\right|^2_{CM}} - 2 + \frac{\left|\overline{v}\right|^2_{CM}}{\left|v^2\right|_{Iso}} \right)}}$$

$$\phi^2_{Respiration} = \left(\frac{\gamma_{Inertial\_Iso}}{\gamma_{Inertial}} \right) = \left(\frac{\gamma_{Inertial\_Iso}}{\gamma_{Super\_Compress}} \right)^2 = \frac{\sqrt{\left(\frac{\left|v^2\right|_{Iso}}{\left|\overline{v}\right|^2_{CM}} - 2 + \frac{\left|\overline{v}\right|^2_{CM}}{\left|v^2\right|_{Iso}} \right)}}{\left(\frac{\left|v^2\right|_{Iso}}{\left|\overline{v}\right|^2_{CM}} - 1 \right)}$$

$$\phi^2_{Respiration} = \sqrt{\phi_{Loco\_Motion}} = \left(\frac{\gamma_{Inertial\_Iso}}{\gamma_{Inertial}} \right) = \left(\frac{\overline{p}_{Inertial\_Net}}{\overline{p}_{Net\_Iso}} \right)$$

$$\phi^2_{Respiration} = \sqrt{\phi_{Loco\_Motion}} = \left(\frac{\left|\overline{v}\right|_{CM}}{\sqrt{\left|v^2\right|_{Iso}}} \right) = \left(\frac{\left|\overline{v}\right|_{CM}}{c_{Sound}} \right) \cdot \sqrt{\frac{\gamma_{Heat}}{3}}$$

Substance	Molecule	Specific Heat @ Constant Pressure cp(kJ/kg K)	Specific Heat @ Constant Volume cv(kJ/kg K)	$(c_p / c_v) = \gamma_{Heat}$
Helium	He	5.19	3.12	1.667
Hydrogen	H_2	14.32	10.16	1.405
Air		1.01	0.718	1.4
Oxygen	O_2	0.919	0.659	1.395
Methane	CH_4	2.22	1.7	1.304
Carbon dioxide	CO_2	0.844	0.655	1.289
Benzene	C_6H_6	1.09	0.99	1.12
Ether		2.01	1.95	1.03

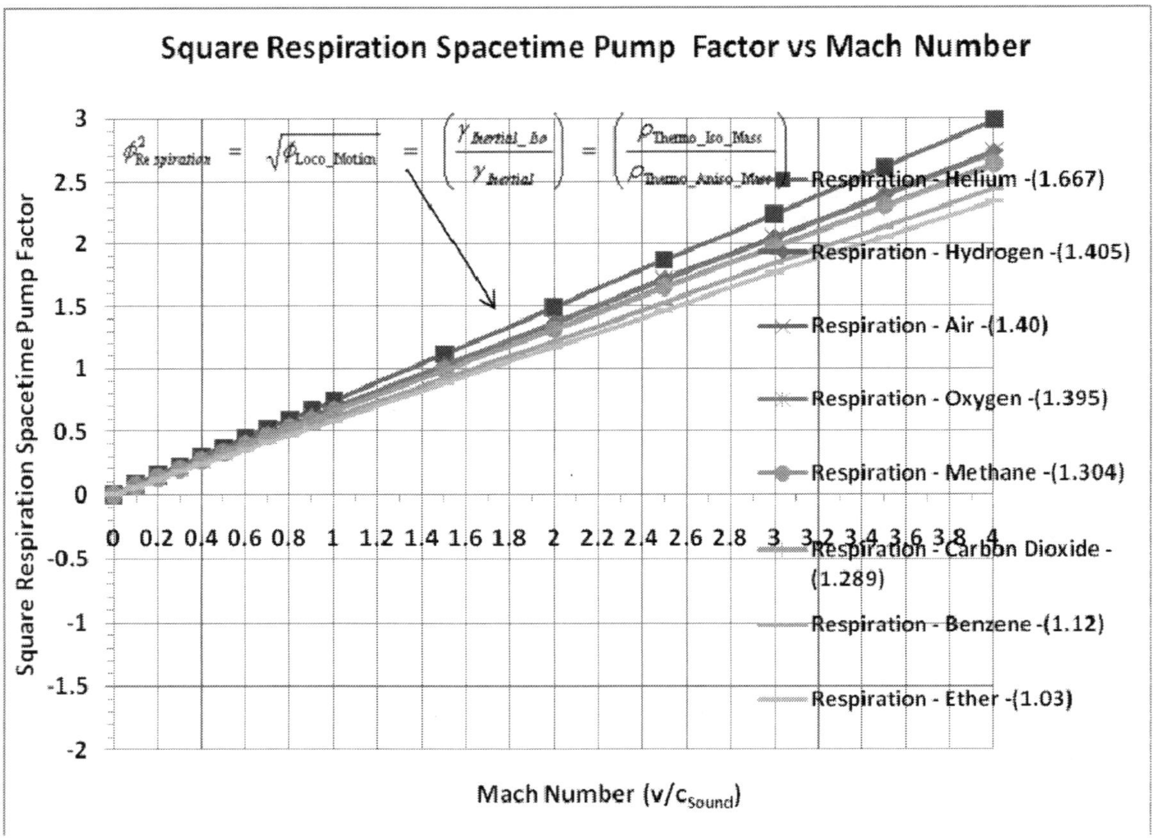

Figure 10.4: Linear - Respiration Spacetime Pump Factor vs. Mach number for various atomic substances.

Chapter 11

The Special Theory of Thermodynamics

(Thermodynamic)
(Spacetime Compressibility)

11.1 Thermodynamic Spacetime Compressibility

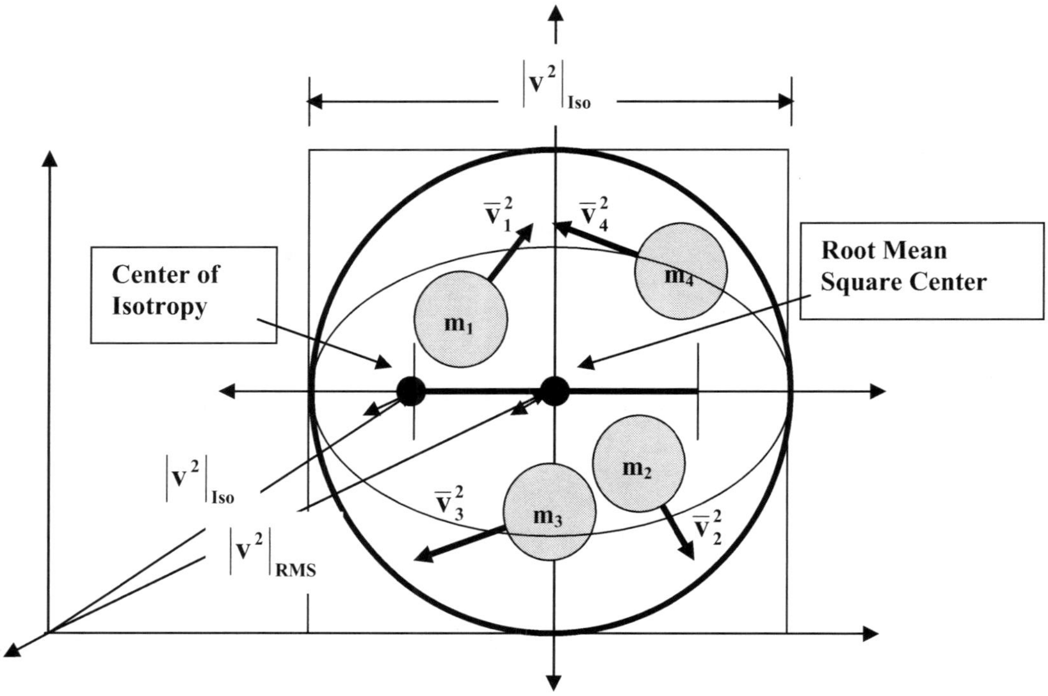

Figure 11.1: Isotropy and the Center of Isotropy.

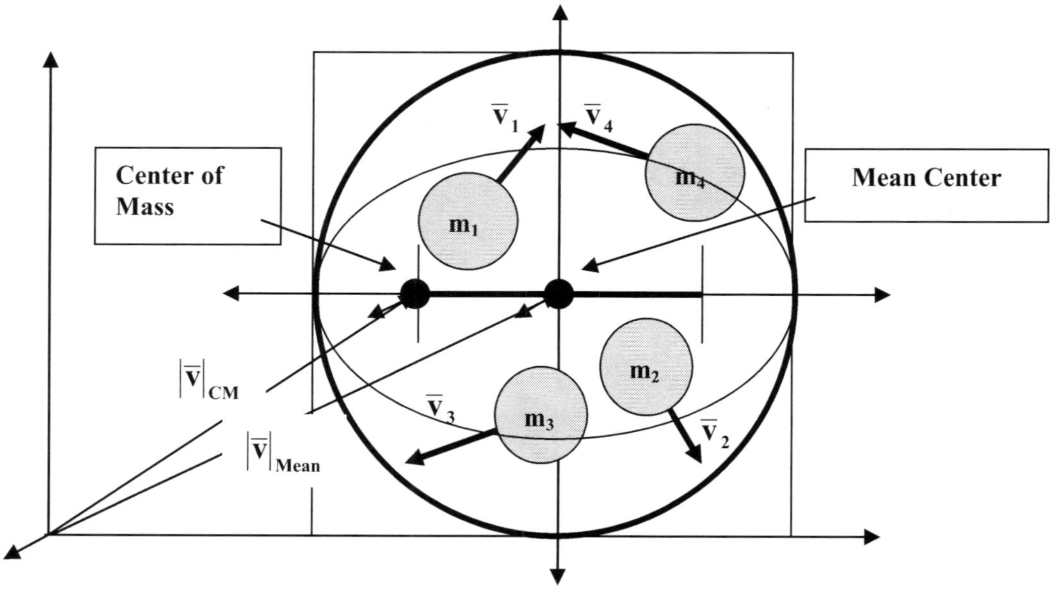

Figure 11.2: Anisotropy and the Center of Mass.

In general Kinetic Heat energy is the collective Omni-directional random motion of mass bodies which interact via collision such as electrons, protons, atoms, molecules, or aetherons. The absolute temperature is a measure of how fast the mass bodies are moving before and after collisions.

Heat energy naturally and spontaneously flows from any high temperature to a lower temperature, according to the Zeroth Law of Thermodynamics. Thus, when a group of fast-moving (warmer) molecules is placed next to a group of slow-moving (cooler) molecules, random collisions will cause the slower molecules to speed up and the faster ones to slow down, thus transferring energy from the warmer group to the cooler group.

In essence a system mass body moves heat energy between the Isotropy and the Anisotropy and the internal and external constituents of the system.

At low Center of Mass Velocities relative to the speed of sound in a specific atomic substance medium, the system mass body behaves like a thermal heat engine moving heat energy from the internal isotropy to the external anisotropy and away from the body; cooling the body and warming the environment.

And at the same time there are low Center of Mass Velocities relative to the speed of sound in a specific atomic substance, the system mass body behaves like a thermal heat pump moving aether heat energy, mass and density from the external anisotropy to the internal isotropy inwards towards the body; condensing the body, rarifying the environment.

Thus the above described isolated system mass body rejects heat energy into the external environment when the internal Isotropy is greater than the external Anisotropy; cooling and condensing the body and warming and rarifying the environment.

11.1

$$\phi^2_{\mathrm{Re}\,spiration} = \left(\frac{\left|\overline{\mathrm{v}}\right|_{\mathrm{CM}}}{\sqrt{\left|\mathrm{v}^2\right|_{\mathrm{Iso}}}} \right) \leq 1$$

$$\phi^2_{\mathrm{Re}\,spiration} = \sqrt{\phi_{\mathrm{Loco\_Motion}}} = \left(\frac{\gamma_{Inertial\_Iso}}{\gamma_{Inertial}} \right) = \left(\frac{\left|\overline{\mathrm{v}}\right|_{\mathrm{CM}}}{\mathrm{c}_{\mathrm{Sound}}} \right) \cdot \sqrt{\frac{\gamma_{Heat}}{3}} \leq 1$$

At high Center of Mass Velocities relative to the speed of sound in a specific atomic substance medium, the system mass body behaves like a thermal heat engine moving heat energy from the external anisotropy to the internal isotropy in towards the body; heating the body and cooling the environment.

And at the same time the high Center of Mass Velocities relative to the speed of sound in a specific atomic substance medium, the system mass body behaves like a thermal heat pump moving aether heat energy, mass and density from the internal isotropy to the external anisotropy away from the body; rarifying the body and condensing the environment.

Thus the above described isolated system mass body absorbs heat energy from the external environment when the internal Isotropy is less than the external Anisotropy; heating and expanding the body and cooling and condensing the environment.

11.2

$$\phi^2_{\text{Respiration}} = \left(\frac{|\overline{\mathrm{v}}|_{\text{CM}}}{\sqrt{|\mathrm{v}^2|_{\text{Iso}}}} \right) \geq 1$$

$$\phi^2_{\text{Respiration}} = \sqrt{\phi_{\text{Loco\_Motion}}} = \left(\frac{\gamma_{\text{Inertial\_Iso}}}{\gamma_{\text{Inertial}}} \right) = \left(\frac{|\overline{\mathrm{v}}|_{\text{CM}}}{\mathrm{c}_{\text{Sound}}} \right) \cdot \sqrt{\frac{\gamma_{\text{Heat}}}{3}} \geq 1$$

Furthermore, a thermal heat pump is found in nature to be a device that extracts heat from one place and transfers it to another. In this theory a Heat pump is used to transfer heat by circulating a substance called Aether into or out of the system mass body.

The Aether is a gaseous heat energy substance that circulates through an isolated system mass body via its Thermodynamic Momentum and Thermodynamic Heat Pump Energy, which acts to absorb, transport, and release heat energy towards or away from mass body.

The Thermodynamic Momentum ($\overline{\mathrm{p}}_{\text{Thermo\_Momentum}}$) like a diaphragm controls the direction of flow of the Aether through the heat pump, and changes the heat pump from heating to cooling mode or vice versa.

The Thermodynamic Momentum ($\overline{\mathrm{p}}_{\text{Thermo\_Momentum}}$) extracts heat from the external ambient environment, and then transfers that heat to either the inside or outside of the isolated system mass body; depending on whether the Anisotropy is greater than or less than the Isotropy.

According to the general wave theory in elastic or compressible mediums, a condensation and rarefaction are necessary to constitute a sound wave. Surely, if a condensation is not produced, there can be no sound wave!

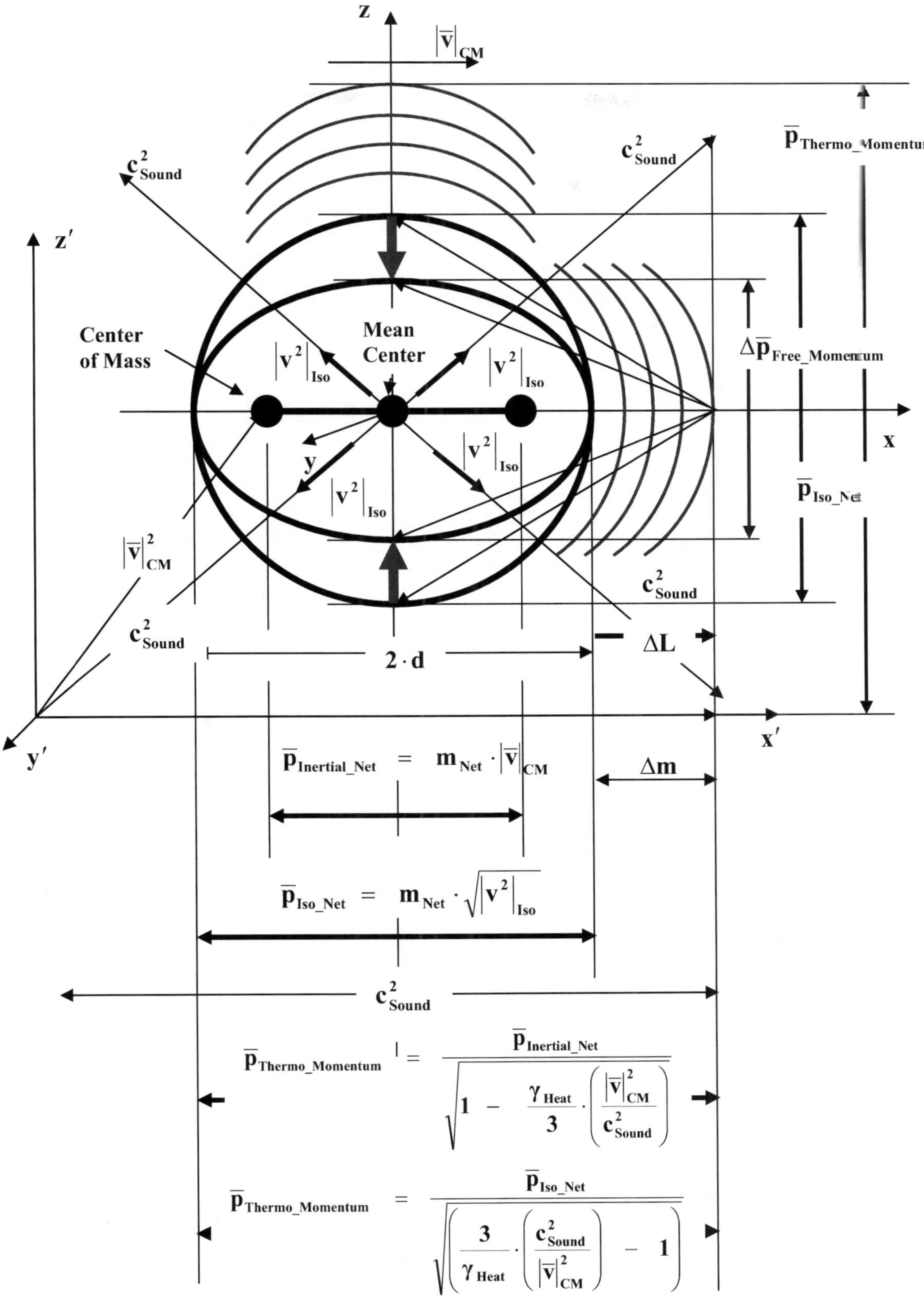

Figure 11.3: A system mass body likewise moves heat energy between the Isotropy and the Anisotropy of the system.

When a common pendulum oscillates in an air filled medium, it tends to form a condensation in front and a rarefaction behind. However, the motion of the pendulum is so slow, and the air so swift and elastic, that it moves away from the front of the pendulum before the air is sensibly condensed; likewise the air fills the space behind the pendulum motion before it can become sensibly dilated.

If that same pendulum were to oscillate at speed that approached the speed of sound in an air, then a condensation in front of and a rarefaction behind the pendulum swing would develop and be sensed. This condensation will manifest in the measurements as additional or surplus resistance inertia, and kinetic energy that must be added to the pendulum system.

From the above description, it can be concluded that a vibrating body, before it can act as a sounding body, must also produce alternate compressions and rarefactions in the air.

If, however, the vibrating body be so small that at each oscillation the surrounding air has time to flow round it, there is at every oscillation a local rearrangement - a local flow and reflow of the air; but the air at a distance is almost wholly unaffected by this.

Based on the above analogies there is unquestionably a resisting limit between motion of a body and compressibility of the medium though which it is moving.

Thus, for slow moving bodies, mobility has no appreciable resistance, and prevents condensation or compression of the air.

But, when the center of mass of those bodies are moving at a speed that is close to the average speed of the constituent particles of the medium, then a resisting limit must be overcome by applying sufficient additional energy, which condenses and compression the particles of that medium which increase or provide surplus, inertia mass, resistance, kinetic energy, momentum, and density.

To the reader I would like to remind you that the External Anisotropy of an isolated system mass body is equally present for a body is at rest or moving with a uniform velocity in a straight line relative to an external frame of reference, in accordance with the First Law of Motion.

The following equations describe the quantities used to predict the Carnot Thermodynamic Heat Engine and Heat Pump for an isolated system mass body and its atmosphere environment surroundings.

Thermodynamic "Pump" Velocity

11.3

$$\left|\overline{v}\right|_{\text{Thermo\_Momentum}} = \frac{\overline{p}_{\text{Thermo\_Momentum}}}{m_{\text{Net}}} = \gamma_{Inertial} \cdot \left|\overline{v}\right|_{\text{CM}} = \gamma_{Inertial\_Iso} \cdot \sqrt{\left|v^2\right|_{\text{Iso}}} \rightarrow m/s$$

Thermodynamic "Pump" Momentum

11.4

$$\overline{p}_{\text{Thermo\_Momentum}} = \gamma_{Inertial} \cdot \overline{p}_{\text{Inertial\_Net}} = \gamma_{Inertial\_Iso} \cdot \overline{p}_{\text{Net\_Iso}} \rightarrow kg \cdot m/s$$

$$\overline{p}_{\text{Thermo\_Momentum}} = \gamma_{Inertial} \cdot \left(m_{\text{Net}} \cdot \left|\overline{v}\right|_{\text{CM}}\right) = \gamma_{Inertial\_Iso} \cdot \left(m_{\text{Net}} \cdot \sqrt{\left|v^2\right|_{\text{Iso}}}\right)$$

$$\overline{p}_{\text{Thermo\_Momentum}} = \frac{\overline{p}_{\text{Inertial\_Net}}}{\sqrt{1 - \frac{\gamma_{Heat}}{3} \cdot \left(\frac{\left|\overline{v}\right|^2_{\text{CM}}}{c^2_{\text{Sound}}}\right)}} = \frac{\overline{p}_{\text{Net\_Iso}}}{\sqrt{\left(\frac{3}{\gamma_{Heat}} \cdot \left(\frac{c^2_{\text{Sound}}}{\left|\overline{v}\right|^2_{\text{CM}}}\right) - 1\right)}}$$

Aphorism 11.1: Relative to an External Observer, the Center of Mass, the Center of Isotropy, and an atomic substance medium, as the **Isotropy is greater than Anisotropy** of the system, the kinetic energy transfer flow direction is outward and the aether flow direction is inward; the result is cooling and condensing of the body, and warming and rarifying of the environment.

$$\phi^2_{Respiration} = \sqrt{\phi_{\text{Loco\_Motion}}} = \left(\frac{\gamma_{Inertial\_Iso}}{\gamma_{Inertial}}\right) = \left(\frac{\left|\overline{v}\right|_{\text{CM}}}{c_{\text{Sound}}}\right) \cdot \sqrt{\frac{\gamma_{Heat}}{3}} \leq 1$$

Aphorism 11.2: Relative to an External Observer, the Center of Mass, the Center of Isotropy, and an atomic substance medium, as the **Isotropy is less than Anisotropy** of the system, the kinetic energy transfer flow direction is inward and the aether flow direction is outward; the result is heating and rarifying of the body, and cooling and condensing of the environment

$$\phi^2_{Respiration} = \sqrt{\phi_{\text{Loco\_Motion}}} = \left(\frac{\gamma_{Inertial\_Iso}}{\gamma_{Inertial}}\right) = \left(\frac{\left|\overline{v}\right|_{\text{CM}}}{c_{\text{Sound}}}\right) \cdot \sqrt{\frac{\gamma_{Heat}}{3}} \geq 1$$

Chapter 12

The Special Theory of Thermodynamics

(Carnot Thermal Anisotropy Engine)

12.1 Carnot Thermal Engine – External (Anisotropy) Inertial Motion Compressibility Factor

In general a mass system body immersed in any environment, including the vacuum of spacetime behaves like a Carnot Thermal Anisotropy Engine.

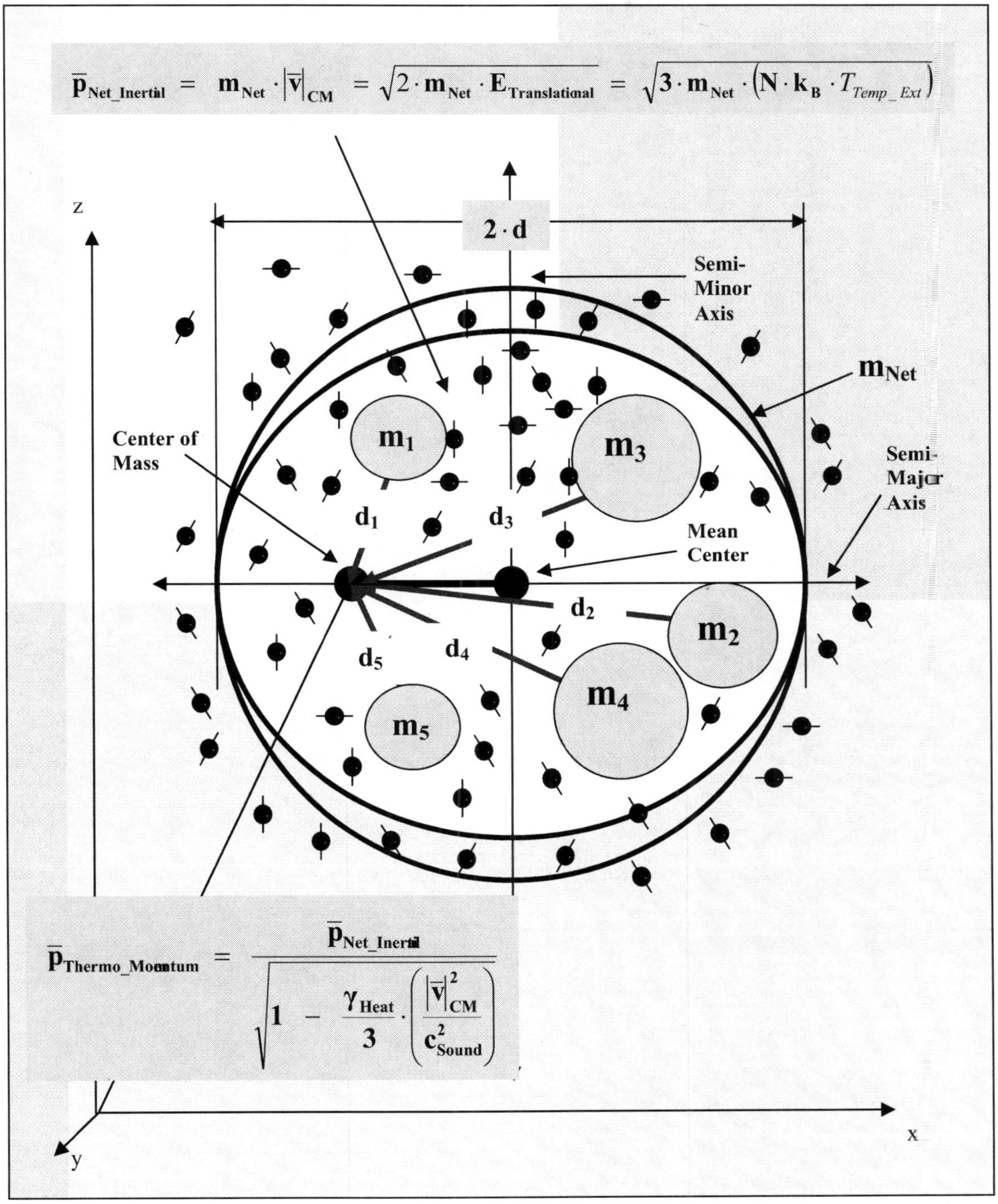

Figure 12.1: External Anisotropy directional uniform inertial rectilinear momentum and the Internal Isotropy uniform inertial rectilinear momentum are measures of the First Law of Motion.

Using the **Carnot Thermal Engine "Anisotropy" Efficiency Factor** the Inertial Anisotropic Rectilinear Momentum, Internal Isotropic Omni-directional Momentum, and Spacetime Compressibility Ratios can be derived.

12.1

$$\eta_{\text{Carnot}} = \frac{1}{\gamma^2_{Inertial}} = \left[\frac{W_{\text{Free\_Work\_Energy}}}{E_{\text{Net\_Iso\_Kinetic}}}\right]$$

$$\eta_{\text{Carnot}} = \frac{1}{\gamma^2_{Inertial}} = \left[\frac{E_{\text{Net\_Iso\_Kinetic}} - E_{\text{Net\_Translational}}}{E_{\text{Net\_Iso\_Kinetic}}}\right] = \frac{\left|v^2\right|_{\text{Iso}} - \left|\bar{v}\right|^2_{\text{CM}}}{\left|v^2\right|_{\text{Iso}}}$$

$$\eta_{\text{Carnot}} = \frac{1}{\gamma^2_{Inertial}} = \left(\frac{E_{\text{Net\_Translational}}}{E_{\text{Net\_Iso\_Kinetic}}}\right)\left(\frac{E_{\text{Net\_Iso\_Kinetic}}}{E_{\text{Net\_Translational}}} - 1\right)$$

$$\eta_{\text{Carnot}} = \frac{1}{\gamma^2_{Inertial}} = \left(\frac{\frac{\left(\bar{p}_{\text{Inertial\_Net}}\right)^2}{2 \cdot m_{\text{Net}}}}{\frac{\bar{p}^2_{\text{Net\_Iso}}}{2 \cdot m_{\text{Net}}}}\right) \cdot \left(\frac{E_{\text{Net\_Iso\_Kinetic}}}{E_{\text{Net\_Translational}}} - 1\right)$$

$$\eta_{\text{Carnot}} = \frac{1}{\gamma^2_{Inertial}} = \left(\frac{\bar{p}^2_{\text{Inertial\_Net}}}{\bar{p}^2_{\text{Net\_Iso}}}\right) \cdot \left(\frac{E_{\text{Net\_Iso\_Kinetic}}}{E_{\text{Net\_Translational}}} - 1\right)$$

Taking the square root of both sides of the above equation yields

$$\frac{1}{\gamma_{Inertial} \cdot \sqrt{\left(\frac{E_{\text{Net\_Iso\_Kinetic}}}{E_{\text{Net\_Translational}}} - 1\right)}} = \left(\frac{\bar{p}_{\text{Inertial\_Net}}}{\bar{p}_{\text{Net\_Iso}}}\right)$$

12.2

$$\boxed{\phi^2_{\text{Re}\,spiration} = \left(\frac{\gamma_{Inertial\_Iso}}{\gamma_{Inertial}}\right) = \left(\frac{\bar{p}_{\text{Inertial\_Net}}}{\bar{p}_{\text{Net\_Iso}}}\right) \rightarrow Unitless}$$

Since a mass system body and its medium environment behave like a heat engine, the isotropy to anisotropic energy transfer between two different parts of an isolated system of differing temperatures, densities, and kinetic energies is also limited or governed by Carnot's theorems.

The **External (Anisotropy) Inertial Motion Compressibility Factor** ($\gamma_{Inertial}$) is a measure of the ratio of the amount of increase or decrease in the Thermodynamic Momentum ($\overline{\mathrm{p}}_{\mathrm{Thermo\_Momentum}}$) relative to the Anisotropy and Center of Mass Momentum ($\overline{\mathrm{p}}_{\mathrm{Inertial\_Net}}$) of an isolated system body.

The increase or decrease in the Thermodynamic Momentum ($\overline{\mathrm{p}}_{\mathrm{Thermo\_Momentum}} = \gamma_{Inertial} \cdot \overline{\mathrm{p}}_{\mathrm{Inertial\_Net}}$) of an isolated system mass body is the measure of the resistance, condensing and accumulation of surplus **External Anisotropy**, inertia mass, density, and anisotropic kinetic energy, which is a result of the **Inertial Locomotion Foreshortening Factor** increases or decreases ($\phi_{\mathrm{Loco\_Motion}} = \left(\dfrac{\left|\overline{\mathrm{v}}\right|^2_{\mathrm{CM}}}{\left|\mathrm{v}^2\right|_{\mathrm{Iso}}} \right)$), and likewise as the **Mach number** increases or decreases ($M_{Mach} = \left(\dfrac{\left|\overline{\mathrm{v}}\right|_{\mathrm{CM}}}{\mathrm{c}_{\mathrm{Sound}}} \right)$).

Aphorism 12.1: Relative to an External Observer, the Mean Center, Center of Mass, and an atomic substance medium, the **External (Anisotropy) Inertial Motion Compressibility Factor** ($\gamma_{Inertial} = \dfrac{1}{\sqrt{1 - \dfrac{\left|\overline{\mathrm{v}}\right|^2_{\mathrm{CM}}}{\left|\mathrm{v}^2\right|_{\mathrm{Iso}}}}}$) of an Isolated System Mass Body is a measure of the accumulation of surplus **External Anisotropy** energy of motion and increases as the **Inertial Locomotion Foreshortening Factor** increases or decreases ($\phi_{\mathrm{Loco\_Motion}} = \left(\dfrac{\left|\overline{\mathrm{v}}\right|^2_{\mathrm{CM}}}{\left|\mathrm{v}^2\right|_{\mathrm{Iso}}} \right) = \dfrac{\gamma_{Heat}}{3} \cdot \left(\dfrac{\left|\overline{\mathrm{v}}\right|^2_{\mathrm{CM}}}{\mathrm{c}^2_{\mathrm{Sound}}} \right)$), and likewise as the Mach number increases or decreases ($M_{Mach} = \left(\dfrac{\left|\overline{\mathrm{v}}\right|_{\mathrm{CM}}}{\mathrm{c}_{\mathrm{Sound}}} \right)$).

Aphorism 12.2: The **External (Anisotropy) Inertial Motion Compressibility Factor** ($\gamma_{Inertial}$) of an isolated dynamic system mass body and its environment is proportional to the inverse square root of the Anisotropic Carnot Thermal Engine Efficiency factor which is a measure of the **Coefficient of Heating** of the "Environment" of the system.

$$\gamma_{Inertial} = \frac{1}{\sqrt{\eta_{\text{Carnot}}}}$$

For an isolated system mass body functioning as a Carnot engine heat pump, the "effectiveness" is the ratio of the energy delivered to the high-temperature reservoir divided by the work required to force the machine around its cycle.

The effectiveness of a heat pump warming the external environment of an isolated dynamic system mass body is called the Coefficient of Heat performance factor ($\text{COP}_{\text{Heating}}$).

While the Coefficient of Heat Performance ($\text{COP}_{\text{Heating}}$) is partly a measure of the efficiency of a heat pump, it is also a measure of the "external" conditions under which it is operating; and is equal to the **Square of the External (Anisotropy) Inertial Motion Compressibility Factor.**

12.3

$$\gamma^2_{Inertial} = \frac{1}{\eta_{\text{Carnot}}} = \left[\frac{\text{E}_{\text{Net\_Iso\_Kinetic}}}{\text{W}_{\text{Work\_Free}}}\right] = \left[\frac{\text{E}_{\text{Net\_Iso\_Kinetic}}}{\text{E}_{\text{Net\_Iso\_Kinetic}} - \text{E}_{\text{Net\_Translational}}}\right]$$

$$\gamma^2_{Inertial} = \left(\frac{\overline{\text{p}}_{\text{Thermo\_Momentum}}}{\overline{\text{p}}_{\text{Inertial\_Net}}}\right)^2 = \left[\frac{High\ \text{Temperature}}{High\ \text{Temperature} - Low\ \text{Temperature}}\right] \rightarrow Unitless$$

External (Anisotropy) Inertial Motion Compressibility Factor

12.4

$$\gamma_{Inertial} = \left(\frac{\overline{p}_{Thermo\_Momentum}}{\overline{p}_{Inertial\_Net}} \right) = \sqrt{\frac{High\ \text{Temperature}}{High\ \text{Temperature} - Low\ \text{Temperature}}}$$

$$\gamma_{Inertial} = \frac{1}{\sqrt{\eta_{Carnot}}} = \sqrt{\left[\frac{E_{Net\_Iso\_Kinetic}}{E_{Net\_Iso\_Kinetic} - E_{Net\_Translational}} \right]}$$

$$\gamma_{Inertial} = \sqrt{\frac{\left| v^2 \right|_{Iso}}{\left| v^2 \right|_{Iso} - \left| \overline{v} \right|_{CM}^2}} = \sqrt{\frac{T_{emp}}{T_{emp} - T_{emp\_Ext}}} \rightarrow Unitless$$

Definition 12.1: Relative to an External Observer, the Mean Center, the Center of Mass, and an atomic substance medium of an isolated system mass body, the **External (Anisotropy) Inertial Motion Compressibility Factor** ($\gamma_{Inertial} = \frac{1}{\sqrt{\eta_{Carnot}}} = \left(\frac{\overline{p}_{Thermo\_Momentum}}{\overline{p}_{Inertial\_Net}} \right)$) is the measure of the condensing and accumulation of surplus external anisotropy, inertia mass, density, and kinetic energy, which is a result of **Inertial Locomotion Foreshortening Factor** increases or decreases ($\phi_{Loco\_Motion} = \left(\frac{\left| \overline{v} \right|_{CM}^2}{\left| v^2 \right|_{Iso}} \right) = \frac{\gamma_{Heat}}{3} \cdot \left(\frac{\left| \overline{v} \right|_{CM}^2}{c_{Sound}^2} \right)$), and likewise as the **Mach number** increases or decreases ($M_{Mach} = \left(\frac{\left| \overline{v} \right|_{CM}}{c_{Sound}} \right)$); and is defined as the ratio of the **Thermodynamic Momentum** ($\overline{p}_{Thermo\_Momentum}$) divided by the **External Anisotropic Inertial Net Linear Momentum** ($\overline{p}_{Inertial\_Net} = m_{Net} \cdot \left| \overline{v} \right|_{CM}$) of an isolated system mass body.

Definition 12.2: Relative to an External Observer, the Mean Center, Center of Mass, and an atomic substance medium of an isolated system mass body, the **External (Anisotropy) Inertial Motion Compressibility Factor** ($\gamma_{Inertial} = \dfrac{1}{\sqrt{\eta_{\text{Carnot}}}} = \left(\dfrac{\overline{\mathrm{p}}_{\text{Thermo\_Momentum}}}{\overline{\mathrm{p}}_{\text{Inertial\_Net}}}\right)$) is the measure of the amount of Coefficient of Heating between the anisotropy and the heat transfer mechanism of the system; and is defined as the inverse square root of the **Carnot Thermal Engine Efficiency Factor** ($\dfrac{1}{\sqrt{\eta_{\text{Carnot}}}} = \sqrt{\left[\dfrac{\mathrm{E}_{\text{Net\_Iso\_Kinetic}}}{\mathrm{E}_{\text{Net\_Iso\_Kinetic}} - \mathrm{E}_{\text{Net\_Translational}}}\right]}$) of an isolated system mass body.

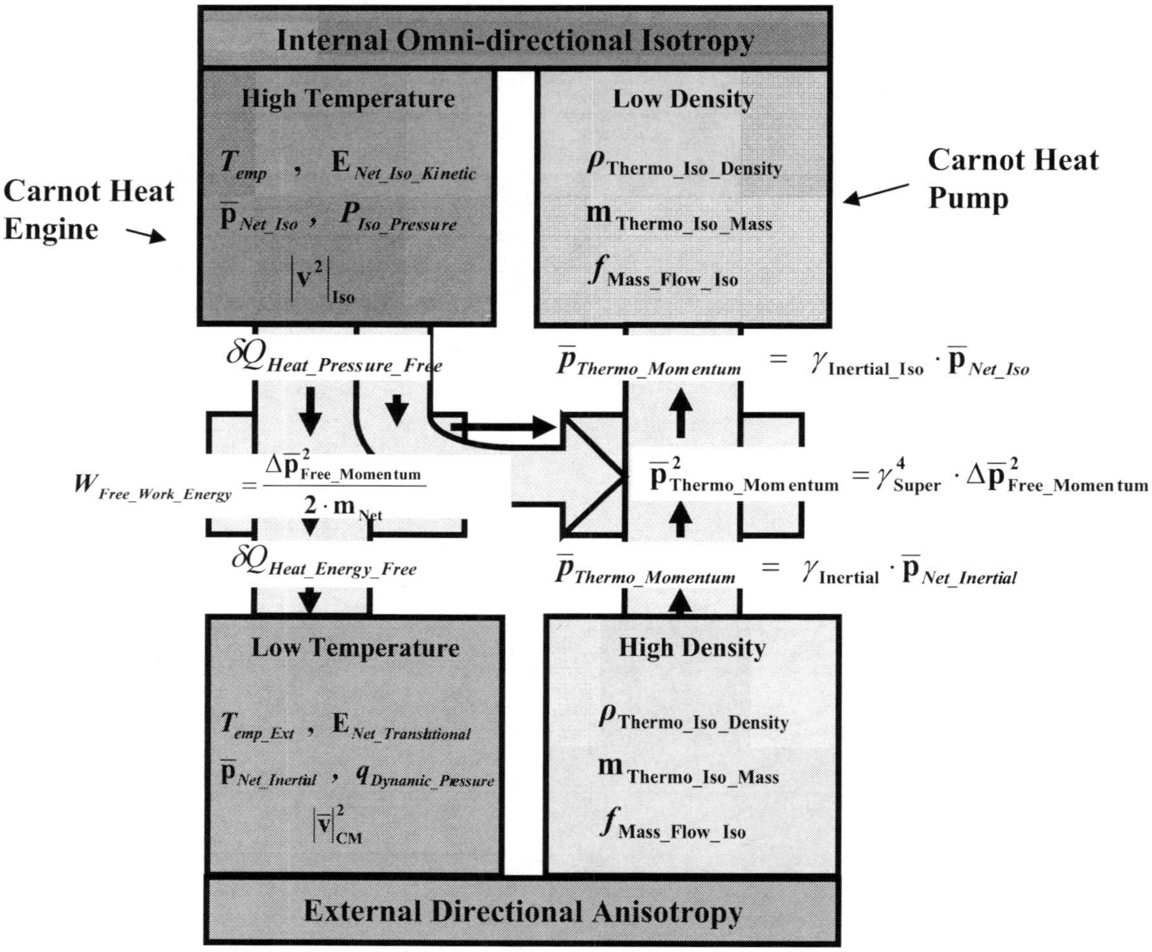

$$\phi_{\text{Respiration}} = \left(\phi_{\text{Loco\_Motion}}\right)^{\frac{1}{4}} \leq 1$$

Figure 12.2: Heat Energy Transfer Device – Kemp Thermal Engine.

External (Anisotropy) Inertial Motion Compressibility Factor

12.5

$$\gamma_{Inertial} = \frac{1}{\sqrt{\eta_{Carnot}}} = \left(\frac{\overline{p}_{Thermo\_Momentum}}{\overline{p}_{Inertial\_Net}}\right) \rightarrow Unitless$$

$$\gamma_{Inertial} = \frac{1}{\sqrt{\eta_{Carnot}}} = \gamma_{Inertial\_Iso} \cdot \left(\frac{\overline{p}_{Net\_Iso}}{\overline{p}_{Inertial\_Net}}\right)$$

$$\gamma_{Inertial} = \frac{\left(\frac{\overline{p}_{Net\_Iso}}{\overline{p}_{Inertial\_Net}}\right)}{\sqrt{\left(\frac{E_{Net\_Iso\_Kinetic}}{E_{Net\_Translational}} - 1\right)}} = \frac{\left(\frac{\overline{p}_{Net\_Iso}}{\overline{p}_{Inertial\_Net}}\right)}{\sqrt{\left(\frac{\left|v^2\right|_{Iso}}{\left|\overline{v}\right|^2_{CM}} - 1\right)}}$$

$$\gamma_{Inertial} = \frac{1}{\sqrt{\eta_{Carnot}}} = \frac{1}{\sqrt{1 - \frac{\left|\overline{v}\right|^2_{CM}}{\left|v^2\right|_{Iso}}}} = \frac{1}{\sqrt{1 - \phi_{Loco\_Motion}}}$$

$$\gamma_{Inertial} = \frac{1}{\sqrt{\eta_{Carnot}}} = \frac{1}{\sqrt{1 - \frac{\gamma_{Heat}}{3} \cdot \left(\frac{\left|\overline{v}\right|^2_{CM}}{c^2_{Sound}}\right)}}$$

$$\gamma_{Inertial} = \sqrt{1 + \gamma^2_{Inertial\_Iso}}$$

External (Anisotropy) Inertial Motion Compressibility Factor

12.6

$$\gamma_{Inertial} = \frac{1}{\sqrt{\eta_{\text{Carnot}}}} = \frac{\gamma^2_{Super\_Compres}}{\gamma_{Inertial\_Iso}} \rightarrow Unitless$$

$$\gamma_{Inertial} = \frac{1}{\sqrt{\eta_{\text{Carnot}}}} = \gamma_{Inertial\_Iso} \cdot \left(\frac{\overline{\text{p}}_{\text{Net\_Iso}}}{\overline{\text{p}}_{\text{Inertial\_Net}}} \right)$$

$$\gamma_{Inertial} = \frac{1}{\sqrt{1 - \frac{\gamma_{Heat}}{3} \cdot \left(\frac{|\overline{\text{v}}|^2_{\text{CM}}}{\text{c}^2_{\text{Sound}}} \right)}} = \frac{\left(\frac{\overline{\text{p}}_{\text{Net\_Iso}}}{\overline{\text{p}}_{\text{Inertial\_Net}}} \right)}{\sqrt{\left(\frac{\text{E}_{\text{Net\_Iso\_Kinetic}}}{\text{E}_{\text{Net\_Translational}}} - 1 \right)}}$$

$$\gamma_{Inertial} = \frac{1}{\sqrt{1 - \frac{|\overline{\text{v}}|^2_{\text{CM}}}{|\text{v}^2|_{\text{Iso}}}}} = \frac{\overline{\text{p}}_{\text{Net\_Iso}}}{\Delta\overline{\text{p}}_{\text{Free\_Momentum}}} = \frac{\overline{\text{p}}_{\text{Thermo\_Momentum}}}{\overline{\text{p}}_{\text{Inertial\_Net}}}$$

$$\gamma_{Inertial} = \frac{1}{\sqrt{\eta_{\text{Carnot}}}} = \frac{1}{\sqrt{1 - \left(\frac{T_{emp\_Ext}}{T_{emp}} \right)}} = \frac{1}{\sqrt{1 - \left(\frac{\text{m}_{\text{Inertia\_Mass}}}{\text{m}_{\text{Net}}} \right)}}$$

External (Anisotropy) Inertial Motion Compressibility Factor

12.7

$$\gamma_{Inertial} = \sqrt{\frac{\left(1 - \frac{1}{\gamma_{Heat}}\right) \cdot \mathrm{m}_{\mathrm{Net}} \cdot C_{Specific\_Pressure} \cdot T_{emp}}{\Delta(PV)_{\mathrm{Free\_Energy}}}} \rightarrow Unitless$$

$$\gamma_{Inertial} = \frac{1}{\sqrt{\eta_{\mathrm{Carnot}}}} = \frac{1}{\sqrt{1 - \phi_{\mathrm{Loco\_Motion}}}} = \sqrt{\frac{\mathrm{m}_{\mathrm{Net}}}{\Delta\mathrm{m}_{\mathrm{Aniso\_WFM}}}}$$

$$\gamma_{Inertial} = \frac{1}{\sqrt{1 - \phi_{\mathrm{Loco\_Motion}}}} = \sqrt{\frac{\mathrm{m}_{\mathrm{Net}} \cdot 3 \cdot \mathrm{N} \cdot \mathrm{k}_{\mathrm{B}} \cdot T_{emp}}{\Delta\overline{\mathrm{p}}^{2}_{\mathrm{Free\_Momentum}}}}$$

$$\gamma_{Inertial} = \frac{1}{\sqrt{\eta_{\mathrm{Carnot}}}} = \frac{1}{\sqrt{1 - \phi_{\mathrm{Loco\_Motion}}}} = \frac{\mathrm{m}_{\mathrm{Net}} \cdot |\mathrm{v}|_{\mathrm{Iso}}}{\Delta\overline{\mathrm{p}}_{\mathrm{Free\_Momentum}}}$$

$$\gamma_{Inertial} = \frac{1}{\sqrt{\eta_{\mathrm{Carnot}}}} = \frac{1}{\sqrt{1 - \phi_{\mathrm{Loco\_Motion}}}} = \sqrt{\frac{\left[1 + \left(\frac{\Delta\mathrm{m}_{\mathrm{Aniso\_WFM}}}{\mathrm{m}_{\mathrm{Net}}}\right)\right]}{\left(\frac{\Delta\mathrm{m}_{\mathrm{Aniso\_WFM}}}{\mathrm{m}_{\mathrm{Net}}}\right)}}$$

$$\gamma_{Inertial} = \frac{1}{\sqrt{1 - \phi_{\mathrm{Loco\_Motion}}}} = \sqrt{\frac{1}{\left(\frac{\Delta\overline{\mathrm{p}}^{2}_{\mathrm{Free\_Momentum}}}{\mathrm{m}_{\mathrm{Net}} \cdot 3 \cdot \mathrm{N} \cdot \mathrm{k}_{\mathrm{B}} \cdot T_{emp\_Ext}}\right)} + 1}$$

External (Anisotropy) Inertial Motion Compressibility Factor

12.8

$$\gamma_{Inertial} = \frac{1}{\sqrt{\eta_{Carnot}}} = \frac{1}{\sqrt{1 - \left(\frac{|\overline{v}|^2_{CM}}{|v^2|_{Iso}}\right)}} \rightarrow Unitless$$

$$\gamma_{Inertial} = \frac{1}{\sqrt{\eta_{Carnot}}} = \frac{1}{\sqrt{1 - \frac{\gamma_{Heat}}{3} \cdot \left(\frac{|\overline{v}|^2_{CM}}{c^2_{Sound}}\right)}}$$

$$\gamma_{Inertial} = \sqrt{\frac{\left[1 + \left(\frac{\Delta(PV)_{Free\_Energy}}{\left(1 - \frac{1}{\gamma_{Heat}}\right) \cdot m_{Net} \cdot C_{Specific\_Pressure} \cdot T_{emp\_Ext}}\right)\right]}{\left(\frac{\Delta(PV)_{Free\_Energy}}{\left(1 - \frac{1}{\gamma_{Heat}}\right) \cdot m_{Net} \cdot C_{Specific\_Pressure} \cdot T_{emp\_Ext}}\right)}}$$

$$\gamma_{Inertial} = \sqrt{\frac{\left[\left(1 - \frac{1}{\gamma_{Heat}}\right) \cdot m_{Net} \cdot C_{Specific\_Pressure} \cdot T_{emp\_Ext} + \Delta(PV)_{Free\_Energy}\right]}{\Delta(PV)_{Free\_Energy}}}$$

External (Anisotropy) Inertial Motion Compressibility Factor

12.9

$$\gamma_{Inertial} = \frac{1}{\sqrt{\eta_{\text{Carnot}}}} = \frac{\gamma^2_{Super\_Compres}}{\gamma_{Inertial\_Iso}} \rightarrow Unitless$$

$$\gamma_{Inertial} = \frac{\gamma^2_{Super\_Compres}}{\gamma_{Inertial\_Iso}} = \frac{1}{\gamma_{Inertial\_Iso}} \cdot \left(\frac{\overline{\mathrm{p}}^2_{\text{Thermo\_Momentum}}}{\overline{\mathrm{p}}_{\text{Net\_Iso}} \cdot \overline{\mathrm{p}}_{\text{Inertial\_Net}}} \right)$$

$$\gamma_{Inertial} = \frac{1}{\sqrt{\left[-\frac{\gamma_{Heat}}{6} \cdot \left(\frac{|\overline{\mathrm{v}}|^2_{\text{CM}}}{\mathrm{c}^2_{\text{Sound}}} \right) \pm \sqrt{1 - \frac{\gamma_{Heat}}{3} \cdot \left(\frac{|\overline{\mathrm{v}}|^2_{\text{CM}}}{\mathrm{c}^2_{\text{Sound}}} \right) + \frac{\gamma^2_{Heat}}{12} \cdot \left(\frac{|\overline{\mathrm{v}}|^4_{\text{CM}}}{\mathrm{c}^4_{\text{Sound}}} \right)} \right]}}$$

$$\gamma_{Inertial} = \frac{1}{\sqrt{\eta_{\text{Carnot}}}} = \frac{1}{\sqrt{1 - \frac{\mathrm{m}_{\text{Net}} \cdot |\overline{\mathrm{v}}|^2_{\text{CM}}}{3 \cdot \left(\mathrm{N} \cdot \mathrm{k_B} \cdot T_{emp} \right)}}}$$

The External (Anisotropy) Inertial Motion Compressibility Factor of an isolated system mass body can also be expressed as the absolute value; which is able to handle the negative values as the anisotropy becomes larger than the isotropy of the system.

12.10

$$\frac{1}{\gamma^2_{Inertial}} = \left\| \left[\frac{E_{Net\_Iso\_Kinetic} - E_{Net\_Translational}}{E_{Net\_Iso\_Kinetic}} \right] \right\| = \left\| \frac{\left| v^2 \right|_{Iso} - \left| \bar{v} \right|^2_{CM}}{\left| v^2 \right|_{Iso}} \right\|$$

External (Anisotropy) Inertial Motion Compressibility Factor

12.11

$$\gamma_{Inertial} = \frac{1}{\sqrt{\eta_{Carnot}}} = \frac{1}{\sqrt{\left\| 1 - \frac{\left| \bar{v} \right|^2_{CM}}{\left| v^2 \right|_{Iso}} \right\|}} \rightarrow \textit{Unitless}$$

$$\gamma_{Inertial} = \frac{1}{\sqrt{\eta_{Carnot}}} = \frac{1}{\sqrt{\left\| 1 - \frac{\gamma_{Heat}}{3} \cdot \left(\frac{\left| \bar{v} \right|^2_{CM}}{c^2_{Sound}} \right) \right\|}}$$

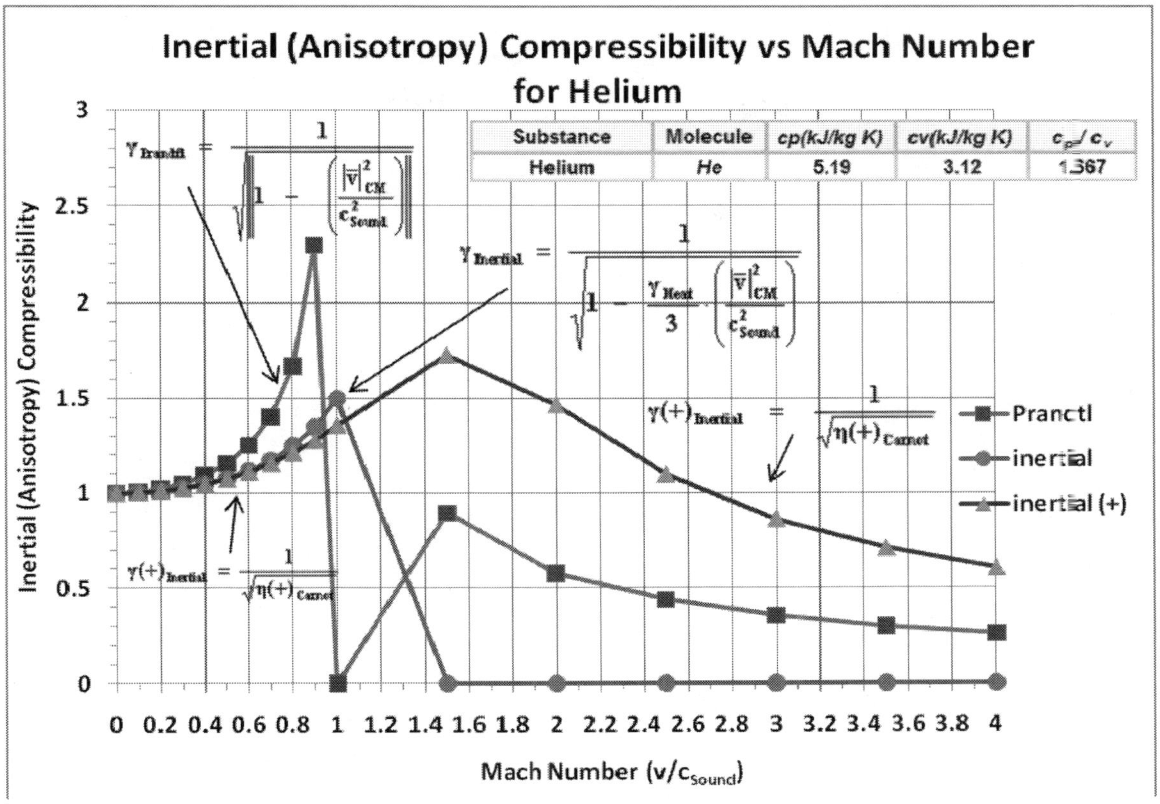

Figure 12.3: Inertial Anisotropy Compressibility vs. Mach number for - Helium.

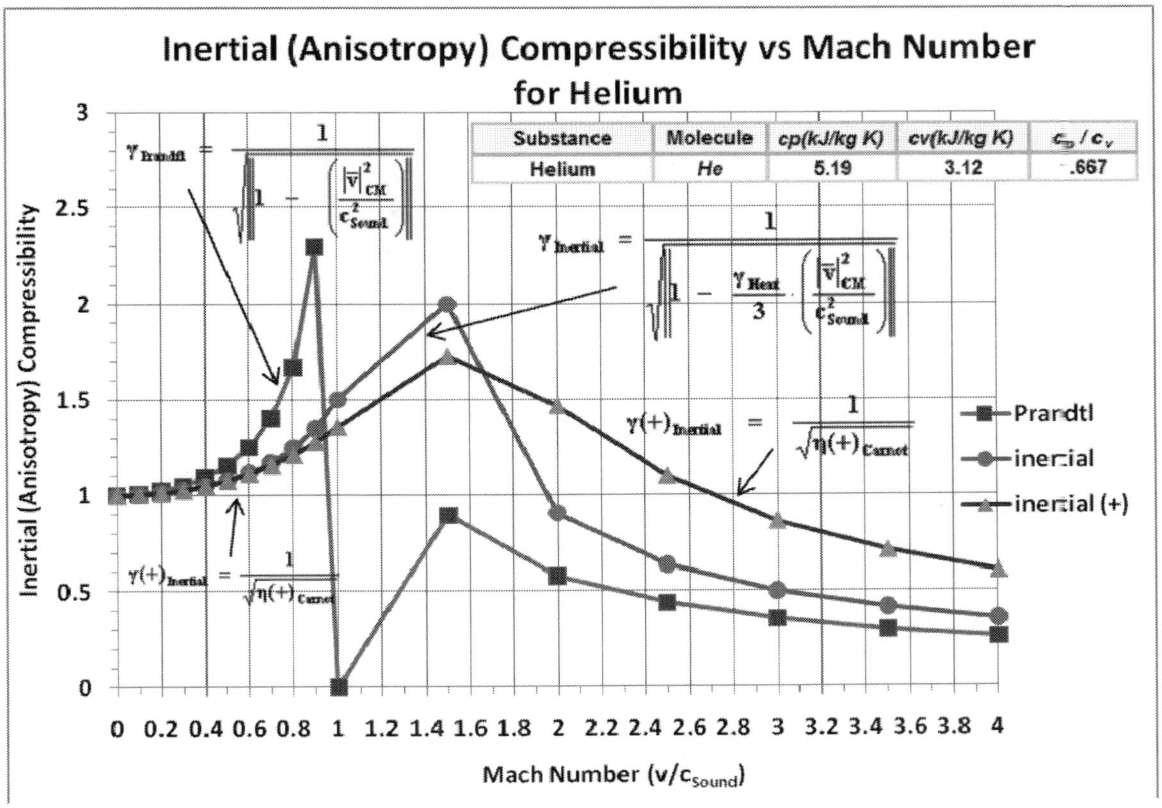

Figure 12.4: Inertial Anisotropy (Absolute) Compressibility vs. Mach number for - Helium.

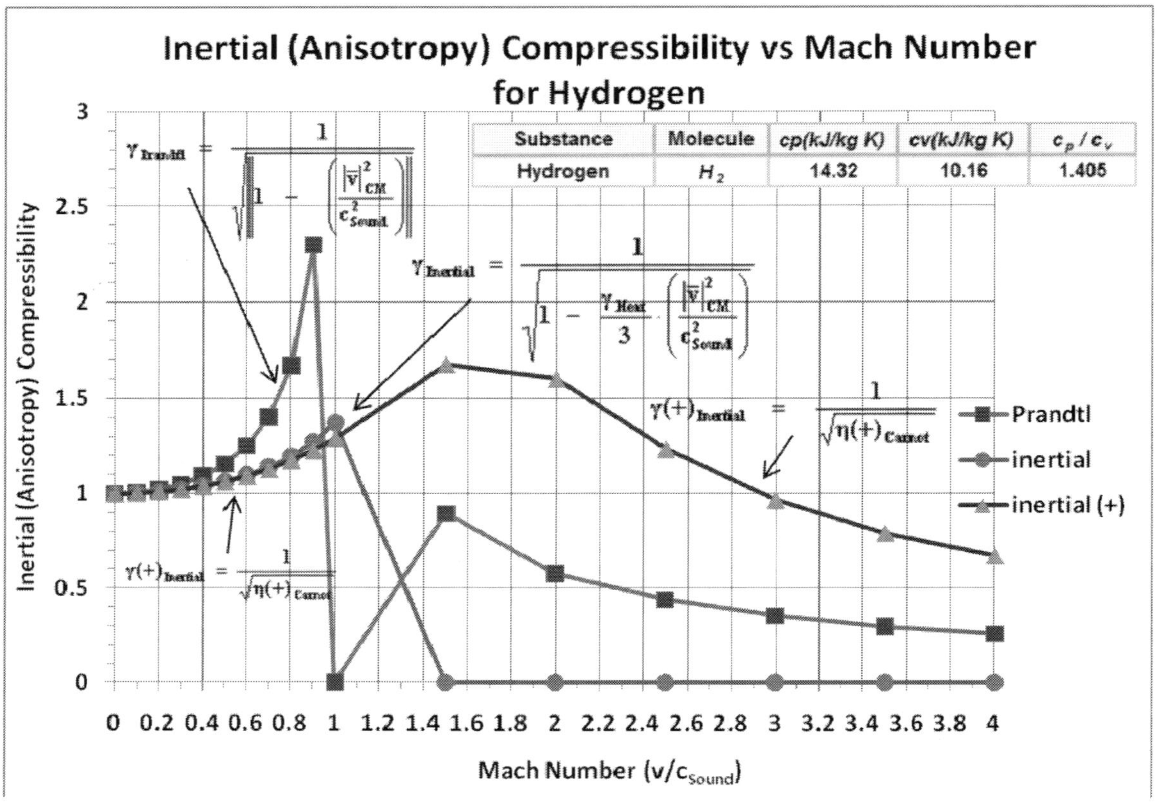

Figure 12.5: Inertial Anisotropy Compressibility vs. Mach number for - Helium.

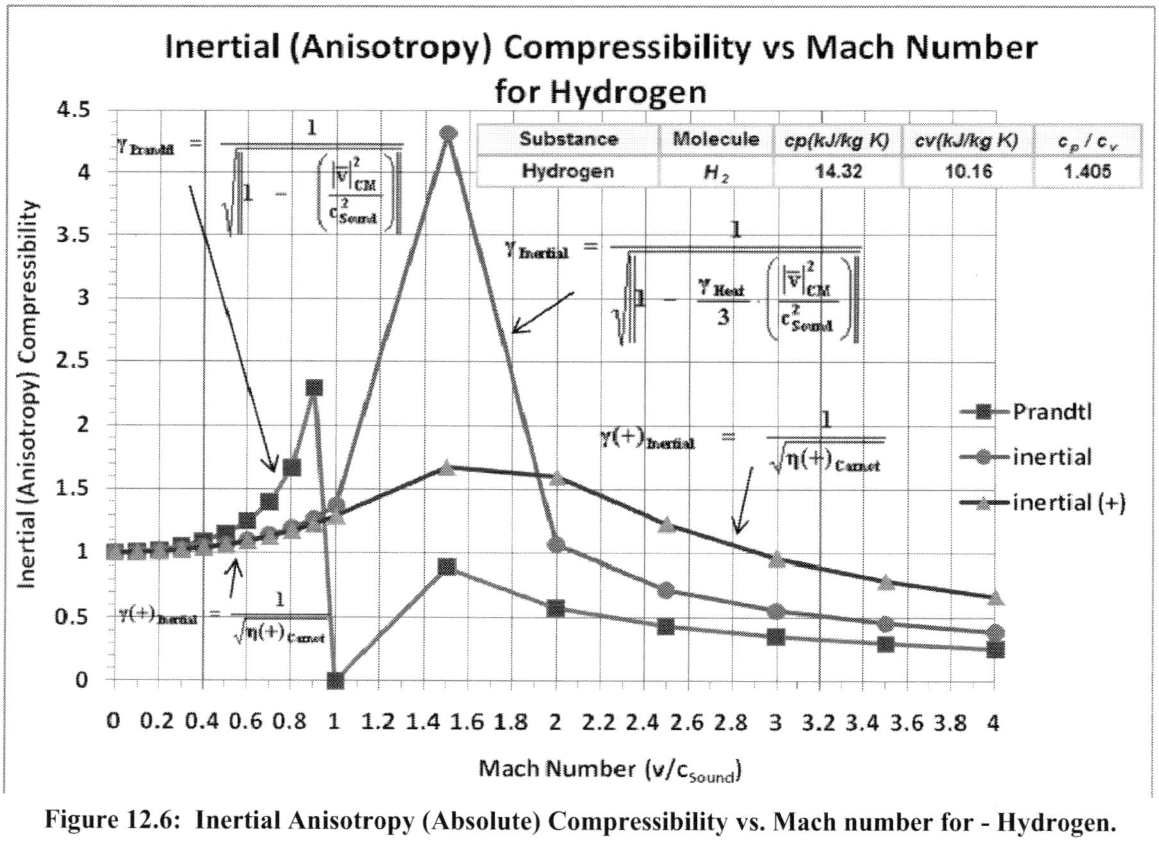

Figure 12.6: Inertial Anisotropy (Absolute) Compressibility vs. Mach number for - Hydrogen.

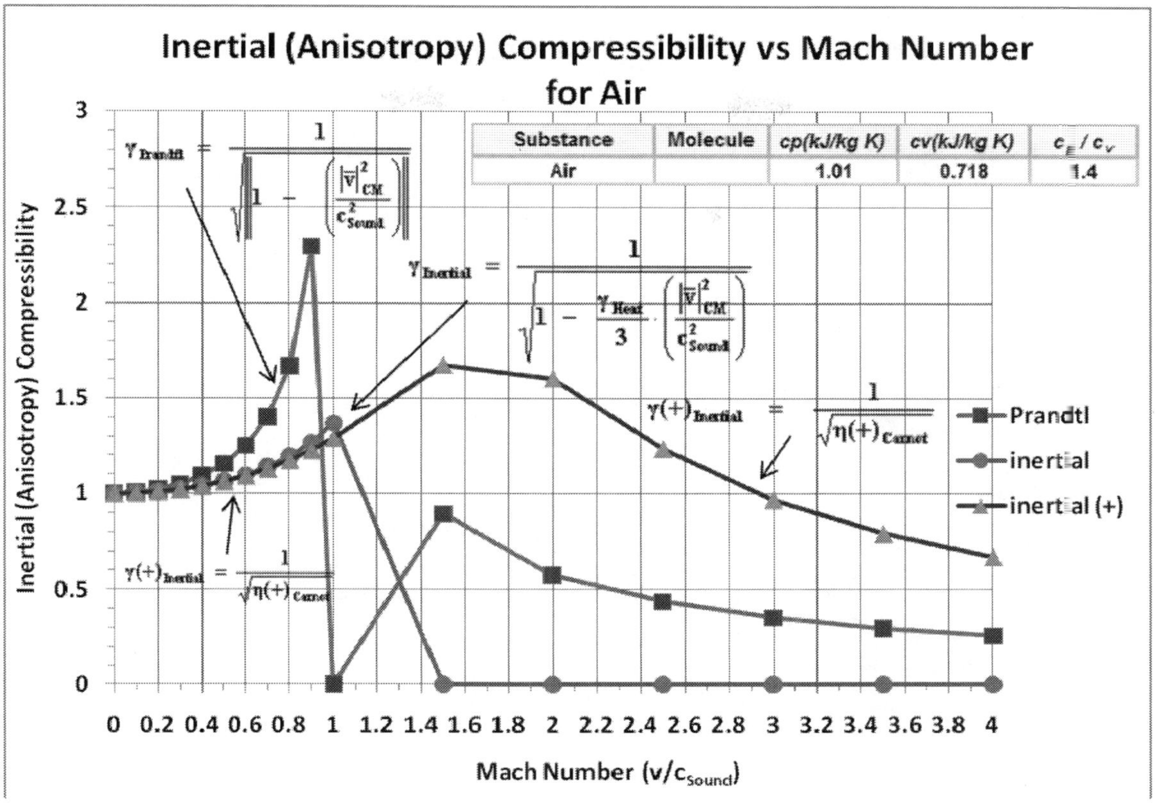

Figure 12.7: Inertial Anisotropy Compressibility vs. Mach number for - Air.

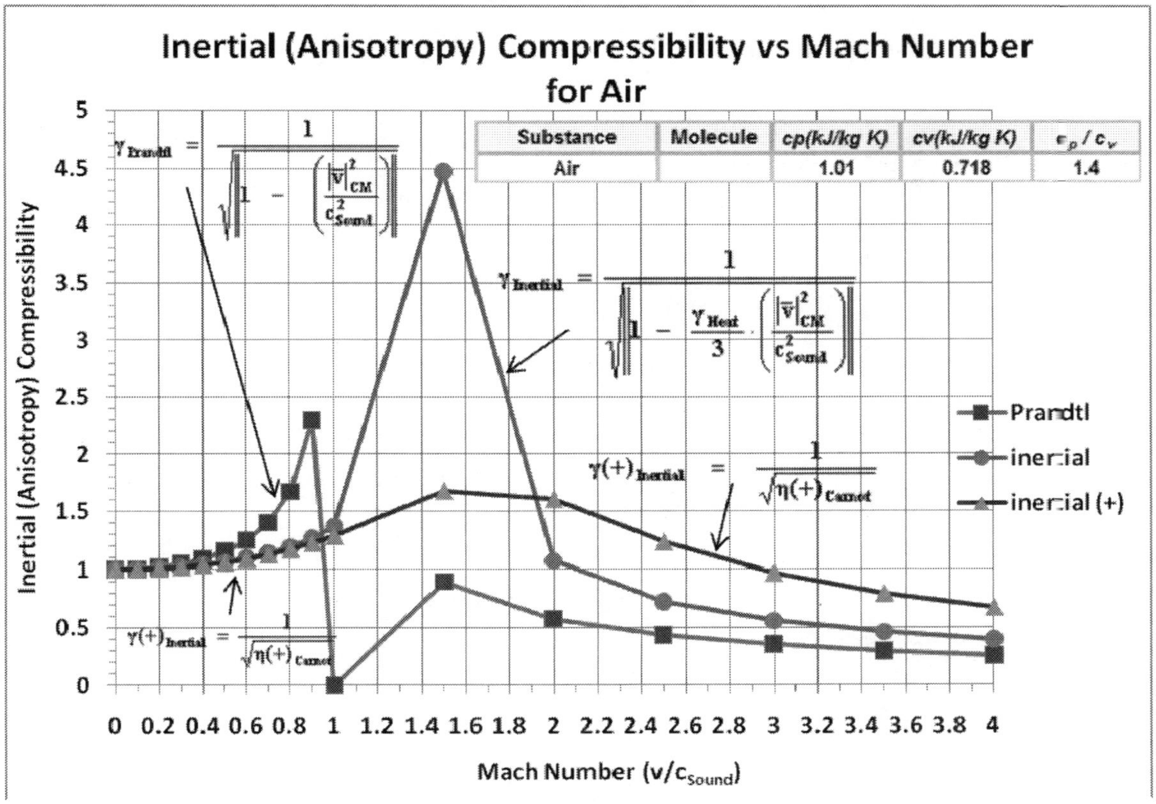

Figure 12.8: Inertial Anisotropy (Absolute) Compressibility vs. Mach number for - Air.

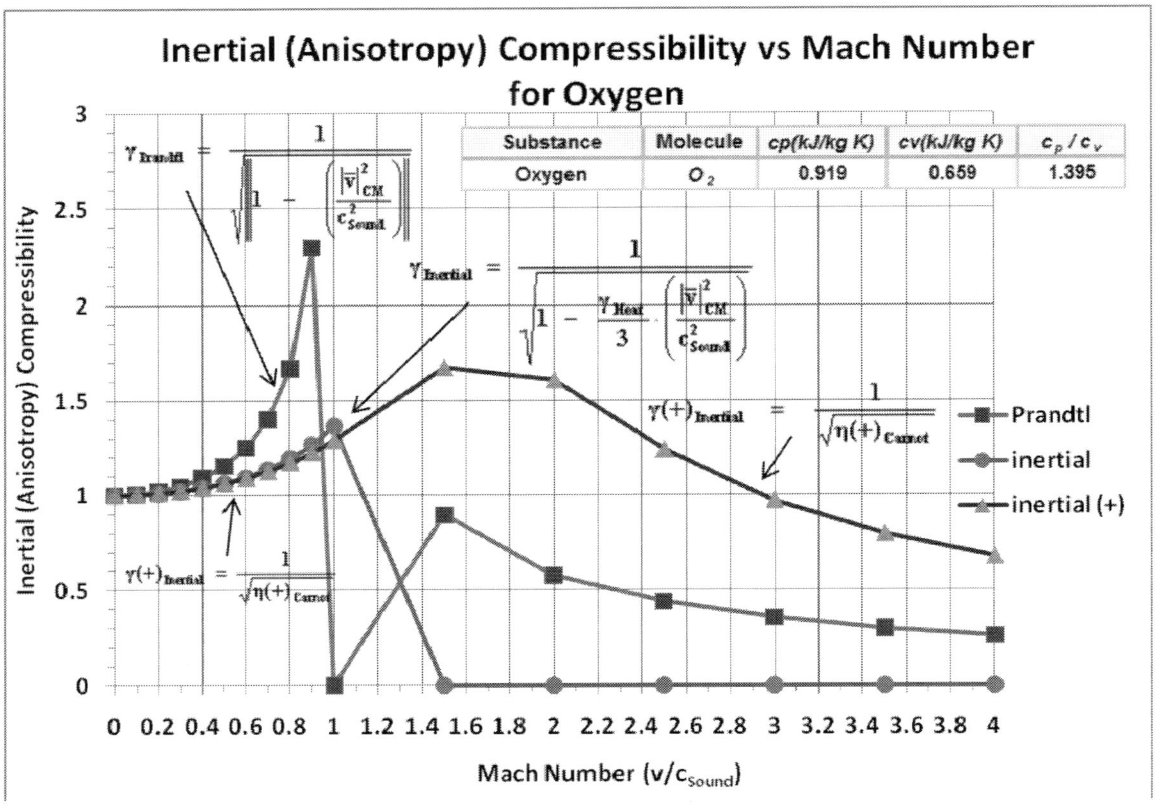

Figure 12.9: Inertial Anisotropy Compressibility vs. Mach number for - Oxygen.

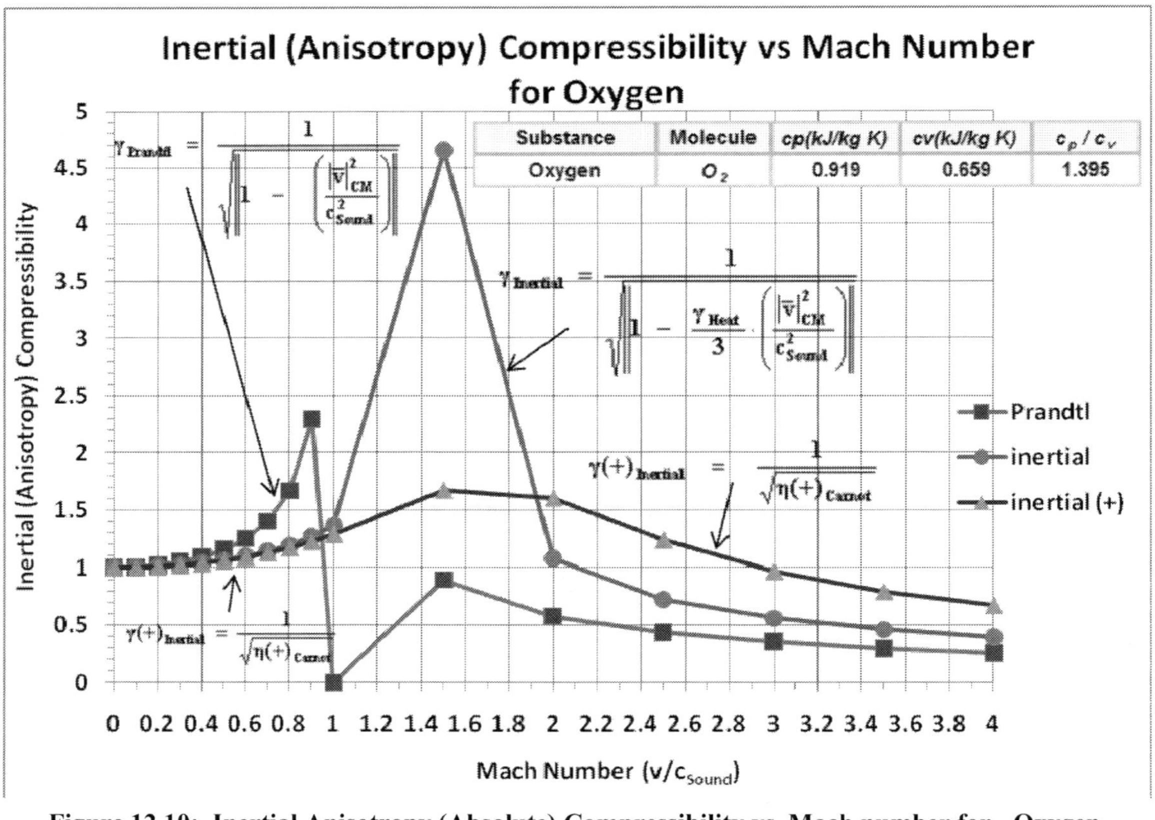

Figure 12.10: Inertial Anisotropy (Absolute) Compressibility vs. Mach number for - Oxygen.

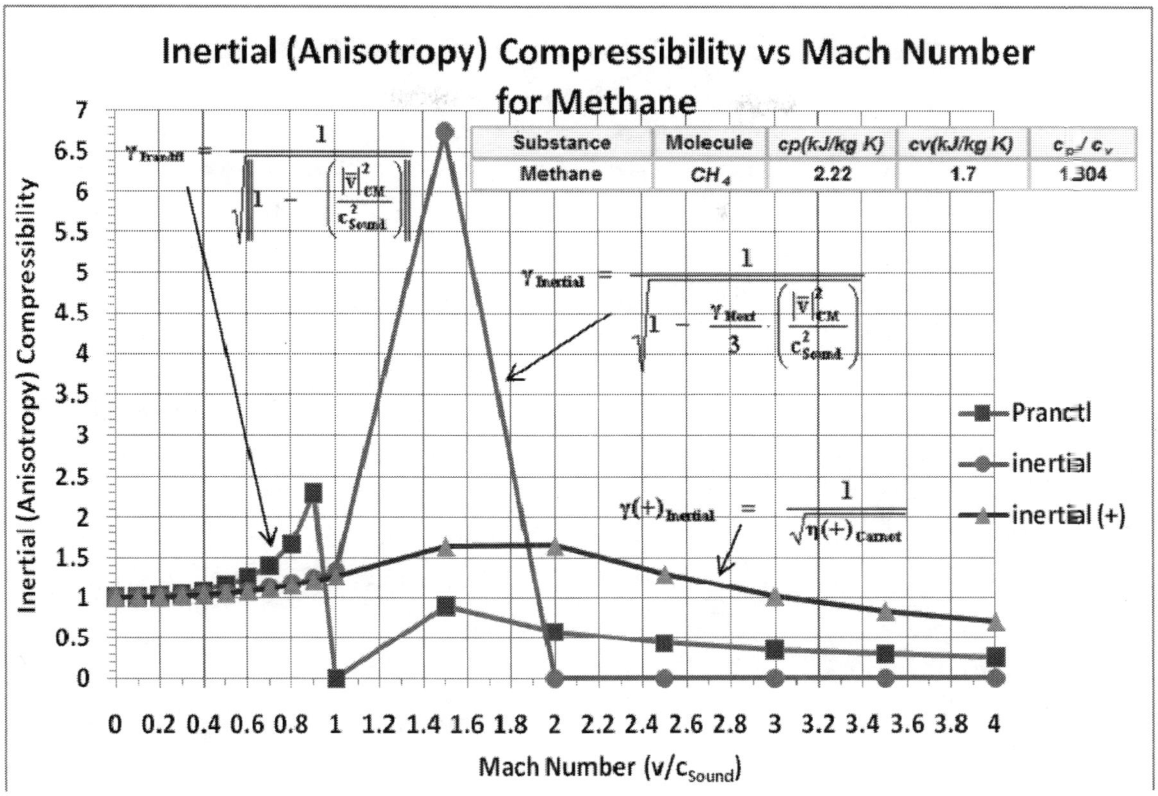

Figure 12.11: Inertial Anisotropy Compressibility vs. Mach number for - Methane.

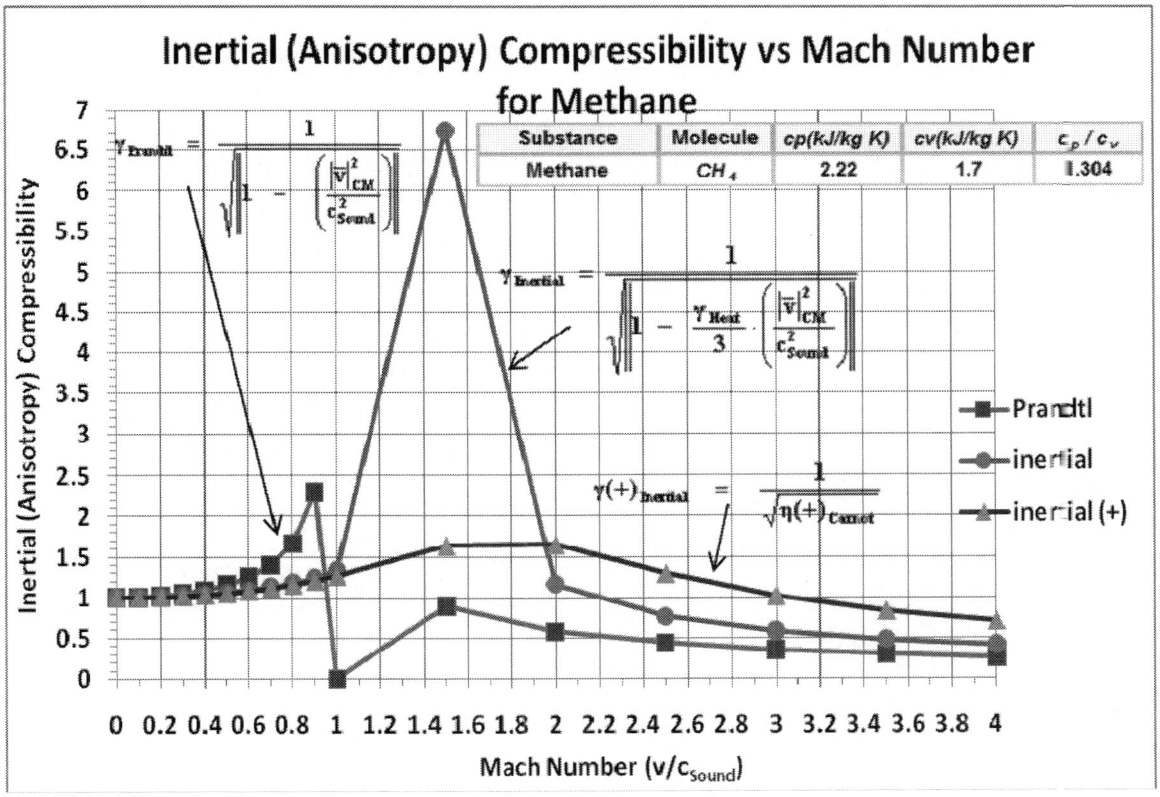

Figure 12.12: Inertial Anisotropy (Absolute) Compressibility vs. Mach number for - Methane.

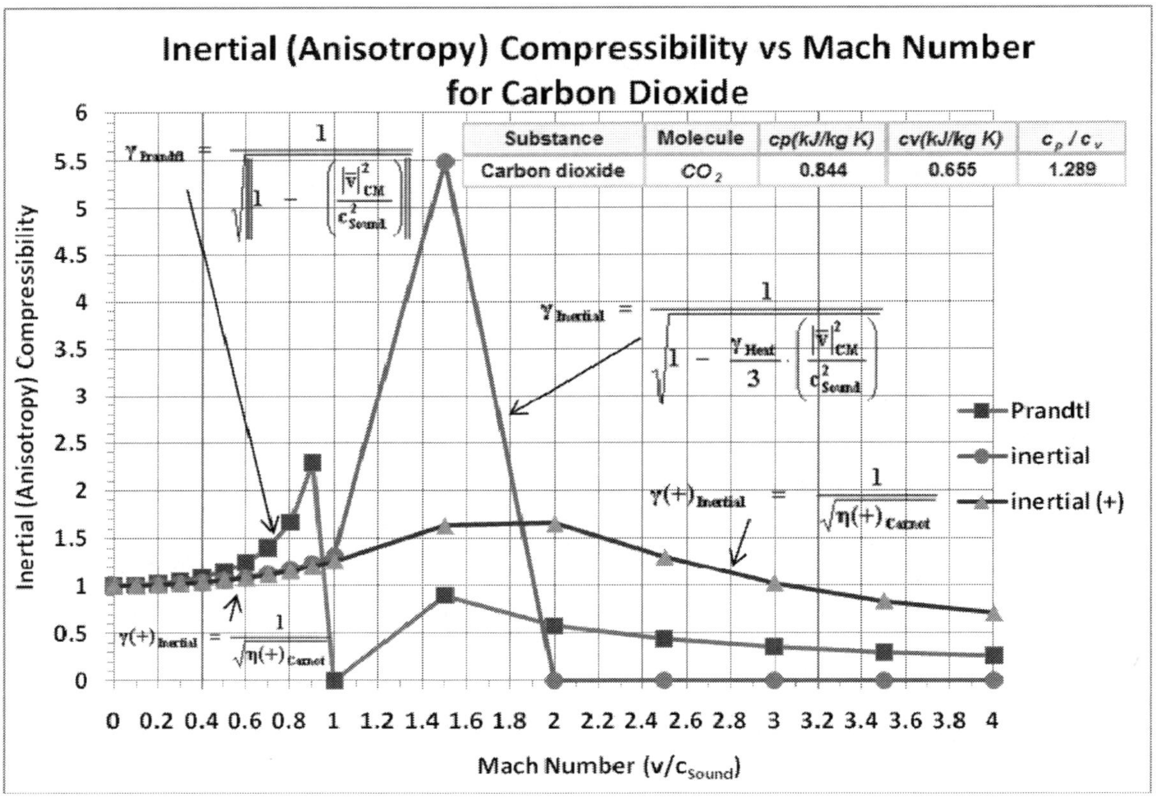

Figure 12.13: Inertial Anisotropy Compressibility vs. Mach number for – Carbon Dioxide.

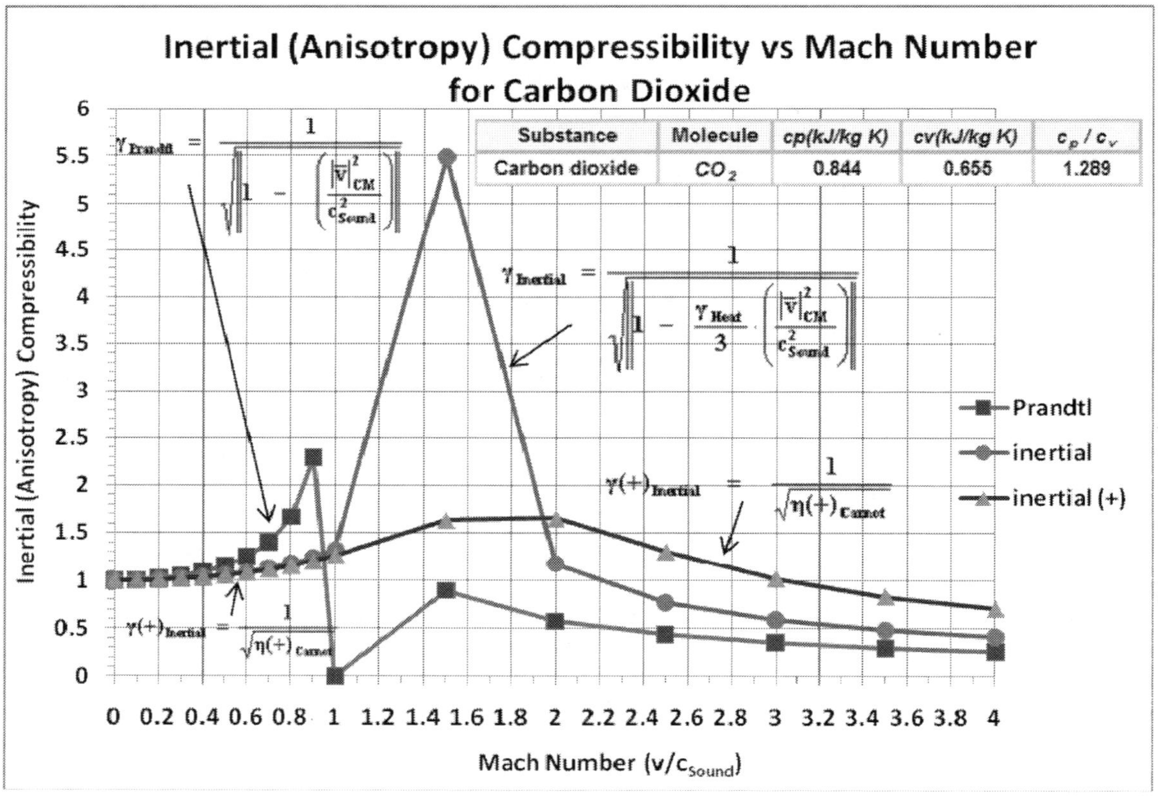

Figure 12.14: Inertial Anisotropy (Absolute) Compressibility vs. Mach number for – Carbon Dioxide.

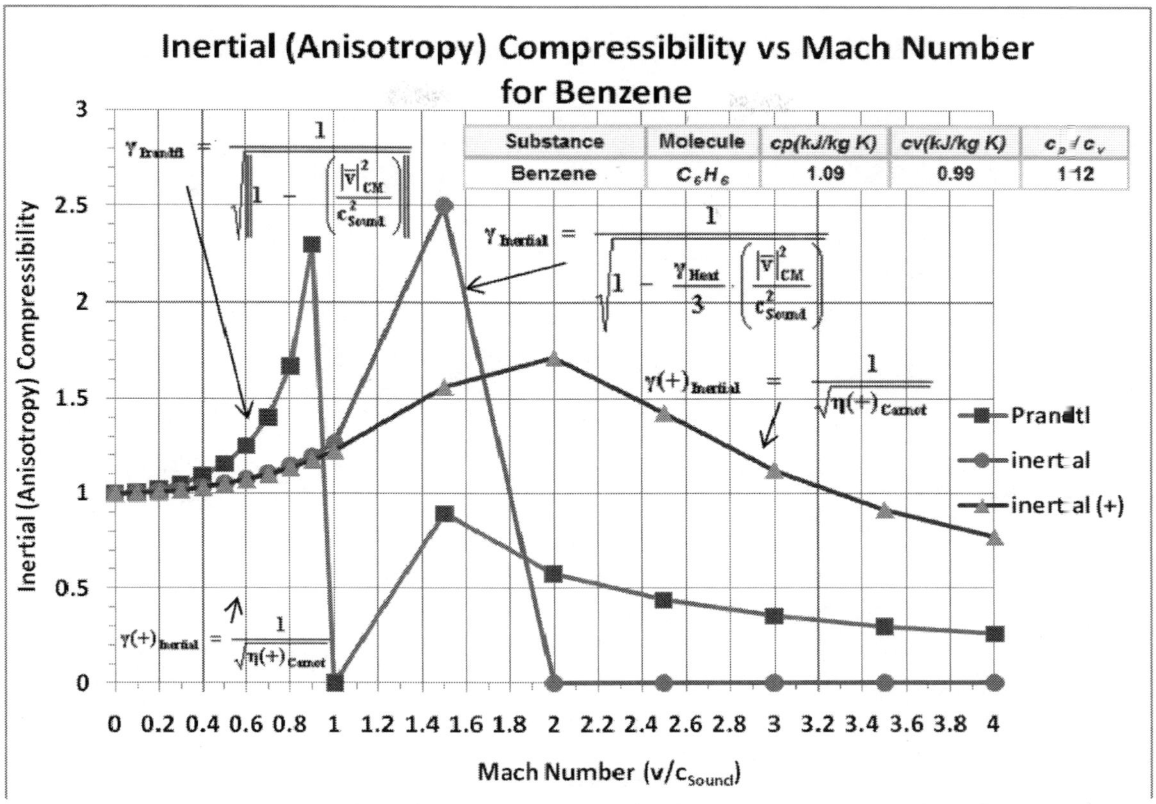

Figure 12.15: Inertial Anisotropy Compressibility vs. Mach number for – Benzene.

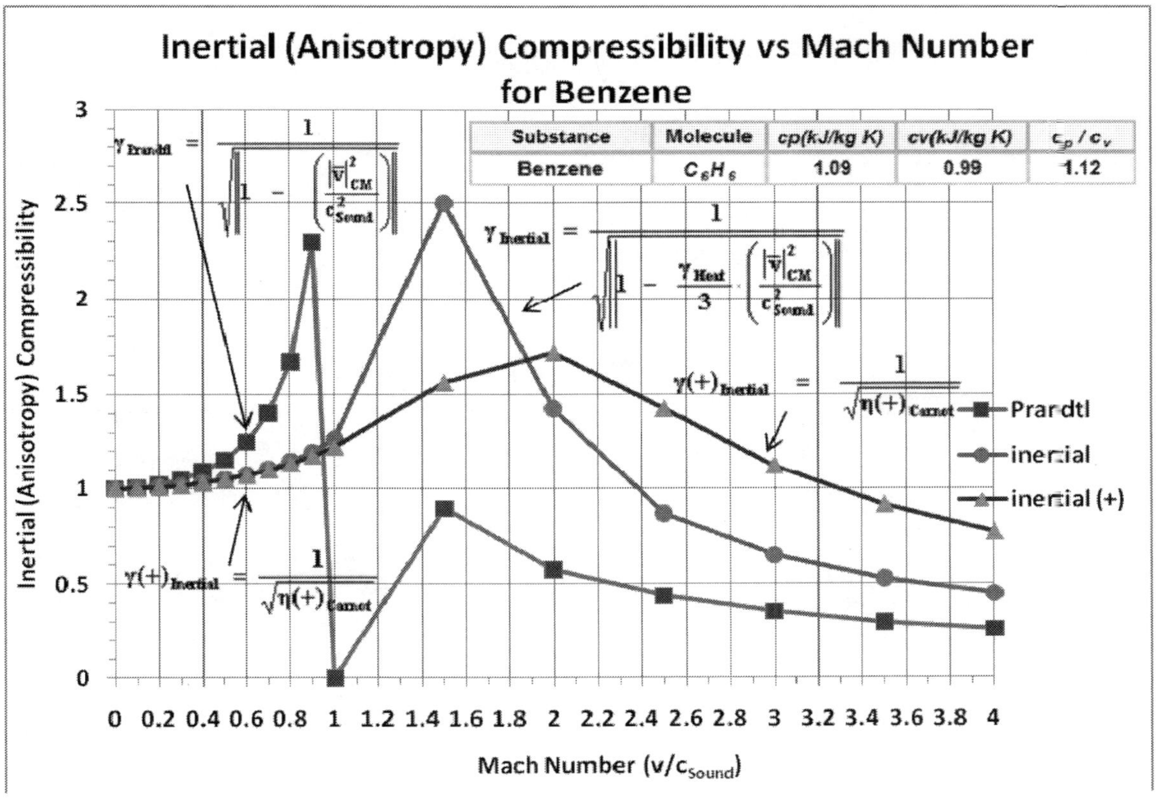

Figure 12.16: Inertial Anisotropy (Absolute) Compressibility vs. Mach number for - Benzene.

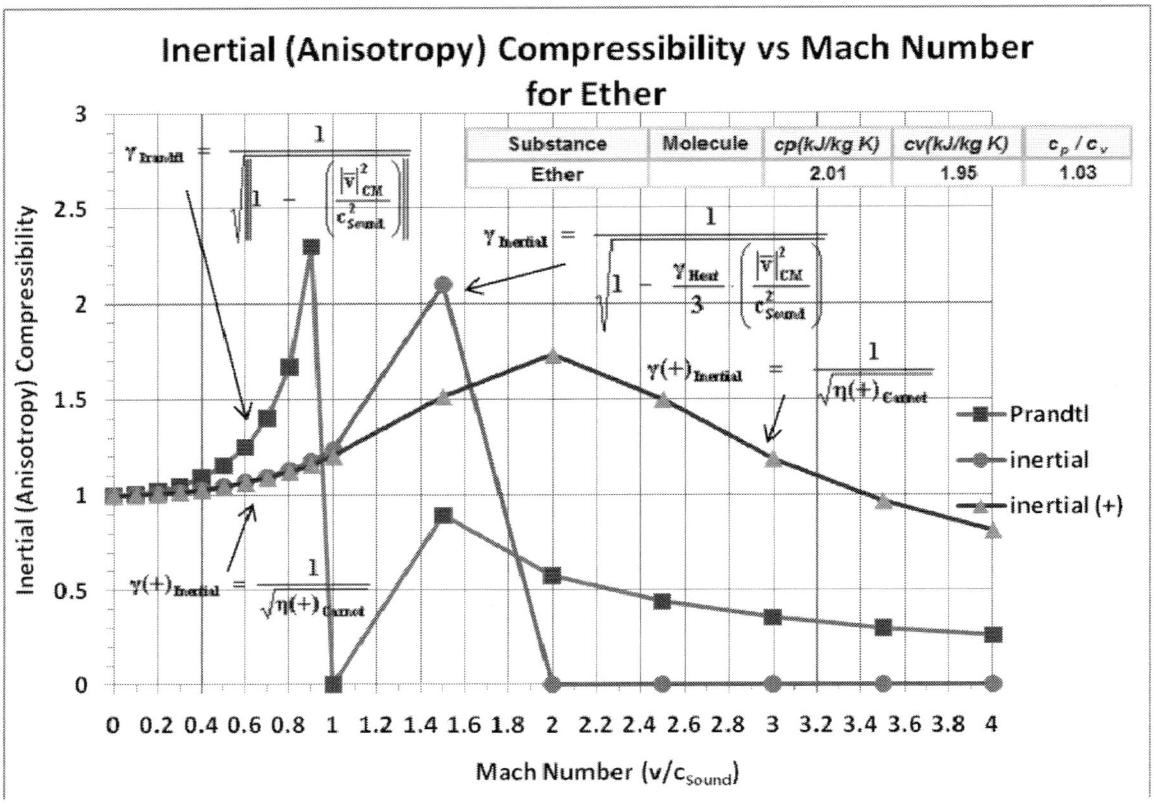

Figure 12.17: Inertial Anisotropy Compressibility vs. Mach number for – Ether

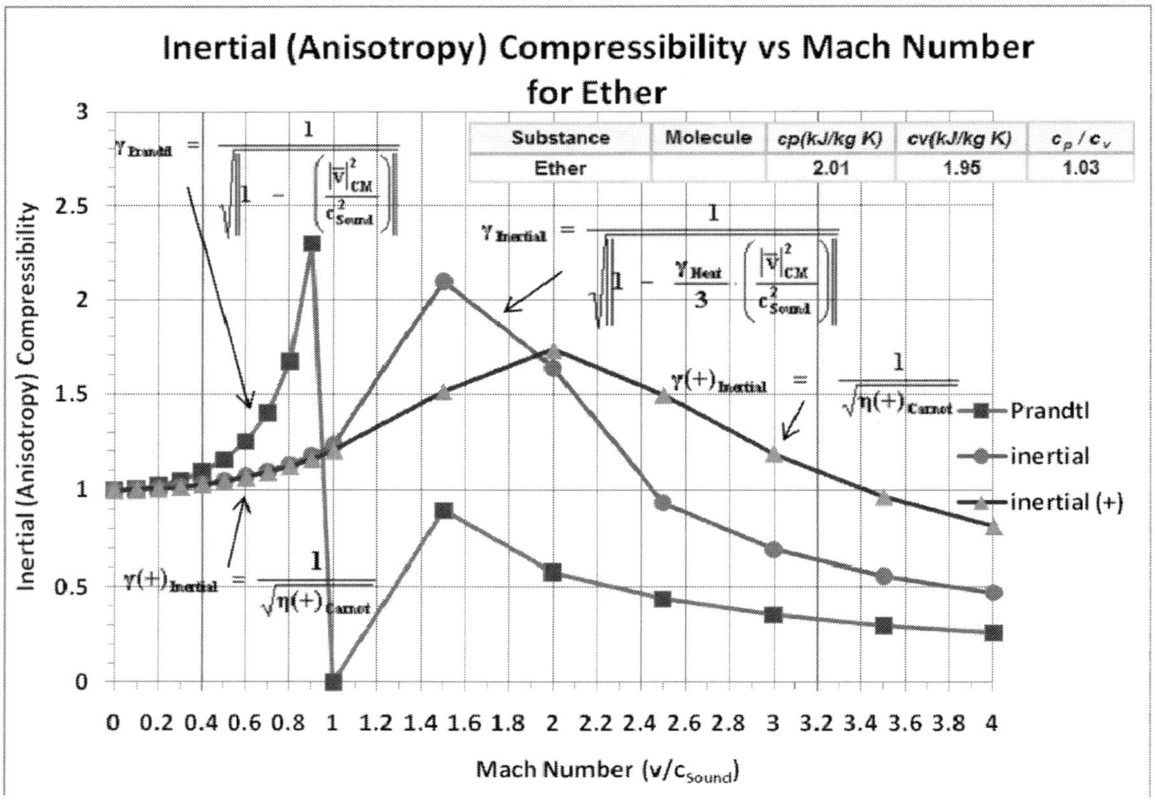

Figure 12.18: Inertial Anisotropy (Absolute) Compressibility vs. Mach number for - Ether.

12.2 Carnot Thermal Anisotropy Engine - Parameters

Thermodynamic Anisotropy – Surplus Density

12.12

$$\rho_{\text{Thermo\_Aniso\_Density}} = \left(\frac{m_{\text{Thermo\_Aniso\_Mass}}}{V_{ol}}\right) \rightarrow {kg}/{m^3}$$

$$\rho_{\text{Thermo\_Aniso\_Density}} = \gamma_{Inertial} \cdot \rho_{\text{Net}} = \frac{\rho_{\text{Net}}}{\sqrt{1 - \left(\frac{|\bar{v}|^2_{CM}}{|v^2|_{Iso}}\right)}}$$

$$\rho_{\text{Thermo\_Aniso\_Density}} = \frac{1}{|\bar{v}|_{CM}} \cdot \left(\frac{\bar{p}_{\text{Thermo\_Momentum}}}{V_{ol}}\right) = \frac{\left(\frac{m_{\text{Net}}}{V_{ol}}\right)}{\sqrt{1 - \frac{\gamma_{Heat}}{3} \cdot \left(\frac{|\bar{v}|^2_{CM}}{c^2_{\text{Sound}}}\right)}}$$

Thermodynamic Anisotropy – Surplus Inertia Mass and Resistance

12.13

$$m_{\text{Thermo\_Aniso\_Mass}} = \gamma_{Inertial} \cdot m_{\text{Net}} \rightarrow kg$$

$$m_{\text{Thermo\_Aniso\_Mass}} = \frac{\bar{p}_{\text{Thermo\_Momentum}}}{|\bar{v}|_{CM}} = \frac{m_{\text{Net}}}{\sqrt{1 - \left(\frac{|\bar{v}|^2_{CM}}{|v^2|_{Iso}}\right)}}$$

$$m_{\text{Thermo\_Aniso\_Mass}} = \frac{m_{\text{Net}}}{\sqrt{1 - \frac{\gamma_{Heat}}{3} \cdot \left(\frac{|\bar{v}|^2_{CM}}{c^2_{\text{Sound}}}\right)}}$$

Thermodynamic Anisotropy – Surplus Mass Flow Rate

Where the Thermodynamic Pump cycle frequency is given by,

12.14

$$f_{Pump\_Frequency} = \left(\frac{1}{T_{Pump\_Cycle\_Period}} \right) \rightarrow {}^{1}\!/_{s}$$

Where the Thermodynamic Pump cycle frequency is given by,

For Light:

12.15

$$\bar{d} = \bar{c}_{Light} \cdot \left(\frac{T_{Pump\_Cycle\_Period}}{2\pi} \right) \rightarrow m$$

For Sound:

12.16

$$\bar{d} = \bar{c}_{Sound} \cdot \left(\frac{T_{Pump\_Cycle\_Period}}{2\pi} \right) \rightarrow m$$

Thermodynamic Anisotropy – Surplus Mass Flow Rate

12.17

$$f_{Mass\_Flow\_Aniso} = \gamma_{Inertial} \cdot f_{Mass\_Flow} \rightarrow {}^{kg}\!/_{s}$$

$$f_{Mass\_Flow\_Aniso} = \left(\frac{m_{Thermo\_Aniso\_Mass}}{T_{Pump\_Cycle\_Period}} \right) = \gamma_{Inertial} \cdot \left(\frac{m_{Net}}{T_{Pump\_Cycle\_Period}} \right)$$

$$f_{Mass\_Flow\_Aniso} = \frac{\bar{p}_{Thermo\_Momentum}}{|\bar{v}|_{CM} \cdot T_{Pump\_Cycle\_Period}} = \frac{\left(\frac{m_{Net}}{T_{Pump\_Cycle\_Period}} \right)}{\sqrt{1 - \frac{\gamma_{Heat}}{3} \cdot \left(\frac{|\bar{v}|_{CM}^{2}}{c_{Sound}^{2}} \right)}}$$

Thermodynamic Anisotropy – Surplus Mass Flow Rate – Light Speed

12.18

$$f_{Mass\_Flow\_Aniso} = \gamma_{Inertial} \cdot f_{Mass\_Flow} \rightarrow {kg}/{s}$$

$$f_{Mass\_Flow\_Aniso} = \frac{\overline{p}_{Thermo\_Momentum}}{2\pi \cdot \overline{d}} \cdot \left(\frac{\overline{c}_{Light}}{|\overline{v}|_{CM}} \right) = \frac{\left(\dfrac{m_{Net} \cdot \overline{c}_{Light}}{2\pi \cdot \overline{d}} \right)}{\sqrt{1 - \dfrac{\gamma_{Heat}}{3} \cdot \left(\dfrac{|\overline{v}|_{CM}^2}{c_{Sound}^2} \right)}}$$

Thermodynamic Anisotropy – Surplus Mass Flow Rate – Sound Speed

12.19

$$f_{Mass\_Flow\_Aniso} = \gamma_{Inertial} \cdot f_{Mass\_Flow} \rightarrow {kg}/{s}$$

$$f_{Mass\_Flow\_Aniso} = \frac{\overline{p}_{Thermo\_Momentum}}{2\pi \cdot \overline{d}} \cdot \left(\frac{\overline{c}_{Sound}}{|\overline{v}|_{CM}} \right) = \frac{\left(\dfrac{m_{Net} \cdot \overline{c}_{Sound}}{2\pi \cdot \overline{d}} \right)}{\sqrt{1 - \dfrac{\gamma_{Heat}}{3} \cdot \left(\dfrac{|\overline{v}|_{CM}^2}{c_{Sound}^2} \right)}}$$

Thermodynamic Anisotropy – Surplus Aether Dynamic Pressure

12.20

$$q_{Aether\_Dynamic\_Pressure} = \gamma_{Inertial} \cdot q_{Dynamic\_Pressure} \rightarrow {kg}/{m \cdot s^2}$$

$$q_{Aether\_Dynamic\_Pressure} = \frac{1}{3} \cdot \left(\frac{\bar{p}_{Thermo\_Momentum}}{V_{ol}} \right) \cdot |\bar{v}|_{CM}$$

$$q_{Aether\_Dynamic\_Pressure} = \frac{1}{3} \cdot \rho_{Thermo\_Aniso\_Density} \cdot |\bar{v}|_{CM}^2 = \frac{1}{3} \cdot \left(\frac{m_{Thermo\_Aniso\_Mass}}{V_{ol}} \right) \cdot |\bar{v}|_{CM}^2$$

$$q_{Aether\_Dynamic\_Pressure} = \frac{\frac{1}{3} \cdot \rho_{Net} \cdot |\bar{v}|_{CM}^2}{\sqrt{1 - \left(\frac{|\bar{v}|_{CM}^2}{|v^2|_{Iso}} \right)}} = \frac{\frac{1}{3} \cdot \left(\frac{m_{Net}}{V_{ol}} \right) \cdot |\bar{v}|_{CM}^2}{\sqrt{1 - \frac{\gamma_{Heat}}{3} \cdot \left(\frac{|\bar{v}|_{CM}^2}{c_{Sound}^2} \right)}}$$

$$q_{Aether\_Dynamic\_Pressure} = \frac{q_{Dynamic\_Pressure}}{\sqrt{1 - \frac{\gamma_{Heat}}{3} \cdot \left(\frac{|\bar{v}|_{CM}^2}{c_{Sound}^2} \right)}}$$

$$q_{Aether\_Dynamic\_Pressure} = \frac{\frac{\phi_{Loco\_Motion}}{\gamma_{Heat}} \cdot \left(\rho_{Net} \cdot c_{Sound}^2 \right)}{\sqrt{1 - \left(\frac{|\bar{v}|_{CM}^2}{|v^2|_{Iso}} \right)}} = \frac{\frac{\phi_{Loco\_Motion}}{\gamma_{Heat}} \cdot \left(\frac{m_{Net} \cdot c_{Sound}^2}{V_{ol}} \right)}{\sqrt{1 - \frac{\gamma_{Heat}}{3} \cdot \left(\frac{|\bar{v}|_{CM}^2}{c_{Sound}^2} \right)}}$$

Thermodynamic Anisotropy – Surplus Aether Sound Bulk Modulus Pressure

12.21

$$B_{\text{Aether\_Pressure\_Aniso}} = \gamma_{Inertial} \cdot B_{\text{Sound\_Bulk\_Pressure}} \rightarrow {kg}/{m \cdot s^2}$$

$$B_{\text{Aether\_Pressure\_Aniso}} = \rho_{\text{Thermo\_Aniso\_Density}} \cdot c^2_{\text{Sound}}$$

$$B_{\text{Aether\_Pressure\_Aniso}} = \left(\frac{m_{\text{Thermo\_Aniso\_Mass}}}{V_{\text{ol}}}\right) \cdot c^2_{\text{Sound}}$$

$$B_{\text{Aether\_Pressure\_Aniso}} = \gamma_{Inertial} \cdot \rho_{\text{Net}} \cdot c^2_{\text{Sound}} = \frac{\rho_{\text{Net}} \cdot c^2_{\text{Sound}}}{\sqrt{1 - \frac{\gamma_{Heat}}{3} \cdot \left(\frac{|\overline{v}|^2_{\text{CM}}}{c^2_{\text{Sound}}}\right)}}$$

$$B_{\text{Aether\_Pressure\_Aniso}} = \left(\frac{\overline{p}_{\text{Thermo\_Momentum}}}{V_{\text{ol}}}\right) \cdot \frac{c^2_{\text{Sound}}}{|\overline{v}|_{\text{CM}}} = \frac{\left(\frac{m_{\text{Net}}}{V_{\text{ol}}}\right) \cdot c^2_{\text{Sound}}}{\sqrt{1 - \left(\frac{|\overline{v}|^2_{\text{CM}}}{|v^2|_{\text{Iso}}}\right)}}$$

$$B_{\text{Aether\_Pressure\_Aniso}} = \frac{\gamma_{Heat}}{\phi_{\text{Loco\_Motion}}} \cdot P_{\text{Aether\_Dynamic\_Pressure}}$$

$$B_{\text{Aether\_Pressure\_Aniso}} = \frac{\gamma_{Heat}}{\phi_{\text{Loco\_Motion}}} \cdot \gamma_{Inertial} \cdot q_{\text{Dynamic\_Pressure}}$$

Thermodynamic Anisotropy – Surplus Aether Light Bulk Modulus Pressure

12.22

$$B_{\text{Aether\_Pressure\_Light\_Aniso}} = \gamma_{Inertial} \cdot B_{\text{Light\_Bulk\_Pressure}} \rightarrow {kg}/{m \cdot s^2}$$

$$B_{\text{Aether\_Pressure\_Light\_Aniso}} = \rho_{\text{Thermo\_Aniso\_Density}} \cdot c^2_{\text{Light}}$$

$$B_{\text{Aether\_Pressure\_Light\_Aniso}} = \left(\frac{m_{\text{Thermo\_Aniso\_Mass}}}{V_{\text{ol}}} \right) \cdot c^2_{\text{Light}}$$

$$B_{\text{Aether\_Pressure\_Light\_Aniso}} = \gamma_{Inertial} \cdot \rho_{\text{Net}} \cdot c^2_{\text{Light}} = \frac{\rho_{\text{Net}} \cdot c^2_{\text{Light}}}{\sqrt{1 - \frac{\gamma_{Heat}}{3} \cdot \left(\frac{|\overline{v}|^2_{\text{CM}}}{c^2_{\text{Sound}}} \right)}}$$

$$B_{\text{Aether\_Pressure\_Light\_Aniso}} = \left(\frac{\overline{p}_{\text{Thermo\_Momentum}}}{V_{\text{ol}}} \right) \cdot \frac{c^2_{\text{Light}}}{|\overline{v}|_{\text{CM}}} = \frac{\left(\frac{m_{\text{Net}}}{V_{\text{ol}}} \right) \cdot c^2_{\text{Light}}}{\sqrt{1 - \left(\frac{|\overline{v}|^2_{\text{CM}}}{|v^2|_{\text{Iso}}} \right)}}$$

Thermodynamic Anisotropy – Surplus Aether Light Energy Content

12.23

$$\left(B_{\text{Aether\_Pressure\_Light\_Aniso}} \cdot \text{V}_{\text{ol}}\right) = \gamma_{Inertial} \cdot \left(B_{\text{Light\_Bulk\_Pressure}} \cdot \text{V}_{\text{ol}}\right) \rightarrow kg \cdot m^2 / s^2$$

$$\left(B_{\text{Aether\_Pressure\_Light\_Aniso}} \cdot \text{V}_{\text{ol}}\right) = \rho_{\text{Thermo\_Aniso\_Density}} \cdot \text{c}^2_{\text{Light}} \cdot \text{V}_{\text{ol}}$$

$$\left(B_{\text{Aether\_Pressure\_Light\_Aniso}} \cdot \text{V}_{\text{ol}}\right) = \text{m}_{\text{Thermo\_Aniso\_Mass}} \cdot \text{c}^2_{\text{Light}}$$

$$\left(B_{\text{Aether\_Pressure\_Light\_Aniso}} \cdot \text{V}_{\text{ol}}\right) = \gamma_{Inertial} \cdot \left(\text{m}_{\text{Net}} \cdot \text{c}^2_{\text{Light}}\right) = \frac{\text{m}_{\text{Net}} \cdot \text{c}^2_{\text{Light}}}{\sqrt{1 - \frac{\gamma_{Heat}}{3} \cdot \left(\frac{\left|\overline{\text{v}}\right|^2_{\text{CM}}}{\text{c}^2_{\text{Sound}}}\right)}}$$

$$\left(B_{\text{Aether\_Pressure\_Light\_Aniso}} \cdot \text{V}_{\text{ol}}\right) = \frac{\overline{\text{p}}_{\text{Thermo\_Momentum}}}{\left(\frac{\left|\overline{\text{v}}\right|_{\text{CM}}}{\text{c}^2_{\text{Light}}}\right)} = \frac{\text{m}_{\text{Net}} \cdot \text{c}^2_{\text{Light}}}{\sqrt{1 - \frac{\gamma_{Heat}}{3} \cdot \left(\frac{\left|\overline{\text{v}}\right|^2_{\text{CM}}}{\text{c}^2_{\text{Sound}}}\right)}}$$

Thermodynamic Anisotropy – Surplus Aether Sound Energy Content

12.24

$$\left(B_{\text{Aether\_Pressure\_Aniso}} \cdot \mathrm{V_{ol}}\right) = \gamma_{Inertial} \cdot \left(B_{\text{Sound\_Bulk\_Pressure}} \cdot \mathrm{V_{ol}}\right) \rightarrow kg \cdot m^2 \big/ s^2$$

$$\left(B_{\text{Aether\_Pressure\_Aniso}} \cdot \mathrm{V_{ol}}\right) = \rho_{\text{Thermo\_Aniso\_Density}} \cdot \mathrm{c}^2_{\text{Sound}} \cdot \mathrm{V_{ol}}$$

$$\left(B_{\text{Aether\_Pressure\_Aniso}} \cdot \mathrm{V_{ol}}\right) = \mathrm{m}_{\text{Thermo\_Aniso\_Mass}} \cdot \mathrm{c}^2_{\text{Sound}}$$

$$\left(B_{\text{Aether\_Pressure\_Aniso}} \cdot \mathrm{V_{ol}}\right) = \gamma_{Inertial} \cdot \left(\mathrm{m}_{\text{Net}} \cdot \mathrm{c}^2_{\text{Sound}}\right) = \frac{\mathrm{m}_{\text{Net}} \cdot \mathrm{c}^2_{\text{Sound}}}{\sqrt{1 - \left(\dfrac{\left|\overline{\mathrm{v}}\right|^2_{\text{CM}}}{\left|\mathrm{v}^2\right|_{\text{Iso}}}\right)}}$$

$$\left(B_{\text{Aether\_Pressure\_Aniso}} \cdot \mathrm{V_{ol}}\right) = \frac{\overline{\mathrm{p}}_{\text{Thermo\_Momentum}}}{\left(\dfrac{\left|\overline{\mathrm{v}}\right|_{\text{CM}}}{\mathrm{c}^2_{\text{Light}}}\right)} = \frac{\mathrm{m}_{\text{Net}} \cdot \mathrm{c}^2_{\text{Sound}}}{\sqrt{1 - \dfrac{\gamma_{Heat}}{3} \cdot \left(\dfrac{\left|\overline{\mathrm{v}}\right|^2_{\text{CM}}}{\mathrm{c}^2_{\text{Sound}}}\right)}}$$

Thermodynamic Anisotropy – Surplus Aether Gravitational Evolutionary In/Out Flow Attraction Rate

12.25

$$K_{Gravity\_Aether\_Aniso} = \gamma_{Inertial} \cdot K_{Gravity} \rightarrow {kg \cdot m^3}/{s^2}$$

$$K_{Gravity\_Aether\_Aniso} = \left(m_{Thermo\_Aniso\_Mass} \cdot G\right) = \gamma_{Inertial} \cdot \left(m_{Net} \cdot G\right)$$

$$K_{Gravity\_Aether\_Aniso} = \left(m_{Thermo\_Aniso\_Mass} \cdot G\right)$$

$$K_{Gravity\_Aether\_Aniso} = \left(\frac{\overline{p}_{Thermo\_Momentum}}{\left|\overline{v}\right|_{CM}}\right) \cdot G = \frac{K_{Gravity}}{\sqrt{1 - \left(\frac{\left|\overline{v}\right|^2_{CM}}{\left|v^2\right|_{Iso}}\right)}}$$

$$K_{Gravity\_Aether\_Aniso} = \left(\frac{\overline{p}_{Thermo\_Momentum}}{\left|\overline{v}\right|_{CM}}\right) \cdot G = \frac{\left(m_{Net} \cdot G\right)}{\sqrt{1 - \frac{\gamma_{Heat}}{3} \cdot \left(\frac{\left|\overline{v}\right|^2_{CM}}{c^2_{Sound}}\right)}}$$

$$K_{Gravity\_Aether\_Aniso} = \frac{\left(\frac{\overline{F}_{Newton\_Gravitation} \cdot d^2}{m_{Test}}\right)}{\sqrt{1 - \left(\frac{\left|\overline{v}\right|^2_{CM}}{\left|v^2\right|_{Iso}}\right)}} = \frac{\left(\frac{\overline{F}_{Self\_Gravitation} \cdot d^2}{m_{Net}}\right)}{\sqrt{1 - \frac{\gamma_{Heat}}{3} \cdot \left(\frac{\left|\overline{v}\right|^2_{CM}}{c^2_{Sound}}\right)}}$$

Chapter 13

The Special Theory of Thermodynamics

(Carnot Thermal Isotropy Engine)

13.1 Carnot Thermal Engine – External (Isotropy) Inertial Motion Compressibility Factor

In general a mass system body immersed in any environment, including the vacuum of spacetime behaves like a Carnot Thermal Anisotropy Engine

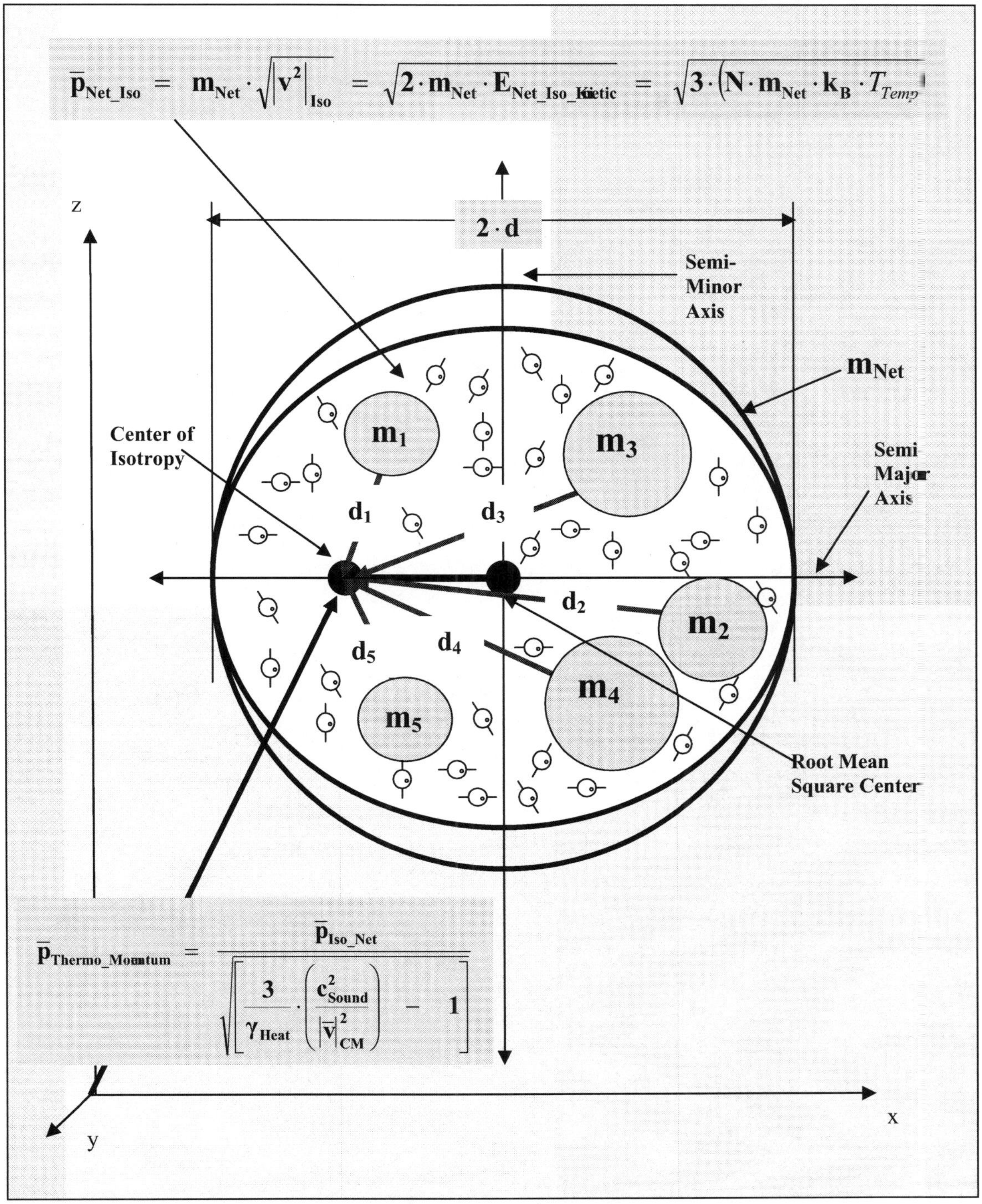

Figure 13.1: Internal Isotropy Omni-directional Center of Isotropy Squared Velocity Inertia

Using the **Carnot Thermal Engine "Isotropy" Efficiency Factor** the Inertial Anisotropic Rectilinear Momentum, Internal Isotropic Omni-directional Momentum, and Spacetime Compressibility Ratios can be derived.

13.1

$$\eta_{\text{Carnot\_Iso}} = \frac{1}{\gamma^2_{Inertial\_Iso}} = \left[\frac{W_{\text{Free\_Work\_Energy}}}{E_{\text{Net\_Translational}}}\right]$$

$$\eta_{\text{Carnot\_Iso}} = \frac{1}{\gamma^2_{Inertial\_Iso}} = \left[\frac{E_{\text{Net\_Iso\_Kinetic}} - E_{\text{Net\_Translational}}}{E_{\text{Net\_Translational}}}\right] = \frac{\left|v^2\right|_{\text{Iso}} - \left|\overline{v}\right|^2_{\text{CM}}}{\left|\overline{v}\right|^2_{\text{CM}}}$$

$$\eta_{\text{Carnot\_Iso}} = \frac{1}{\gamma^2_{Inertial\_Iso}} = \left(\frac{E_{\text{Net\_Iso\_Kinetic}}}{E_{\text{Net\_Translational}}}\right)\cdot\left(1 - \frac{E_{\text{Net\_Translational}}}{E_{\text{Net\_Iso\_Kinetic}}}\right)$$

$$\eta_{\text{Carnot\_Iso}} = \frac{1}{\gamma^2_{Inertial\_Iso}} = \left(\frac{\dfrac{\overline{p}^2_{\text{Net\_Iso}}}{2\cdot m_{\text{Net}}}}{\dfrac{\left(\overline{p}_{\text{Inertial\_Net}}\right)^2}{2\cdot m_{\text{Net}}}}\right)\cdot\left(1 - \frac{E_{\text{Net\_Translational}}}{E_{\text{Net\_Iso\_Kinetic}}}\right)$$

$$\eta_{\text{Carnot\_Iso}} = \frac{1}{\gamma^2_{Inertial\_Iso}} = \left(\frac{\overline{p}^2_{\text{Net\_Iso}}}{\overline{p}^2_{\text{Inertial\_Net}}}\right)\cdot\left(1 - \frac{E_{\text{Net\_Translational}}}{E_{\text{Net\_Iso\_Kinetic}}}\right)$$

Taking the square root of both sides of the above equation yields

$$\frac{1}{\gamma_{Inertial\_Iso}\cdot\sqrt{\left(1 - \dfrac{E_{\text{Net\_Translational}}}{E_{\text{Net\_Iso\_Kinetic}}}\right)}} = \left(\frac{\overline{p}_{\text{Net\_Iso}}}{\overline{p}_{\text{Inertial\_Net}}}\right)$$

13.2

$$\phi^2_{\text{Re}\,spiration} = \left(\frac{\gamma_{Inertial}}{\gamma_{Inertial\_Iso}}\right) = \left(\frac{\overline{p}_{\text{Net\_Iso}}}{\overline{p}_{\text{Inertial\_Net}}}\right) \rightarrow \textit{Unitless}$$

Since a mass system body and its medium environment behave like a heat engine, the isotropy to anisotropic energy transfer between two different parts of an isolated system of differing temperatures and kinetic energies is also limited or governed by Carnot's theorems.

The increase or decrease in the Thermodynamic Momentum ($\overline{p}_{\text{Thermo\_Momentum}} = \gamma_{Inertial\_Iso} \cdot \overline{p}_{\text{Net\_Iso}}$) of an isolated system mass body is also the measure of the resistance, condensing and accumulation of surplus **Internal Isotropy**, inertia mass, density, and internal kinetic energy, which is a result the **Inertial Locomotion Foreshortening Factor** increases or decreases ($\phi_{\text{Loco\_Motion}} = \left(\frac{|\overline{v}|^2_{\text{CM}}}{|v^2|_{\text{Iso}}} \right)$), and likewise as the **Mach number** increases or decreases ($M_{Mach} = \left(\frac{|\overline{v}|_{\text{CM}}}{c_{\text{Sound}}} \right)$).

The **Internal (Isotropy) Inertial Motion Compressibility Factor** ($\gamma_{Inertial\_Iso}$) is a measure of the ratio of the amount of increase or decrease in the Thermodynamic Momentum ($\overline{p}_{\text{Thermo\_Momentum}}$) relative to the Isotropy and Center of Isotropy Momentum ($\overline{p}_{\text{Net\_Iso}}$) of an isolated system body.

Aphorism 13.1: Relative to an External Observer, the Root Mean Square Center, Center of Isotropy, and an atomic substance medium, the **Internal (Isotropy) Inertial Motion Compressibility Factor (**$\gamma_{Inertial\_Iso} = \frac{1}{\sqrt{\left(\frac{|v^2|_{\text{Iso}}}{|\overline{v}|^2_{\text{CM}}} - 1 \right)}}$**)** of an Isolated System Mass Body is the accumulation of surplus **Internal Isotropy** energy of motion and increases as the **Inertial Locomotion Foreshortening Factor** increases or decreases ($\phi_{\text{Loco\_Motion}} = \left(\frac{|\overline{v}|^2_{\text{CM}}}{|v^2|_{\text{Iso}}} \right) = \frac{\gamma_{Heat}}{3} \cdot \left(\frac{|\overline{v}|^2_{\text{CM}}}{c^2_{\text{Sound}}} \right)$), and likewise as the Mach number increases or decreases ($M_{Mach} = \left(\frac{|\overline{v}|_{\text{CM}}}{c_{\text{Sound}}} \right)$).

Aphorism 13.2: The **Internal (Isotropy) Inertial Motion Compressibility Factor (**$\gamma_{Inertial\_Iso}$**)** of an isolated dynamic system mass body and its environment is proportional to the inverse square root of the Isotropic Carnot Pump Efficiency factor which is a measure of the **Coefficient of Cooling** of the "Environment" of the system.

$$\gamma_{Inertial\_Iso} = \frac{1}{\sqrt{\eta_{\text{Carnot\_Iso}}}}$$

For an isolated system mass body functioning as a Carnot refrigerator, the "effectiveness" is the ratio of the energy removed from the low-temperature reservoir divided by the work required to force the machine around its cycle.

The effectiveness of a refrigerator cooling the external environment of a isolated dynamic system mass body is called the Coefficient of Cooling performance factor ($COP_{Cooling}$).

While the Coefficient of Cooling Performance ($COP_{Cooling}$) is partly a measure of the efficiency of a refrigerator, it is also a measure of the "external" conditions under which it is operating; and is equal to the **Square of the Internal (Isotropy) Inertial Motion Compressibility Factor.**

13.3

$$\gamma^2_{Inertial\_Iso} = \frac{1}{\eta_{\text{Carnot\_Iso}}} = \left[\frac{E_{\text{Net\_Translational}}}{W_{\text{Work\_Free}}}\right] = \left[\frac{E_{\text{Net\_Translational}}}{E_{\text{Net\_Iso\_Kinetic}} - E_{\text{Net\_Translational}}}\right]$$

$$\gamma^2_{Inertial\_Iso} = \left(\frac{\bar{p}_{\text{Thermo\_Momentum}}}{\bar{p}_{\text{Net\_Iso}}}\right)^2 = \left[\frac{Low\ \text{Temperature}}{High\ \text{Temperature} - Low\ \text{Temperature}}\right]$$

Internal (Isotropy) Inertial Motion Compressibility Factor

13.4

$$\gamma_{Inertial\_Iso} = \left(\frac{\overline{p}_{Thermo\_Momentum}}{\overline{p}_{Net\_Iso}}\right) = \sqrt{\frac{Low\ \text{Temperature}}{High\ \text{Temperature} - Low\ \text{Temperature}}}$$

$$\gamma_{Inertial\_Iso} = \frac{1}{\sqrt{\eta_{Carnot\_Iso}}} = \sqrt{\left[\frac{E_{Net\_Translational}}{E_{Net\_Iso\_Kinetic} - E_{Net\_Translational}}\right]}$$

$$\gamma_{Inertial\_Iso} = \sqrt{\frac{\left|\overline{v}\right|_{CM}^{2}}{\left|v^2\right|_{Iso} - \left|\overline{v}\right|_{CM}^{2}}} = \sqrt{\frac{T_{emp\_Ext}}{T_{emp} - T_{emp\_Ext}}}$$

Definition 13.1: Relative to an External Observer, the Root Mean Square Center, Center of Isotropy, and an atomic substance medium of an isolated system mass body, the **Internal (Isotropy) Inertial Motion Compressibility Factor** ($\gamma_{Inertial\_Iso} = \left(\frac{\overline{p}_{Thermo\_Momentum}}{\overline{p}_{Net\_Iso}}\right)$) is the measure of the condensing and accumulation of surplus Internal Isotropy inertia mass, density, and kinetic energy, which is a result the **Inertial Locomotion Foreshortening Factor** increases or decreases ($\phi_{Loco\_Motion} = \left(\frac{\left|\overline{v}\right|_{CM}^{2}}{\left|v^2\right|_{Iso}}\right)$), and likewise as the **Mach number** increases or decreases ($M_{Mach} = \left(\frac{\left|\overline{v}\right._{CM}}{c_{Sound}}\right)$); and is defined as the ratio of the **Thermodynamic Momentum** ($\overline{p}_{Thermo\_Momentum}$) divided by the **Internal Isotropic Inertial/Aether Omni-directional Center of Isotropy Net Momentum** ($\overline{p}_{Net\_Iso} = m_{Net} \cdot \sqrt{\left|v^2\right|_{Iso}}$) of an isolated system mass body.

Definition 13.2: Relative to an External Observer, the Root Mean Square Center, Center of Isotropy, and an atomic substance medium of an isolated system mass body, the **Internal (Isotropy) Inertial Motion Compressibility Factor** ($\gamma_{Inertial\_Iso} = \left(\frac{\overline{p}_{Thermo\_Momentum}}{\overline{p}_{Net\_Iso}}\right)$) is the measure of the amount of Coefficient of Cooling between the isotropy and the heat transfer mechanism of the system; and is defined the inverse square root of the **Carnot Thermal Heat Pump Factor** ($\frac{1}{\sqrt{\eta_{Carnot\_Iso}}} = \sqrt{\left[\frac{E_{Net\_Translational}}{E_{Net\_Iso\_Kinetic} - E_{Net\_Translational}}\right]}$) of an isolated system mass body.

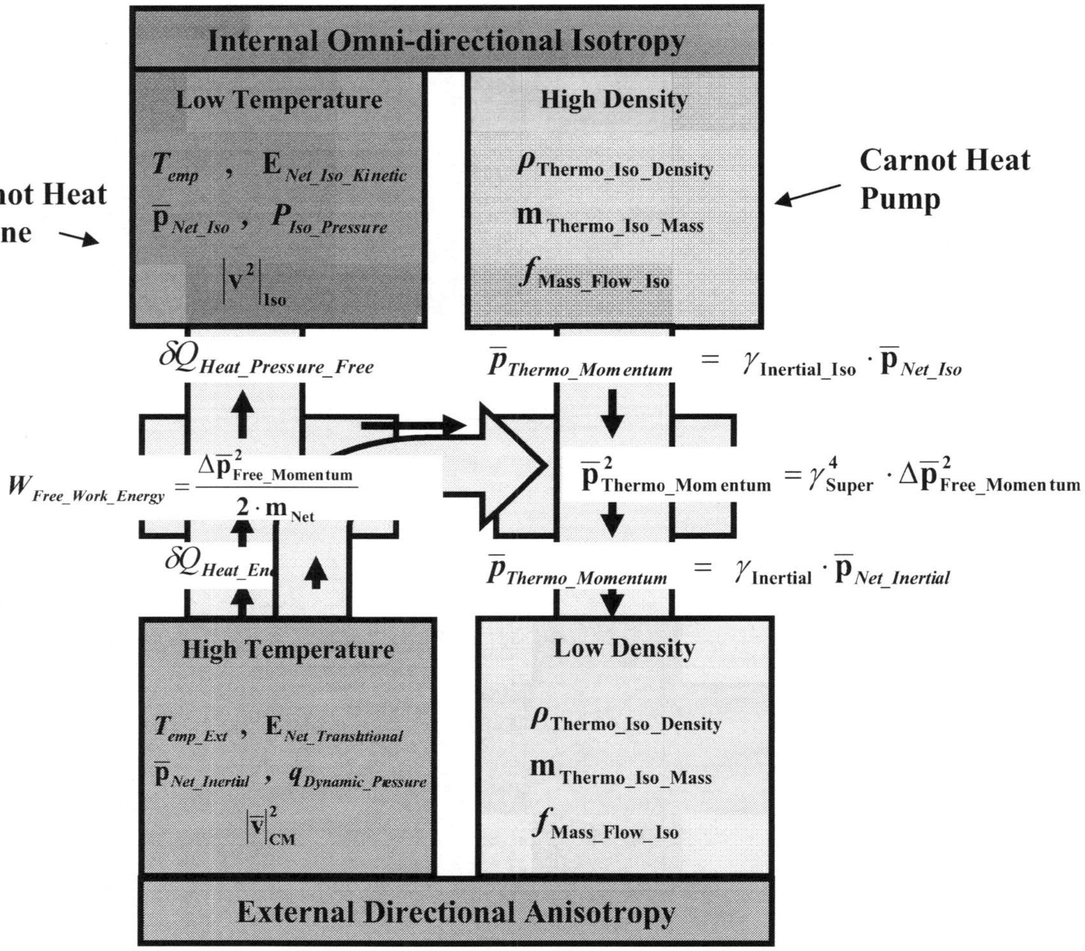

$$\phi_{Respiration} = \left(\phi_{Loco\_Motion}\right)^{\frac{1}{4}} \geq 1$$

Figure 13.2: Heat Energy Transfer Device – Kemp Thermal Engine.

Internal (Isotropy) Inertial Motion Compressibility Factor

13.5

$$\gamma_{Inertial\_Iso} = \left(\frac{\overline{p}_{\text{Thermo\_Momentum}}}{\overline{p}_{\text{Net\_Iso}}} \right) \rightarrow \textit{Unitless}$$

$$\gamma_{Inertial\_Iso} = \gamma_{Inertial} \cdot \left(\frac{\overline{p}_{\text{Inertial\_Net}}}{\overline{p}_{\text{Net\_Iso}}} \right) = \gamma_{Inertial} \cdot \left(\frac{|\overline{v}|_{\text{CM}}}{\sqrt{|v^2|_{\text{Iso}}}} \right)$$

$$\gamma_{Inertial\_Iso} = \frac{1}{\sqrt{\left(\frac{E_{\text{Net\_Iso\_Kinetic}}}{E_{\text{Net\_Translational}}} - 1 \right)}} = \frac{1}{\sqrt{\left(\frac{|v^2|_{\text{Iso}}}{|\overline{v}|^2_{\text{CM}}} - 1 \right)}}$$

$$\gamma_{Inertial\_Iso} = \frac{\gamma^2_{Super\_Compres}}{\gamma_{Inertial}} = \frac{1}{\gamma_{Inertial}} \cdot \left(\frac{\overline{p}^2_{\text{Thermo\_Momentum}}}{\overline{p}_{\text{Net\_Iso}} \cdot \overline{p}_{\text{Inertial\_Net}}} \right)$$

$$\gamma_{Inertial\_Iso} = \frac{1}{\sqrt{\eta_{\text{Carnot\_Iso}}}} = \frac{1}{\sqrt{\left(\frac{3}{\gamma_{Heat}} \cdot \left(\frac{c^2_{\text{Sound}}}{|\overline{v}|^2_{\text{CM}}} \right) - 1 \right)}}$$

$$\gamma_{Inertial\_Iso} = \frac{1}{\sqrt{\left(\frac{1}{\phi_{\text{Loco\_Motion}}} - 1 \right)}} = \frac{1}{\sqrt{\left(\frac{3}{\gamma_{Heat}} \cdot \left(\frac{1}{M_{Mach}} \right) - 1 \right)}}$$

$$\gamma_{Inertial\_Iso} = \sqrt{\gamma^2_{Inertial} - 1}$$

The Internal (Isotropy) Inertial Motion Compressibility Factor of an isolated system mass body can also be expressed as the absolute value; which is able to handle the negative values as the anisotropy becomes larger than the isotropy of the system.

$$\frac{1}{\gamma^2_{Inertial\_Iso}} = \left\| \left[\frac{E_{Net\_Iso\_Kinetic} - E_{Net\_Translational}}{E_{Net\_Translational}} \right] \right\| = \left\| \frac{\left|v^2\right|_{Iso} - \left|\overline{v}\right|^2_{CM}}{\left|\overline{v}\right|^2_{CM}} \right\|$$

Internal (Isotropy) Inertial Motion Compressibility Factor

13.6

$$\gamma_{Inertial\_Iso} = \frac{1}{\sqrt{\left\| \frac{\left|v^2\right|_{Iso}}{\left|\overline{v}\right|^2_{CM}} - 1 \right\|}} \rightarrow \textit{Unitless}$$

$$\gamma_{Inertial\_Iso} = \frac{1}{\sqrt{\left\| \frac{3}{\gamma_{Heat}} \cdot \left(\frac{c^2_{Sound}}{\left|\overline{v}\right|^2_{CM}} \right) - 1 \right\|}}$$

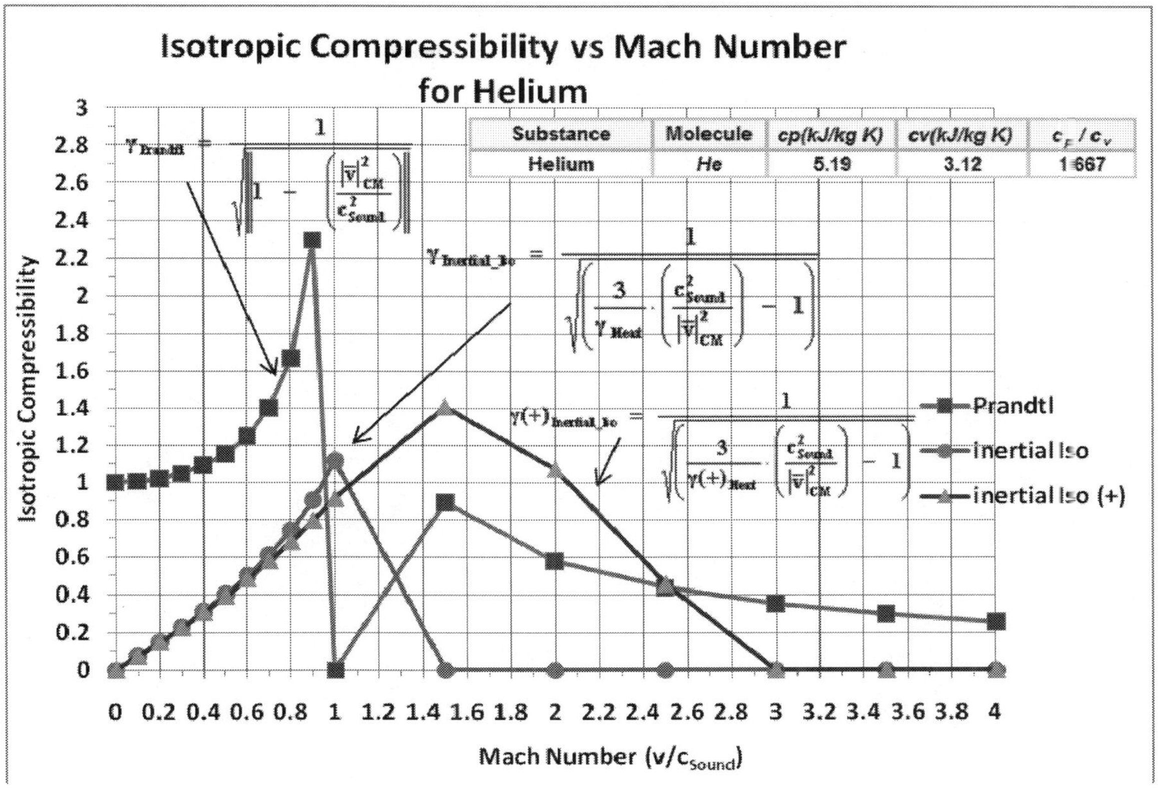

Figure 13.3: Isotropic Compressibility vs. Mach number for - Helium.

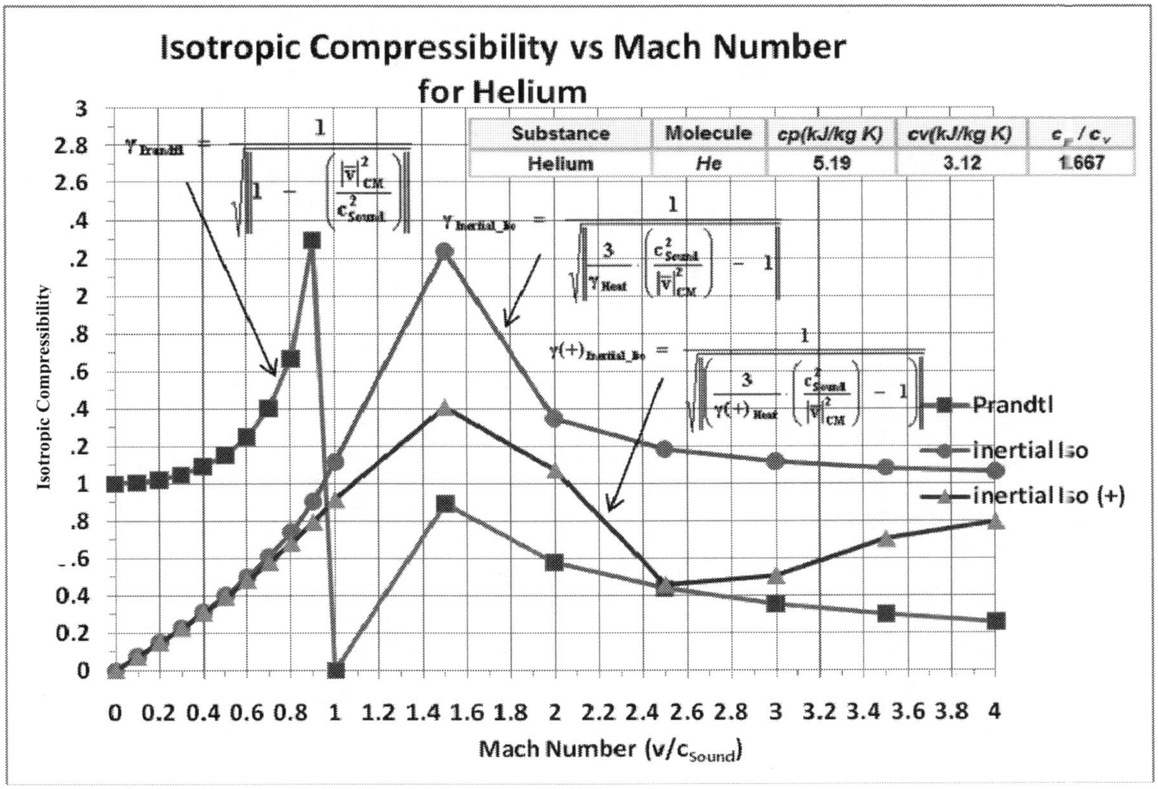

Figure 13.4: Isotropic (Absolute) Compressibility vs. Mach number for - Helium.

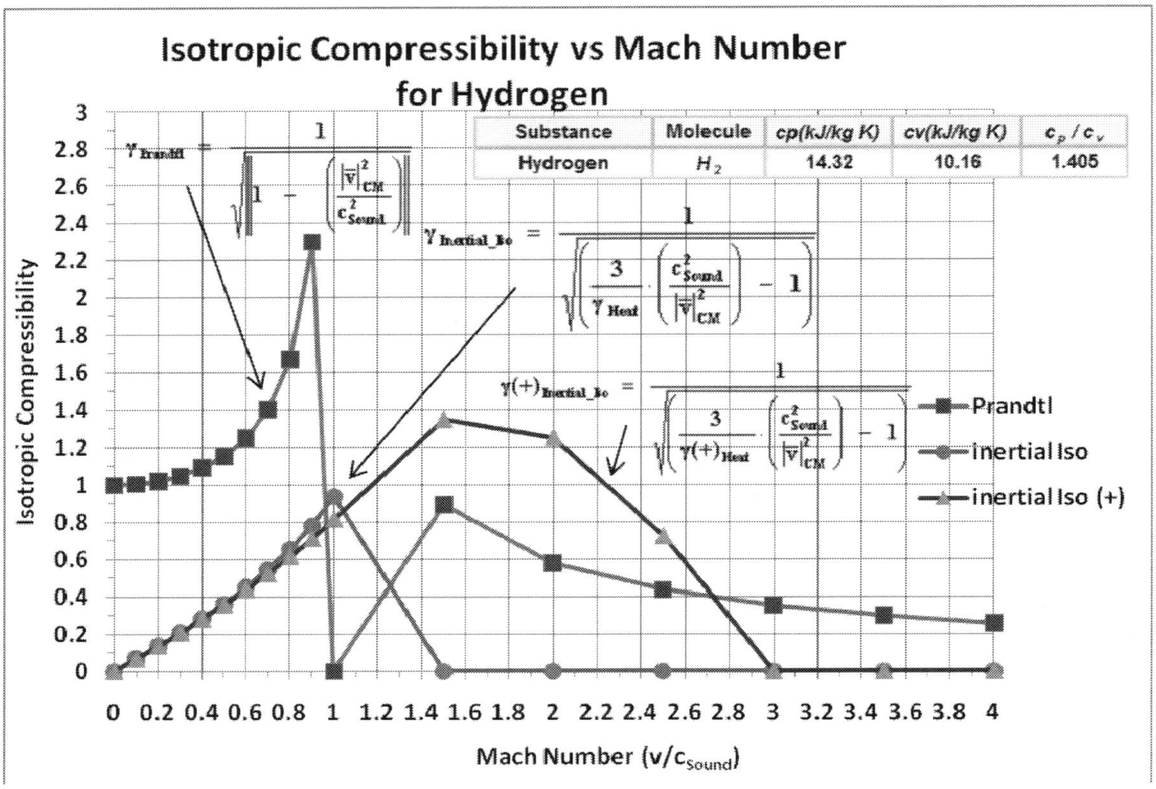

Figure 13.5: Isotropic Compressibility vs. Mach number for - Hydrogen.

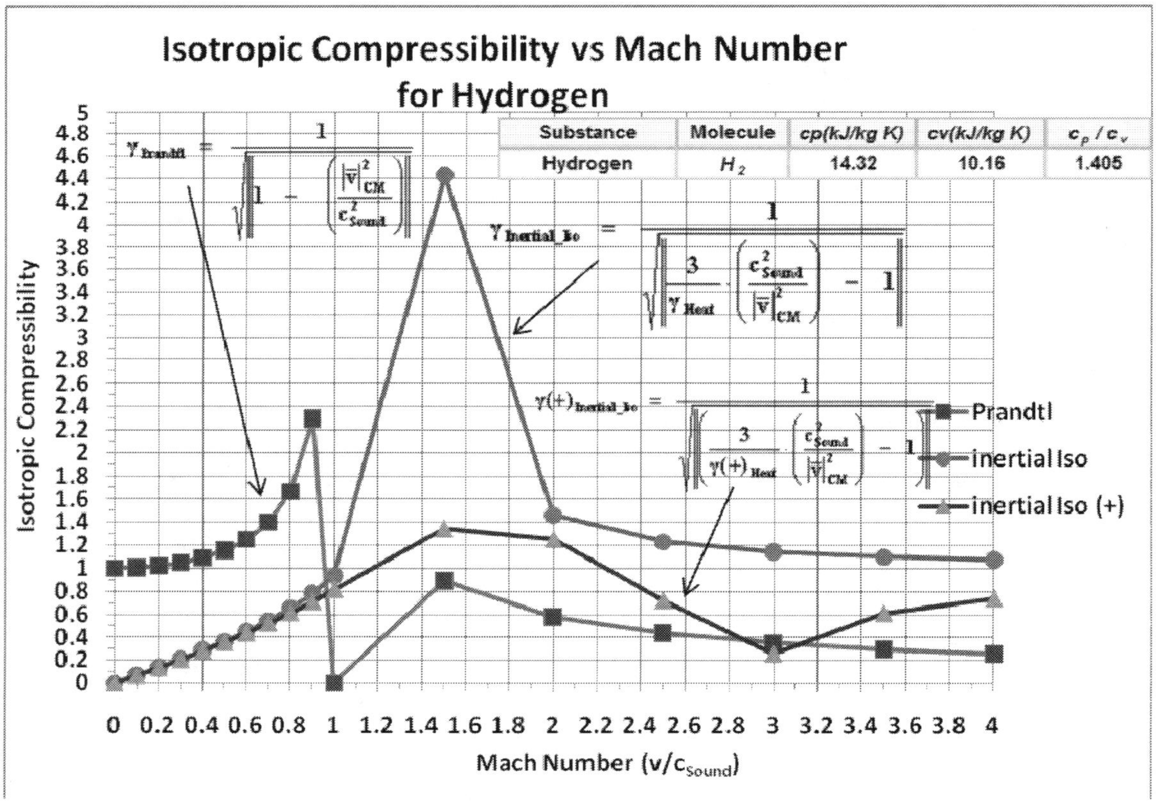

Figure 13.6: Isotropic (Absolute) Compressibility vs. Mach number for - Hydrogen.

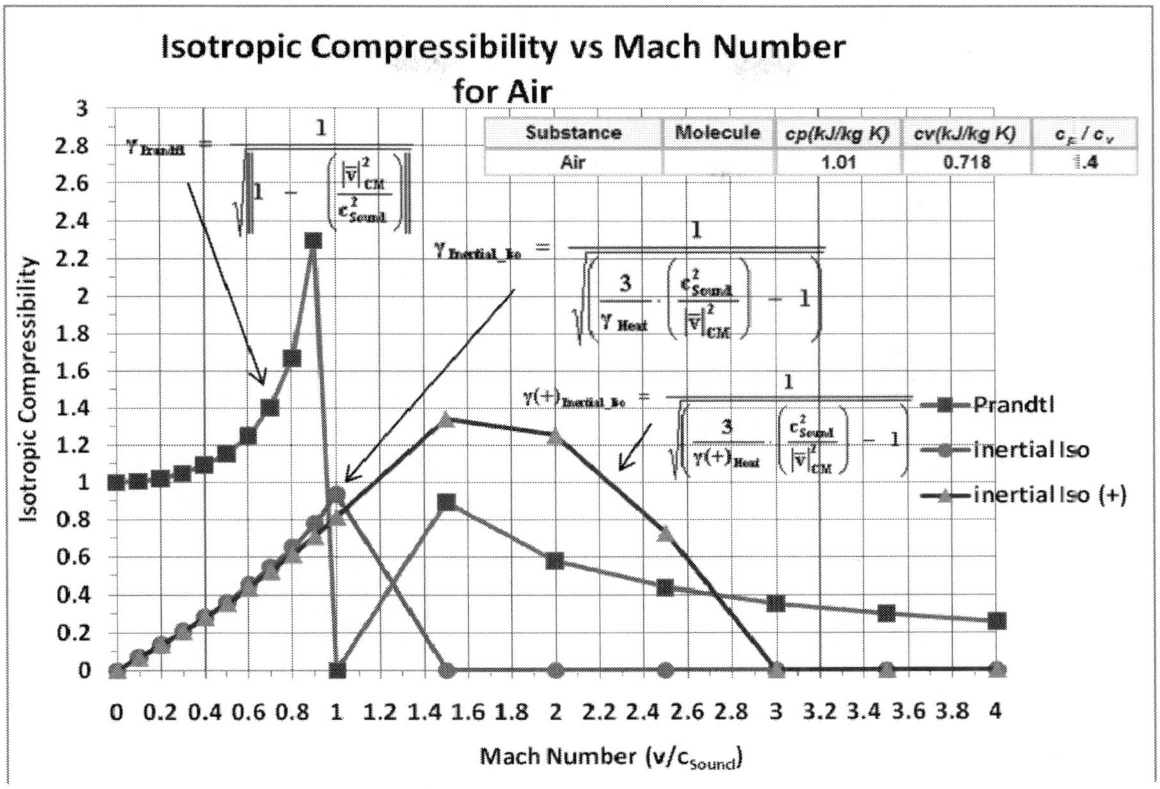

Figure 13.7: Isotropic Compressibility vs. Mach number for - Air.

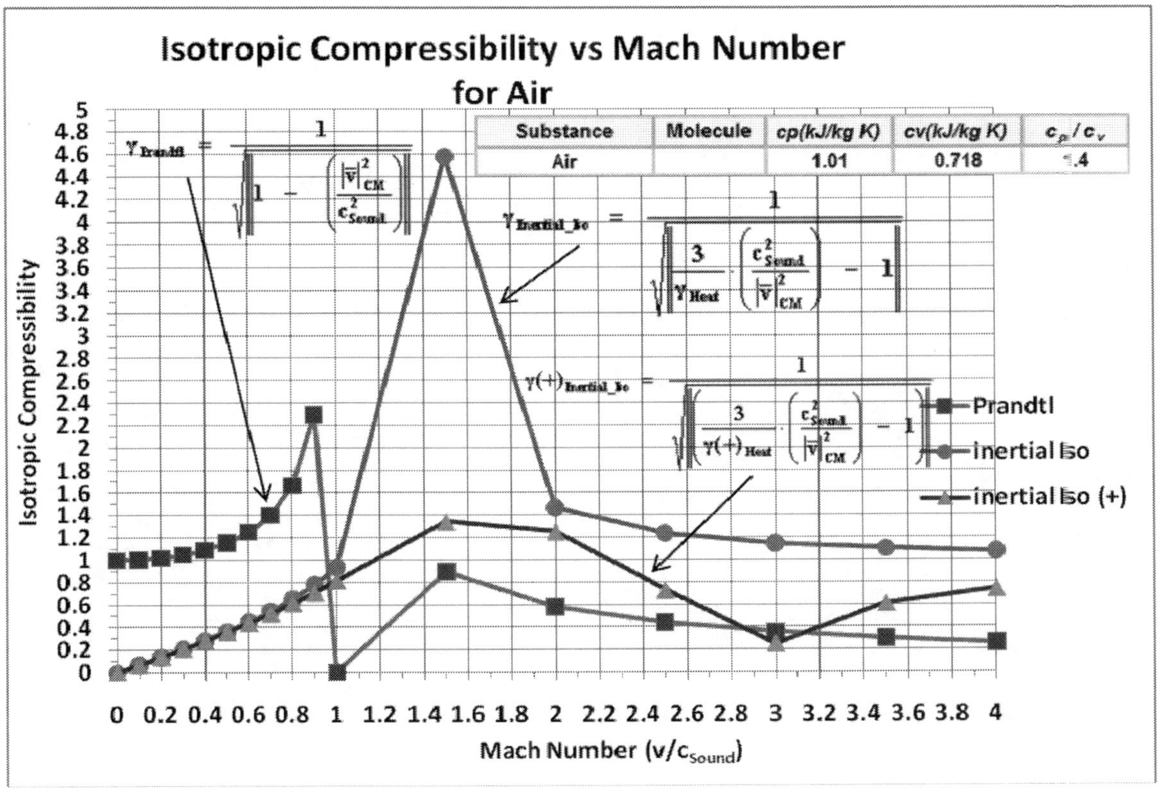

Figure 13.8: Isotropic (Absolute) Compressibility vs. Mach number for - Air.

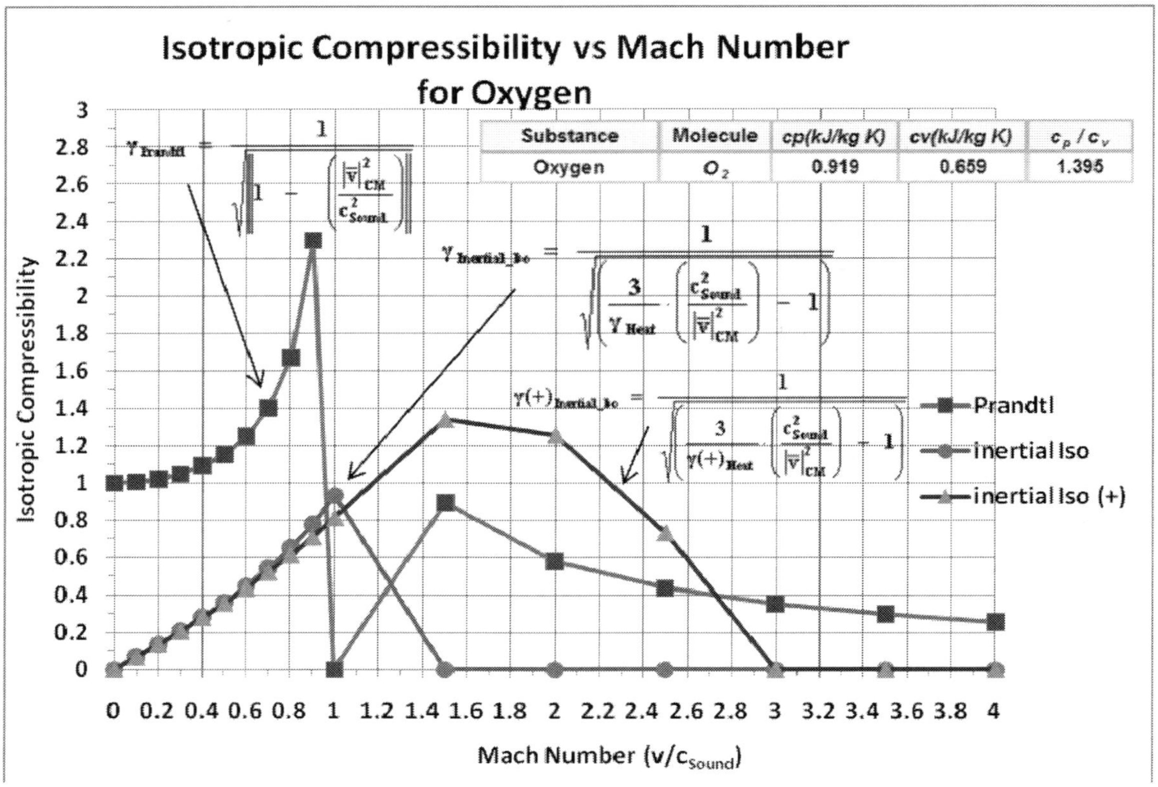

Figure 13.9: Isotropic (Absolute) Compressibility vs. Mach number for - Oxygen.

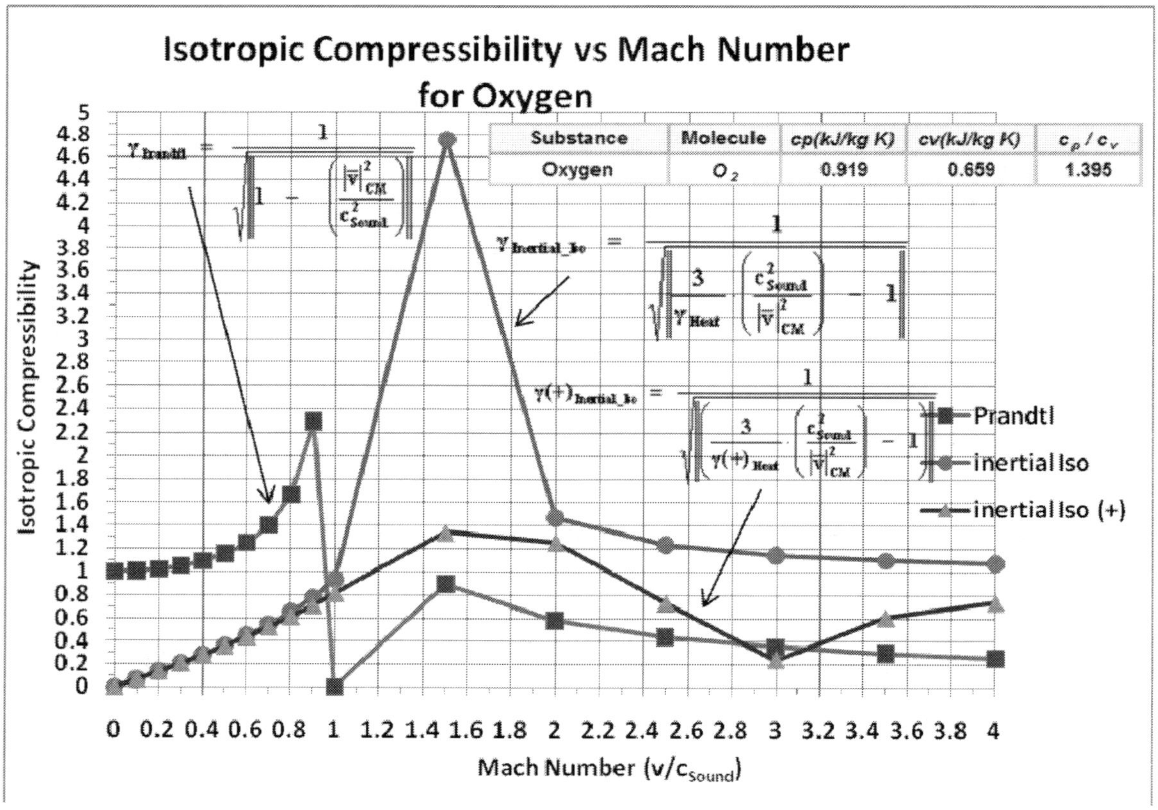

Figure 13.10: Isotropic (Absolute) Compressibility vs. Mach number for - Oxygen.

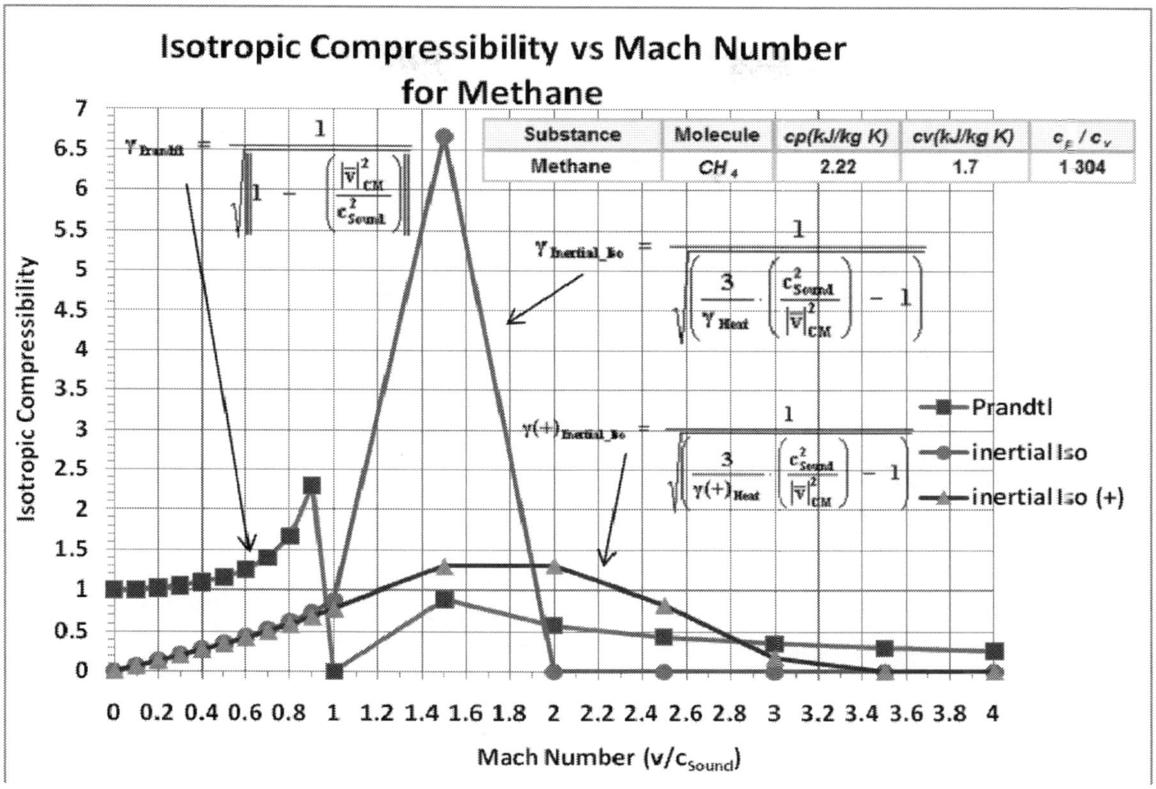

Figure 13.11: Isotropic Compressibility vs. Mach number for - Methane.

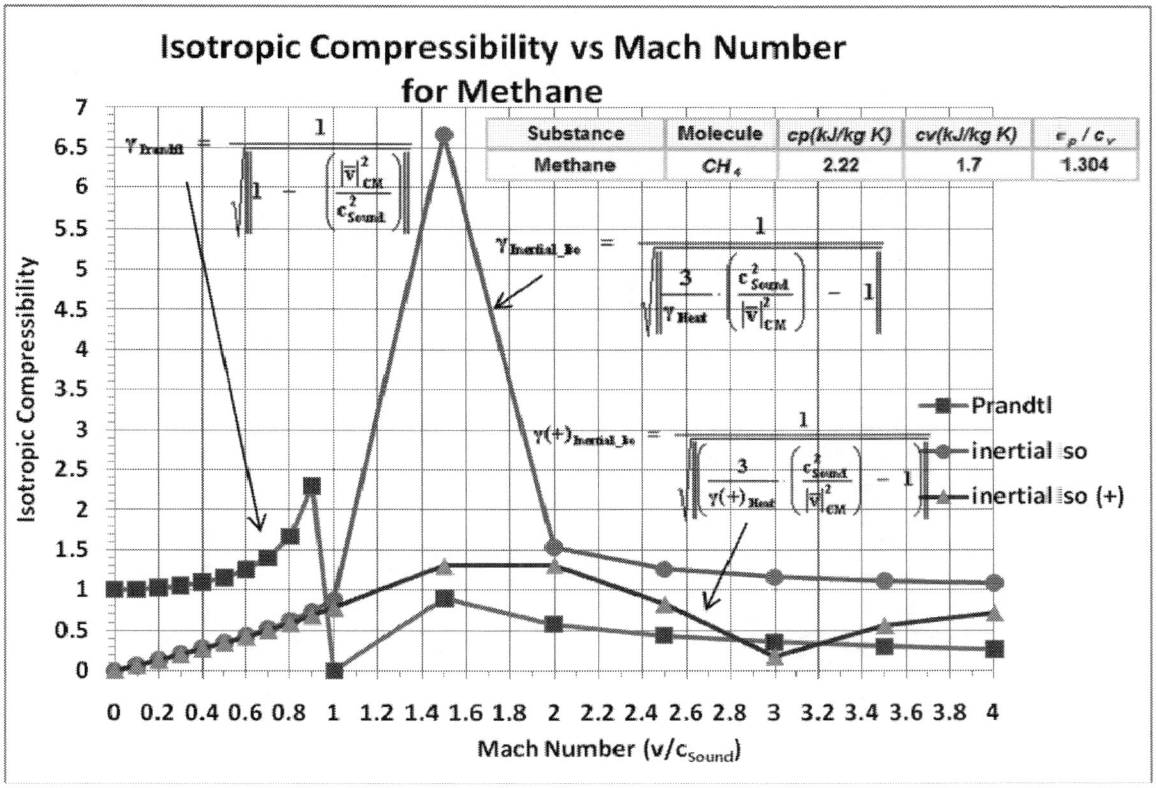

Figure 13.12: Isotropic (Absolute) Compressibility vs. Mach number for - Methane

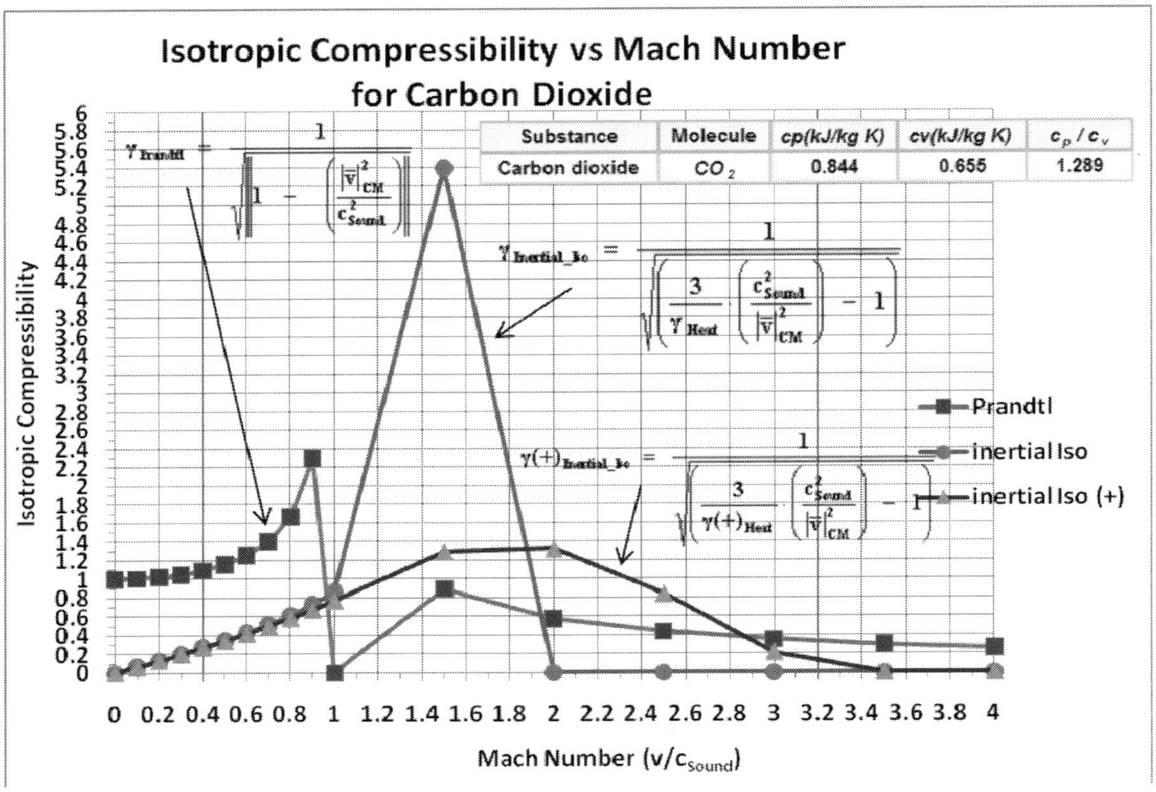

Figure 13.13: Isotropic Compressibility vs. Mach number for – Carbon Dioxide.

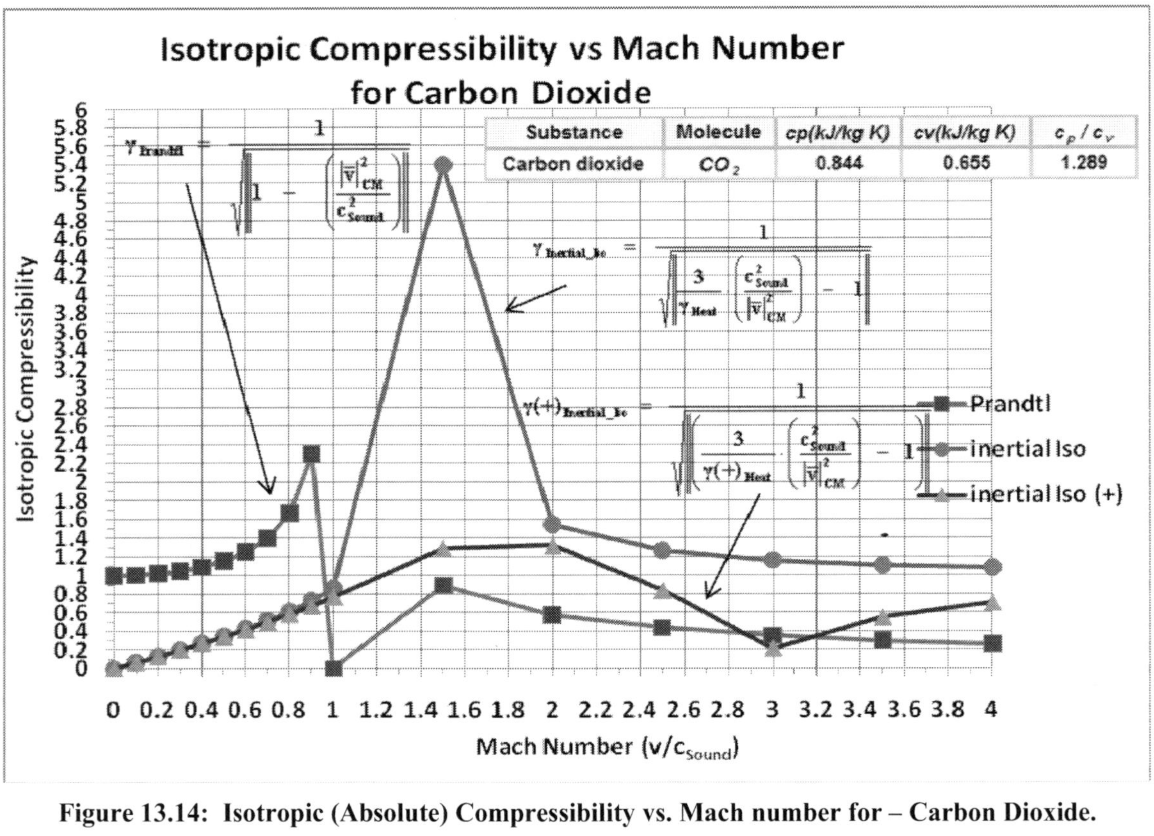

Figure 13.14: Isotropic (Absolute) Compressibility vs. Mach number for – Carbon Dioxide.

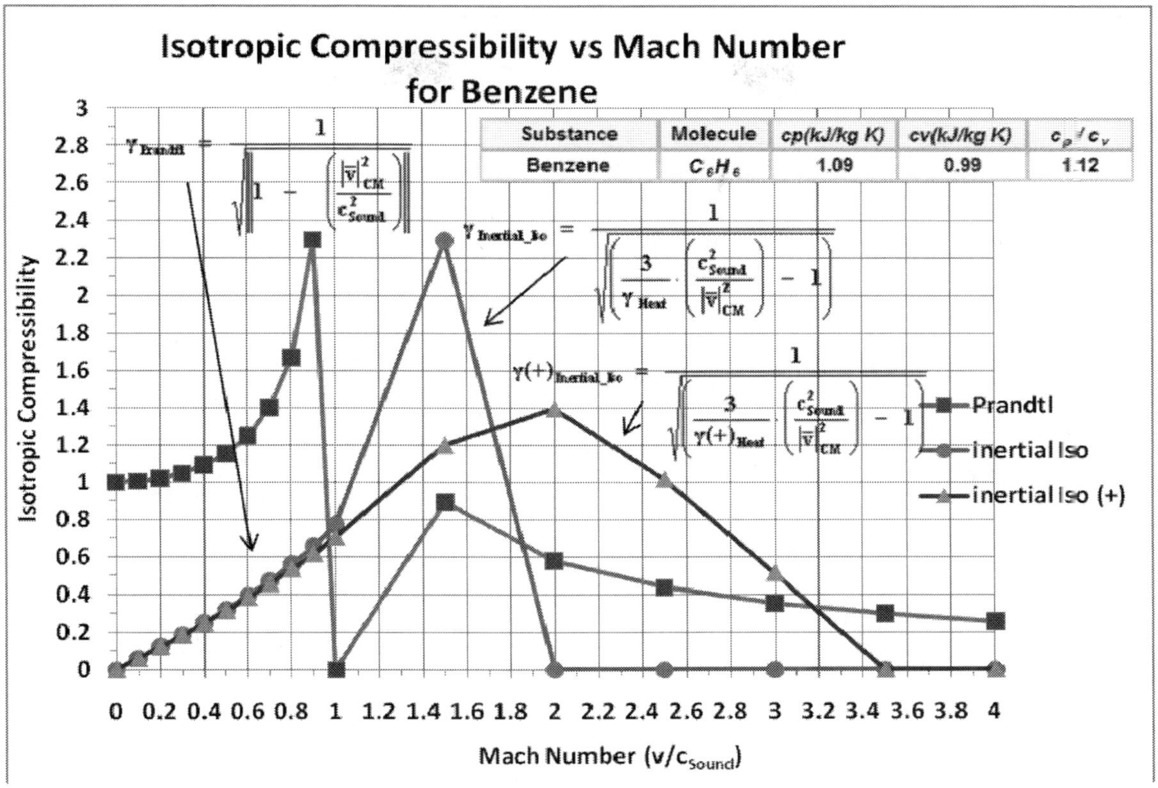

Figure 13.15: Isotropic Compressibility vs. Mach number for – Benzene.

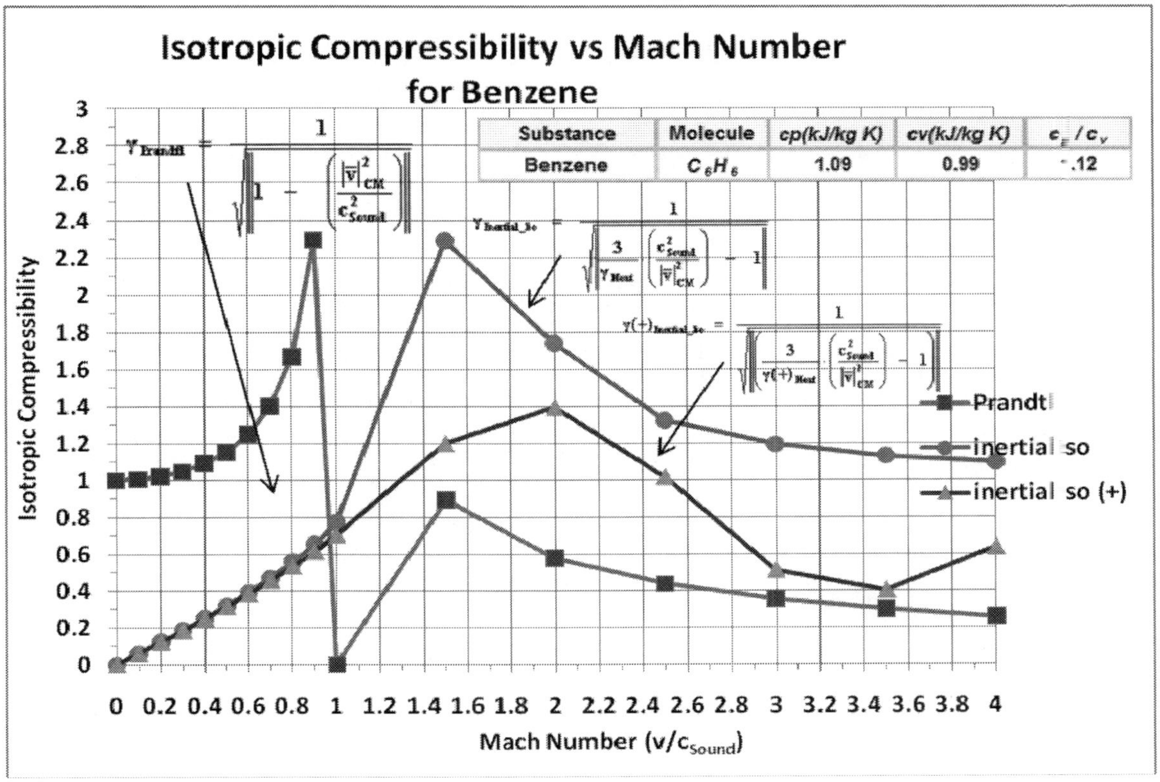

Figure 13.16: Isotropic (Absolute) Compressibility vs. Mach number for – Benzene.

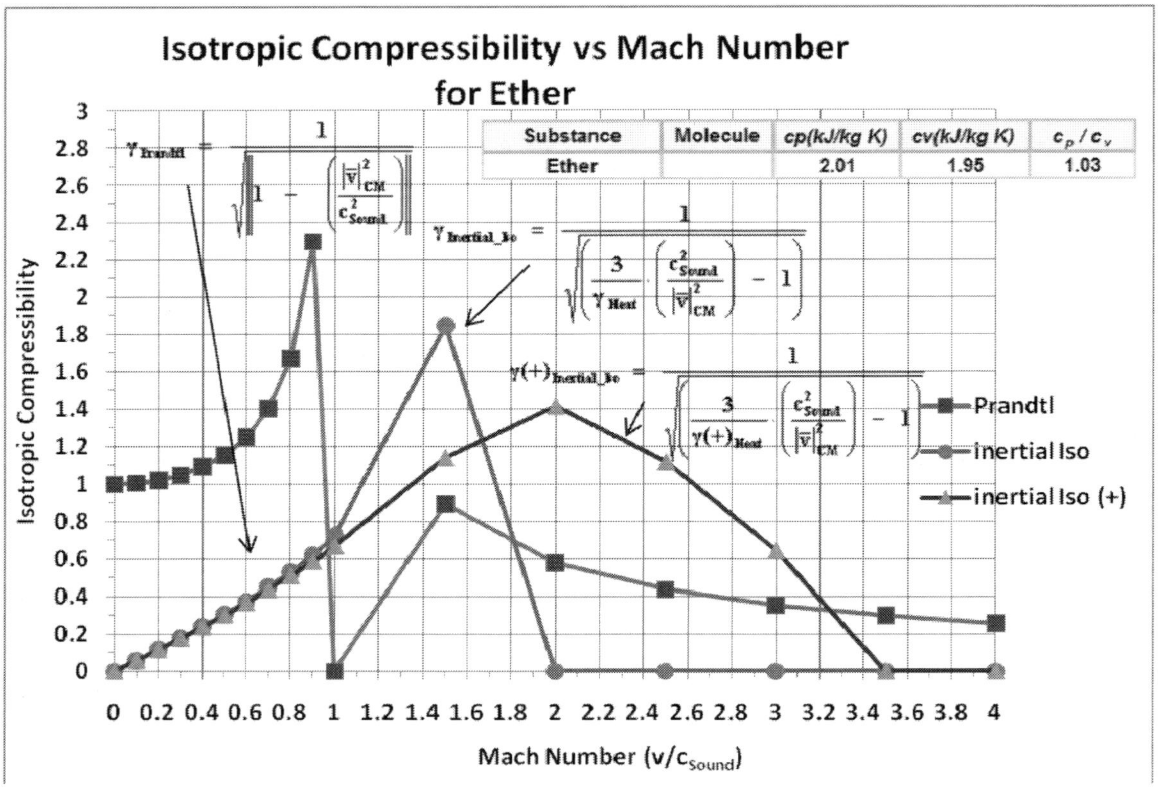

Figure 13.17: Isotropic Compressibility vs. Mach number for – Ether.

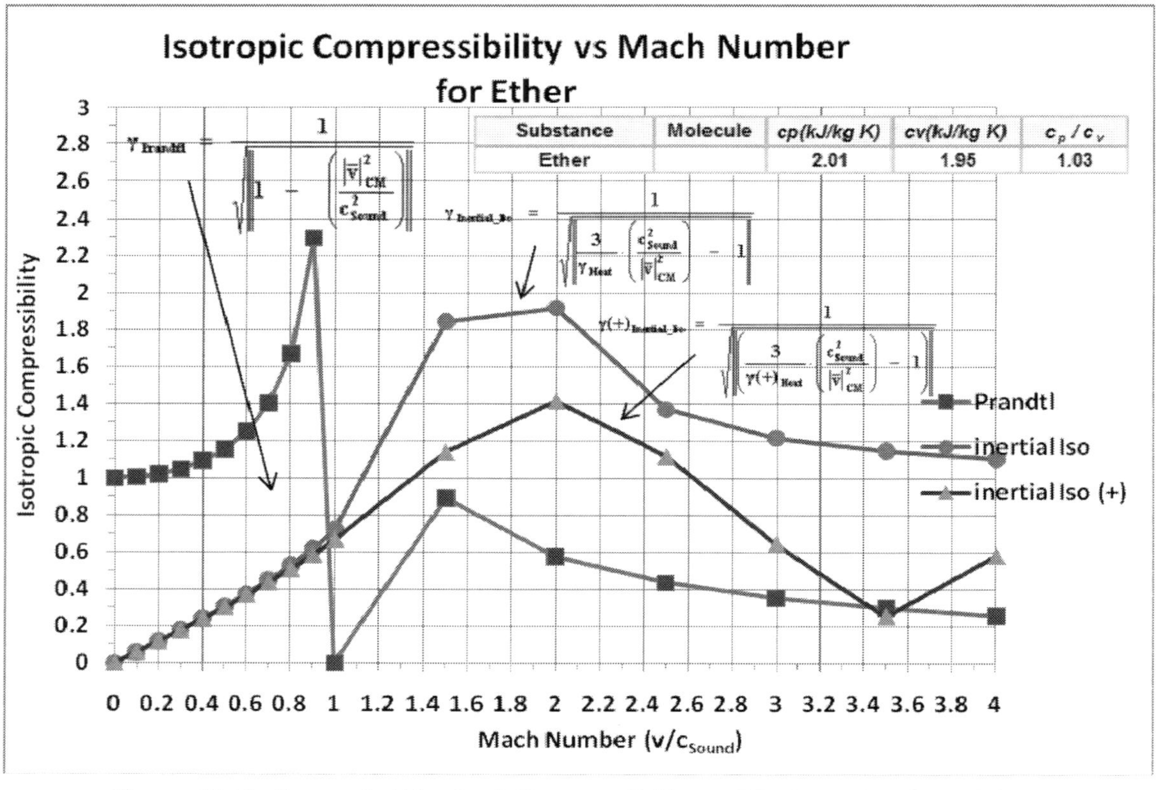

Figure 13.18: Isotropic (Absolute) Compressibility vs. Mach number for – Ether.

13.2 Carnot Thermal Isotropy Engine - Parameters

Thermodynamic Isotropy – Surplus Density

13.7

$$\rho_{\text{Thermo\_Iso\_Density}} = \frac{\text{m}_{\text{Thermo\_Iso\_Mass}}}{\text{V}_{\text{ol}}} \rightarrow {kg}/{m^3}$$

$$\rho_{\text{Thermo\_Iso\_Density}} = \gamma_{Inertial\_Iso} \cdot \rho_{\text{Net}} = \frac{\rho_{\text{Net}}}{\sqrt{\left(\left(\frac{\left|\text{v}^2\right|_{\text{Iso}}}{\left|\overline{\text{v}}\right|^2_{\text{CM}}}\right) - 1\right)}}$$

$$\rho_{\text{Thermo\_Iso\_Density}} = \frac{\overline{\text{p}}_{\text{Thermo\_Momentum}}}{\text{V}_{\text{ol}} \cdot \sqrt{\left|\text{v}^2\right|_{\text{Iso}}}} = \frac{\left(\frac{\text{m}_{\text{Net}}}{\text{V}_{\text{ol}}}\right)}{\sqrt{\left(\frac{3}{\gamma_{Heat}} \cdot \left(\frac{\text{c}^2_{\text{Sound}}}{\left|\overline{\text{v}}\right|^2_{\text{CM}}}\right) - 1\right)}}$$

Thermodynamic Isotropy – Surplus Inertia Mass and Resistance

13.8

$$\text{m}_{\text{Thermo\_Iso\_Mass}} = \gamma_{Inertial\_Iso} \cdot \text{m}_{\text{Net}} \rightarrow kg$$

$$\text{m}_{\text{Thermo\_Iso\_Mass}} = \frac{\overline{\text{p}}_{\text{Thermo\_Momentum}}}{\sqrt{\left|\text{v}^2\right|_{\text{Iso}}}} = \frac{\text{m}_{\text{Net}}}{\sqrt{\left(\left(\frac{\left|\text{v}^2\right|_{\text{Iso}}}{\left|\overline{\text{v}}\right|^2_{\text{CM}}}\right) - 1\right)}}$$

$$\text{m}_{\text{Thermo\_Iso\_Mass}} = \frac{\text{m}_{\text{Net}}}{\sqrt{\left(\frac{3}{\gamma_{Heat}} \cdot \left(\frac{\text{c}^2_{\text{Sound}}}{\left|\overline{\text{v}}\right|^2_{\text{CM}}}\right) - 1\right)}}$$

Thermodynamic Isotropy – Surplus Mass Flow Rate

Where the Thermodynamic Pump cycle frequency is given by,

13.9

$$f_{Pump\_Frequency} = \left(\frac{1}{T_{Pump\_Cycle\_Period}} \right) \rightarrow \frac{1}{s}$$

Where the Thermodynamic Pump cycle frequency is given by,

13.10

For Light:

13.11

$$\bar{d} = \bar{c}_{Light} \cdot \left(\frac{T_{Pump\_Cycle\_Period}}{2\pi} \right) \rightarrow m$$

For Sound:

13.12

$$\bar{d} = \bar{c}_{Sound} \cdot \left(\frac{T_{Pump\_Cycle\_Period}}{2\pi} \right) \rightarrow m$$

Thermodynamic Isotropy – Surplus Mass Flow Rate

13.13

$$f_{Mass\_Flow\_Iso} = \gamma_{Inertial\_Iso} \cdot f_{Mass\_Rate} \rightarrow \frac{kg}{s}$$

$$f_{Mass\_Flow\_Iso} = \left(\frac{m_{Thermo\_Iso\_Mass}}{T_{Pump\_Cycle\_Period}} \right) = \gamma_{Inertial\_Iso} \cdot \left(\frac{m_{Net}}{T_{Pump\_Cycle\_Period}} \right)$$

$$f_{Mass\_Flow\_Iso} = \frac{\bar{p}_{Thermo\_Momentum}}{T_{Pump\_Cycle\_Period} \cdot \sqrt{\left|v^2\right|_{Iso}}} = \frac{\left(\frac{m_{Net}}{T_{Pump\_Cycle\_Period}} \right)}{\sqrt{\left(\frac{3}{\gamma_{Heat}} \cdot \left(\frac{c_{Sound}^2}{\left|\bar{v}\right|_{CM}^2} \right) - 1 \right)}}$$

Thermodynamic Isotropy – Surplus Mass Flow Rate – Light Speed

13.14

$$f_{Mass\_Flow\_Iso} = \gamma_{Inertial\_Iso} \cdot f_{Mass\_Rate} \rightarrow kg/s$$

$$f_{Mass\_Flow\_Iso} = \frac{\overline{p}_{Thermo\_Momentum}}{2\pi \cdot \overline{d}} \cdot \left(\frac{\overline{c}_{Light}}{\sqrt{\left|v^2\right|_{Iso}}} \right) = \frac{\left(\frac{m_{Net} \cdot \overline{c}_{Light}}{2\pi \cdot \overline{d}} \right)}{\sqrt{\left(\frac{3}{\gamma_{Heat}} \cdot \left(\frac{c^2_{Sound}}{\left|\overline{v}\right|^2_{CM}} \right) - 1 \right)}}$$

Thermodynamic Isotropy – Surplus Mass Flow Rate – Sound Speed

13.15

$$f_{Mass\_Flow\_Iso} = \gamma_{Inertial\_Iso} \cdot f_{Mass\_Rate} \rightarrow kg/s$$

$$f_{Mass\_Flow\_Iso} = \frac{\overline{p}_{Thermo\_Momentum}}{2\pi \cdot \overline{d}} \cdot \left(\frac{\overline{c}_{Sound}}{\sqrt{\left|v^2\right|_{Iso}}} \right) = \frac{\left(\frac{m_{Net} \cdot \overline{c}_{Sound}}{2\pi \cdot \overline{d}} \right)}{\sqrt{\left(\frac{3}{\gamma_{Heat}} \cdot \left(\frac{c^2_{Sound}}{\left|\overline{v}\right|^2_{CM}} \right) - 1 \right)}}$$

Thermodynamic Isotropy – Surplus Aether Static Pressure

13.16

$$P_{Aether\_Pressure\_Static} = \gamma_{Inertial\_Iso} \cdot P_{Iso\_Pressure} \rightarrow {}^{kg}/_{m \cdot s^2}$$

$$P_{Aether\_Pressure\_Static} = \frac{1}{3} \cdot \left(\frac{\bar{p}_{Thermo\_Momentum}}{V_{ol}} \right) \cdot \sqrt{\left| v^2 \right|_{Iso}}$$

$$P_{Aether\_Pressure\_Static} = \frac{1}{3} \cdot \rho_{Thermo\_Iso\_Density} \cdot \left| v^2 \right|_{Iso} = \frac{1}{3} \cdot \left(\frac{m_{Thermo\_Iso\_Mass}}{V_{ol}} \right) \cdot \left| v^2 \right|_{Iso}$$

$$P_{Aether\_Pressure\_Static} = \frac{\frac{1}{3} \cdot \rho_{Net} \cdot \left| v^2 \right|_{Iso}}{\sqrt{\left(\left(\frac{\left| v^2 \right|_{Iso}}{\left| \bar{v} \right|^2_{CM}} \right) - 1 \right)}} = \frac{\frac{1}{3} \cdot \left(\frac{m_{Net}}{V_{ol}} \right) \cdot \left| v^2 \right|_{Iso}}{\sqrt{\left(\frac{3}{\gamma_{Heat}} \cdot \left(\frac{c^2_{Sound}}{\left| \bar{v} \right|^2_{CM}} \right) - 1 \right)}}$$

$$P_{Aether\_Pressure\_Static} = \frac{P_{Iso\_Pressure}}{\sqrt{\left(\frac{3}{\gamma_{Heat}} \cdot \left(\frac{c^2_{Sound}}{\left| \bar{v} \right|^2_{CM}} \right) - 1 \right)}}$$

$$P_{Aether\_Pressure\_Static} = \frac{\frac{1}{\gamma_{Heat}} \cdot \left(\rho_{Net} \cdot c^2_{Sound} \right)}{\sqrt{\left(\left(\frac{\left| v^2 \right|_{Iso}}{\left| \bar{v} \right|^2_{CM}} \right) - 1 \right)}} = \frac{\frac{1}{\gamma_{Heat}} \cdot \left(\frac{m_{Net} \cdot c^2_{Sound}}{V_{ol}} \right)}{\sqrt{\left(\frac{3}{\gamma_{Heat}} \cdot \left(\frac{c^2_{Sound}}{\left| \bar{v} \right|^2_{CM}} \right) - 1 \right)}}$$

Thermodynamic Isotropy – Surplus Aether Sound Bulk Modulus Pressure

13.17

$$B_{\text{Aether\_Pressure\_Iso}} = \gamma_{Inertial\_Iso} \cdot B_{\text{Sound\_Pressure}} \rightarrow {kg}/{m \cdot s^2}$$

$$B_{\text{Aether\_Pressure\_Iso}} = \rho_{\text{Thermo\_Iso\_Density}} \cdot c^2_{\text{Sound}}$$

$$B_{\text{Aether\_Pressure\_Iso}} = \left(\frac{m_{\text{Thermo\_Iso\_Mass}}}{V_{\text{ol}}} \right) \cdot c^2_{\text{Sound}}$$

$$B_{\text{Aether\_Pressure\_Iso}} = \gamma_{Inertial\_Iso} \cdot \rho_{\text{Net}} \cdot c^2_{\text{Sound}} = \frac{\rho_{\text{Net}} \cdot c^2_{\text{Sound}}}{\sqrt{\left(\left(\frac{\left| v^2 \right|_{\text{Iso}}}{\left| \overline{v} \right|^2_{\text{CM}}} \right) - 1 \right)}}$$

$$B_{\text{Aether\_Pressure\_Iso}} = \left(\frac{\overline{p}_{\text{Thermo\_Momentum}}}{V_{\text{ol}}} \right) \cdot \frac{c^2_{\text{Sound}}}{\sqrt{\left| v^2 \right|_{\text{Iso}}}} = \frac{\left(\frac{m_{\text{Net}}}{V_{\text{ol}}} \right) \cdot c^2_{\text{Sound}}}{\sqrt{\left(\frac{3}{\gamma_{Heat}} \cdot \left(\frac{c^2_{\text{Sound}}}{\left| \overline{v} \right|^2_{\text{CM}}} \right) - 1 \right)}}$$

$$B_{\text{Aether\_Pressure\_Iso}} = \gamma_{Heat} \cdot B_{\text{Aether\_Pressure\_Static}}$$

$$B_{\text{Aether\_Pressure\_Iso}} = \gamma_{Heat} \cdot \gamma_{Inertial\_Iso} \cdot P_{\text{Iso\_Pressure}}$$

Thermodynamic Isotropy – Surplus Aether Light Bulk Modulus Pressure

13.18

$$B_{\text{Aether\_Pressure\_Light\_Iso}} = \gamma_{Inertial\_Iso} \cdot B_{\text{Light\_Bulk\_Pressure}} \rightarrow {kg}/{m \cdot s^2}$$

$$B_{\text{Aether\_Pressure\_Light\_Iso}} = \rho_{\text{Thermo\_Iso\_Density}} \cdot c_{\text{Light}}^2$$

$$B_{\text{Aether\_Pressure\_Light\_Iso}} = \left(\frac{m_{\text{Thermo\_Iso\_Mass}}}{V_{\text{ol}}} \right) \cdot c_{\text{Light}}^2$$

$$B_{\text{Aether\_Pressure\_Light\_Iso}} = \gamma_{Inertial\_Iso} \cdot \rho_{\text{Net}} \cdot c_{\text{Light}}^2 = \frac{\rho_{\text{Net}} \cdot c_{\text{Light}}^2}{\sqrt{\left(\left(\frac{\left|v^2\right|_{\text{Iso}}}{\left|\overline{v}\right|_{\text{CM}}^2} \right) - 1 \right)}}$$

$$B_{\text{Aether\_Pressure\_Light\_Iso}} = \left(\frac{\overline{p}_{\text{Thermo\_Momentum}}}{V_{\text{ol}}} \right) \cdot \frac{c_{\text{Light}}^2}{\sqrt{\left|v^2\right|_{\text{Iso}}}} = \frac{\left(\frac{m_{\text{Net}}}{V_{\text{ol}}} \right) \cdot c_{\text{Light}}^2}{\sqrt{\left(\frac{3}{\gamma_{Heat}} \cdot \left(\frac{c_{\text{Sound}}^2}{\left|\overline{v}\right|_{\text{CM}}^2} \right) - 1 \right)}}$$

Thermodynamic Isotropy – Surplus Aether Sound Energy Content

13.19

$$\left(B_{\text{Aether\_Pressure\_Iso}} \cdot \mathrm{V}_{\text{ol}}\right) \;=\; \gamma_{Inertial\_Iso} \cdot \left(B_{\text{Sound\_Pressure}} \cdot \mathrm{V}_{\text{ol}}\right) \;\rightarrow\; kg \cdot m^2 \big/ s^2$$

$$\left(B_{\text{Aether\_Pressure\_Iso}} \cdot \mathrm{V}_{\text{ol}}\right) \;=\; \rho_{\text{Thermo\_Iso\_Density}} \cdot \mathrm{c}^2_{\text{Sound}} \cdot \mathrm{V}_{\text{ol}}$$

$$\left(B_{\text{Aether\_Pressure\_Iso}} \cdot \mathrm{V}_{\text{ol}}\right) \;=\; \mathrm{m}_{\text{Thermo\_Iso\_Mass}} \cdot \mathrm{c}^2_{\text{Sound}}$$

$$\left(B_{\text{Aether\_Pressure\_Iso}} \cdot \mathrm{V}_{\text{ol}}\right) \;=\; \gamma_{Inertial\_Iso} \cdot \mathrm{m}_{\text{Net}} \cdot \mathrm{c}^2_{\text{Sound}} \;=\; \frac{\mathrm{m}_{\text{Net}} \cdot \mathrm{c}^2_{\text{Sound}}}{\sqrt{\left(\left(\dfrac{\left|\mathrm{v}^2\right|_{\text{Iso}}}{\left|\overline{\mathrm{v}}\right|^2_{\text{CM}}}\right) - 1\right)}}$$

$$\left(B_{\text{Aether\_Pressure\_Iso}} \cdot \mathrm{V}_{\text{ol}}\right) \;=\; \overline{\mathrm{p}}_{\text{Thermo\_Momentum}} \cdot \left(\frac{\mathrm{c}^2_{\text{Sound}}}{\sqrt{\left|\mathrm{v}^2\right|_{\text{Iso}}}}\right) \;=\; \frac{\mathrm{m}_{\text{Net}} \cdot \mathrm{c}^2_{\text{Sound}}}{\sqrt{\left(\dfrac{3}{\gamma_{Heat}} \cdot \left(\dfrac{\mathrm{c}^2_{\text{Sound}}}{\left|\overline{\mathrm{v}}\right|^2_{\text{CM}}}\right) - \right)}}$$

$$\left(B_{\text{Aether\_Pressure\_Iso}} \cdot \mathrm{V}_{\text{ol}}\right) \;=\; \gamma_{Heat} \cdot \left(B_{\text{Aether\_Pressure\_Static}} \cdot \mathrm{V}_{\text{ol}}\right)$$

$$\left(B_{\text{Aether\_Pressure\_Iso}} \cdot \mathrm{V}_{\text{ol}}\right) \;=\; \gamma_{Heat} \cdot \gamma_{Inertial\_Iso} \cdot \left(\mathrm{P}_{\text{Iso\_Pressure}} \cdot \mathrm{V}_{\text{ol}}\right)$$

Thermodynamic Isotropy – Surplus Aether Light Energy Content

13.20

$$\left(B_{\text{Aether\_Pressure\_Light\_Iso}} \cdot \text{V}_{\text{ol}}\right) = \gamma_{Inertial\_Iso} \cdot \left(B_{\text{Light\_Bulk\_Pressure}} \cdot \text{V}_{\text{ol}}\right) \rightarrow kg \cdot m^2 / s^2$$

$$\left(B_{\text{Aether\_Pressure\_Light\_Iso}} \cdot \text{V}_{\text{ol}}\right) = \rho_{\text{Thermo\_Iso\_Density}} \cdot \text{c}^2_{\text{Light}} \cdot \text{V}_{\text{ol}}$$

$$\left(B_{\text{Aether\_Pressure\_Light\_Iso}} \cdot \text{V}_{\text{ol}}\right) = \text{m}_{\text{Thermo\_Iso\_Mass}} \cdot \text{c}^2_{\text{Light}}$$

$$\left(B_{\text{Aether\_Pressure\_Light\_Iso}} \cdot \text{V}_{\text{ol}}\right) = \gamma_{Inertial\_Iso} \cdot \text{m}_{\text{Net}} \cdot \text{c}^2_{\text{Light}} = \frac{\text{m}_{\text{Net}} \cdot \text{c}^2_{\text{Light}}}{\sqrt{\left(\frac{3}{\gamma_{Heat}} \cdot \left(\frac{\text{c}^2_{\text{Sound}}}{|\overline{\text{v}}|^2_{\text{CM}}}\right) - 1\right)}}$$

$$\left(B_{\text{Aether\_Pressure\_Light\_Iso}} \cdot \text{V}_{\text{ol}}\right) = \overline{\text{p}}_{\text{Thermo\_Momentum}} \cdot \left(\frac{\text{c}^2_{\text{Light}}}{\sqrt{|\text{v}^2|_{\text{Iso}}}}\right) = \frac{\text{m}_{\text{Net}} \cdot \text{c}^2_{\text{Light}}}{\sqrt{\left(\left(\frac{|\text{v}^2|_{\text{Iso}}}{|\overline{\text{v}}|^2_{\text{CM}}}\right) - 1\right)}}$$

Thermodynamic Isotropy – Surplus Aether Gravitational Evolutionary In/Out Flow Attraction Rate

13.21

$$K_{Gravity\_Aether\_Iso} = \gamma_{Inertial\_Iso} \cdot K_{Gravity} \rightarrow kg \cdot m^3 / s^2$$

$$K_{Gravity\_Aether\_Iso} = \left(m_{Thermo\_Iso\_Mass} \cdot G\right) = \gamma_{Inertial\_Iso} \cdot \left(m_{Net} \cdot G\right)$$

$$K_{Gravity\_Aether\_Iso} = \left(m_{Thermo\_Iso\_Mass} \cdot G\right)$$

$$K_{Gravity\_Aether\_Iso} = \left(\frac{\overline{p}_{Thermo\_Momentum}}{\sqrt{\left|v^2\right|_{Iso}}}\right) \cdot G = \frac{K_{Gravity}}{\sqrt{\left(\left(\frac{\left|v^2\right|_{Iso}}{\left|\overline{v}\right|_{CM}^2}\right) - 1\right)}}$$

$$K_{Gravity\_Aether\_Iso} = \left(\frac{\overline{p}_{Thermo\_Momentum}}{\sqrt{\left|v^2\right|_{Iso}}}\right) \cdot G = \frac{\left(m_{Net} \cdot G\right)}{\sqrt{\left(\frac{3}{\gamma_{Heat}} \cdot \left(\frac{c_{Sound}^2}{\left|\overline{v}\right|_{CM}^2}\right) - 1\right)}}$$

$$K_{Gravity\_Aether\_Iso} = \frac{\left(\frac{\overline{F}_{Newton\_Gravitation} \cdot d^2}{m_{Test}}\right)}{\sqrt{\left(\left(\frac{\left|v^2\right|_{Iso}}{\left|\overline{v}\right|_{CM}^2}\right) - 1\right)}} = \frac{\left(\frac{\overline{F}_{Self\_Gravitation} \cdot d^2}{m_{Net}}\right)}{\sqrt{\left(\frac{3}{\gamma_{Heat}} \cdot \left(\frac{c_{Sound}^2}{\left|\overline{v}\right|_{CM}^2}\right) - 1\right)}}$$

Chapter 14

The Special Theory of Thermodynamics

(Relativistic Aether Motion)
&
(Super Compressibility)

14.1 Universal Geometric Mean Theorem – Super Compressibility Factor

Universal Geometric Mean Theorem

$$(Mean)^2 = (Max)\cdot(Min)$$

$$(Mean) = \sqrt{(Max)\cdot(Min)}$$

Universal Geometric Mean Theorem – Square Super Compressibility Factor

14.1

$$\gamma^2_{Super\_Compress} = \gamma_{Inertial}\cdot\gamma_{Inertial\_Iso} \rightarrow Unitless$$

Universal Geometric Mean Theorem – Super Compressibility Factor

14.2

$$\gamma_{Super\_Compress} = \sqrt{\gamma_{Inertial}\cdot\gamma_{Inertial\_Iso}} \rightarrow Unitless$$

Universal Geometric Mean Ratio

14.3

$$\frac{(Min)}{(Mean)} = \frac{(Mean)}{(Max)} = \sqrt{\frac{(Min)}{(Max)}} = \left(\frac{(Mean) - (Min)}{(Max) - (Mean)}\right)$$

Universal Geometric Mean Ratio – Respiration Spacetime Factor

14.4

$$\frac{\gamma_{Inertial\_Iso}}{\gamma_{Super\_Compress}} = \frac{\gamma_{Super\_Compress}}{\gamma_{Inertial}} = \sqrt{\frac{\gamma_{Inertial\_Iso}}{\gamma_{Inertial}}} = \left(\frac{\gamma_{Super\_Compress} - \gamma_{Inertial\_Iso}}{\gamma_{Inertial} - \gamma_{Super\_Compress}}\right)$$

Universal Geometric Mean Ratio – Respiration Spacetime Factor

14.5

$$\phi_{Respiration} = \sqrt{\frac{\gamma_{Inertial\_Iso}}{\gamma_{Inertial}}} = \sqrt{\frac{\rho_{Thermo\_Iso\_Mass}}{\rho_{Thermo\_Aniso\_Mass}}} \rightarrow Unitless$$

$$\phi_{Respiration} = \sqrt{\frac{\gamma_{Inertial\_Iso}}{\gamma_{Inertial}}} = \frac{\gamma_{Super\_Compress}}{\gamma_{Inertial}} = \frac{\sqrt{1 - \frac{|\overline{v}|^2_{CM}}{|v^2|_{Iso}}}}{\left(\frac{|v^2|_{Iso}}{|\overline{v}|^2_{CM}} - 2 + \frac{|\overline{v}|^2_{CM}}{|v^2|_{Iso}}\right)^{\frac{1}{4}}}$$

$$\phi_{Respiration} = \sqrt{\frac{\gamma_{Inertial\_Iso}}{\gamma_{Inertial}}} = \frac{\gamma_{Inertial\_Iso}}{\gamma_{Super\_Compress}} = \frac{\left(\frac{|v^2|_{Iso}}{|\overline{v}|^2_{CM}} - 2 + \frac{|\overline{v}|^2_{CM}}{|v^2|_{Iso}}\right)^{\frac{1}{4}}}{\sqrt{\left(\frac{|v^2|_{Iso}}{|\overline{v}|^2_{CM}} - 1\right)}}$$

$$\phi_{Respiration} = \left(\phi_{Loco\_Motion}\right)^{\frac{1}{4}} = \sqrt{\frac{\gamma_{Inertial\_Iso}}{\gamma_{Inertial}}} = \sqrt{\left(\frac{\overline{p}_{Inertial\_Net}}{\overline{p}_{Net\_Iso}}\right)}$$

$$\phi_{Respiration} = \left(\phi_{Loco\_Motion}\right)^{\frac{1}{4}} = \sqrt{\frac{|\overline{v}|_{CM}}{\sqrt{|v^2|_{Iso}}}} = \sqrt{\left(\frac{\overline{p}_{Inertial\_Net}}{\overline{p}_{Net\_Iso}}\right)}$$

$$\phi_{Respiration} = \sqrt{\left(\frac{|\overline{v}|_{CM}}{c_{Sound}}\right) \cdot \sqrt{\frac{\gamma_{Heat}}{3}}} = \sqrt{\left(\frac{\overline{p}_{Inertial\_Net}}{\overline{p}_{Net\_Iso}}\right)}$$

14.2 Square Super Compressibility Factor

For an isolated system mass body and its environment, the **Square Super Compressibility Factor** ($\gamma^2_{Super\_Compress}$) is a measure of the amount of work required to move the aether of the system from the interior to the exterior and vice versa of the system. The aether mass spontaneously flows from regions of high density aether-mass-energy to regions of low density aether-mass-energy.

The energy transfer mechanism for moving aether-mass-energy through a system mass body is the Thermodynamic Momentum, and Thermodynamic Pump Kinetic Energy.

This, thermodynamic pump kinetic energy accumulates surplus aether mass-energy in either the interior or exterior of the mass system, depending on the value of either the **Inertial Locomotion Foreshortening Factor** ($\phi_{\text{Loco\_Motion}} = \left(\dfrac{\left|\overline{\text{v}}\right|^2_{\text{CM}}}{\left|\text{v}^2\right|_{\text{Iso}}}\right) = \dfrac{\gamma_{Heat}}{3}\cdot\left(\dfrac{\left|\overline{\text{v}}\right|^2_{\text{CM}}}{\text{c}^2_{\text{Sound}}}\right)$) , or the **Mach number** ($M_{Mach} = \left(\dfrac{\left|\overline{\text{v}}\right|_{\text{CM}}}{\text{c}_{\text{Sound}}}\right)$).

The **Square Super Compressibility Factor** ($\gamma^2_{Super\_Compress}$) is a measure of the work-energy required to pump aether mass-energy based on density differences between the net inertial mass body and the atmosphere environment in which it is immersed.

This aether density difference is controlled by the Thermodynamic Momentum Pump Energy operates in a direction that is directly opposite to the pressure difference created by Anisotropic Translational and Isotropic Omni-directional kinetic energies.

The **Square Super Compressibility Factor** ($\gamma^2_{Super\_Compress}$) is also a measure of the ratio of the amount of increase or decrease in the Thermodynamic Momentum ($\overline{\text{p}}_{\text{Thermo\_Momentum}}$) relative to the Aether-Kinematic Work "Free" Momentum ($\Delta\overline{\text{p}}_{\text{Free\_Momentum}}$), and likewise relative to the Conic Inertial Momentum ($\overline{\text{p}}_{\text{Conic\_Momentum}}$) of an isolated system body.

Square Super Compressibility Factor (Absolute Value Method)			
	Mach Number $(M_{Mach} = \left(\frac{\lvert\overline{V}\rvert_{CM}}{c_{Sound}}\right))$	***Locomotion Foreshortening Factor*** $(\phi_{Loco\_Motion} = \frac{\gamma_{Heat}}{3} \cdot \left(\frac{\lvert\overline{V}\rvert^2_{CM}}{c^2_{Sound}}\right))$	***Square Super Compressibility Factor*** $(\gamma^2_{Super\_Compress})$
$(\gamma^2_{Super\_Compress})$ **Minimum Value**	0	0	$\gamma^2_{Super\_Compress} = 0$
$(\gamma^2_{Super\_Compress})$ **Greater Than Minimum Value**	1.0	$\frac{\gamma_{Heat}}{3}$	$\gamma^2_{Super\_Compress} = \frac{1}{\sqrt{\left(\frac{3}{\gamma_{Heat}} - 2 + \frac{\gamma_{Heat}}{3}\right)}}$
$(\gamma^2_{Super\_Compress})$ **Maximum Value**	1.5	$\frac{3 \cdot \gamma_{Heat}}{4}$	$\gamma^2_{Super\_Compress} = \frac{1}{\sqrt{\left(\frac{1}{\gamma_{Heat}} \cdot \left(\frac{4}{3}\right) - 2 + \left(\frac{3}{4}\right) \cdot \gamma_{Heat}\right)}}$
$(\gamma^2_{Super\_Compress})$ **Less Than Maximum Value**	2.0	$\frac{4 \cdot \gamma_{Heat}}{3}$	$\gamma^2_{Super\_Compress} = \frac{1}{\sqrt{\left(\frac{1}{\gamma_{Heat}} \cdot \left(\frac{3}{4}\right) - 2 + \left(\frac{4}{3}\right) \cdot \gamma_{Heat}\right)}}$

Aphorism 14.1: Relative to an External Observer, the Center of Isotropy, the Center of Mass, and an atomic substance medium, the **Super Compressibility Factor**

$$(\gamma^2_{Super\_Compress} = \frac{1}{\sqrt{\left(\frac{3}{\gamma_{Heat}} \cdot \left(\frac{c^2_{Sound}}{\lvert\overline{V}\rvert^2_{CM}}\right) - 2 + \frac{\gamma_{Heat}}{3} \cdot \left(\frac{\lvert\overline{V}\rvert^2_{CM}}{c^2_{Sound}}\right)\right)}})$$

of an Isolated System Mass Body is the measure of the amount of work required to accumulate surplus Isotropy energy of motion and surplus Anisotropy energy of motion internal and external to the system, and increases or decreases as the **Inertial Locomotion Foreshortening Factor** increases or decreases $(\phi_{Loco\_Motion} = \left(\frac{\lvert\overline{V}\rvert^2_{CM}}{\lvert V^2\rvert_{Iso}}\right) = \frac{\gamma_{Heat}}{3} \cdot \left(\frac{\lvert\overline{V}\rvert^2_{CM}}{c^2_{Sound}}\right))$, and likewise as the **Mach number** increases or decreases ($M_{Mach} = \left(\frac{\lvert\overline{V}\rvert_{CM}}{c_{Sound}}\right)$).

Square Super Compressibility Factor (Specific Value Method)			
	Mach Number ($M_{Mach} = \left(\frac{\lvert\overline{v}\rvert_{CM}}{c_{Sound}}\right)$)	**Locomotion Foreshortening Factor** ($\phi_{Loco\_Motion} = \frac{\gamma_{Heat}}{3} \cdot \left(\frac{\lvert\overline{v}\rvert^2_{CM}}{c^2_{Sound}}\right)$)	**Square Super Compressibility Factor** ($\gamma^2_{Super\_Compress}$)
($\gamma^2_{Super\_Compress}$) Minimum Value	0	0	$\gamma^2_{Super\_Compress} = 0$
($\gamma^2_{Super\_Compress}$) Greater Than Minimum Value	1.0	$\frac{\gamma_{Heat}}{3}$	$\gamma^2_{Super\_Compress} = \frac{\sqrt{\frac{\gamma_{Heat}}{3}}}{\left(1 - \frac{\gamma_{Heat}}{3}\right)}$
($\gamma^2_{Super\_Compress}$) Maximum Value	1.5	$\frac{3 \cdot \gamma_{Heat}}{4}$	$\gamma^2_{Super\_Compress} = \frac{\left(\frac{3}{2}\right) \cdot \sqrt{\frac{\gamma_{Heat}}{3}}}{\left(1 - \gamma_{Heat} \cdot \left(\frac{3}{4}\right)\right)}$
($\gamma^2_{Super\_Compress}$) Less Than Maximum Value	2.0	$\frac{4 \cdot \gamma_{Heat}}{3}$	$\gamma^2_{Super\_Compress} = \frac{2 \cdot \sqrt{\frac{\gamma_{Heat}}{3}}}{\left(1 - \frac{4 \cdot \gamma_{Heat}}{3}\right)}$

Definition 14.1: The **Square Super Compressibility Factor** ($\gamma^2_{Super\_Compress} = \frac{\left(\frac{\lvert\overline{v}\rvert_{CM}}{c_{Sound}}\right) \cdot \sqrt{\frac{\gamma_{Heat}}{3}}}{\left(1 - \frac{\gamma_{Heat}}{3} \cdot \left(\frac{\lvert\overline{v}\rvert^2_{CM}}{c^2_{Sound}}\right)\right)}$) is a measure of the amount of work required to accumulate surplus aether and transfer the energy form the interior isotropy to the exterior anisotropy and vice versa between a system mass body and its environment and increases or decreases as the **Inertial Locomotion Foreshortening Factor** increases or decreases ($\phi_{Loco\_Motion} = \left(\frac{\lvert\overline{v}\rvert^2_{CM}}{\lvert v^2\rvert_{Iso}}\right)$), and likewise as the **Mach number** increases or decreases ($M_{Mach} = \left(\frac{\lvert\overline{v}\rvert_{CM}}{c_{Sound}}\right)$).

Definition 14.2: The **Square Super Compressibility Factor** ($\gamma^2_{Super\_Compress} = \gamma_{Inertial} \cdot \gamma_{Inertial\_Iso} = \frac{1}{\sqrt{\eta_{\text{Carnot}} \cdot \eta_{\text{Carnot\_Iso}}}}$) is a measure of the amount of work required to accumulate surplus aether and transfer the energy form the interior isotropy to the exterior anisotropy and vice versa between a system mass body and its environment, defined as the inverse square root of the product of the **Anisotropic Carnot Thermodynamic Efficiency factor** (η_{Carnot}) multiplied by the **Isotropic Carnot Efficiency factor** ($\eta_{\text{Carnot\_Iso}}$); and likewise is equal to the product of the **Anisotropic Carnot Thermodynamic Efficiency factor** ($\gamma_{Inertial}$) multiplied by the **Isotropic Carnot Efficiency factor** ($\gamma_{Inertial\_Iso}$).

Definition 14.3: Relative to an External Observer, the Center of Isotropy, the Center of Mass, and an atomic substance medium of an isolated system mass body, the **Square Super Compressibility Factor** ($\gamma^2_{Super\_Compress} = \left(\frac{\overline{\text{p}}_{\text{Thermo\_Momentum}}}{\Delta\overline{\text{p}}_{\text{Free\_Momentum}}}\right) = \left(\frac{\overline{\text{p}}^2_{\text{Thermo\_Momentum}}}{\overline{\text{p}}^2_{\text{Conic\_Momentum}}}\right)$) Body is the measure of the amount of work required to accumulate surplus Isotropy energy of motion and surplus Anisotropy energy of motion between a mass body and its atmospheric environment and increases or decreases as the **Inertial Locomotion Foreshortening Factor** increases or decreases ($\phi_{\text{Loco\_Motion}} = \left(\frac{\left|\overline{\text{v}}\right|^2_{\text{CM}}}{\left|\text{v}^2\right|_{\text{Iso}}}\right)$), and likewise as the **Mach number** increases or decreases ($M_{Mach} = \left(\frac{\left|\overline{\text{v}}\right|_{\text{CM}}}{\text{c}_{\text{Sound}}}\right)$); and is defined as the ratio of the **Thermodynamic Momentum** ($\overline{\text{p}}_{\text{Thermo\_Momentum}}$) divided by the **Aether-Kinematic Work "Free" Momentum** ($\Delta\overline{\text{p}}_{\text{Free\_Momentum}} = \sqrt{3 \cdot \text{N} \cdot \text{k}_{\text{B}} \cdot \text{m}_{\text{Net}} \cdot \left[T_{Temp} - T_{Temp\_Ext}\right]}$) of an isolated system mass body.

Definition 14.4: Relative to an External Observer, the Center of Isotropy, the Center of Mass, and an atomic substance medium of an isolated system mass body, the **Square Super Compressibility Factor** ($\gamma^2_{Super\_Compress} = \frac{1}{\sqrt{\eta_{\text{Carnot}} \cdot \eta_{\text{Carnot\_Iso}}}} = \sqrt{\left[\frac{\text{E}_{\text{Net\_Translational}} \cdot \text{E}_{\text{Net\_Iso\_Kinetic}}}{\text{W}^2_{\text{Work\_Free}}}\right]}$) is a measure of the **Coefficient of Entropy** between a system mass body and its environment defined as the square root of the product of the **External Anisotropic Inertial Rectilinear Translational Kinetic Energy** ($\text{E}_{\text{Net\_Translational}}$) multiplied by the **Internal Isotropic Omni-directional Kinetic Energy** ($\text{E}_{\text{Net\_Iso\_Kinetic}}$) divided by the **Square of the Aether-Kinematic Work "Free" Energy** ($\text{W}^2_{\text{Work\_Free}}$) of an isolated system mass body.

Square Super Compressibility Factor

14.6

$$\gamma^2_{Super\_Compress} = \gamma_{Inertial} \cdot \gamma_{Inertial\_Iso} = \frac{1}{\sqrt{\eta_{\text{Carnot}} \cdot \eta_{\text{Carnot\_Iso}}}} \rightarrow Unitless$$

$$\gamma^2_{Super\_Compress} = \sqrt{\frac{\textit{Low}\text{ Temperature} \cdot \textit{High}\text{ Temperature}}{\left(\textit{High}\text{ Temperature} - \textit{Low}\text{ Temperature}\right)^2}}$$

$$\gamma^2_{Super\_Compress} = \sqrt{\left[\frac{\text{E}_{\text{Net\_Translational}} \cdot \text{E}_{\text{Net\_Iso\_Kinetic}}}{\text{W}^2_{\text{Work\_Free}}}\right]}$$

$$\gamma^2_{Super\_Compress} = \frac{1}{\sqrt{\eta_{\text{Carnot}} \cdot \eta_{\text{Carnot\_Iso}}}} = \sqrt{\left[\frac{\text{E}_{\text{Net\_Translational}} \cdot \text{E}_{\text{Net\_Iso\_Kinetic}}}{\left(\text{E}_{\text{Net\_Iso\_Kinetic}} - \text{E}_{\text{Net\_Translational}}\right)^2}\right]}$$

$$\gamma^2_{Super\_Compress} = \left(\frac{\overline{\text{p}}_{\text{Thermo\_Momentum}}}{\Delta\overline{\text{p}}_{\text{Free\_Momentum}}}\right) = \left(\frac{\overline{\text{p}}^2_{\text{Thermo\_Momentum}}}{\overline{\text{p}}^2_{\text{Conic\_Momentum}}}\right)$$

Square Super Compressibility Factor

14.7

$$\gamma^2_{Super\_Compress} = \sqrt{\frac{\left|\overline{\mathrm{v}}\right|^2_{\mathrm{CM}} \cdot \left|\mathrm{v}^2\right|_{\mathrm{Iso}}}{\left(\left|\mathrm{v}^2\right|_{\mathrm{Iso}} - \left|\overline{\mathrm{v}}\right|^2_{\mathrm{CM}}\right)^2}} \rightarrow Unitless$$

$$\gamma^2_{Super\_Compress} = \sqrt{\frac{T_{emp\_Ext} \cdot T_{emp}}{\left(T_{emp} - T_{emp\_Ext}\right)^2}}$$

$$\gamma^2_{Super\_Compress} = \sqrt{\frac{1}{\Delta S_{Entropy\ Free\_Surroundings}} \cdot \left(\frac{\Delta(PV)_{\mathrm{Free\_Energy}}}{\Delta T_{Free}}\right)}$$

$$\gamma^2_{Super\_Compress} = \sqrt{\frac{1}{\Delta S_{Entropy\ Free\_Surroundings}} \cdot \left(\frac{\left[\delta Q_{\mathrm{Heat\_Pressure\_Free}} - \delta Q_{\mathrm{Heat\_Energy\_Free}}\right]}{\left(T_{emp} - T_{emp\_Ext}\right)}\right)}$$

Square Super Compressibility Factor

14.8

$$\gamma^2_{Super\_Compress} = \gamma_{Inertial} \cdot \gamma_{Inertial\_Iso} \rightarrow Unitless$$

$$\gamma^2_{Super\_Compress} = \gamma^2_{Inertial} \cdot \left(\frac{\overline{p}_{Inertial\_Net}}{\overline{p}_{Net\_Iso}} \right) = \gamma^2_{Inertial} \cdot \left(\frac{\left|\overline{v}\right|_{CM}}{\sqrt{\left|v^2\right|_{Iso}}} \right)$$

$$\gamma^2_{Super\_Compress} = \gamma^2_{Inertial\_Iso} \cdot \left(\frac{\overline{p}_{Net\_Iso}}{\overline{p}_{Inertial\_Net}} \right) = \gamma^2_{Inertial\_Iso} \cdot \left(\frac{\sqrt{\left|v^2\right|_{Iso}}}{\left|\overline{v}\right|_{CM}} \right)$$

$$\gamma^2_{Super\_Compress} = \frac{\left(\frac{\overline{p}_{Net\_Iso}}{\overline{p}_{Inertial\_Net}} \right)}{\left(\frac{E_{Net\_Iso\_Kinetic}}{E_{Net\_Translational}} - 1 \right)} = \frac{\left(\frac{\overline{p}_{Net\_Iso}}{\overline{p}_{Inertial\_Net}} \right)}{\left(\frac{\left|v^2\right|_{Iso}}{\left|\overline{v}\right|^2_{CM}} - 1 \right)}$$

$$\gamma^2_{Super\_Compress} = \gamma_{Inertial} \cdot \gamma_{Inertial\_Iso} = \left(\frac{\overline{p}^2_{Thermo\_Momentum}}{\overline{p}_{Net\_Iso} \cdot \overline{p}_{Inertial\_Net}} \right)$$

Square Super Compressibility Factor

14.9

$$\gamma^2_{Super\_Compress} = \gamma_{Inertial} \cdot \gamma_{Inertial\_Iso} = \left(\frac{\overline{p}^2_{\mathrm{Thermo\_Momentum}}}{\overline{p}_{\mathrm{Net\_Iso}} \cdot \overline{p}_{\mathrm{Inertial\_Net}}} \right) \rightarrow Unitless$$

$$\gamma^2_{Super\_Compress} = \gamma_{Inertial} \cdot \gamma_{Inertial\_Iso} = \left(\frac{\overline{p}_{\mathrm{Thermo\_Momentum}}}{\Delta\overline{p}_{\mathrm{Free\_Momentum}}} \right)$$

$$\gamma^2_{Super\_Compress} = \left(\frac{1}{\sqrt{1 - \dfrac{\left|\overline{v}\right|^2_{\mathrm{CM}}}{\left|v^2\right|_{\mathrm{Iso}}}}} \right) \cdot \left(\frac{1}{\sqrt{\left(\dfrac{\left|v^2\right|_{\mathrm{Iso}}}{\left|\overline{v}\right|^2_{\mathrm{CM}}} - 1 \right)}} \right)$$

$$\gamma^2_{Super\_Compress} = \frac{1}{\sqrt{\left(\dfrac{\left|v^2\right|_{\mathrm{Iso}}}{\left|\overline{v}\right|^2_{\mathrm{CM}}} - 2 + \dfrac{\left|\overline{v}\right|^2_{\mathrm{CM}}}{\left|v^2\right|_{\mathrm{Iso}}} \right)}}$$

$$\gamma^2_{Super\_Compress} = \frac{1}{\sqrt{\left(\dfrac{3}{\gamma_{Heat}} \cdot \left(\dfrac{c^2_{\mathrm{Sound}}}{\left|\overline{v}\right|^2_{\mathrm{CM}}} \right) - 2 + \dfrac{\gamma_{Heat}}{3} \cdot \left(\dfrac{\left|\overline{v}\right|^2_{\mathrm{CM}}}{c^2_{\mathrm{Sound}}} \right) \right)}}$$

Square Super Compressibility Factor

14.10

$$\gamma^2_{Super\_Compress} = \gamma^2_{Inertial} \cdot \left(\frac{\overline{p}_{Inertial\_Net}}{\overline{p}_{Net\_Iso}} \right) = \gamma^2_{Inertial} \cdot \left(\frac{\left|\overline{v}\right|_{CM}}{\sqrt{\left|v^2\right|_{Iso}}} \right) \rightarrow Unitless$$

$$\gamma^2_{Super\_Compress} = \frac{\left(\dfrac{\left|\overline{v}\right|_{CM}}{\sqrt{\left|v^2\right|_{Iso}}} \right)}{\left(1 - \dfrac{\left|\overline{v}\right|^2_{CM}}{\left|v^2\right|_{Iso}} \right)} = \frac{\left(\dfrac{\left|\overline{v}\right|_{CM}}{c_{Sound}} \right) \cdot \sqrt{\dfrac{\gamma_{Heat}}{3}}}{\left(1 - \dfrac{\gamma_{Heat}}{3} \cdot \left(\dfrac{\left|\overline{v}\right|^2_{CM}}{c^2_{Sound}} \right) \right)}$$

$$\gamma^2_{Super\_Compress} = \gamma^2_{Inertial\_Iso} \cdot \left(\frac{\overline{p}_{Net\_Iso}}{\overline{p}_{Inertial\_Net}} \right) = \gamma^2_{Inertial\_Iso} \cdot \left(\frac{\sqrt{\left|v^2\right|_{Iso}}}{\left|\overline{v}\right|_{CM}} \right)$$

$$\gamma^2_{Super\_Compress} = \frac{\left(\dfrac{\sqrt{\left|v^2\right|_{Iso}}}{\left|\overline{v}\right|_{CM}} \right)}{\left(\dfrac{\left|v^2\right|_{Iso}}{\left|\overline{v}\right|^2_{CM}} - 1 \right)} = \frac{\left(\dfrac{c_{Sound}}{\left|\overline{v}\right|_{CM}} \right) \cdot \sqrt{\dfrac{3}{\gamma_{Heat}}}}{\left(\dfrac{3}{\gamma_{Heat}} \cdot \left(\dfrac{c^2_{Sound}}{\left|\overline{v}\right|^2_{CM}} \right) - 1 \right)}$$

14.3 Super Compressibility Factor

For an isolated system mass body and its environment, the **Super Compressibility Factor** ($\gamma_{Super\_Compress}$) is a measure of the amount of work required to move the aether of the system from the interior to the exterior and vice versa of the system. The aether mass spontaneously flows from regions of high density aether mass to regions of low density mass.

The energy transfer mechanism for moving aether mass and energy through a system mass body is the Thermodynamic Momentum, and Thermodynamic Pump Kinetic Energy.

This, thermodynamic pump kinetic energy accumulates surplus aether mass-energy in either the interior or exterior of the mass system, depending on the value of either the **Inertial Locomotion Foreshortening Factor** ($\phi_{\text{Loco\_Motion}} = \left(\frac{|\overline{\text{v}}|^2_{\text{CM}}}{|\text{v}^2|_{\text{Iso}}}\right) = \frac{\gamma_{Heat}}{3} \cdot \left(\frac{|\overline{\text{v}}|^2_{\text{CM}}}{\text{c}^2_{\text{Sound}}}\right)$), or the **Mach number** ($M_{Mach} = \left(\frac{|\overline{\text{v}}|_{\text{CM}}}{\text{c}_{\text{Sound}}}\right)$).

The **Super Compressibility Factor** ($\gamma_{Super\_Compress}$) is also a measure of the ratio of the amount of increase or decrease in the Thermodynamic Momentum ($\overline{\text{p}}_{\text{Thermo\_Momentum}}$) relative to the Aether-Kinematic Work "Free" Momentum ($\Delta\overline{\text{p}}_{\text{Free\_Momentum}}$), and likewise relative to the Conic Inertial Momentum ($\overline{\text{p}}_{\text{Conic\_Momentum}}$) of an isolated system body.

Super Compressibility Factor Specific Values			
	Mach Number $\left(M_{Mach} = \left(\frac{\lvert\overline{\text{v}}\rvert_{\text{CM}}}{\text{c}_{\text{Sound}}} \right) \right)$	**Locomotion Foreshortening Factor** $\left(\phi_{\text{Loco\_Motion}} = \frac{\gamma_{Heat}}{3} \cdot \left(\frac{\lvert\overline{\text{v}}\rvert^2_{\text{CM}}}{\text{c}^2_{\text{Sound}}} \right) \right)$	**Super Compressibility Factor** $\left(\gamma_{Super\_Compress} \right)$
$\left(\gamma_{Super\_Compress} \right)$ **Minimum Value**	0	0	$\gamma_{Super\_Compress} = 0$
$\left(\gamma_{Super\_Compress} \right)$ **Greater Than Minimum Value**	1.0	$\frac{\gamma_{Heat}}{3}$	$\gamma_{Super\_Compress} = \frac{1}{\left(\frac{3}{\gamma_{Heat}} - 2 + \frac{\gamma_{Heat}}{3} \right)^{\frac{1}{4}}}$
$\left(\gamma_{Super\_Compress} \right)$ **Maximum Value**	1.5	$\frac{3 \cdot \gamma_{Heat}}{4}$	$\gamma_{Super\_Compress} = \frac{1}{\left(\frac{1}{\gamma_{Heat}} \cdot \left(\frac{4}{3} \right) - 2 + \left(\frac{3}{4} \right) \cdot \gamma_{Heat} \right)^{\frac{1}{4}}}$
$\left(\gamma_{Super\_Compress} \right)$ **Less Than Maximum Value**	2.0	$\frac{4 \cdot \gamma_{Heat}}{3}$	$\gamma_{Super\_Compress} = \frac{1}{\left(\frac{1}{\gamma_{Heat}} \cdot \left(\frac{3}{4} \right) - 2 + \left(\frac{4}{3} \right) \gamma_{Heat} \right)^{\frac{1}{4}}}$

Aphorism 14.2: Relative to an External Observer, the Center of Isotropy, the Center of Mass, and an atomic substance medium, the **Super Compressibility Factor** $\left(\gamma_{Super\_Compress} = \frac{1}{\left(\frac{3}{\gamma_{Heat}} \cdot \left(\frac{\text{c}^2_{\text{Sound}}}{\lvert\overline{\text{v}}\rvert^2_{\text{CM}}} \right) - 2 + \frac{\gamma_{Heat}}{3} \cdot \left(\frac{\lvert\overline{\text{v}}\rvert^2_{\text{CM}}}{\text{c}^2_{\text{Sound}}} \right) \right)^{\frac{1}{4}}} \right)$ of an Isolated System Mass Body is the measure of the amount of work required to accumulate surplus Isotropy energy of motion and surplus Anisotropy energy of motion and increases or decreases as the **Inertial Locomotion Foreshortening Factor** increases or decreases $\left(\phi_{\text{Loco\_Motion}} = \left(\frac{\lvert\overline{\text{v}}\rvert^2_{\text{CM}}}{\lvert \text{v}^2 \rvert_{\text{Iso}}} \right) = \frac{\gamma_{Heat}}{3} \cdot \left(\frac{\lvert\overline{\text{v}}\rvert^2_{\text{CM}}}{\text{c}^2_{\text{Sound}}} \right) \right)$, and likewise as the Mach number increases or decreases $\left(M_{Mach} = \left(\frac{\lvert\overline{\text{v}}\rvert_{\text{CM}}}{\text{c}_{\text{Sound}}} \right) \right)$.

Definition 14.5: The **Super Compressibility Factor** ($\gamma_{Super\_Compress} = \sqrt{\gamma_{Inertial} \cdot \gamma_{Inertial\_Iso}} = \dfrac{1}{(\eta_{\text{Carnot}} \cdot \eta_{\text{Carnot\_Iso}})^{\frac{1}{4}}}$) is a measure of the amount of work required to accumulate surplus aether and transfer the energy form the interior isotropy to the exterior anisotropy and vice versa between a system mass body and its environment, defined as the inverse raised to the fourth root of the product of the **Anisotropic Carnot Thermodynamic Efficiency factor** (η_{Carnot}) multiplied by the **Isotropic Carnot Efficiency factor** ($\eta_{\text{Carnot\_Iso}}$); and likewise is equal to the product of the **Anisotropic Carnot Thermodynamic Efficiency factor** ($\gamma_{Inertial}$) multiplied by the **Isotropic Carnot Efficiency factor** ($\gamma_{Inertial\_Iso}$).

Definition 14.6: Relative to an External Observer, the Center of Isotropy, the Center of Mass, and an atomic substance medium of an isolated system mass body, the **Super Compressibility Factor** ($\gamma_{Super\_Compress} = \sqrt{\left(\dfrac{\overline{\text{p}}_{\text{Thermo\_Momentum}}}{\Delta\overline{\text{p}}_{\text{Free\_Momentum}}}\right)} = \left(\dfrac{\overline{\text{p}}_{\text{Thermo\_Momentum}}}{\overline{\text{p}}_{\text{Conic\_Momentum}}}\right)$) Body is the measure of the amount of work required to accumulate surplus Isotropy energy of motion and surplus Anisotropy energy of motion between a mass body and its atmospheric environment and increases as the **Inertial Locomotion Foreshortening Factor** increases or decreases ($\phi_{\text{Loco\_Motion}} = \left(\dfrac{|\overline{\text{v}}|^2_{\text{CM}}}{|\text{v}^2|_{\text{Iso}}}\right)$), and likewise as the **Mach number** increases or decreases ($M_{Mach} = \left(\dfrac{|\overline{\text{v}}|_{\text{CM}}}{\text{c}_{\text{Sound}}}\right)$); and is defined as the ratio of the **Thermodynamic Momentum** ($\overline{\text{p}}_{\text{Thermo\_Momentum}}$) divided by the **Conic Inertial Momentum** ($\overline{\text{p}}_{\text{Conic\_Momentum}} = \gamma_{Super\_Compress} \cdot \Delta\overline{\text{p}}_{\text{Free\_Momentum}}$) of an isolated system mass body.

Definition 14.7: Relative to an External Observer, the Center of Isotropy, the Center of Mass, and an atomic substance medium of an isolated system mass body, the **Square Super Compressibility Factor** ($\gamma_{Super\_Compress} = \frac{1}{\left(\eta_{\mathrm{Carnot}} \cdot \eta_{\mathrm{Carnot\_Iso}}\right)^{\frac{1}{4}}} = \left[\frac{\mathrm{E}_{\mathrm{Net\_Translational}} \cdot \mathrm{E}_{\mathrm{Net\_Iso\_Kinetic}}}{\mathrm{W}^2_{\mathrm{Work\_Free}}}\right]^{\frac{1}{4}}$) is a measure of the **Coefficient of Entropy** between a system mass body and its environment defined as the fourth root of the product of the **External Anisotropic Inertial Rectilinear Translational Kinetic Energy** ($\mathrm{E}_{\mathrm{Net\_Translational}}$) multiplied by the **Internal Isotropic Omni-directional Kinetic Energy** ($\mathrm{E}_{\mathrm{Net\_Iso\_Kinetic}}$) divided by the **Square of the Aether-Kinematic Work "Free" Energy** ($\mathrm{W}^2_{\mathrm{Work\_Free}}$) of an isolated system mass body.

Super Compressibility Factor

14.11

$$\gamma_{Super\_Compress} = \sqrt{\gamma_{Inertial} \cdot \gamma_{Inertial\_Iso}} = \frac{1}{\left(\eta_{\text{Carnot}} \cdot \eta_{\text{Carnot\_Iso}}\right)^{\frac{1}{4}}} \rightarrow Unitless$$

$$\gamma_{Super\_Compress} = \left[\frac{\text{Low Temperature} \cdot \textit{High}\ \text{Temperature}}{\left(\textit{High}\ \text{Temperature} - \text{Low Temperature}\right)^2}\right]^{\frac{1}{4}}$$

$$\gamma_{Super\_Compress} = \frac{1}{\left(\eta_{\text{Carnot}} \cdot \eta_{\text{Carnot\_Iso}}\right)^{\frac{1}{4}}} = \left[\frac{\text{E}_{\text{Net\_Translational}} \cdot \text{E}_{\text{Net\_Iso\_Kinetic}}}{\left(\text{E}_{\text{Net\_Iso\_Kinetic}} - \text{E}_{\text{Net\_Translational}}\right)^2}\right]^{\frac{1}{4}}$$

$$\gamma_{Super\_Compress} = \sqrt{\left(\frac{\overline{\text{p}}_{\text{Thermo\_Momentum}}}{\Delta\overline{\text{p}}_{\text{Free\_Momentum}}}\right)} = \left(\frac{\overline{\text{p}}_{\text{Thermo\_Momentum}}}{\overline{\text{p}}_{\text{Conic\_Momentum}}}\right)$$

$$\gamma_{Super\_Compress} = \left[\frac{\left|\overline{\text{v}}\right|^2_{\text{CM}} \cdot \left|\text{v}^2\right|_{\text{Iso}}}{\left(\left|\text{v}^2\right|_{\text{Iso}} - \left|\overline{\text{v}}\right|^2_{\text{CM}}\right)^2}\right]^{\frac{1}{4}} = \left[\frac{T_{emp\_Ext} \cdot T_{emp}}{\left(T_{emp} - T_{emp\_Ext}\right)^2}\right]^{\frac{1}{4}}$$

$$\gamma_{Super\_Compress} = \left[\frac{1}{\Delta S_{Entropy\ Free\_Surroundings}} \cdot \left(\frac{\Delta(PV)_{\text{Free\_Energy}}}{\Delta T_{Free}}\right)\right]^{\frac{1}{4}}$$

$$\gamma_{Super\_Compress} = \left[\frac{1}{\Delta S_{Entropy\ Free\_Surroundings}} \cdot \left(\frac{\left[\delta Q_{\text{Heat\_Pressure\_Free}} - \delta Q_{\text{Heat\_Energy\_Free}}\right]}{\left(T_{emp} - T_{emp\_Ext}\right)}\right)\right]^{\frac{1}{4}}$$

Super Compressibility Factor

14.12

$$\gamma_{Super\_Compress} = \sqrt{\gamma_{Inertial} \cdot \gamma_{Inertial\_Iso}} \rightarrow Unitless$$

$$\gamma_{Super\_Compress} = \gamma_{Inertial} \cdot \sqrt{\left(\frac{\overline{p}_{Inertial\_Net}}{\overline{p}_{Net\_Iso}}\right)} = \gamma_{Inertial} \cdot \sqrt{\left(\frac{\left|\overline{v}\right|_{CM}}{\sqrt{\left|v^2\right|_{Iso}}}\right)}$$

$$\gamma_{Super\_Compress} = \gamma_{Inertial\_Iso} \cdot \sqrt{\left(\frac{\overline{p}_{Net\_Iso}}{\overline{p}_{Inertial\_Net}}\right)} = \gamma_{Inertial\_Iso} \cdot \sqrt{\left(\frac{\sqrt{\left|v^2\right|_{Iso}}}{\left|\overline{v}\right|_{CM}}\right)}$$

$$\gamma_{Super\_Compress} = \frac{\sqrt{\left(\frac{\overline{p}_{Net\_Iso}}{\overline{p}_{Inertial\_Net}}\right)}}{\sqrt{\left(\frac{E_{Net\_Iso\_Kinetic}}{E_{Net\_Translational}} - 1\right)}} = \frac{\sqrt{\left(\frac{\overline{p}_{Net\_Iso}}{\overline{p}_{Inertial\_Net}}\right)}}{\sqrt{\left(\frac{\left|v^2\right|_{Iso}}{\left|\overline{v}\right|^2_{CM}} - 1\right)}}$$

$$\gamma_{Super\_Compress} = \sqrt{\gamma_{Inertial} \cdot \gamma_{Inertial\_Iso}} = \sqrt{\left(\frac{\overline{p}^2_{Thermo\_Momentum}}{\overline{p}_{Net\_Iso} \cdot \overline{p}_{Inertial\_Net}}\right)}$$

Super Compressibility Factor

14.13

$$\gamma_{Super\_Compress} = \sqrt{\gamma_{Inertial} \cdot \gamma_{Inertial\_Iso}} = \sqrt{\left(\frac{\overline{p}^{2}_{Thermo\_Momentum}}{\overline{p}_{Net\_Iso} \cdot \overline{p}_{Inertial\_Net}} \right)} \rightarrow Unitless$$

$$\gamma_{Super\_Compress} = \sqrt{\gamma_{Inertial} \cdot \gamma_{Inertial\_Iso}} = \sqrt{\left(\frac{\overline{p}_{Thermo\_Momentum}}{\Delta\overline{p}_{Free\_Momentum}} \right)}$$

$$\gamma_{Super\_Compress} = \sqrt{\left(\frac{1}{\sqrt{1 - \frac{|\overline{v}|^{2}_{CM}}{|v^{2}|_{Iso}}}} \right) \cdot \left(\frac{1}{\sqrt{\left(\frac{|v^{2}|_{Iso}}{|\overline{v}|^{2}_{CM}} - 1 \right)}} \right)}$$

$$\gamma_{Super\_Compress} = \frac{1}{\left(\frac{|v^{2}|_{Iso}}{|\overline{v}|^{2}_{CM}} - 2 + \frac{|\overline{v}|^{2}_{CM}}{|v^{2}|_{Iso}} \right)^{\frac{1}{4}}}$$

$$\gamma_{Super\_Compress} = \frac{1}{\left(\frac{3}{\gamma_{Heat}} \cdot \left(\frac{c^{2}_{Sound}}{|\overline{v}|^{2}_{CM}} \right) - 2 + \frac{\gamma_{Heat}}{3} \cdot \left(\frac{|\overline{v}|^{2}_{CM}}{c^{2}_{Sound}} \right) \right)^{\frac{1}{4}}}$$

Super Compressibility Factor

14.14

$$\gamma_{Super\_Compress} = \gamma_{Inertial} \cdot \sqrt{\left(\frac{\overline{p}_{Inertial\_Net}}{\overline{p}_{Net\_Iso}}\right)} = \gamma_{Inertial} \cdot \sqrt{\left(\frac{|\overline{v}|_{CM}}{\sqrt{|v^2|_{Iso}}}\right)} \rightarrow Unitless$$

$$\gamma_{Super\_Compress} = \frac{\sqrt{\left(\frac{|\overline{v}|_{CM}}{\sqrt{|v^2|_{Iso}}}\right)}}{\sqrt{\left(1 - \frac{|\overline{v}|^2_{CM}}{|v^2|_{Iso}}\right)}} = \frac{\sqrt{\left(\frac{|\overline{v}|_{CM}}{c_{Sound}}\right) \cdot \sqrt{\frac{\gamma_{Heat}}{3}}}}{\sqrt{\left(1 - \frac{\gamma_{Heat}}{3} \cdot \left(\frac{|\overline{v}|^2_{CM}}{c^2_{Sound}}\right)\right)}}$$

$$\gamma_{Super\_Compress} = \gamma_{Inertial\_Iso} \cdot \sqrt{\left(\frac{\overline{p}_{Net\_Iso}}{\overline{p}_{Inertial\_Net}}\right)} = \gamma_{Inertial\_Iso} \cdot \sqrt{\left(\frac{\sqrt{|v^2|_{Iso}}}{|\overline{v}|_{CM}}\right)}$$

$$\gamma_{Super\_Compress} = \frac{\sqrt{\left(\frac{\sqrt{|v^2|_{Iso}}}{|\overline{v}|_{CM}}\right)}}{\sqrt{\left(\frac{|v^2|_{Iso}}{|\overline{v}|^2_{CM}} - 1\right)}} = \frac{\sqrt{\left(\frac{c_{Sound}}{|\overline{v}|_{CM}}\right) \cdot \sqrt{\frac{3}{\gamma_{Heat}}}}}{\sqrt{\left(\frac{3}{\gamma_{Heat}} \cdot \left(\frac{c^2_{Sound}}{|\overline{v}|^2_{CM}}\right) - 1\right)}}$$

Super Compressibility Factor

14.15

$$\gamma_{Super\_Compress} = \left[\frac{1}{\Delta S_{Entropy\ Free\_Surroundings}} \cdot \left(\frac{\Delta (PV)_{\text{Free\_Energy}}}{\Delta T_{Free}} \right) \right]^{\frac{1}{4}} \rightarrow Unitless$$

$$\gamma_{Super\_Compress} = \left[\frac{1}{\Delta S_{Entropy\ Free\_Surroundings}} \cdot \left(\frac{\left[\delta Q_{\text{Heat\_Pressure\_Free}} - \delta Q_{\text{Heat\_Energy\_Free}} \right]}{\left(T_{emp} - T_{emp\_Ext} \right)} \right) \right]^{\frac{1}{4}}$$

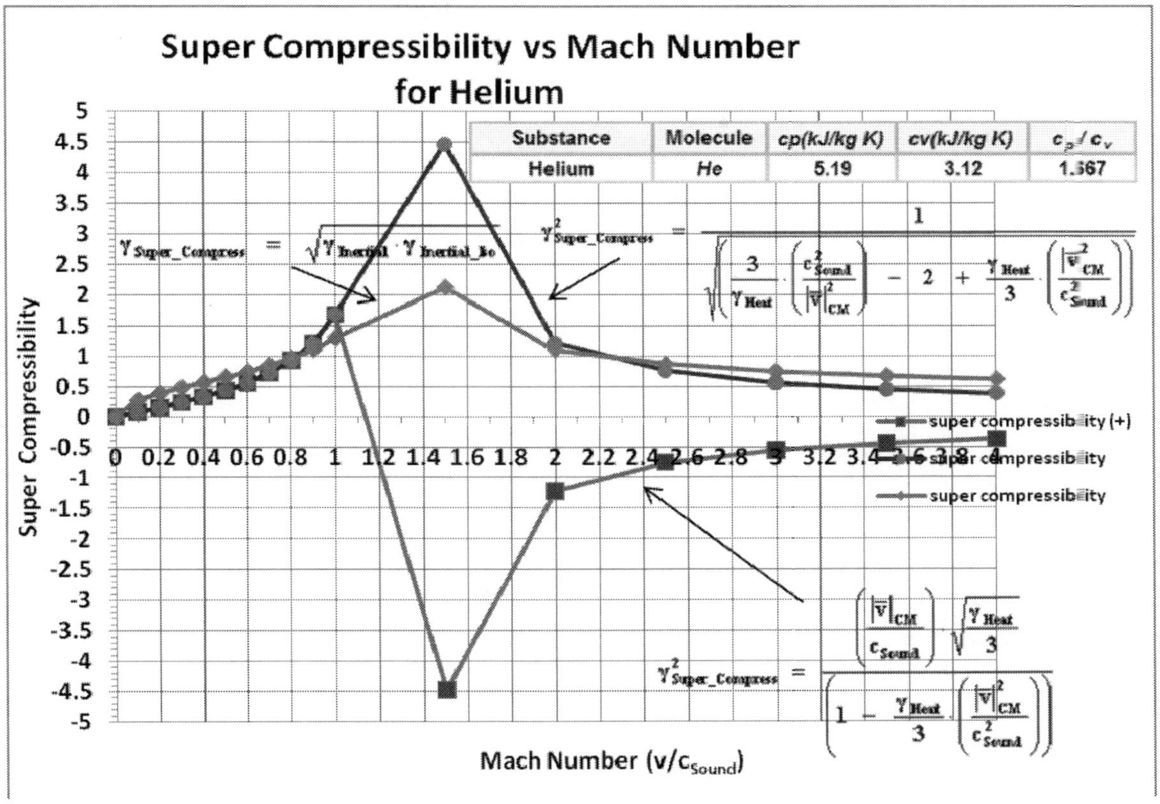

Figure 14.1: Super Compressibility vs. Mach number for – Helium.

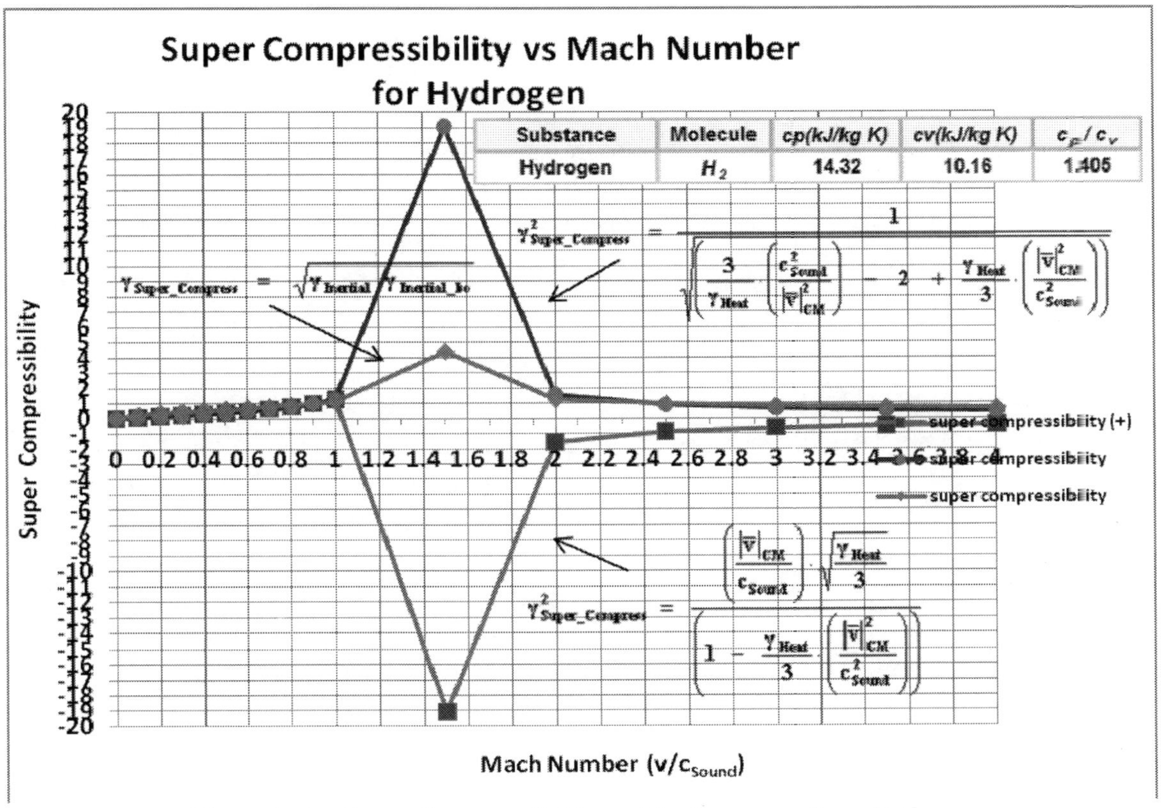

Figure 14.2: Super Compressibility vs. Mach number for – Hydrogen.

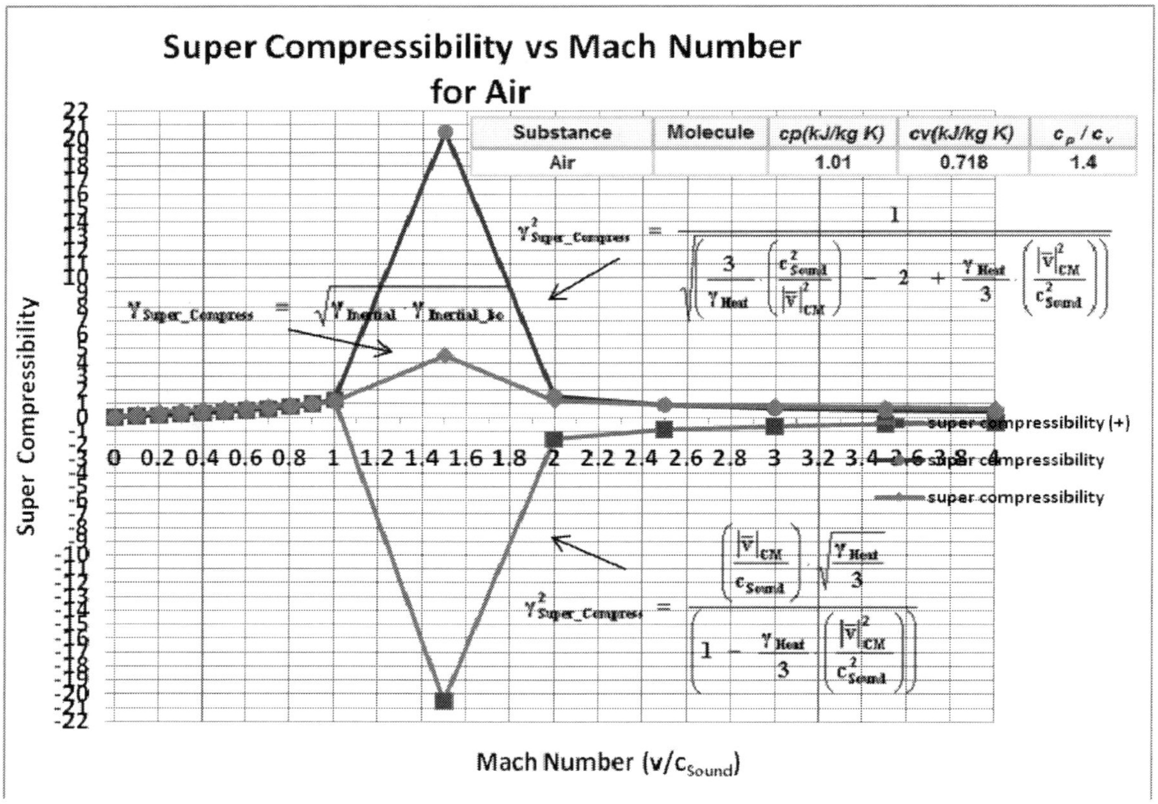

Figure 14.3: Super Compressibility vs. Mach number for – Air.

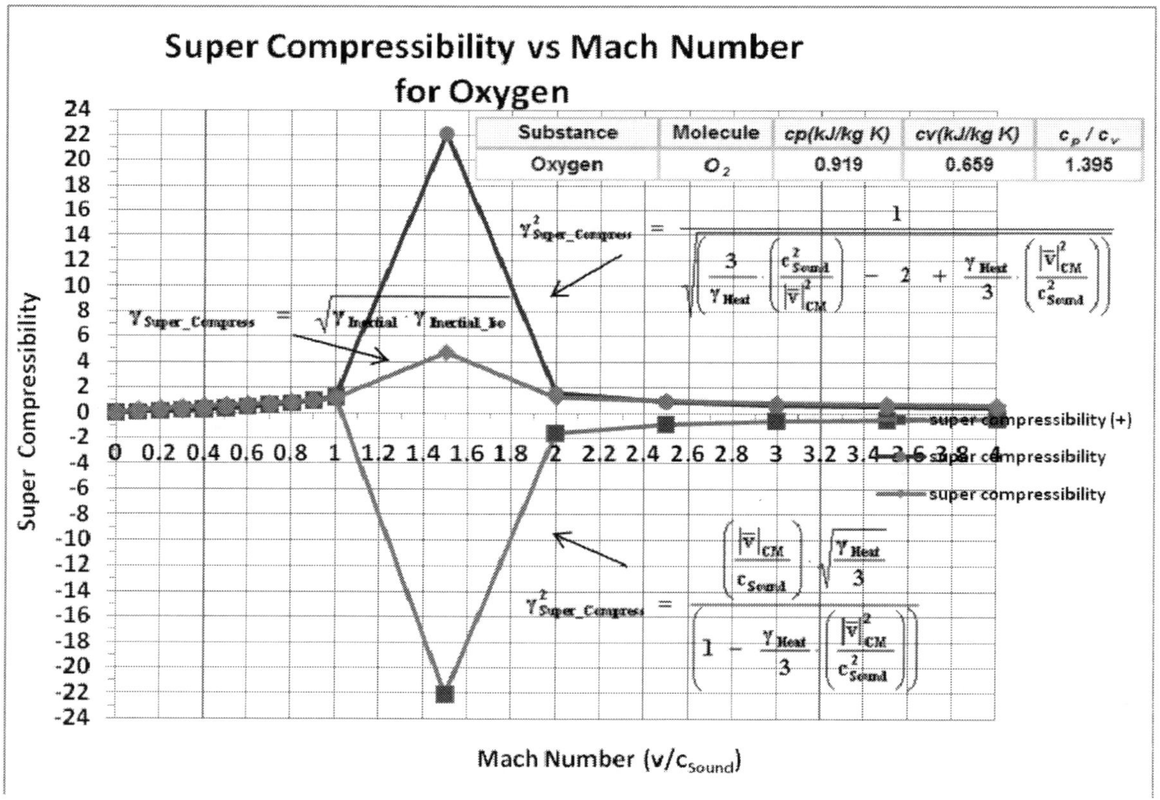

Figure 14.4: Super Compressibility vs. Mach number for – Oxygen.

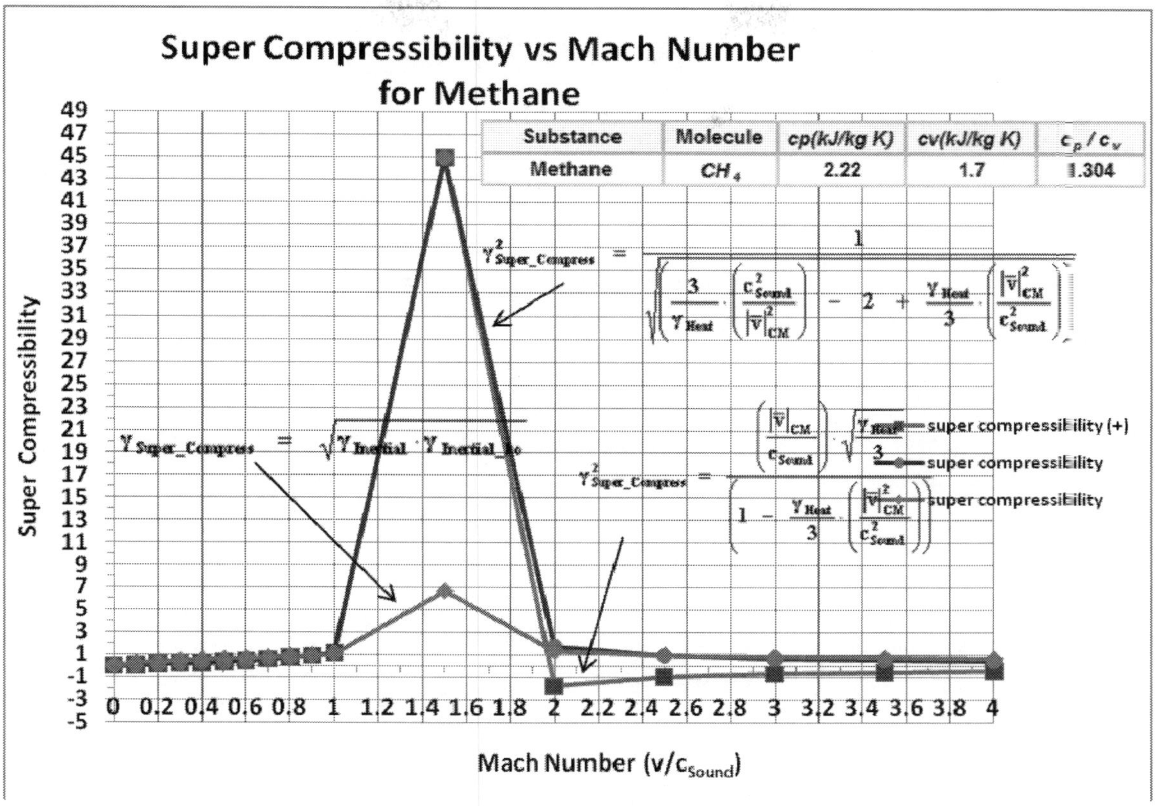

Figure 14.5: Super Compressibility vs. Mach number for – Methane.

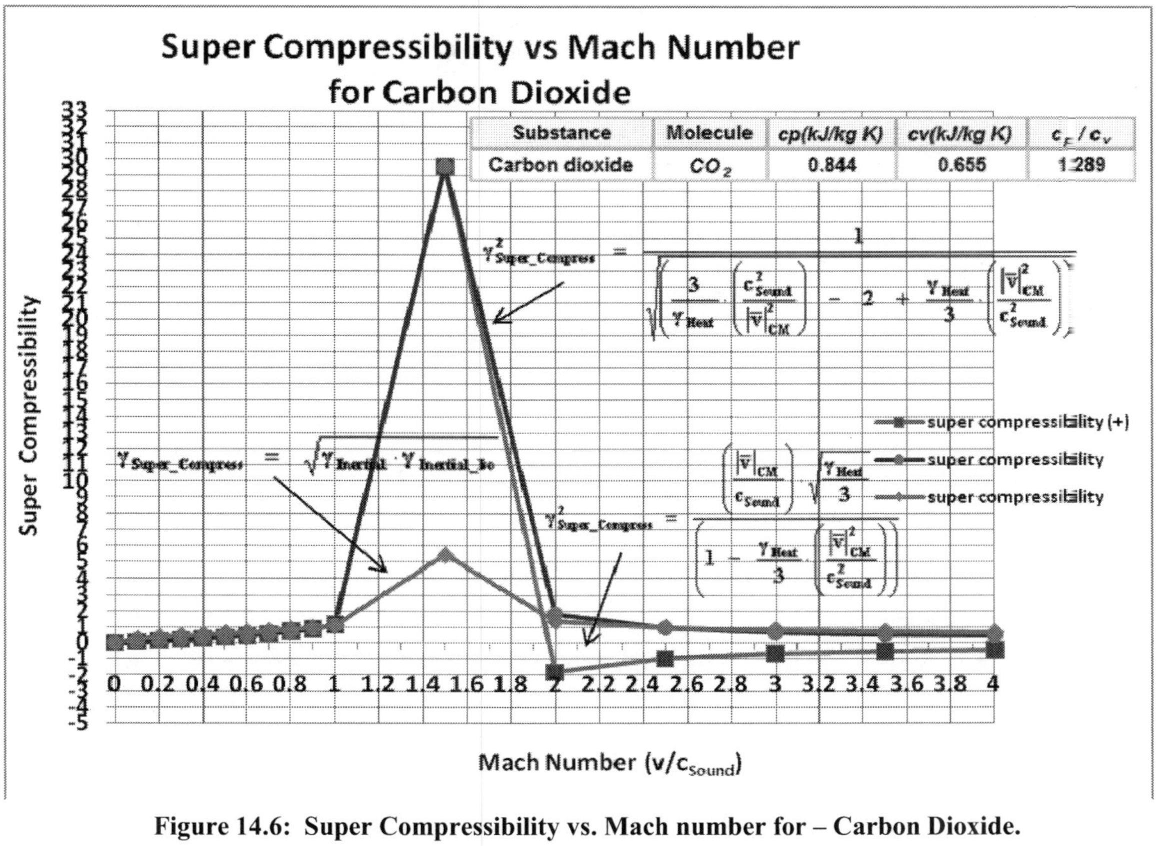

Figure 14.6: Super Compressibility vs. Mach number for – Carbon Dioxide.

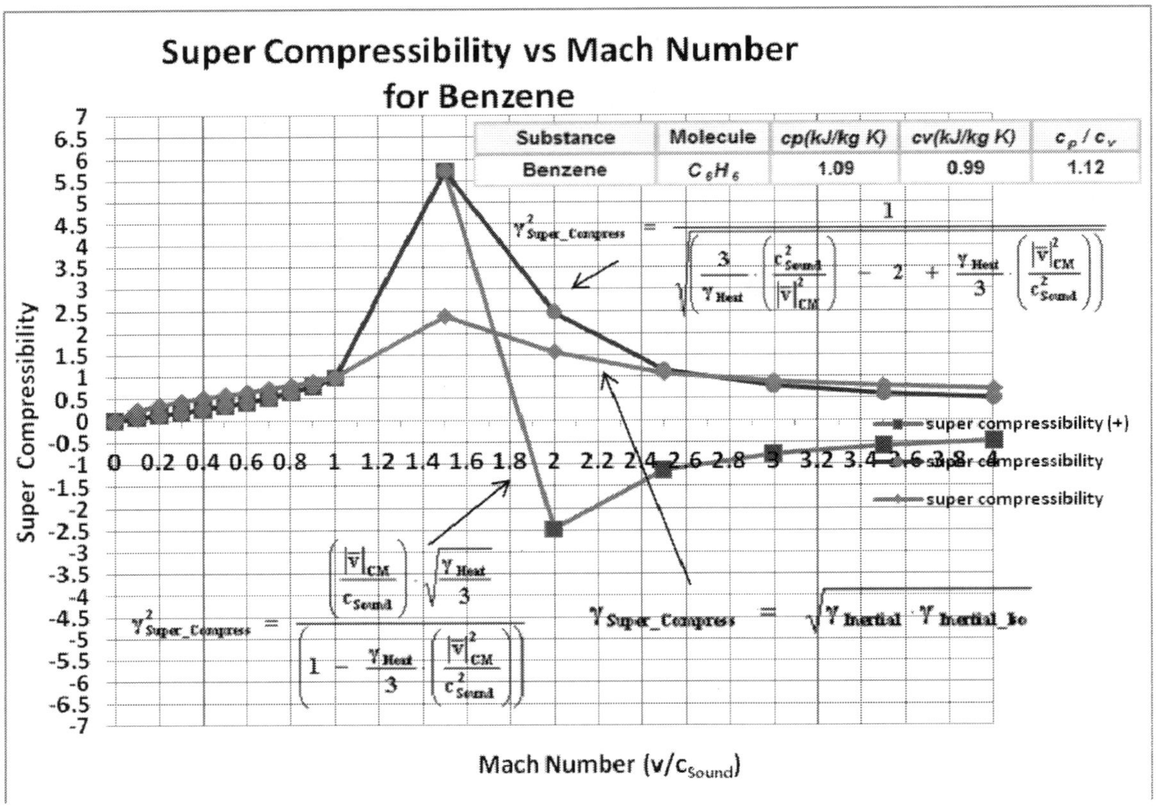

Figure 14.7: Super Compressibility vs. Mach number for – Hydrogen.

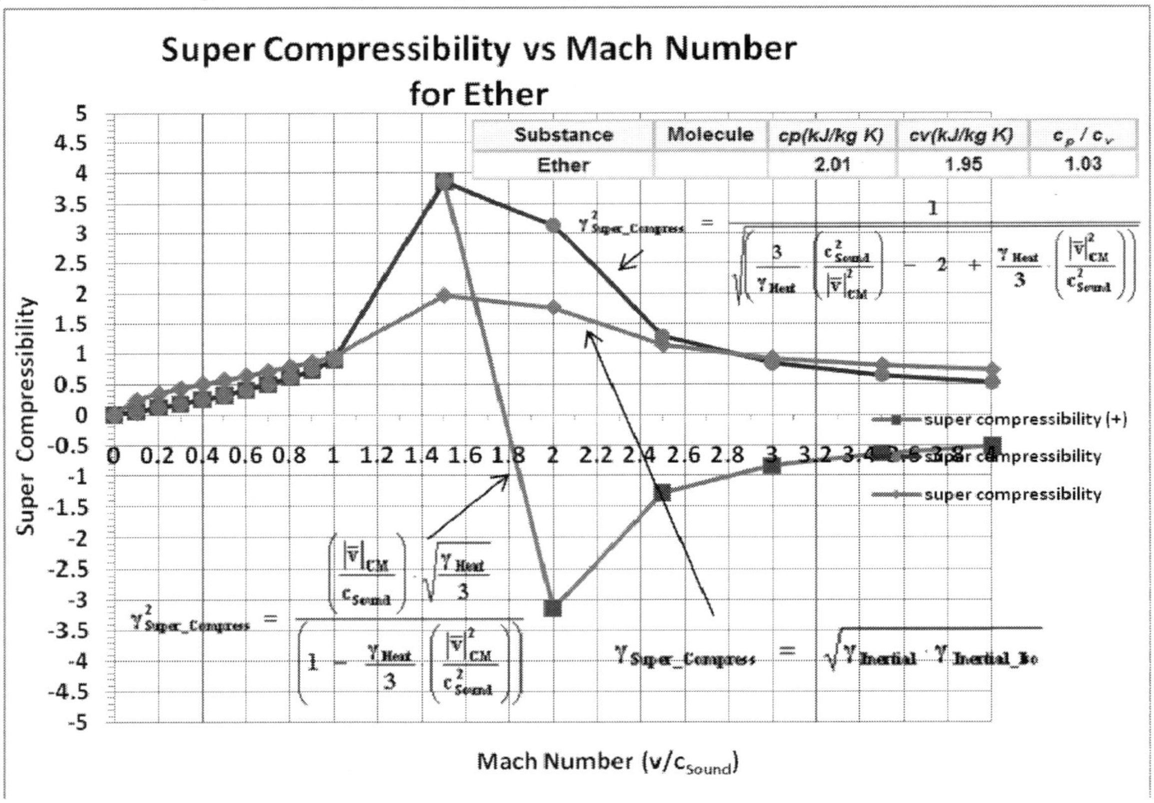

Figure 14.8: Super Compressibility vs. Mach number for – Ether.

14.4 Carnot Thermal Engine/Pump - Parameters

Thermodynamic Super Surplus Pump Density

14.16

$$\rho_{\text{Thermo\_Pump\_Density}} = \gamma_{Super\_Compress} \cdot \rho_{\text{Net}} \rightarrow {kg}/{m^3}$$

$$\rho_{\text{Thermo\_Pump\_Density}} = \left(\frac{\mathrm{m}_{\text{Thermo\_Super\_Mass}}}{\mathrm{V}_{\text{ol}}}\right) = \frac{\overline{\mathrm{p}}_{\text{Thermo\_Momentum}}}{\mathrm{V}_{\text{ol}} \cdot \sqrt{\left|\overline{\mathrm{v}}\right|_{\text{CM}}} \cdot \sqrt{\left|\mathrm{v}^2\right|_{\text{Iso}}}}$$

$$\rho_{\text{Thermo\_Pump\_Density}} = \frac{\rho_{\text{Net}}}{\left(\frac{3}{\gamma_{Heat}} \cdot \left(\frac{\mathrm{c}^2_{\text{Sound}}}{\left|\overline{\mathrm{v}}\right|^2_{\text{CM}}}\right) - 2 + \frac{\gamma_{Heat}}{3} \cdot \left(\frac{\left|\overline{\mathrm{v}}\right|^2_{\text{CM}}}{\mathrm{c}^2_{\text{Sound}}}\right)\right)^{\frac{1}{4}}}$$

$$\rho_{\text{Thermo\_Pump\_Density}} = \frac{\left(\frac{\mathrm{m}_{\text{Net}}}{\mathrm{V}_{\text{ol}}}\right)}{\left(\frac{3}{\gamma_{Heat}} \cdot \left(\frac{\mathrm{c}^2_{\text{Sound}}}{\left|\overline{\mathrm{v}}\right|^2_{\text{CM}}}\right) - 2 + \frac{\gamma_{Heat}}{3} \cdot \left(\frac{\left|\overline{\mathrm{v}}\right|^2_{\text{CM}}}{\mathrm{c}^2_{\text{Sound}}}\right)\right)^{\frac{1}{4}}}$$

Thermodynamic Super Surplus Inertia Mass and Resistance

14.17

$$m_{\text{Thermo\_Super\_Mass}} = \gamma_{Super\_Compress} \cdot m_{\text{Net}} \quad \rightarrow kg$$

$$m_{\text{Thermo\_Super\_Mass}} = \frac{\bar{p}_{\text{Thermo\_Momentum}}}{\sqrt{\left|\bar{v}\right|_{\text{CM}} \cdot \sqrt{\left|v^2\right|_{\text{Iso}}}}}$$

$$m_{\text{Thermo\_Super\_Mass}} = \frac{m_{\text{Net}}}{\left(\frac{3}{\gamma_{Heat}} \cdot \left(\frac{c_{\text{Sound}}^2}{\left|\bar{v}\right|_{\text{CM}}^2}\right) - 2 + \frac{\gamma_{Heat}}{3} \cdot \left(\frac{\left|\bar{v}\right|_{\text{CM}}^2}{c_{\text{Sound}}^2}\right)\right)^{\frac{1}{4}}}$$

Thermodynamic Super Surplus Mass Flow Rate

Where the Thermodynamic Pump cycle frequency is given by,

$$f_{\text{Thermo\_Pump\_Frequency}} = \left(\frac{1}{T_{\text{Pump\_Cycle\_Period}}} \right) \rightarrow 1/s$$

Thermodynamic Super Surplus Mass Flow Rate

14.18

$$f_{Mass\_Flow\_Rate\_Super} = \gamma_{Super\_Compress} \cdot f_{Mass\_Flow} \rightarrow kg/s$$

$$f_{Mass\_Flow\_Rate\_Super} = \left(\frac{m_{\text{Thermo\_Iso\_Mass}}}{T_{\text{Pump\_Cycle\_Period}}} \right) = \gamma_{Super\_Compress} \cdot \left(\frac{m_{\text{Net}}}{T_{\text{Pump\_Cycle\_Period}}} \right)$$

$$f_{Mass\_Flow\_Rate\_Super} = \frac{\overline{p}_{\text{Thermo\_Momentum}}}{T_{\text{Pump\_Cycle\_Period}} \cdot \sqrt{\left|\overline{v}\right|_{\text{CM}} \cdot \sqrt{\left|v^2\right|_{\text{Iso}}}}}$$

$$f_{Mass\_Flow\_Rate\_Super} = \frac{\left(\frac{m_{\text{Net}}}{T_{\text{Pump\_Cycle\_Period}}} \right)}{\left(\frac{3}{\gamma_{Heat}} \cdot \left(\frac{c_{\text{Sound}}^2}{\left|\overline{v}\right|_{\text{CM}}^2} \right) - 2 + \frac{\gamma_{Heat}}{3} \cdot \left(\frac{\left|\overline{v}\right|_{\text{CM}}^2}{c_{\text{Sound}}^2} \right) \right)^{\frac{1}{4}}}$$

Thermodynamic Isotropy — Super Surplus Aether Gravitational Evolutionary In/Out Flow Attraction Rate

14.19

$$\mathrm{K}_{\text{Gravity\_Super}} = \gamma_{Super\_Compress} \cdot \mathrm{K}_{\text{Gravity}} \rightarrow {kg \cdot m^3}/{s^2}$$

$$\mathrm{K}_{\text{Gravity\_Super}} = \mathrm{m}_{\text{Thermo\_Super\_Mass}} \cdot \mathrm{G} = \gamma_{Super\_Compress} \cdot \left(\mathrm{m}_{\text{Net}} \cdot \mathrm{G}\right)$$

$$\mathrm{K}_{\text{Gravity\_Super}} = \mathrm{m}_{\text{Thermo\_Super\_Mass}} \cdot \mathrm{G} = \frac{\overline{\mathrm{p}}_{\text{Thermo\_Momentum}} \cdot \mathrm{G}}{\sqrt{\left|\overline{\mathrm{v}}\right|_{\text{CM}} \cdot \sqrt{\left|\mathrm{v}^2\right|_{\text{Iso}}}}}$$

$$\mathrm{K}_{\text{Gravity\_Super}} = \mathrm{m}_{\text{Thermo\_Super\_Mass}} \cdot \mathrm{G} = \frac{\left(\mathrm{m}_{\text{Net}} \cdot \mathrm{G}\right)}{\left(\frac{3}{\gamma_{Heat}} \cdot \left(\frac{\mathrm{c}^2_{\text{Sound}}}{\left|\overline{\mathrm{v}}\right|^2_{\text{CM}}}\right) - 2 + \frac{\gamma_{Heat}}{3} \cdot \left(\frac{\left|\overline{\mathrm{v}}\right|^2_{\text{CM}}}{\mathrm{c}^2_{\text{Sound}}}\right)\right)^{\frac{1}{4}}}$$

$$\mathrm{K}_{\text{Gravity\_Super}} = \mathrm{m}_{\text{Thermo\_Super\_Mass}} \cdot \mathrm{G} = \frac{\left(\frac{\overline{\mathrm{F}}_{\text{Self\_Gravitation}} \cdot \mathrm{d}^2}{\mathrm{m}_{\text{Net}}}\right)}{\left(\frac{3}{\gamma_{Heat}} \cdot \left(\frac{\mathrm{c}^2_{\text{Sound}}}{\left|\overline{\mathrm{v}}\right|^2_{\text{CM}}}\right) - 2 + \frac{\gamma_{Heat}}{3} \cdot \left(\frac{\left|\overline{\mathrm{v}}\right|^2_{\text{CM}}}{\mathrm{c}^2_{\text{Sound}}}\right)\right)^{\frac{1}{4}}}$$

$$\mathrm{K}_{\text{Gravity\_Super}} = \mathrm{m}_{\text{Thermo\_Super\_Mass}} \cdot \mathrm{G} = \frac{\left(\frac{\overline{\mathrm{F}}_{\text{Newton\_Gravitation}} \cdot \mathrm{d}^2}{\mathrm{m}_{\text{Test}}}\right)}{\left(\frac{3}{\gamma_{Heat}} \cdot \left(\frac{\mathrm{c}^2_{\text{Sound}}}{\left|\overline{\mathrm{v}}\right|^2_{\text{CM}}}\right) - 2 + \frac{\gamma_{Heat}}{3} \cdot \left(\frac{\left|\overline{\mathrm{v}}\right|^2_{\text{CM}}}{\mathrm{c}^2_{\text{Sound}}}\right)\right)^{\frac{1}{4}}}$$

Chapter 15

The Special Theory of Thermodynamics

(Thermodynamic Spacetime) (Compressibility Conservation)

15.1 Conservation of Thermodynamic Spacetime Compressibility Factors *(Non-Linear Compressibility)*

Conservation of Thermodynamic Spacetime Square Compressibility Factor (Non-Linear Compressibility)

15.1

$$\left[\gamma^2_{Inertial} + \gamma^2_{Inertial\_Iso}\right] = \left[\gamma^4_{Inertial} - \gamma^4_{Inertial\_Iso}\right] \rightarrow Unitless$$

Square External (Anisotropy) Inertial Motion Compressibility Factor

15.2

$$\gamma^2_{Inertial} = \left[\gamma^4_{Inertial} - \gamma^4_{Super\_Compress}\right] \rightarrow Unitless$$

$$\gamma^2_{Inertial} = \left[1 + \gamma^2_{Inertial\_Iso}\right]$$

$$\gamma^2_{Inertial} = \frac{1}{\left(1 - \dfrac{\left|\overline{v}\right|^2_{CM}}{\left|v^2\right|_{Iso}}\right)} = \frac{1}{\left(1 - \dfrac{\gamma_{Heat}}{3} \cdot \left(\dfrac{\left|\overline{v}\right|^2_{CM}}{c^2_{Sound}}\right)\right)}$$

Square Internal (Isotropy) Inertial Motion Compressibility Factor

15.3

$$\gamma^2_{Inertial\_Iso} = \left[\gamma^4_{Super\_Compress} - \gamma^4_{Inertial\_Iso}\right] \rightarrow Unitless$$

$$\gamma^2_{Inertial\_Iso} = \left[\gamma^2_{Inertial} - 1\right]$$

$$\gamma^2_{Inertial\_Iso} = \frac{1}{\left(\dfrac{\left|v^2\right|_{Iso}}{\left|\overline{v}\right|^2_{CM}} - 1\right)} = \frac{1}{\left(\dfrac{3}{\gamma_{Heat}} \cdot \left(\dfrac{c^2_{Sound}}{\left|\overline{v}\right|^2_{CM}}\right) - 1\right)}$$

Fourth Power of the Super Compressibility Factor

15.4

$$\gamma^{4}_{Super\_Compress} = \left[\gamma^{4}_{Inertial} - \gamma^{2}_{Inertial}\right] \rightarrow Unitless$$

$$\gamma^{4}_{Super\_Compress} = \left[\gamma^{4}_{Inertial\_Iso} + \gamma^{2}_{Inertial\_Iso}\right]$$

$$\gamma^{4}_{Super\_Compress} = \frac{1}{\left(\frac{3}{\gamma_{Heat}} \cdot \left(\frac{c^{2}_{Sound}}{\left|\overline{v}\right|^{2}_{CM}}\right) - 2 + \frac{\gamma_{Heat}}{3} \cdot \left(\frac{\left|\overline{v}\right|^{2}_{CM}}{c^{2}_{Sound}}\right)\right)}$$

15.2 Conservation of Thermodynamic Spacetime Compressibility *(Linear Compressibility)*

Conservation of Thermodynamic Spacetime Square Compressibility Factor (Non-Linear Compressibility)

15.5

$$1 = \left[\gamma^2_{Inertial} - \gamma^2_{Inertial\_Iso}\right] = \left[\frac{1}{\eta_{\text{Carnot}}} - \frac{1}{\eta_{\text{Carnot\_Iso}}}\right]$$

$$1 = \left(\gamma_{Inertial} - \gamma_{Inertial\_Iso}\right)\cdot\left(\gamma_{Inertial} + \gamma_{Inertial\_Iso}\right) \rightarrow Unitless$$

Conservation of Thermodynamic Spacetime Compressibility Factor (Linear Compressibility)

15.6

$$\frac{1}{\left(\gamma_{Inertial} + \gamma_{Inertial\_Iso}\right)} = \left(\gamma_{Inertial} - \gamma_{Inertial\_Iso}\right) \rightarrow Unitless$$

Conservation of Thermodynamic Spacetime Square Compressibility Factor (Linear Compressibility)

15.7

$$\frac{1}{\left(\gamma_{Inertial} - \gamma_{Inertial\_Iso}\right)} = \left(\gamma_{Inertial} + \gamma_{Inertial\_Iso}\right) \rightarrow Unitless$$

Conservation of Thermodynamic Spacetime Square Compressibility Factor (Linear Compressibility)

15.8

$$\frac{1}{\left(\gamma_{Inertial} + \gamma_{Inertial\_Iso}\right)} = \left(\gamma_{Inertial} - \gamma_{Inertial\_Iso}\right) \rightarrow Unitless$$

$$\frac{1}{\left(\gamma_{Inertial} + \gamma_{Inertial\_Iso}\right)} = \left(\left(\frac{\overline{p}_{Thermo\_Momentum}}{\overline{p}_{Inertial\_Net}}\right) - \left(\frac{\overline{p}_{Thermo\_Momentum}}{\overline{p}_{Net\_Iso}}\right)\right)$$

$$\frac{1}{\left(\gamma_{Inertial} + \gamma_{Inertial\_Iso}\right)} = \overline{p}_{Thermo\_Momentum} \cdot \left(\frac{1}{\overline{p}_{Inertial\_Net}} - \frac{1}{\overline{p}_{Net\_Iso}}\right)$$

$$\frac{1}{\left(\gamma_{Inertial} + \gamma_{Inertial\_Iso}\right)} = \overline{p}_{Thermo\_Momentum} \cdot \left(\frac{\overline{p}_{Net\_Iso} - \overline{p}_{Inertial\_Net}}{\overline{p}_{Inertial\_Net} \cdot \overline{p}_{Net\_Iso}}\right)$$

$$\frac{1}{\left(\gamma_{Inertial} + \gamma_{Inertial\_Iso}\right)} = \overline{p}_{Thermo\_Momentum} \cdot \left(\frac{\overline{p}_{Net\_Iso} - \overline{p}_{Inertial\_Net}}{\overline{p}_{Thermo\_Momentum} \cdot \Delta\overline{p}_{Free\_Momentum}}\right)$$

Conservation of Thermodynamic Spacetime Square Compressibility Factor (Linear Compressibility)

15.9

$$\frac{1}{\left(\gamma_{Inertial} + \gamma_{Inertial\_Iso}\right)} = \left(\gamma_{Inertial} - \gamma_{Inertial\_Iso}\right) \rightarrow Unitless$$

$$\frac{1}{\left(\gamma_{Inertial} + \gamma_{Inertial\_Iso}\right)} = \left(\frac{\overline{p}_{Net\_Iso} - \overline{p}_{Inertial\_Net}}{\Delta\overline{p}_{Free\_Momentum}}\right)$$

$$\frac{1}{\left(\gamma_{Inertial} + \gamma_{Inertial\_Iso}\right)} = \left(\frac{\overline{p}_{Net\_Iso} - \overline{p}_{Inertial\_Net}}{\sqrt{\overline{p}^{2}_{Net\_Iso} - \overline{p}^{2}_{Inertial\_Net}}}\right)$$

$$\frac{1}{\left(\gamma_{Inertial} + \gamma_{Inertial\_Iso}\right)} = \left(\sqrt{\frac{\overline{p}_{Net\_Iso} - \overline{p}_{Inertial\_Net}}{\overline{p}_{Net\_Iso} + \overline{p}_{Inertial\_Net}}}\right)$$

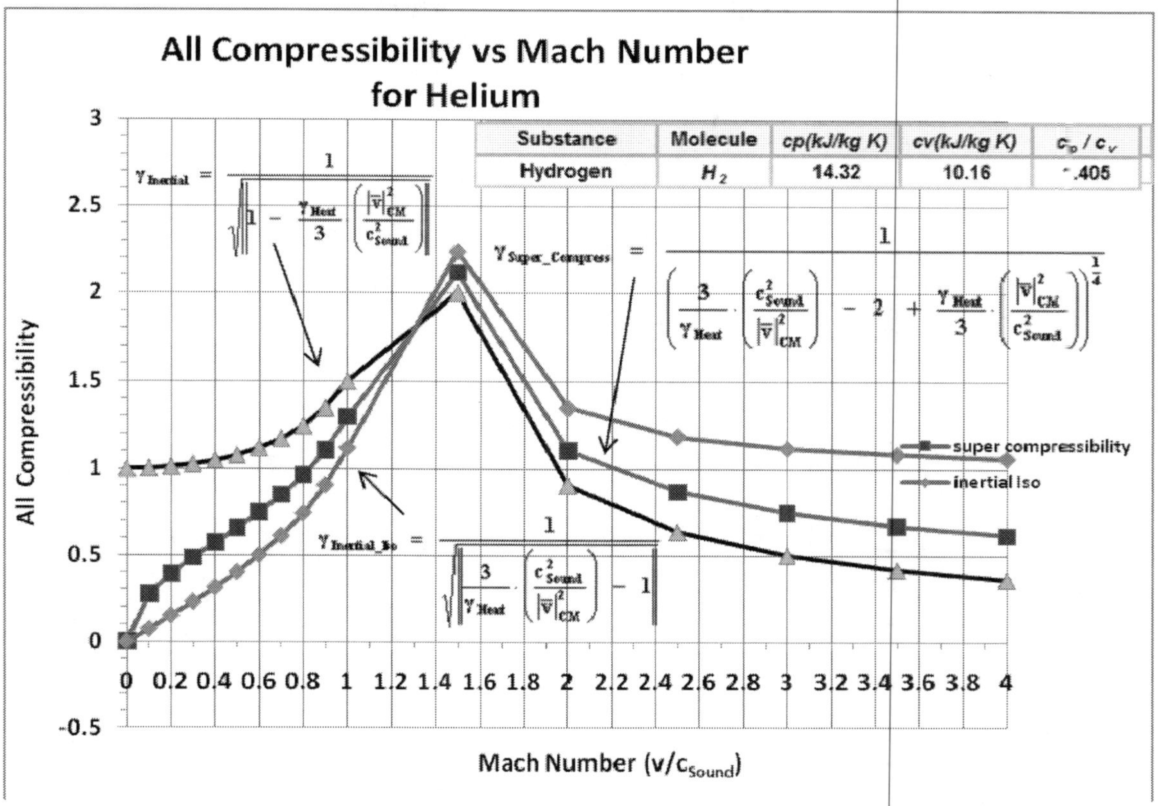

Figure 15.1: All Compressibility vs. Mach number for – Helium

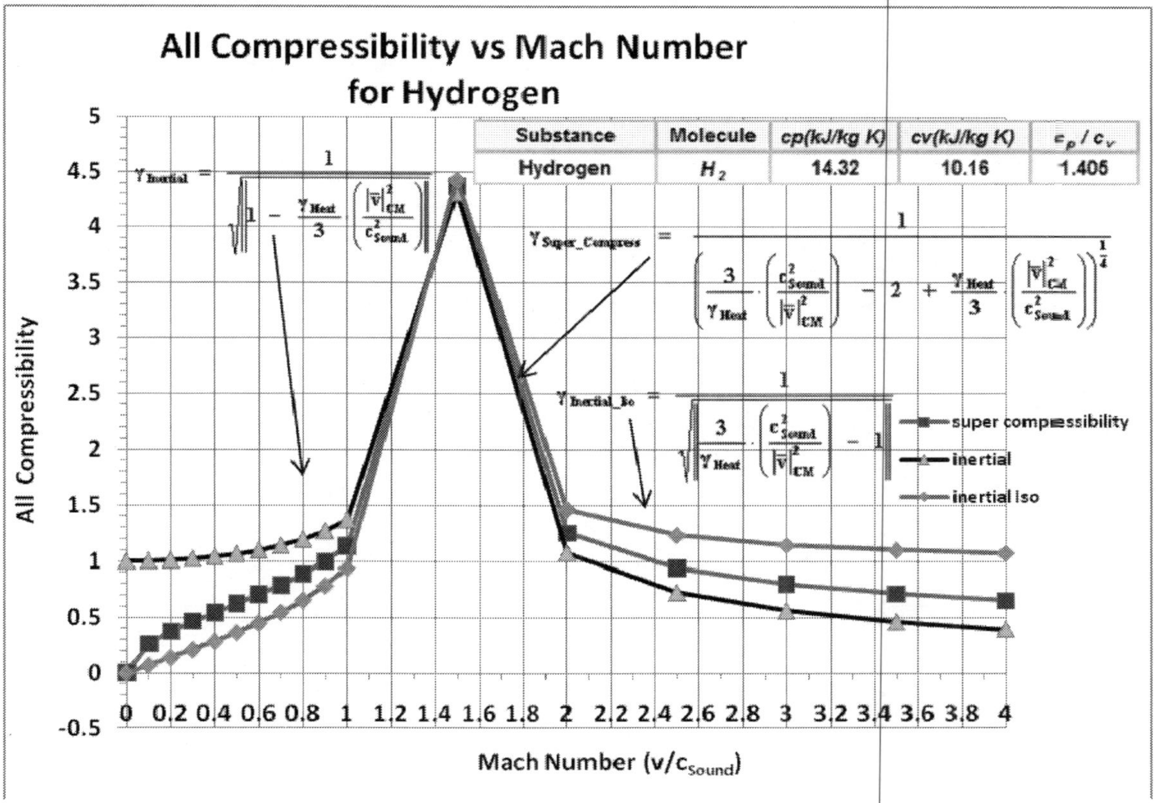

Figure 15.2: All Compressibility vs. Mach number for – Hydrogen.

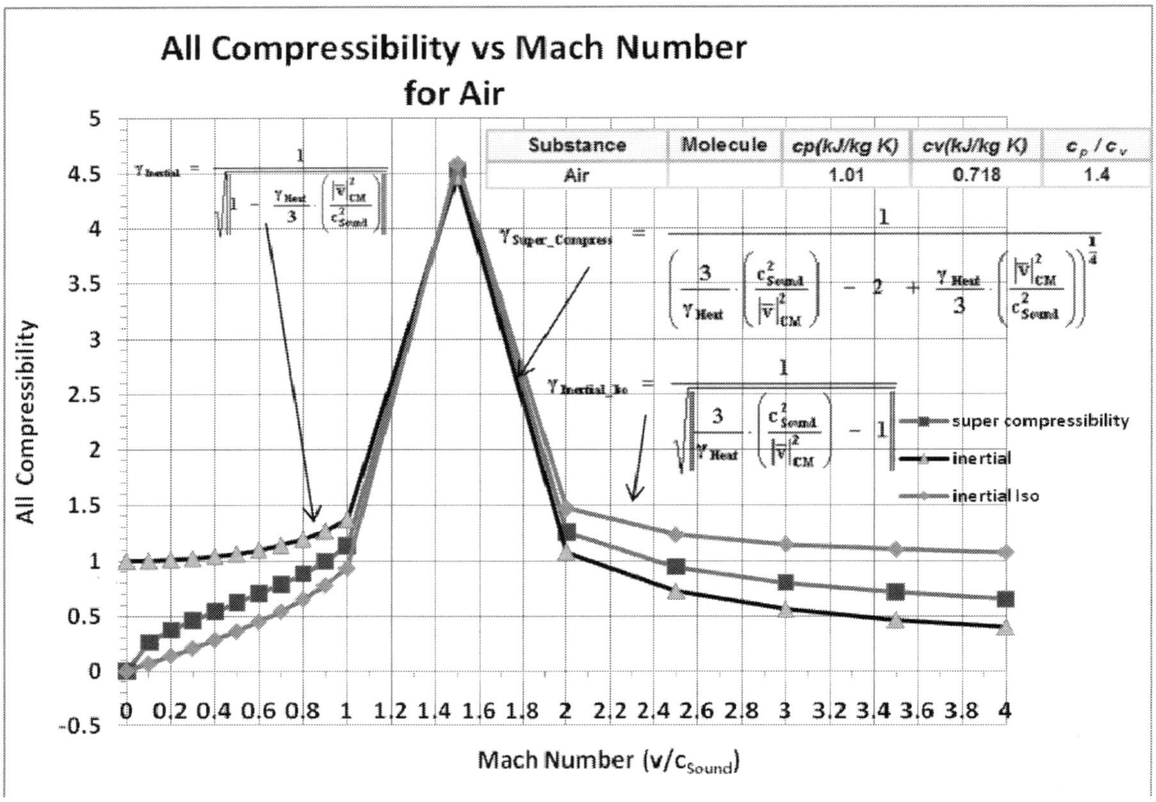

Figure 15.3: All Compressibility vs. Mach number for – Air.

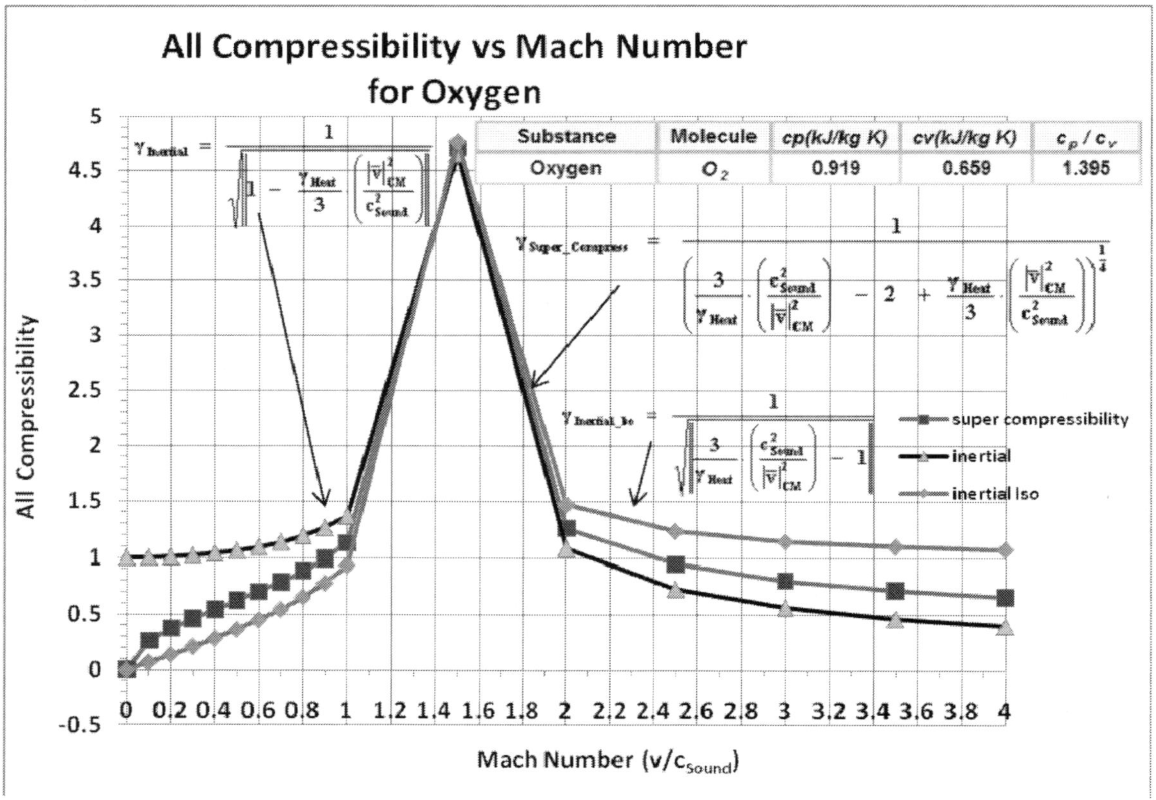

Figure 15.4: All Compressibility vs. Mach number for – Oxygen.

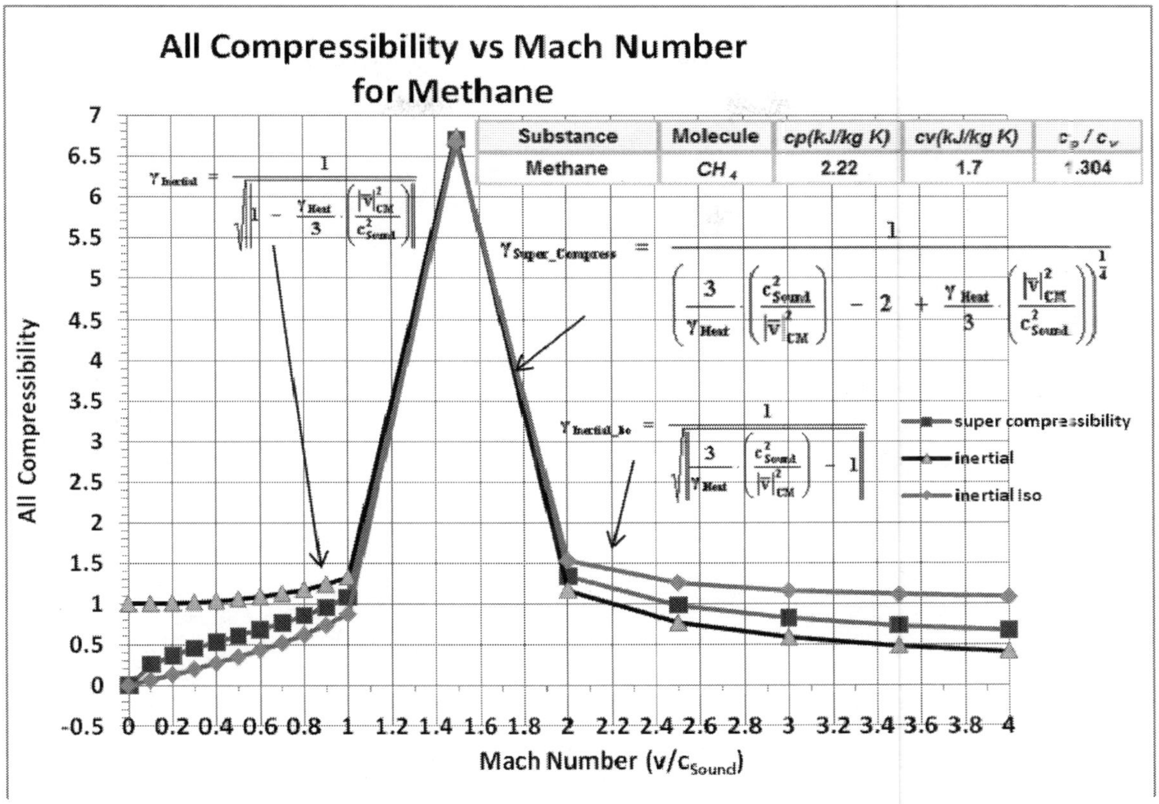

Figure 15.5: All Compressibility vs. Mach number for – Methane.

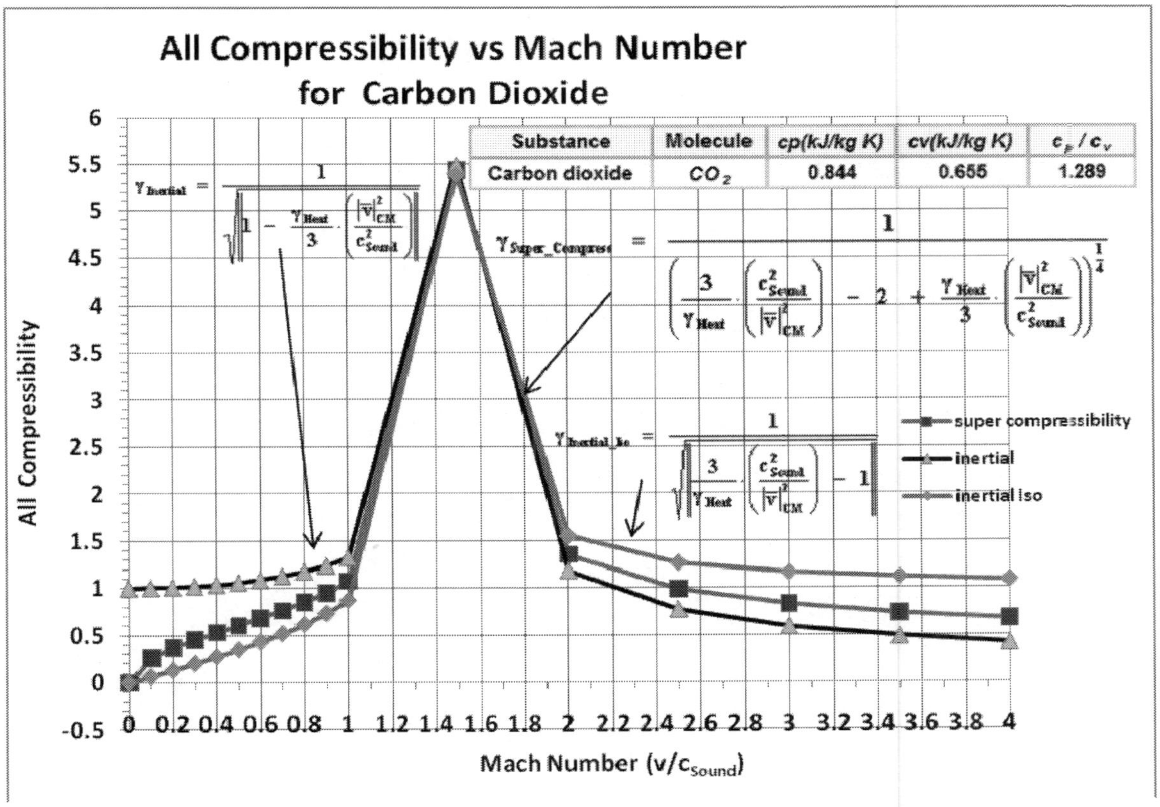

Figure 15.6: All Compressibility vs. Mach number for – Carbon Dioxide.

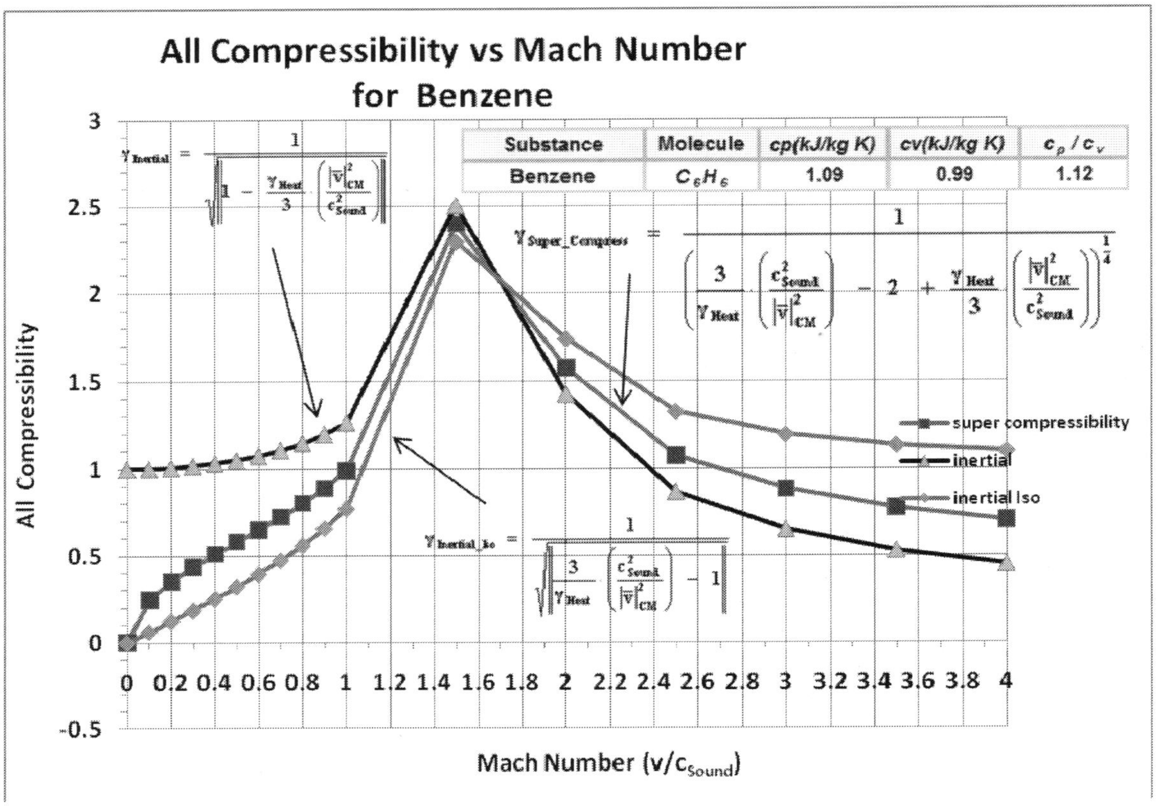

Figure 15.7: All Compressibility vs. Mach number for – Benzene.

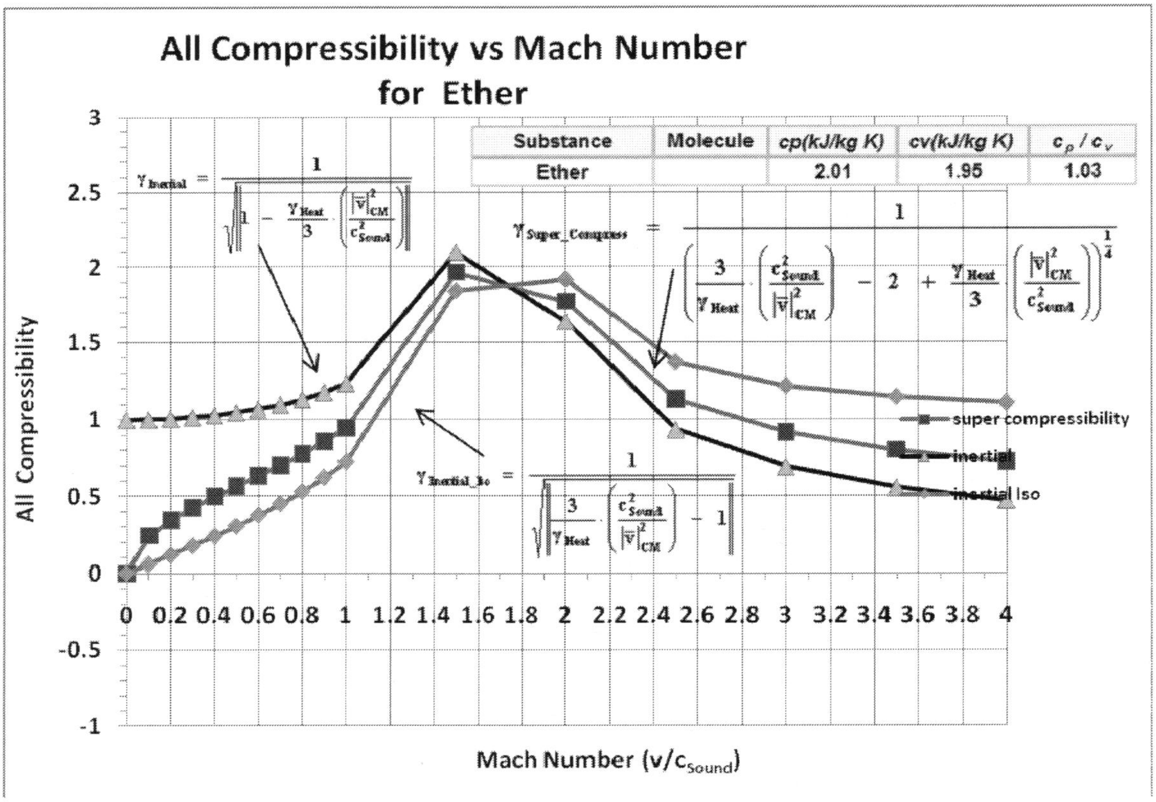

Figure 15.8: All Compressibility vs. Mach number for – Ether.

15.3 Conservation of Thermodynamic Spacetime Compressibility – For Various Physical Quantities

Conservation of Thermodynamic Spacetime Compressibility (Linear Compressibility — Net Inertial Mass)

15.10

$$\frac{m_{Net}}{\left(\gamma_{Inertial} + \gamma_{Inertial\_Iso}\right)} = m_{Net} \cdot \left(\gamma_{Inertial} - \gamma_{Inertial\_Iso}\right) \rightarrow kg$$

Conservation of Thermodynamic Spacetime Compressibility (Linear Compressibility — Net Inertial Mass Density)

15.11

$$\frac{\rho_{Net}}{\left(\gamma_{Inertial} + \gamma_{Inertial\_Iso}\right)} = \rho_{Net} \cdot \left(\gamma_{Inertial} - \gamma_{Inertial\_Iso}\right) \rightarrow {kg}/{m^3}$$

Conservation of Thermodynamic Spacetime Compressibility (Linear Compressibility — Sound Bulk Modulus Pressure)

15.12

$$\frac{\rho_{Net} \cdot c_{Sound}^2}{\left(\gamma_{Inertial} + \gamma_{Inertial\_Iso}\right)} = \left(\rho_{Net} \cdot c_{Sound}^2\right) \cdot \left(\gamma_{Inertial} - \gamma_{Inertial\_Iso}\right) \rightarrow {kg}/{m \cdot s^2}$$

Conservation of Thermodynamic Spacetime Compressibility (Linear Compressibility — Light Bulk Modulus Pressure)

15.13

$$\frac{\rho_{Net} \cdot c_{Light}^2}{\left(\gamma_{Inertial} + \gamma_{Inertial\_Iso}\right)} = \left(\rho_{Net} \cdot c_{Light}^2\right) \cdot \left(\gamma_{Inertial} - \gamma_{Inertial\_Iso}\right) \rightarrow {kg}/{m \cdot s^2}$$

Conservation of Thermodynamic Spacetime Compressibility (Linear Compressibility – Gravitational Acceleration)

15.14

$$\frac{\left(\frac{m_{Net} \cdot G}{d^2}\right)}{\left(\gamma_{Inertial} + \gamma_{Inertial\_Iso}\right)} = \left(\frac{m_{Net} \cdot G}{d^2}\right) \cdot \left(\gamma_{Inertial} - \gamma_{Inertial\_Iso}\right) \rightarrow {}^{m}\!/_{s^2}$$

Conservation of Thermodynamic Spacetime Compressibility (Linear Compressibility – Aether-Kinematic Work "Free" Momentum)

15.15

$$\frac{\Delta \overline{p}_{Free\_Momentum}}{\left(\gamma_{Inertial} + \gamma_{Inertial\_Iso}\right)} = \Delta \overline{p}_{Free\_Momentum} \cdot \left(\gamma_{Inertial} - \gamma_{Inertial\_Iso}\right) \rightarrow {}^{kg \cdot m}\!/_{s}$$

Epilogue

To the reader, I hope that you enjoyed this unique look into matter, energy and the aether. In essence this second volume (2) of the Super Principia Mathematica – The Rage to Master, is a Special Theory of Thermodynamics.

I hope that the reader enjoyed this work, I consider it art, as well as science. I earnestly ask that everything be read with an open mind and that the shortcomings in any of the subjects addressed, which are new concepts, may be not so much reprehended as investigated, and kindly supplemented, by new endeavors of my readers.

You may have noticed that the subjects addressed assumed that the reader already has an understanding of Universal Kinetic Theory as presented in Super Principia Volume (1) The First Law of Motion – Inertial Motion. These subjects and similar are described in greater detail in the subsequent volumes of the Super Principia Mathematic – The Rage to Master, such as:

- Volume - The First Law of Motion – Inertial Motion
- Volume - The Laws of Non Inertial Motion
- Volume - The Special Theory of Thermodynamics
- Volume - The General Theory of Thermodynamics
- Volume - The Special Theory of Relativity
- Volume - The General Theory of Relativity

To the reader, because of the invention of Hubble type telescope, optics, and digital signal processing techniques we are entering a new paradigm in our understanding into the workings of the firmament in the created universe; which opens the door for a need to probe deeper into the understanding of the workings of nature that makes these writings possible.

Robert Kemp
June 2010

Table of Common Quantity Symbols			
	Quantity Name	**Quantity Symbol**	**Units**
1	**Adiabatic Isentropic Thermodynamic Field Entropy – Production**	$\Delta S_{Entropy\,Free\_Production}$	$kg \cdot m^2 / s^2 \cdot K$
2	**Adiabatic Isentropic Thermodynamic Field Entropy – Surroundings**	$\Delta S_{Entropy\,Free\_Surroundings}$	$kg \cdot m^2 / s^2 \cdot K$
3	**Adiabatic Isentropic Thermodynamic Field Free Aether Entropy – System**	$\Delta S_{Entropy\,Free\_Energy}$	$kg \cdot m^2 / s^2 \cdot K$
4	**Aether Anisotropy mass flow rate**	$f_{Mass\_Flow\_Aniso}$	kg/s
5	**Aether Isotropy mass flow rate**	$f_{Mass\_Flow\_Iso}$	kg/s
6	**Aether Kinematic Work "Free" Energy – Carnot Heat Engine**	$\mathrm{W}_{\mathrm{Free\_Work\_Energy}}$	$kg \cdot m^2 / s^2$
7	**Aether-Kinematic Work "Free" Momentum**	$\Delta \overline{\mathrm{p}}_{\mathrm{Free\_Momentum}}$	$kg \cdot m / s$
8	**Anisotropic – Aether-Kinematic Work "Free" Momentum Difference**	$\Delta \overline{\mathrm{p}}_{\mathrm{Aniso\_Momentum}}$	$kg \cdot m / s$
9	**Anisotropic Aerodynamic Pressure**	$\mathrm{q}_{\mathrm{Dynamic\_Pressure}}$	$kg / m \cdot s^2$
10	**Anisotropic Aether Kinematic Working Fluid Mass**	$\Delta \mathrm{m}_{\mathrm{Aniso\_WFM}}$	kg
11	**Anisotropic Inertial Arithmetic Mean Center Rest Momentum**	$\lvert \overline{\mathrm{p}} \rvert_{\mathrm{Inertial\_Rest}}$	$kg \cdot m / s$
12	**Average - Vibrational Energy of Quantized Linear Harmonic Oscillator**	$\in_{\mathrm{Thermal\_Energy}}$	$kg \cdot m^2 / s^2$
13	**Avogadro's number**	$\mathrm{N}_{\mathrm{Avogadro}}$	*Unitless*
14	**Carnot Thermal Engine "Anisotropy" Efficiency Factor**	η_{Carnot}	*Unitless*
15	**Carnot Thermal Engine "Isotropy" Efficiency Factor**	$\eta_{\mathrm{Carnot\_Iso}}$	*Unitless*
16	**Conic Inertial Momentum**	$\overline{\mathrm{p}}_{\mathrm{Conic\_Momentum}}$	$kg \cdot m / s$
17	**External Anisotropic Center of Mass Squared Velocity Inertia**	$\lvert \overline{\mathrm{v}} \rvert_{\mathrm{CM}}^2$	m^2 / s^2
18	**External Anisotropic Inertial Translational Kinetic Energy**	$\mathrm{E}_{\mathrm{Net\_Translational}}$	$kg \cdot m^2 / s^2$
19	**External Anisotropic Uniform Directional Rectilinear Inertial Momentum**	$\overline{\mathrm{p}}_{\mathrm{Inertial\_Net}}$	$kg \cdot m / s$
20	**External Anisotropic Uniform Rectilinear Center of Mass Velocity**	$\lvert \overline{\mathrm{v}} \rvert_{\mathrm{CM}}$	m / s

21	**Ideal Gas Energy**	$P_{Iso\_Pressure} \cdot V_{ol}$	$kg \cdot m^2 / s^2$
22	**Inertial Motion Compressibility Factor**	$\gamma_{Inertial}$	*Unitless*
23	**Inertial Net Mass Density**	ρ_{Net}	kg / m^3
24	**Inertial Squared Velocity Inertia of Body (1)**	$\overline{v}_1^2$	m^2 / s^2
25	**Inertial Squared Velocity Inertia of Body (2)**	$\overline{v}_2^2$	m^2 / s^2
26	**Internal "Inherent Inertia" Mass**	$m_{Inertia\_Mass}$	kg
27	**Internal Isotropic Omni-Directional Center of Isotropy Squared Velocity Inertia**	$\left\|v^2\right\|_{Iso}$	m^2 / s^2
28	**Internal Isotropic Omni-directional Speed**	$\left\|v\right\|_{Iso} = \sqrt{\left\|v^2\right\|_{Iso}}$	m / s
29	**Internal Isotropic Uniform Omni-directional Inertial Momentum**	$\overline{p}_{Net\_Iso}$	$kg \cdot m / s$
30	**Internal Net Isotropic Omni-directional Kinetic Energy**	$E_{Net\_Iso\_Kinetic}$	$kg \cdot m^2 / s^2$
31	**Isotropic "Dark Matter" Halo Mass**	m_{Dark_Matter}	kg
32	**Isotropic — Aether-Kinematic Work "Free" Momentum Difference**	$\Delta\overline{p}_{Iso\_Momentum}$	$kg \cdot m / s$
33	**Isotropic Aether Kinematic Working Fluid Mass**	Δm_{Iso_WFM}	kg
34	**Isotropic Aether Squared Velocity Inertia of Body (1)**	$\overline{v}_{A1}^2$	m^2 / s^2
35	**Isotropic Aether Squared Velocity Inertia of Body (2)**	$\overline{v}_{A2}^2$	m^2 / s^2
36	**Isotropic Aether Squared Velocity Inertia of Body (2) – "After State"**	$\left(\overline{v}'^2_{A2}\right)_{Iso}$	m^2 / s^2
37	**Isotropic Inertial Motion Compressibility Factor**	$\gamma_{Inertial\_Iso}$	*Unitless*
38	**Isotropic Static Pressure**	$P_{Iso\_Pressure}$	$kg / m \cdot s^2$
39	**Kemp's Thermodynamic Efficiency Factor**	η_{Kemp}	*Unitless*
40	**Lorentz Compressibility Factor**	$\gamma_{Lorentz}$	*Unitless*
41	**Mach Number**	$\beta_{Mach} = M_{Mach} = \left(\frac{\left\|\overline{v}\right\|_{CM}}{c_{Sound}}\right)$	*Unitless*
42	**Macroscopic Gas Energy-Temperature Constant**	R_{Gas}	$kg \cdot m^2 / s^2 \cdot K$
43	**Microscopic Boltzmann Gas Energy-Temperature Constant**	k_B	$kg \cdot m^2 / s^2 \cdot K$
44	**Net Inertial Mass**	m_{Net}	kg

45	**Net Inertial Rest Mass of the Atom**	$\mathrm{m}_{\mathrm{Net\_Atom}}$	kg
46	**Number of Mass Units**	N	*Unitless*
47	**Number of Moles**	$\mathrm{n}_{\mathrm{Moles}}$	mol
48	**Prandtl-Glauert Atomic Substance - Spacetime Compressibility Factor**	$\gamma_{\Pr andtl}$	*Unitless*
49	**Specific Heat at constant Pressure**	$C_{Specific\_\Pr essure}$	$m^2 / s^2 \cdot K$
50	**Specific Heat at constant Temperature**	$C_{Specific\_Temperature}$	$m^2 / s^2 \cdot K$
51	**Specific Heat at constant Volume**	$C_{Specific\_Heat}$	$m^2 / s^2 \cdot K$
52	**Specific Heat Capacity Adiabatic Index Ratio**	γ_{Heat}	*Unitless*
53	**Speed/Velocity of Light**	$\mathrm{c}_{\mathrm{Light}}$	m/s
54	**Speed/Velocity of Sound**	$\mathrm{c}_{\mathrm{Sound}}$	m/s
55	**Squared Total Momentum Conservation**	$\dfrac{\overline{\mathrm{p}}^2_{\mathrm{Total\_Momentum}}}{2}$	$kg^2 \cdot m^2 / s^2$
56	**Super Carnot Efficiency Factor**	$\eta_{\mathrm{Canot\_Super}}$	*Unitless*
57	**Super Compressibility Factor**	$\gamma_{Super\_Compress}$	*Unitless*
58	**Thermodynamic "Pump" Momentum**	$\overline{\mathrm{p}}_{\mathrm{Thermo\_Momentum}}$	$kg \cdot m / s$
59	**Thermodynamic Aether Free Energy Temperature Difference**	ΔT_{Free}	K
60	**Thermodynamic Anisotropy – Surplus Aether Gravitational Evolutionary In/Out Flow Attraction Rate**	$\mathrm{K}_{\mathrm{Gravity\_Aether\_Aniso}}$	$kg \cdot m^3 / s^2$
61	**Thermodynamic Anisotropy – Surplus Aether Light Bulk Modulus Pressure**	$B_{\mathrm{Aether\_Pressure\_Light\_Aniso}}$	$kg \cdot / m \cdot s^2$
62	**Thermodynamic Anisotropy – Surplus Aether Light Energy Content**	$\left(B_{\mathrm{Aether\_Pressure\_Light\_Aniso}} \cdot \mathrm{V}_{\mathrm{ol}}\right)$	$kg \cdot m^2 / s^2$
63	**Thermodynamic Anisotropy – Surplus Aether Sound Bulk Modulus Pressure**	$B_{\mathrm{Aether\_Pressure\_Aniso}}$	$kg \cdot / m \cdot s^2$
64	**Thermodynamic Anisotropy – Surplus Aether Sound Energy Content**	$\left(B_{\mathrm{Aether\_Pressure\_Aniso}} \cdot \mathrm{V}_{\mathrm{ol}}\right)$	$kg \cdot m^2 / s^2$
65	**Thermodynamic Anisotropy Surplus Aether Dynamic Pressure**	$\mathrm{q}_{\mathrm{Aether\_Dynamic\_Pressure}}$	$kg \cdot / m \cdot s^2$
66	**Thermodynamic Anisotropy Surplus Density**	$\rho_{\mathrm{Thermo\_Aniso\_Density}}$	kg / m^3

67	**Thermodynamic Anisotropy Surplus Inertia Mass and Resistance**	$m_{Thermo\_Aniso\_Mass}$	kg
68	**Thermodynamic Available Energy**	$H_{Enthalpy\_Energy}$	$kg \cdot m^2 / s^2$
69	**Thermodynamic Enthalpy Energy**	$H_{Enthalpy\_Energy}$	$kg \cdot m^2 / s^2$
70	**Thermodynamic Entropic "Light" Work/Energy**	$\overline{F}_{Entropic\_Light\_Force}$	$kg \cdot m / s^2$
71	**Thermodynamic Entropic "Sound" Work/Energy**	$\overline{F}_{Entropic\_Sound\_Force}$	$kg \cdot m / s^2$
72	**Thermodynamic Entropic Force**	$\overline{F}_{Entropic\_Force}$	$kg \cdot m / s^2$
73	**Thermodynamic External Anisotropic Aerodynamic Temperature**	T_{emp_Ext}	K
74	**Thermodynamic Field Irreversible Heat "Free Aether" Pressure Potential Energy Transfer**	$\delta Q_{Heat\_Pressure\_Free}$	$kg \cdot m^2 / s^2$
75	**Thermodynamic Field Irreversible Heat "Free Aether" Volume Potential Energy Transfer**	$\delta Q_{Heat\_Energy\_Free}$	$kg \cdot m^2 / s^2$
76	**Thermodynamic Field Irreversible Heat "Free Aether" Work Energy Transfer**	$\Delta(PV)_{Free\_Energy}$	$kg \cdot m^2 / s^2$
77	**Thermodynamic Field Irreversible Heat "Free Aether" Work Force**	$F_{Thermo\_Free\_Force}$	$kg \cdot m / s^2$
78	**Thermodynamic Heat Death Temperature**	T_{Death_Temp}	K
79	**Thermodynamic Internal Energy**	$U_{Internal\_Energy}$	$kg \cdot m^2 / s^2$
80	**Thermodynamic Internal Isotropic Omni-directional Absolute Temperature**	T_{emp}	K
81	**Thermodynamic Isotropic Aether Temperature**	T_{Aether_Temp}	K
82	**Thermodynamic Isotropy – Super Surplus Aether Gravitational Evolutionary In/Out Flow Attraction Rate**	$K_{Gravity\_Super}$	$kg \cdot m^3 / s^2$
83	**Thermodynamic Isotropy – Surplus Aether Gravitational Evolutionary In/Out Flow Attraction Rate**	$K_{Gravity\_Aether\_Iso}$	$kg \cdot m^3 / s^2$
84	**Thermodynamic Isotropy – Surplus Aether Light Bulk Modulus Pressure**	$B_{Aether\_Pressure\_Light\_Iso}$	$kg / m \cdot s^2$
85	**Thermodynamic Isotropy – Surplus Aether Light Energy Content**	$(B_{Aether\_Pressure\_Light\_Iso} \cdot V_{ol})$	$kg \cdot m^2 / s^2$
86	**Thermodynamic Isotropy – Surplus Aether Sound Bulk Modulus Pressure**	$B_{Aether\_Pressure\_Iso}$	$kg / m \cdot s^2$
87	**Thermodynamic Isotropy – Surplus Aether Sound Energy Content**	$(B_{Aether\_Pressure\_Iso} \cdot V_{ol})$	$kg \cdot m^2 / s^2$
88	**Thermodynamic Isotropy Surplus Aether Static Pressure**	$P_{Aether\_Pressure\_Static}$	$kg / m \cdot s^2$

89	**Thermodynamic Isotropy Surplus Density**	$\rho_{\text{Thermo\_Iso\_Density}}$	kg/m^3
90	**Thermodynamic Isotropy Surplus Inertia Mass and Resistance**	$\text{m}_{\text{Thermo\_Iso\_Mass}}$	kg
91	**Thermodynamic Pump Work Energy**	$\text{W}_{\text{Thermo\_Pump\_Energy}}$	$kg \cdot m^2/s^2$
92	**Thermodynamic Super Surplus Density**	$\rho_{\text{Thermo\_Pump\_Density}}$	kg/m^3
93	**Thermodynamic Super Surplus Inertia Mass and Resistance**	$\text{m}_{\text{Thermo\_Super\_Mass}}$	kg
94	**Thermodynamic Super Surplus Mass Flow Rate**	$f_{Mass\_Flow\_Rate\_Super}$	kg/s
95	**Thermodynamic Temperature Difference**	ΔT_{Temp}	K
96	**Thermodynamic Zero Point Energy Temperature**	$T_{emp\,Zero\_Point}$	K
97	**Total Aether Kinematic Working Fluid Mass**	$\Delta\text{m}_{\text{Net\_WFM}}$	kg
98	**Total Kinetic Energy Conservation**	$\text{E}_{\text{Total\_Kinetic}}$	$kg \cdot m^2/s^2$
99	**Universal Geometric Mean Ratio – Respiration Spacetime Factor**	$\phi^2_{\text{Re}spiration}$	$Unitless$
100	**Universal Geometric Mean Ratio – Anisotropy to Isotropy Inertial Locomotion Foreshortening Factor**	$\phi_{\text{Loco\_Motion}} = \left(\frac{\lvert\vec{\text{v}}\rvert^2_{\text{CM}}}{\lvert\text{v}^2\rvert_{\text{Iso}}}\right) = \left(\frac{T_{emp\_Ext}}{T_{emp}}\right)$	$Unitless$
101	**Volume**	V_{ol}	m^3
102	**Zero Point Energy — Ground State Electromagnetic Isotropic Velocity**	$\lvert\text{v}\rvert_{\text{Iso}Ground}$	m/s

Table of Universal Constants

Constant's Name	Symbol	Constant Value	Units
Stephan Boltzmann's Constant	$\sigma_{Stephan}$	5.670400474E-08	$kg / s^3 \cdot K^{\circ 4}$
Wien's Maximum Displacement (spectrum) Law	$b_{Wien\_Constant}$	2.8977684963805E-03	$m \cdot K^{\circ}$
Thermal Excitation Energy Ratio (Max Peak)	$\psi_{Temp\_Quantization}$	4.9651142317443	Unit-less
Gas Constant (R)	R	8.314472471E+00	$kg \cdot m^2 / s^2 \cdot K^{\circ}$
Boltzmann's Energy Constant	k_B	1.380650400E-23	$kg \cdot m^2 / s^2 \cdot K^{\circ}$
Avogadro's Number (Na)	Na	6.022141790E+23	Unit-less
Permittivity of Free Space	ε_0	8.854187817E-12	$C_{oulomb}^2 \cdot s / kg \cdot m^3$
Atomic Mass	m_{Atomic}	1.660538782E-27	kg
Electromagnetic Saturation Constant	$\Psi_{\in}$	3.6617975320614E-26	$C_{oulomb} \cdot s / m$
Speed of Light	c_{Light}	2.997924580E+08	m / s
Zero-Point Energy Ground State Velocity	$\lvert v \rvert_{Iso\,Ground}$	2.187691254052E+06	m / s
Planck's Constant	h_{Planck}	6.626068960E-34	$kg \cdot m^2 / s$
Planck's Constant for Rotation	$\hbar_{Planck}$	1.054571628E-34	$kg \cdot m^2 / s$
Graviton Constant	$h_{Graviton}$	9.080099839E-32	$kg \cdot m^2 / s$
Graviton Constant for Rotation	$\hbar_{Graviton}$	1.445142773E-32	$kg \cdot m^2 / s$
Electronic Charge	$e = q_{Electric}$	1.602176487E-19	C_{oulomb}
Permeability of Free Space	μ_0	1.256637061E-06	m / C_{oulomb}^2
Electric Field Evolution Syntropy	$\Upsilon_{Electric\,0}$	2.307077128E-28	$kg \cdot m^3 / s^2$

Electromagnetic Field Evolution Syntropy (bound)	$\Upsilon_{\hbar\_Bound}$	**3.161526206E-26**	$kg \cdot m^3 / s^2$
Electromagnetic Field Evolution Syntropy (unbound)	$\Upsilon_{\in 0}$	**1.986445500E-25**	$kg \cdot m^3 / s^2$
Aether Gravitation Field Evolution Syntropy	$\Upsilon_{Graviton\,0}$	**2.722145449E-23**	$kg \cdot m^3 / s^2$
Bulk Structure Constant	$\gamma_{Bulk\_Structure}$	**1.370359997E+02**	**Unit-less**
Fine Structure Constant	$\alpha_{\mathrm{Fine\_Structure}}$	**7.29735253730775E-03**	**Unit-less**
Vacuum Adiabatic Heat Capacity Ratio	γ_{Vacuum_Heat}	**5.63365956285694E+04**	**Unit-less**
Universal Gravitational Constant	**G**	**6.6742800E-11**	$m^3 / kg \cdot s^2$
Electromagnetic/Gravitational Vacuum Linear Mass Density	$\mu_{\mathrm{L\_Electromagnetic\_Density}}$	**6.73297478332358E+26**	kg / m
Aether Vacuum Linear Mass Density	$\mu_{\mathrm{L\_Aether\_Density}}$	**3.58540024021747E+22**	kg / m
Dark Vacuum Force	$\mathrm{F}_{\mathrm{Dark\_Force}}$	**3.0256479774082E+43**	$kg \cdot m / s^2$
Aether Evolutionary Flow Rate	$\mathrm{K}_{\mathrm{Aether}}$	**1.45261526648399E-18**	m^3 / s^2
Plank mass	$\mathrm{m}_{\mathrm{Planck}}$	**2.176437408E-08**	kg
Plank length	$\mathrm{d}_{\mathrm{Planck}}$	**1.61625246E-35**	m

Super Principia Mathematica – Index

Super Principia Mathematica – Index

Super Principia Mathematica – Index

Super Principia Mathematica – Index

Super Principia Mathematica – Index

Super Principia Mathematica – Index

Super Principia Mathematica – Index

Super Principia Mathematica – Index

Autobiography - About the Author
Robert Louis Kemp (1966 – Present)

I was born in Detroit Michigan on October 29, 1966 during a full moon. I grew up in Detroit until the age of ten (10) when in 1976 my family moved to Los Angeles/Compton California. I graduated high school in summer of 1984, and three days later I enrolled in college at Tuskegee University in Tuskegee Alabama as an Electrical Engineer with a Physics Minor.

Three years later, I landed my first job as an engineer at the Jet Propulsion Laboratory in Pasadena, CA, in the summer of 1987. My grandfather introduced me to a senior Engineer working at the laboratory.

While enrolled as an Electrical Engineering Student at Tuskegee University, in March of 1988, I got the physics bug; and I started working in physics all through the summer of 1988.

Then, in the fall of 1989, I was led by the Holy Spirit within, to drop out of school for six months; thus I locked myself in a room and studied only physics and the Bible. And for a total of two years all I did was study physics and the Bible.

Those first six months eventually turned into roughly twenty one years total.

I continued studying at the university and graduated from Tuskegee with a Masters Degree in the spring of 1994; where I wrote a Thesis titled: "The Principle of Photon Inertia." In that paper I prove that photons have a mass equivalent to inertia, matter, and energy.

Later that fall of 1994, I wrote a paper titled: The quantization of Electromagnetic Change. In that paper I tied James Maxwell's electromagnetic equations to Albert Einstein and Max Planck's quantum of electromagnetic energy concept. Those papers can be found circulating the internet.

While I was in enrolled in Tuskegee University, the Jet Propulsion Laboratory allowed me to come home and work as a Propulsion Systems Engineer, any time I was home from school.

I worked for Jet Propulsion Laboratory for a total of eight summer/years conducting research, doing satellite data analysis, analyzing computer simulations of interstellar bodies using Orbital Mechanics techniques, and Radio Frequency signal analysis of interstellar space objects.

When I graduated from Tuskegee University in the fall of 1994, I left Jet Propulsion Laboratory to work for Hughes Aircraft Company as an Aircraft Radar and Satellite Systems Engineer.

While studying physics, I always entertained designing and building the "Flying Car," and made many designs over the years. Then in the year 1995, I contacted Moller International, and met with the President of the Company, and showed him some of my flying car designs. He rewarded me by offering me a job making half of what I was making at Hughes Aircraft Company.

But because I really wanted to work on the flying Car, I left Hughes Aircraft Company to work for Moller International as Chief Engineer in charge of the Flight Control System design of the (M200, M400 and Aerobot) Vertical Take-Off and Landing and "Flying Car" aircrafts.

When the Moller Corporation did not win a major contract that would have pumped more money into the company, I left a year later to return to Hughes Aircraft Company.

I returned to Hughes Aircraft in 1996, and in 1997 they were bought out by Raytheon Systems Corporation. For the Raytheon Systems Corporation, I worked as a systems engineer on the F-14, F-15, F-18, and Global Hawk fighter aircraft radar systems.

In the year 1997 I got the physics bug again, and from 1997 through 1999 in my spare time all I did was study mainly orbital mechanics, and rotation. I eventually wrote a book on the subject of rotation that I never published.

While working as radar systems engineer and studying physics in my spare time, I also picked up a third job and started teaching Mathematics in the year 1999 for the University of Phoenix Southern California Campus.

I continue to teach mathematics to this present day, and prior to the writing of this book I have taught at the University of Phoenix for a total of 10 years.

In the year 2001 I stopped working on physics, because I was tired of working on physics and working three jobs. (Engineer, Teacher, Physics Writer). When I stopped studying physics instead of resting, I started studying software and web site design.

A year later in the year 2002 I left Raytheon to work for the Disney Corporation as a software computer programmer in web site design; however, I got in on the tail end of the Dot Com "Bang"; and experienced the Dot Com "Bust."

I left the Disney Corporation and web site design, to head back into aerospace; and landed a job working for the Northrop Grumman Corporation Aircraft Avionics Division in 2004, working on the X-47B Naval Unmanned Air Combat Vehicle.

Two years later, in the year 2006 I was promoted to Algorithm Development and System Design Verification Manager of the F-35 Fighter Aircraft Program at the Northrop Grumman Corporation.

In June 2007, I got the physics bug again, and resigned from my management job, sold my house and moved into a one bed room condo, and returned to physics.

I started working on completing this book that I started 21 years ago. And for the last three years, on my own free time, after work, and on weekends, night and day, day and night, I spent a total of three years of blood, sweat and tears creating this work.

I know that there are some readers that would ask, why does Kemp not have a PhD? The truth is reader, I did not want to waste years trying to convince others of my ideas, or doing research for someone else, when my own personal research required that same enormous time.

I hope that the reader enjoys this work, I consider it art, as well as science. I earnestly ask that everything be read with an open mind and that the shortcomings in any of the subjects addressed, which are new concepts, may be not so much reprehended as investigated, and kindly supplemented, by new endeavors of my readers.

For me, the mathematics of physics, are the tools that God gave man that he may understand, describe, and predict the great works of God's created universe.

Robert Louis Kemp
June 7, 2010

Notes

Made in the USA
Monee, IL
12 November 2026

1e7d3696-ceb4-42ca-94b4-08680fe597acR01